AF342086

APPLIED STUDIES OF COASTAL AND MARINE ENVIRONMENTS

Edited by **Maged Marghany**

Applied Studies of Coastal and Marine Environments

http://dx.doi.org/10.5772/60743
Edited by Maged Marghany

Contributors

Humood Naser, Shamim A Dar, K.F Khan, Saif A Khan, Gemma Aiello, Lubna Alam, Yuanzhi Zhang, Azhar Mashiatullah, Nasir Ahmad, Riffat Mahmood Qureshi, Magdalena Bełdowska, Dong Jiang, Jingying Fu, Mengmeng Hao, Patricia Popoola, Nicholus Malatji, Ojo Sunday Fayomi, Francesca Dezi, Fabrizio Gara, Davide Roia, Gustavo Lopez, Iolanda Lisi, Marcello Di Risio, Paolo De Girolamo, Massimo Gabellini, Rintaro Ono, Sofia Koukina, Nikolay Lobus, Valeriy Peresypkin, Olga Dara, Olawale Popoola, Gabriel Farotade, Maged Marghany

CBS, Edition **2018**

Published by InTech

Janeza Trdine 9, 51000 Rijeka, Croatia

Notice

Publishing Process Manager Iva Simčić

Technical Editor SPi Global

Cover Designer InTech Design Team

Additional hard copies can be obtained from orders@intechopen.com

Applied Studies of Coastal and Marine Environments, Edited by Maged Marghany
p. cm.
Print ISBN 978-953-51-2548-8
Online ISBN 978-953-51-2549-

Contents

Preface

Since World War II, scientists have paid a great attention on oceanic exploration. Ocean revolutionary began in the mid-twentieth century, which is one of the supreme periods of exploration of the oceans. Since 1950s, oceanic explorations have momentously grown. Oceanography data mobilized from oceans guided to the innovation of immense mountain series on the ocean floor, essentially enfolding the Earth. With this regard, the British National Antarctic Expedition, which is known as the Discovery Expedition, was lunched. The Discovery Expedition spent 3 years cruising across the world oceans and documented a lot of significant ground information regarding biology, zoology, geology, meteorology and magnetism.

Recently, coastal oceanography and marine environment are stressed because of a rapid development of technologies and in industries. Under these circumstances, coastal oceanography and marine environment are damaged due to harmful effects of pollution. Consequently, understanding of physics, chemistry, geology, biology, atmospheric sciences and, in many cases, mathematical models is required for comprehensive understanding of the coastal and marine environment sustainability. Without the mathematical and numerical models, oceanographers cannot study the complexity of coastal oceanography and marine environment. Indeed, this can assist to forecast, offset and minimize the coastal disaster and marine pollution impacts.

A broad, thoughtful coastal oceanography and marine environment cannot be built without a comprehensive understanding of geology, biology, chemistry, marine ecosystem, physics, mathematics, numerical analysis and computer modelling. Some of oceanography academic and research institutes, however, have a narrow point of view regarding coastal oceanography and marine environment. Those academic institutes are using only sophisticated equipment for data collection from coastal waters. Ocean data observations are not end work of scientists and oceanographers because they need to go beyond in situ data collections by developing models and solving the boundary condition problems. The low quality of oceanography academic institutes can lead to hiding of many of the cities which are located along the coastline due to sea level raising. For my point of view, oceanography scientists must be graduated from faculty of oceanography under different aspects such as physical oceanography, marine chemistry and marine biology. For instance, in some academic institutes are just appointed civil engineering or biology graduates to be expert on physical oceanography.

It is a hard task to cover all the aspects of coastal and marine studies in this book. Instead, this book is a collection of a number of high-quality and comprehensive work on coastal and marine environment. I have explained briefly the concept and significance of coastal oceanography in the Introduction chapter. In fact, there are many of communities, for instance, in

Egypt, where coastal oceanography studies are totally ignored because they do not know the great value of coastal information, which can bloom the country's economy. Consequently, the following high-quality book chapters cover a wide range of coastal and marine environment studies. This book has an Introductory Chapter, followed by 15 chapters. Chapters 2 and 3 are devoted to coastal geological sedimentation and its impacts on marine environment. Consequently, Chapter 4 investigates neo-tectonic movement in the Pearl River Delta. Different aspects of the coastal pollution and its impacts are addressed in Chapter 5 through Chapter 13. Furthermore, coastal management is also discussed in Chapter 14, and monitoring the coastal environment using remote sensing and GIS techniques is reported in Chapter 15. Finally, Chapter 16 addresses the human history of maritime exploitation and adaptation process to coastal and marine environments. It is important to investigate the history of maritime exploitation and adaptation to environment coastal zone to learn how to explore the oceans.

I wish to convey my appreciation to all authors who have contributed to this book. Without their intense commitment, this book would not have become such a precious piece of knowledge. I am also grateful to the InTech editorial team, specially Ms. Iva Simcic, who has afforded the opportunity to publish this book.

Maged Marghany,Ph.D
Geospatial Information Science Research
Centre, Faculty of Engineering
University Putra Malaysia
Serdang, Selangor, Malaysia

Introduction

Chapter 1

Introduction to Coastal and Marine Environment Concepts and Significances

Maged Marghany and Shattri Mansor

Additional information is available at the end of the chapter

http://dx.doi.org/10.5772/64621

1. Introduction

Oceanography is the investigation of features found in coastal water. It extends from shallow estuaries to deep oceans, everywhere sea water can be found. This involves a broad scope of issues, beginning with the physical aspects of coastal dynamics and coastal sedimentation to the seafloor's geology and ecosystems comprised by marine life. Consequently, the study of the ocean is always a challenging and intriguing process when considering its size and global expanse. Features of oceans are complex and challenge for human beings to comprehend. All life and climatic changes depend on ocean dynamic fluctuations. There is no doubt that oceans are important to life on this planet. It is not easy to understand oceans because there are many factors which control their features. Scientists cannot study these factors individually as this will deliver imperfect information. For example, the chemical components of water shape the nature of which organisms inhabit the ocean. Consecutively, marine organisms supply deposits to shape the ocean floor's geology.

2. Subdisciplines of oceanography

There are four major sub-disciplines of oceanography. These broader subjects are (i) marine chemistry, (ii) marine geology, (iii) marine biology and (iv) physical oceanography.

2.1. Marine chemistry

This is the examination of oceanic chemistry. With regard to this, the ocean contains practically all the element which exist in the periodic table. Therefore, the dynamic changes of these

---

chemical elements can be studied within the Earth's oceans. Scientists study the dynamic cycling of these elements in both oceans and adjacent land. In addition, scientists identify advanced cycles - known as a biogeochemical cycles, the cycling of substances or substance turnover. These are considered routes by which a chemical substances interchanges in the biosphere (i.e., biotic and lithosphere), atmosphere and hydrosphere (i.e., abiotic partitions of the Earth [3]).

The term "biogeochemical" expresses the involvement of biological, geological and chemical factors in ocean studies. For instance, the distribution of carbon, oxygen, nitrogen, phosphorus, calcium and water, etc., within the physical and biological system – this is known as a biogeochemical cycle [3].

2.2. Marine geology

Marine geology studies the early and recent structure of the ocean floor. It also explores the heritage of the dynamic changes and the development of underwater features within the landscape. Marine geology correspondingly investigates the chemical and physical properties of ocean floor sediments and rocks. In addition, it implies geochemical, geophysical, sedimentological and paleontological examination of the ocean floor and coastal zones. Thus, marine geology has extremely robust bonds to physical oceanography (e.g., alongshore currents which are induced because the oblique wave propagation can cause sediment transport along coastal waters [9]).

The study of the ocean floor is a corner stone to predicate the Earth's climate changes. The investigation of climate and weather patterns over hundreds of millions of years is a study known as paleoclimatology - under ocean floor sediments can be collected using resolution drills taking sediment-core samples[7, 9].

2.3. Marine biology

Marine biology investigates exactly how every one of the sub-fields of oceanography operate either discretely or simultaneously to affect the circulation of animals and marine plants along with what purpose marine organisms perform relative to their environment. Marine biology additionally emphasizes the way species fit with conservational fluctuations (e.g., increased water temperature, artificial and natural instabilities, and pollution). A natural turbulence can be caused by the explosion of an underwater volcano or a hurricane, whilst an oil spill and overfishing represent an artificial disturbance [9, 11].

Marine biotechnology is a new technology which is expanding the opportunity to investigate in which way marine resources can be expended to develop medical industry, ecological inventions, and food technology.

2.4. Physical oceanography

Physical oceanography models the dynamic relationship between the ocean's physical properties, the atmosphere, and the coast and seafloor. The ocean's physical properties include temperature, salinity, and density variations. The ocean's dynamic components are waves, current, and tides. Physical oceanographers additionally investigate the manner by which the ocean interacts with the Earth's atmosphere to generate climate and weather systems [2, 7, 12].

Under this circumstance, physical oceanography theories and models must extremely be instigated to investigate the mystery of flight MH370 [1, 17]. The observational procedures which are taught to undergraduates of physical oceanography do not work in this case. Indeed, standard and modified models are required to verify the information from the *Inmarsat* satellite. In fact, there are many researchers who just use physical oceanography models and do not really understand how the models operate. Physical oceanography scientists who have been correctly trained have deep background knowledge of physics, mathematics, numerical analysis, finite element models, and modelling and can verify information delivered by any source [6]. To date, some marine institutes with a low quality of physical oceanography research are unable to explain, scientifically, how flight MH370 vanished in the southern Indian Ocean. It is believed that researchers are not even able to solve the 1-D mathematical coastal water flow equation! In addition, these researchers are blind users of software, and in situ ocean observations in the water territories without perfect understanding of the principle theories. Ocean data observations do not represent the end work of scientists - they need to go beyond in situ data collection by developing models and solving boundary condition problems.

For instance, the turbulent water flow throughout the southern top of Africa - known as the Agulhas Current - is a portion of a superior "ocean conveyor belt" which distributes the water across the globe. The Agulhas Current is based on currents, wind, and water density changes. Scientists have defined a new finding, named the Agulhas leakage. This represents the increments of water flow from the southern Indian Ocean into the Atlantic Ocean. It may well be that flight MH370 debris moved to the Atlantic Ocean through the Agulhas leakage [7, 8].

Modern physical oceanography studies predict that the ocean conveyor belt will be slow down, something which could cause dramatic changes to weather and climate patterns, and consequently sea level rise due to the melting of ice caps. This melting could reduce the salinity of ocean water and change its biogeochemical matrix.

3. The significance of oceanography and coastal studies

The main question which can of course be raised is why ocean and coastal studies are important to our lives? This book covers some significant aspects of coastal studies. There are countless benefits to be gained from coastal studies which may represent solutions to many economic crises around the world. Due to the huge coverage of the oceans, which are approximately 70% of the Earth's surface, the ocean plays a tremendous role for transportation, natural resources, weather and climate, biodiversity, and the economy[2, 8].

3.1. Transportation

Cities which are located on the coast and have harbors always have great benefits, however, they are exposed to pollution from time to time [2, 7].

3.2. Natural resources

The ocean floor is rich in oil and natural gases - not only for benthic organisms. The ocean floor and continental shelf contains numerous minerals. However, 50% of nations cannot discover the wealth of resources stored in their continental shelves and ocean floors [2, 8, 11].

3.3. Weather and climate

The oceans interact in tandem with the atmosphere and influence worldwide climate and weather. Whilst the air circulates over the warm ocean, it escalates because of warming. When it cools, the condensation induces precipitation. For instance, the Gulf Stream, synchronized withwarm air flowing toward northern Europe, makes winters tolerable [5, 12, 16].

3.4. Biodiversity

A wide number of discrete species of organisms have an extraordinary biodiversity due to the existence of salt marshes, coral reefs, estuaries, seagrass beds, and mangrove [15]. The brackish water of estuaries is key for fish and others marine organisms. Furthermore, coral reefs sustain, along with estuaries, approximately 75% of living marine organisms and fish. With this regard, reefs and estuaries represent shelter for fish because of the accumulation of food and nutrients both required for feeding and nursing [14]. Moreover, birds reside along the sandy shores which represent a source of food – e.g., burrowing worms and fiddler crabs [4, 5, 8, 11, 13].

3.5. Economy

The significance of the ocean cannot be deliberated easily. Economic income is a function of shipping, marine biomedical products, fish industries, and oil and gas mining. Furthermore, beautiful beaches and islands are a corner stone for income as they attract a large number of tourists. In general coastal studies are important to policy makers, allowing them to minimize the effects of marine pollution to ecologically sensitive zones along the shorelines. Hence, pollution-free zones will attract more tourists and boost national economies [8].

3.6. Coastal engineering

Coastal engineers also required information about ocean and coastal dynamics before handing out offshore construction contracts. Furthermore, coastal studies can sustain shoreline changes and assist decision-makers in understanding the mechanisms of sediment transport in order to avoid erosion and sedimentation problems. In addition, information about salinity and temperature can be used to avoid the problem of corrosion to ships, pipelines, and coastal structures containing iron [3, 8].

4. Outline of this book

This book examines the impact of physical oceanography parameters on several applications which range from mercury circulation, corrosion due to salinity, and coastal sedimentation to monitoring the coastal environment using geographical information systems (GIS) and remote sensing. Furthermore, the marine ecology study in this book covers a broad spectrum of clues about the physico-chemical properties of coastal waters. The significance of this study is the integration between different aspects of oceanography with wind monsoon pattern changes. Consequently, salinity and phytoplankton are the main factors that impact zooplankton growth and proliferation [10,11]. Most of the work represented in this book is based on accurate in-situ observationIndeed, the British National Antarctic Expedition which is known as the *Discovery Expedition* spent three years cruising across the world's oceans documenting significant ground information regarding biology, zoology, geology, meteorology, and magnetism. Furthermore, the expedition discovered the existence of the only snow-free Antarctic valleys.

This book contains 16 chapters, as follows:

Chapter 2 [Geological Evolution of Coastal and Marine Environments off the Campania Continental Shelf Through Marine Geological Mapping - The Example of the Cilento Promontory]. This chapter presents the geological evolution of coastal and marine environments offshore the Cilento Promontory, Italy, through marine geological mapping.

Chapter 3 [Engineering Tools for the Estimation of Dredging- Induced Sediment Resuspension and Coastal Environmental Management]. This chapter introduces predictive techniques used to estimate the suspended solid concentration arising from dredges with different mechanisms of sediment release and assesses the spatial and temporal variability of the resulting plume in estuarine and coastal areas.

Chapter 4 [Neo-tectonic Movement in the Pearl River Delta (PRD) Region of China and Its Effects on the Coastal Sedimentary Environment]. This chapter presents the Late Quaternary neo-tectonic movement in the Pearl River Delta (PRD) region of China and its effects on the coastal sedimentary environment. Furthermore, it discusses the evolution of the PRD formation and its controlling factors on the coastal sedimentary environment.

Chapter 5 [Review of Mercury Circulation Changes in the Coastal Zone of Southern Baltic Sea]. This chapter reviews how a surge in mercury concentration in marine organisms is not always linked to an increased inflow of anthropogenic mercury, directly influenced by humans, but may also be related to climate change occurring in a given area, the type of catchment, as well as to the predominance of either marine or terrestrial air masses.

Chapter 6 [Computatonal Analysis of System and Design Parameters of Electrodeposition for Marine Applications]. This chapter investigates marine corrosion in coastal water. In addition, it also addresses factors which induce marine corrosion.

Chapter 7 [Fabrication and Properties of Zinc Composite Coatingsfor Mitigation of Corrosion in Coastal and Marine Zone.] This chapter shows how electrodeposited zinc composite coatings have high potential to serve as protective films against corrosion for components and pieces that are made of mild steel in coastal and marine environments.

Chapter 8 [Dynamic Characterization of Open-ended Pipe Piles in Marine Environment]. This chapter introduces experimental investigations used to study dynamically characterized open-ended pipe piles in the marine environment.

Chapter 9 [Relationship between Bulk Metal Concentration and Bioavailability in Tropic Estuarine Sediments]. This chapter investigates the abundance, distribution, and bioavailability of major and trace elements in the surface sediments of the Cai River estuary, Russia.

Chapter 10 [Stable Isotope Techniques to Address Coastal Marine Pollution]. This chapter explores stable isotopes of carbon (δ^{13}C), sulfur (δ^{34}S), oxygen (δ^{18}O), hydrogen (δ^{2}H), nitrogen (δ^{15}N), and a radioactive isotope of hydrogen (tritium) in order to study coastal pollution. This is new approach is compared to classic methods, such as heavy metals.

Chapter 11 [Investigation of Po-210 and Heavy Metal Concentration in Seafood Due to Coal Burning – Case Study in Malaysia]. This chapter investigates the level of Po-210 and heavy metals in marine organisms from the coastal area of Kapar and makes comparisons with other places around the world.

Chapter 12 [Atmospheric Pollution Causes Deterioration of Sweeteners of Treats and Decreases Competitiveness in the Food Industry of Coastal Baja California, Mexico]. This chapter presents the impact of atmospheric pollution on the food industry along the coastal zone of Mexico.

Chapter 13 [Depositional Environment of Phosphorites of the Sonrai Basin, Lalitpur District, Uttar Pradesh, India]. This chapters evaluates the governing factors involved during the process of phosphorite formation, and deposition. In addition, it investigates the marine mineralogy and chemistry of the various rocks.

Chapter 14 [Management of Marine Protected Zones – Case Study of Bahrain, Arabian Gulf]. This chapter explores the ecological and legal contexts of marine protected areas (MPAs) in Bahrain, and evaluates their effectiveness.

Chapter 15 [Monitoring the Coastal Environment Using Remote Sensing and GIS Techniques]. This is only the chapter in this book which involves remote sensing and GIS techniques in order to monitor the coastal environment.

Chapter 16 [Human History of Maritime Exploitation and Adaptation Process to Coastal and Marine Environments – A View from the Case of Wallacea and the Pacific]. This chapter investigates the historical maritime exploitation and adaption of the environmental coastal zone.

Author details

Maged Marghany* and Shattri Mansor

*Address all correspondence to: magedupm@hotmail.com

Geospatial Information Science Research Centre, Faculty of Engineering, University Putra Malaysia, Serdang, Selangor, Malaysia

References

[1] Marghany M. (2014). Developing genetic algorithm for surveying of MH370 flight in Indian Ocean using altimetry satellite data. 35th Asian conference of remote sensing, at Nay Pyi Taw, Mynamar, 27-31 October 2014. a-a-r-s.org/acrs/administrator/components/com.../OS-081%20.pdf.

[2] Defant, Albert. "Physical oceanography; volume 2." (1961).

[3] Riley, John Price, and Roy Chester, eds. *Chemical oceanography*. Elsevier, 2013.

[4] Philip, N. "Kinetic control of dissolved phosphate in natural rivers and estuaries: A primer on the phosphate buffer mechanism1."*OCEANOGRAPHY* 33.4 -Part 2 (1988).

[5] Officer, Charles B. "Physical Oceanography Of Estuarines." (2013).

[6] Thomson, Richard E., and William J. Emery. *Data analysis methods in physical oceanography*. Newnes, 2014.

[7] Tomczak, Matthias, and J. Stuart Godfrey. *Regional oceanography: an introduction*. Elsevier, 2013.

[8] Dietrich, Günter, et al. "General oceanography." (1980).

[9] Carpenter, K. E. "An introduction to the oceanography, geology, biogeography, and fisheries of the tropical and subtropical western and central Pacific." *FAO Species Identification Guide for Fishery Purposes. The Living Marine Resources of the Western Central Pacific, FAO, Rome* (1998): 117.

[10] Marghany, M. "Linear algorithm for salinity distribution modelling from MODIS data." *Geoscience and Remote Sensing Symposium, 2009 IEEE International, IGARSS 2009*. Vol. 3. IEEE, 2009.

[11] Legendre, L_, and S. Demers. "Towards dynamic biological oceanography and limnology." *Canadian Journal of Fisheries and Aquatic Sciences* 41.1 (1984): 2-19.

[12] Navon, I. M. "Practical and theoretical aspects of adjoint parameter estimation and identifiability in meteorology and oceanography." *Dynamics of Atmospheres and Oceans* 27.1 (1998): 55-79.

[13] Doney, Scott C. "Oceanography: Plankton in a warmer world." *Nature*444.7120 (2006): 695-696.

[14] Cury, Philippe Maurice, et al. "Ecosystem oceanography for global change in fisheries." *Trends in Ecology & Evolution* 23.6 (2008): 338-346.

[15] Cao, Yong, D. Dudley Williams, and Nancy E. Williams. "How important are rare species in aquatic community ecology and bioassessment?."*OCEANOGRAPHY* 43.7 (1998).

[16] Stabeno, P. J., N. A. Bond, A. J. Hermann, N. B. Kachel, C. W. Mordy, and J. E. Overland. "Meteorology and oceanography of the Northern Gulf of Alaska." *Continental Shelf Research* 24, no. 7 (2004): 859-897.

[17] Marghany, M. "Intelligent optimization system for uncertainty MH370 Debris Detection." 36th Asian conference of remote sensing, acrs2015.ccgeo.info/proceedings/TH4-5-6.pdf. Crowne Plaza Manila Galleria in Metro Manila, Philippines, October 19, 2015 - October 23, 2015.

Coastal Geological Sedimentation

Geological Evolution of Coastal and Marine Environments off the Campania Continental Shelf Through Marine Geological Mapping - The Example of the Cilento Promontory

Gemma Aiello and Ennio Marsella

Additional information is available at the end of the chapter

http://dx.doi.org/10.5772/61738

Abstract

The geological evolution of coastal and marine environments offshore the Cilento Prom‐
ontory through marine geological mapping is discussed here. The marine geological map
n. 502 "Agropoli," located offshore the Cilento Promontory (southern Italy), is described
and put in regional geologic setting. The study area covers water depths ranging between
30 and 200 m isobaths. The geologic map has been constructed in the frame of a research
program financed by the National Geological Survey of Italy (CARG Project), finalized to
the construction of an up-to date cartography of the Campania region. Geological and ge‐
ophysical data on the continental shelf and slope offshore the southern Campania region
have been acquired in an area bounded northward by the Gulf of Salerno and southward
by the Gulf of Policastro. A high-resolution multibeam bathymetry has permitted the
construction of a digital elevation model (DEM). Sidescan sonar profiles have also been
collected and interpreted, and their merging with bathymetric data has allowed for the
realization of the base for the marine geologic cartography. The calibration of geophysical
data has been attempted through sea-bottom samples. The morpho-structures and the
seismic sequences overlying the outcrops of acoustic basement reported in the carto‐
graphic representation have been studied in detail using single-channel seismics. The in‐
terpretation of seismic profiles has been a support for the reconstruction of the
stratigraphic and structural setting of the Quaternary continental shelf successions and
the outcrops of rocky acoustic basement in correspondence to the Licosa Cape morpho‐
structural high. These areas result from the seaward prolongation of the stratigraphic and
structural units, widely cropping out in the surrounding emerged sector of the Cilento
Promontory. The cartographic approach is based on the recognition of laterally coeval
depositional systems, interpreted in the frame of system tracts of the Late Quaternary
depositional sequence.

Keywords: Marine geological maps, seismic stratigraphy, Agropoli offshore, Cilento
Promontory, Southern Tyrrhenian Sea

1. Introduction

The geological evolution of coastal and marine environments in selected areas of the Tyrrhenian offshore (Cilento Promontory and Naples Bay, Campania, Italy) is highlighted through the construction of marine geological maps. The geological map n. 502 "Agropoli" (scale 1:50.000, Cilento Promontory) includes the coastal sector offshore the Cilento Promontory between the Agropoli and Agnone towns (southern Campania, Italy) and the surrounding marine areas. Sidescan sonar data have been calibrated by sea-bottom samples. The geological structures overlying the outcrops of acoustic basement have been investigated using subbottom Chirp profiles. The interpretation of Chirp data lends support for the reconstruction of the stratigraphic and structural setting of the Quaternary continental shelf successions and correlations to the outcrops of rocky acoustic basement in the Licosa Cape high. These areas result from the seaward prolongation of the stratigraphic and structural units, widely cropping out in the surrounding emerged sector of the Cilento Promontory(Southern Tyrrhenian Sea) [1; 2].

The techniques of sequence stratigraphy have been applied in the construction of geologic maps, according to the guidelines of marine geologic cartography [3; 4]. Local unconformities have been recognized at the top of pockets of coarse-grained materials, deposited in depressions or channels carving the top of the acoustic basement. The polycyclic nature of several unconformities is suggested by terraced surfaces localized at different water depths and recognized on seismic profiles. Several phases of emersion and erosion coupled to the development of marine terraces and progressive flooding occurred, as suggested by the seismic interpretation. The glacio-eustatic sea-level fluctuations during the Late Quaternary and the Pleistocenic uplift affecting the whole area represented the main control factors of marine-terraced surfaces [5; 6].

In the frame of the research convention for the realization of the geological map n. 502 "Agropoli," the research projects co-financed in the CARG Project financed by the Region Campania, Sector of Soil Defence, Geothermy and Geotechnics are worth mentioning. These projects mainly include the research project GEOSED (financed by the Italian Ministry of the University and Scientific and Technologic Research) for geomorphologic and sedimentologic researches on continental shelf of southern Italy. The GEOSED project, which is complete, includes the geologic and morphologic survey at the scale 1:25.000 of selected areas of continental shelf of southern Italy (Capri Island, Punta Campanella, Ischia Island, Li Galli islands, and coastal belt surrounding the Naples and Pozzuoli towns). In the frame of this project, marine geophysical data have been recorded, mainly including multibeam bathymetry, around the Licosa Cape, allowing for the realization of a high-resolution digital elevation model (DEM) of the marine area surrounding the Licosa Cape [7].

The GEOSED acoustic project and swath bathymetric surveys carried out around the Licosa Cape and the Castellabate littoral (Salerno) have allowed to evaluate in detail the main environmental characters of the sea bottoms, furnishing a term of comparison for future observation in terms of environmental monitoring. These surveys have been realized in the frame of the GEOSED project (geomorphologic and sedimentologic study in selected areas of the continental shelf of southern Italy) finalized to the acquisition of geological knowledge on the sea bottom and on the first subbottom in relationship to the management of the coastal belt. The used instruments, in dotation to the R/V Thetis (Consiglio Nazionale delle Ricerche (CNR), Italy), included a lateral scanning probe (100–500 kHz digital Towfish – 2260NV) linked to a computerized station (ISIS Sonar Triton Elics) for sidescan sonar surveys and a multibeam system SEABAT 8111 Tx/Rx (Thales Inc.) with an operative frequency of 100 kHz. The performed surveys have allowed to individuate the morphological characteristics (terraces, vertical surfaces, grade of steepness, and drainage axes). The sea-bottom lineaments (ripples, current lineations, fishery lineations, and sea-bottom undulations), the occurrence of algal grasses (*Posidonia oceanica*), the processes of gravity instability, and the sedimentary dynamics at the sea bottom have allowed tracing the sediment distribution in terms of acoustic facies. The preliminary analysis of aerial photographs and official cartography has furnished the necessary hints for a geologic setting of adjacent emerged areas, allowing the establishment of land–sea correlations, a necessary element for a correct approach to the study and the management of coastal belts.

During the start-up phase of the research project of realization of a geologic map, a detailed collection of bibliographic data existing on the area of the Cilento Promontory and surrounding offshore has been carried out. The bibliographic research has contributed to delineate the state of the art of previous knowledge on the study of marine area and on the geologic knowledge on the adjacent emerged sectors (Cilento Promontory and Bulgheria Mt.). The collection of bibliographic data has evidenced that, while a good scientific knowledge exists on the emerged areas surrounding the geological map n. 502 "Agropoli," including part of the Cilento Promontory, in particular the stratigraphic and structural aspects, due to the contribution of institutions and research settings operative in Campania from several tens of days (in particular some research groups pertaining to the Earth Science Department of the University of Naples "Federico II"), a poor knowledge exists on the recent geologic setting of the marine areas surrounding the Agropoli map, to whom concerns the sea bottom and subbottom geology. This knowledge is due to a few articles of marine geology carried out by the Istituto di Geologia Marina, CNR, Bologna (now ISMAR-CNR, Sezione di Geologia Marina di Bologna, Italy) and by the research group coordinated by Prof. T.S. Pescatore, Department of Earth Sciences, University of Naples "Federico II," Italy.

Some papers will be recalled here in order to better focus on the research themes of this chapter [8- 13]. Marghany [8] has focused on a three-dimensional coastal deformation based on interferometry synthetic aperture radar (InSAR). Conventional InSAR procedures have been implemented to three repeat passes of ENVISAT (Environmental Satellite) ASAR (Advanced Synthetic Aperture Radar) data. The InSAR technique has allowed to produce detailed maps of earth's surface deformation or digital terrain models through the implementation of the single-look complex synthetic aperture radar (SAR) images, which are received from two or more separate antennas. Moreover, the three-dimensional sorting reliability algorithm (3D-SRA) has been implemented by using the phase unwrapping technique. The aforementioned algorithm has been used to eliminate the phase decorrelation impact from the interferograms. The integration of the 3D-SRA algorithm with the phase unwrapping has produced accurate models of three-dimensional models of coastal deformation. Marghany [9] has presented a new approach to the 3D spit simulation using a differential synthetic aperture interferometry

(DInSAR). The conventional procedures of this interferometry have been implemented to three repeat passes of RADARSAT-1 SAR fine mode data. A new application, based on the fuzzy B-spline algorithm, has been implemented with the phase unwrapping technique. This analysis has shown that the performance of the DInSAR method using fuzzy B-spline is better than the DInSAR method. The integration of the fuzzy B-spline with the phase unwrapping has produced detailed three-dimensional geomorphological reconstructions. Marghany [10] has developed a new approach for the simulation of wave refraction patterns in airborne radar data. The analysis has shown that the wave refraction patterns can be simulated from AIRSAR and POLSAR data with a convergence and divergence spectra energy of 0.84 and 0.4 m^2 respectively. The modification of conventional azimuth cutoff algorithm can be used to retrieve significant wave height in Cvv-band data under circumstances of wave transformation using first-order partial differential equations (PDEs). Marghany [11] has presented a three-dimensional model of coastal deformation using the sorting reliability algorithm of ENVISAT (Environmental Satellite) interferometric synthetic aperture radar. The InSAR procedures have determined discontinuous interferogram patterns because of the high decorrelation. On the contrary, the 3D-sorting reliability algorithm has generated a 3D shoreline deformation with a bias of 0.08 m, lower than the ground measurements and the InSAR method. The 3D-SR algorithm has been used to solve the problem of decorrelation and has produced accurate 3D coastal deformation using ENVISAT (Environmental Satellite) ASAR (Advanced Synthetic Aperture Radar) data. Marghany (2014c) has studied the simulation of tsunami impact on the salinity of the sea surface in the coastal waters of Banda Aceh (Indonesia) using the least square algorithm from MODIS satellite data. The analysis has shown significant variations in the values of the SSS pro during and after the tsunami event. The maximum salinity of the seawaters observed after the tsunami event was 38 psu, higher than the salinities observed during and after the tsunami event. The 2004 tsunami had a significant impact on the sea surface salinity because of the high-sediment deposit concentrations, which added more salts and minerals to the coastal waters of Banda Aceh. The study of Marghany [13] was aimed at modeling the three-dimensional shoreline change rates using the differential interferometric synthetic aperture radar (DInSAR) technique. Nonetheless, the decorrelation plays a significant role to control the accuracy of three-dimensional object reconstruction using the aforementioned technique. This problem has been solved through the implementation of the multichannel MAP height-estimator algorithm within the ENVISAT ASAR data. The developed method has been applied to the example of the shoreline of Johor (Malaysia). The obtained results have shown that the accurate rate of shoreline change can be obtained with a root mean square error (RMSE) of 0.05 m using a multichannel height-estimator algorithm.

2. Geologic setting

2.1. The Eastern Tyrrhenian margin between the Gulf of Salerno and the Gulf of Policastro

The origin of the sedimentary basins and their subsidence along the Campania sector of the Eastern Tyrrhenian margin have been the subject of several studies based on field geological observations and on seismic and well data onshore and offshore [14- 39].

The interpretation of multichannel seismic data pertaining to the "Zone E" (Ministry commercial seismics), together with the subsidence analysis carried out on lithostratigraphic data of deep wells, has shown notable differences in the long-term trends expected for the subsidence in the Salerno Basin [40], allowing for the recognition of late stages of compressional tectonics in the evolution of the sedimentary basin. This compressional tectonics, showing an age between the Late Pliocene and the Early Pleistocene, has involved the Campania offshore, between the Sorrento Peninsula and the Palinuro Cape. In correspondence with the opening of the Tyrrhenian basin, the extensional and strike-slip tectonics has been developed in this area since the Late Miocene, mainly after the compressional phases which have constructed the western sectors of the southern Apennines. The seismic profiles have demonstrated that the tectonic deformation was verified along a main system of listric normal faults, having an NE–SW trending, controlling the present-day setting of half-graben of the Salerno Basin. The compressional tectonic phase, ranging in age from Late Pliocene to Early Pleistocene, having an NNW–SSE trending for the maximum horizontal shortening, has provoked a reactivation of preexisting systems of normal faults having an NE–SW trending and has been responsible for processes of tectonic inversion in the Salerno Basin. This has provoked the deposition of sedimentary sequences characterized by shallowing upward at the top of the compressional structures during their formation, as well as the deformation and the flexural uplift for the subaerial exposure of the flanks of the basin (Mesozoic carbonate reliefs together with the basin filling of the Early Pleistocene) and the subsidence in the central sector of the basin. According to the geologic interpretation of Sacchi et al. [40], the compression had a main role in enhancing the tectonic uplift and the subsidence; the tectonic phases recognized seaward may be tentatively related to the neotectonic evidence documented landward. After the Late Pleistocene, a new and generalized tectonic subsidence started in the Salerno Basin and around the southern rim of the Picentini Mountains, toward the inner part of the Sele Plain. The tectonic deformation and the erosional truncations in the Late Pleistocene sequences in the southern Salerno offshore, as well as the local tectonic uplift of the deposits of the latest Pleistocene several meters above the present-day sea level, suggest that the compressional events have been active since recent times. Hints of Late Neogene to Quaternary compressional tectonics in extensional domains have been already evidenced in other sites of the Mediterranean and North Atlantic regions; the evidence of these compressional phases is not directly related to thrust and belt deformation, which may be characterized by regional extension, may be related with main intraplate stresses.

The continental shelf between the Gulf of Gaeta and the Cilento Promontory represents the seaward prolongation of the alluvial coastal plains bounding the Tyrrhenian sector of the Apenninic chain, which has been individuated after strong rates of tectonic subsidence in a regime of continental stretching and extension. These alluvial plains (Campania Plain, Sarno Plain, and Sele Plain) are bounded toward NE by the inner reliefs of the Apenninic chain, whose continuity is interrupted by structural highs having an NE–SW trending and by volcanic complexes (Phlegrean Fields and Somma-Vesuvius). The sedimentary filling consists of clastic marine and continental deposits, alternating, both in the Campania Plain and in the Sarno Plain, with abundant volcanic complexes. The age of this filling is mainly Pleistocene, although a precise dating of the first sequences overlying the acoustic basement is lacking [5]. The acoustic basement is represented by Meso-Cenozoic carbonates (Campania-Lucania platform) [1]and by deformed Cenozoic sequences and related flysch deposits (Cilento Flysch; Sicilide Units, Liguride Units) [41]. The southern boundary of the Campania Plain is represented by the volcanic district of the Phlegrean Fields, which has been active for the past 50 ky [42].

The Tyrrhenian margin of Campania is characterized by the occurrence of marine areas undergoing a strong subsidence during the Plio-Quaternary and site of strong sedimentary fillings ("peri-Tyrrhenian basins"), such as the Terracina and Gaeta basins, the Capri basin, the Gulf of Naples, the Salerno Valley, and the Sapri and Paola basins. Under the Plio-Quaternary sedimentary cover, the continental margin of Campania is characterized by the occurrence of inner tectonic units of the Apenninic chain, resulting from the seaward elongment of the corresponding units cropping out in the coastal sectors of southern Apennines [1].

Bartole [16] has formulated a regional synthesis on the tectonic setting of the Campania continental margin based on seismic and well data, where four main seismo-stratigraphic units have been recognized.

1. Upper post-orogenic sequence, composed of Early–Middle Pliocene sediments overlying a regional unconformity;

2. Lower post-orogenic sequence, ranging in age from Late Miocene to Early Pliocene;

3. Chaotic sequence, formed by strongly deformed Cenozoic sediments and related flysch deposits (Sicilide Units, Liguride Units, Frosinone Flysch, Cilento Flysch Auct.)

4. Deep carbonate sequence, forming the local acoustic basement, composed of Mesozoic carbonate platform units, widely cropping out onshore in the coastal sectors of central-southern Apennines.

According to the interpretation of Bartole et al. [17] and Bartole [16] in the Campania Tyrrhenian margin, both orogenic structures (Palmarola-Terracina overthrust, Zannone-Volturno overthrust, Salerno Valley, Sele line, Palinuro structure) directly related to the tectonic phases of orogenic transport of the Apenninic chain (Aquitanian–Langhian, Tortonian, Messinian, and Early Pliocene) [1; 43] and post-orogenic structures produced by Plio-Pleistocene extensional tectonics have been distinguished.

Main orogenic structures, distinguished by these authors, in the offshore among the Gaeta and the Campania Plains are the Palmarola-Terracina overthrust, which is a deformational belt Messinian in age and is NE–SW trending, related to right strike-slip faulting during the Early Pliocene and the Zannone-Volturno overthrust, an N-verging thrust of the Middle Miocene having an E–W trending. The two described compressional trends seem to evidence a tectonic transport toward N–NE of tectonic units referred as Sicilide Units.

The orogenic lineaments recognized in the offshore between the Gulf of Salerno and the Gulf of Policastro are the Salerno Valley, the Sele line, and the Palinuro structure. The Salerno Valley is represented by a deep depression having a WSW–ENE elongment, bounded northward by the normal faults downthrowing Mesozoic carbonates of the Sorrento Peninsula and filled with a Plio-Quaternary sequence up to 3300 m thickness (Mina 1 well); [44].

The neotectonic movements responsible for the uplift of southern Apennines start from the Early Pliocene and explicate through some extensional phases since the Middle–Late Pleistocene. The Quaternary neotectonics has played a fundamental role in constructing the present-day morphology of the Campania-Latium Tyrrhenian margin. This is demonstrated, for example, by the Quaternary marine and continental sedimentation of the Campania coastal plains, reaching a thickness of 3000 m and 1500 m in the Volturno Plain and the Sele Plain, respectively [43].

The structures having an Apenninic trending characterize the continental slope areas extending from the Pontine Islands (Ventotene slope) to the Cilento shoreline. The structural lineaments having a counter-Apenninic trending are widespread in the offshore between the Gulf of Salerno and the Sorrento Peninsula, where they have originated parallel structures having an NE–SW trending, giving rise to a horst-and-graben submarine topography.

The half-graben sedimentary basins which characterize the Campania margin are interpreted, as a general rule, as the result of two extensional events, Pleistocene in age. The first event, which was developed during the Early Pleistocene along normal faults having an NW–SE trending, has allowed for the formation of coastal depressions of the Campania Plain, of the Sele Plain, and of the Volturno Plain. The second event, superimposed on an NE–SW trending fault system, at the boundary between the Early Pleistocene and the Middle Pleistocene, has produced the formation of the Naples Bay depression, producing half-graben-type asymmetrical structures, which characterize the whole Campania continental margin.

The offshore between the Sorrento Peninsula and the Cilento Promontory, hosting the structural depression of the Salerno Bay, is characterized by a continental shelf relatively wide and extended (up to 25 km) westward of the Cilento shoreline, with a thin sedimentary cover. Southward of the Sorrento Peninsula a reduced continental shelf and steep continental slopes occur, bounding the structural depression of the Salerno Basin. The individuation and the tectonic setting of the Salerno Basin have been mainly controlled by the Capri-Sorrento Peninsula master fault, showing a WSW–ENE trending (with an average throw of 1.5 km), bounding southward the structural high of the Sorrento Peninsula, where Mesozoic carbonate platform deposits widely crop out. Previous studies in the Salerno and Cilento offshore have mainly regarded the stratigraphy and the sedimentology of the Quaternary deposits [19; 45]. Minor attention has been paid to the structural setting of the area intensively explored by Italian oil companies [44; 20].

A Mesozoic carbonate platform unit, occurring in the subsurface of the Gulf of Salerno, has been recognized by deep exploration wells Mina 1, Milena 1, and Margherita mare 1 (VIDEPI Project; www.socgeol.it) [44] and is composed of limestones and dolomitic limestones, Cretaceous in age. The maximum thickness of the unit is in the order of 1300 m at the Milena 1 well. This unit is correlated with the Taburno-Picentini and Alburno-Cervati units, respectively, cropping out to the north and to the south of the fault zone known as the Sele line [1; 46; 41; 40; 21; 22]. This unit represents the carbonate acoustic basement on seismic profiles.

A siliciclastic unit, Miocene in age, unconformably overlying the Meso-Cenozoic carbonate unit has been drilled by deep lithostratigraphic wells only in the Cilento offshore southward of the Sele line and has been correlated with the Liguride Units and the related foredeep deposits (Cilento Flysch), cropping out onshore in the Cilento Promontory [41].

The filling of the Salerno Basin starts (at the lithostratigraphic well Mina 1) [44] with the deposition of shales and marls ranging in age from the Late Tortonian and the Messinian, whose thickness is about 600 m, overlying a succession of marly shales and marls, with intercalations of sands and conglomerates, grading upward into marly shales with intercalations of thin sands (Pleistocene). The overall thickness of the Pleistocene filling of the Salerno Basin based on offshore well data ranges from 1500 to 2000 m. A regional unconformity, probably Messinian in age, related to a non-depositional/erosional hiatus (the Pliocene is completely lacking), characterizes the base of the Pleistocene sequences in the Cilento offshore (lithostratigraphic wells Milena 1 and Margherita mare 1)[44].

2.2. The Salerno Basin: Tectonic setting and seismic stratigraphy

The Salerno Basin (or Salerno Valley) is a Pleistocene half-graben, whose individuation and tectonic setting have been controlled by the normal fault Capri-Sorrento Peninsula, showing an average throw of about 1500 m [17; 16; 21; 22; 47]. This fault has controlled the origin of the Salerno canyon, showing active erosional processes and synsedimentary tectonics. The acoustic basement is composed of Mesozoic carbonate platform units (lithostratigraphic wells Mina 1, Milena 1, and Margherita mare 1) and overlying Miocene siliciclastic units related to the Liguride Units and foredeep deposits (Cilento Flysch), cropping out onshore in the Cilento Promontory [1; 46; 41].

The Salerno Basin shows an example of complex Quaternary filling of a tectonically controlled sedimentary basin, recording the interactions among the effects of eustatic sea-level oscillations and tectonic activity in the source region coupled with the tectonic deformation in the depositional area, lasting up to recent times. High sedimentary supply, combined with restrict sediment dispersal, has produced the deposition of a thick Quaternary succession, locally exceeding 3 km of thickness. The basin filling is composed of 1) marls and marly shales with intercalations of sands and conglomerates and 2) marly shales with intercalations of thin sands, Pleistocene in age. The average thickness of Pleistocene filling of the Gulf of Salerno based on offshore well data ranges between 1500 and 2000 m. In the Cilento offshore (Milena 1 and Margherita mare 1 wells), a regional unconformity, probably Messinian in age, related with a non-depositional and erosional hiatus (the Pliocene is completely lacking) characterizes the base of Pleistocene sequences.

Interesting evidence on the tectonic setting and the seismic stratigraphy of the Salerno Basin has been outlined by multichannel seismic profiles collected by IAMC-CNR of Naples, Italy. The line drawing of the seismic profile SAM7a (Cilento offshore) is represented in Fig. 1 [48] and discussed here. The seismic profile SAM7a runs from the Salerno offshore starting from the Amalfi town and shows a couple of conjugate listric normal faults, which is NNW–SSE dipping. These faults, indicative of strong releases of clastic multilayer, converge at several stratigraphic layers. In particular, some faults appear related to a master fault, NE–SW oriented (Capo D'Orso fault).

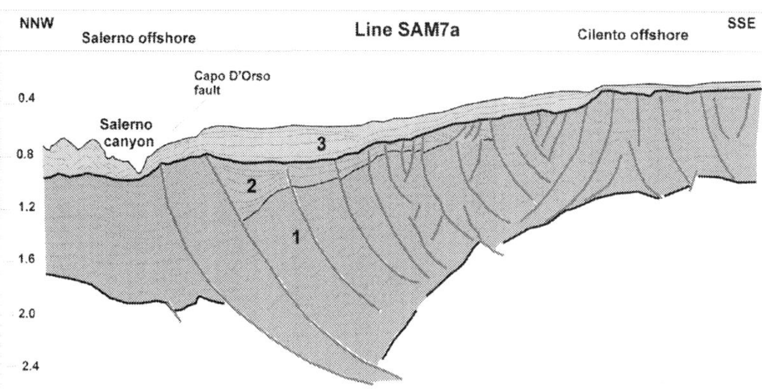

Figure 1. Line drawing of the seismic profile SAM7a (Cilento offshore; modified after Aiello et al., 2009)

This seismic profile runs in the shallowest sector of the Salerno Valley, which, unlike the distal areas where the depositional processes prevail during the Late Quaternary, and appears to be struck by recent erosional processes, enabling the formation of the Salerno canyon. A main unconformity overlies the seismic sequences involved by extensional tectonics in correspondence to normal faults; it also marks the beginning of the deposition of relatively undeformed seismic sequences, forming a thick prograding wedge thickening from the platform margin toward the slope and toward the Salerno canyon. The prograding wedge has been supplied by the Sele river during the Late Quaternary.

The seismic profile SAM5 (Fig. 2) [48], showing a WNW–ESE trending (perpendicular to the Tyrrhenian shoreline), runs perpendicularly to the line SAM7 and has explored the central sector of the Salerno Basin where depositional processes, of basin filling, prevail. The tectonic setting, highlighted on the basis of geological interpretation of seismic profiles, is consistent with an extensional tectonic style, complicated by the occurrence of wide antiformal structures with ENE–WSW axes, involving the Pleistocene seismic sequences.

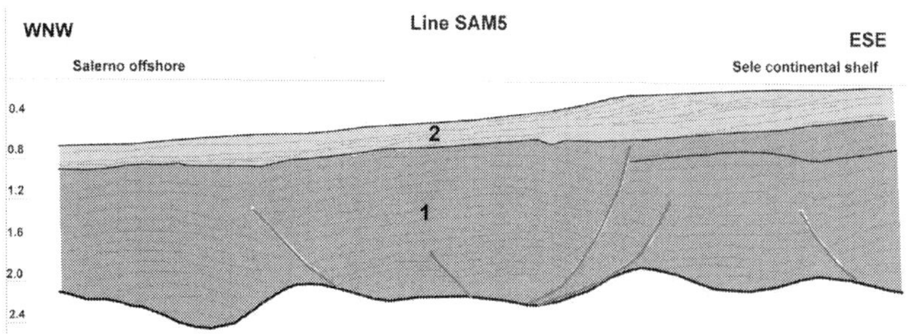

Figure 2. Line drawing of the seismic profile SAM5 (Salerni Basin; modified after Aiello et al., 2009).

The seismic interpretation suggests that these structures are probably linked to rollover mechanisms, resulting from the envelopment of listric normal faults, shown by N–S-trending seismic profiles. A main tectonic unconformity marks the occurrence of a prograding wedge, relatively undeformed, which has been supplied by the Sele river mouth.

The stratigraphic architecture of the prograding wedge is well developed in continental shelf areas, where a stacking pattern of seismic sequences probably corresponding to a fourth-order glacio-eustatic cyclicity has been identified. The offlap breaks of the prograding clinoforms appear well preserved and indicate an aggradational component in the basal parasequences of the wedge and a more pronounced progradational component in the upper parasequences. A progressive seaward shifting of the parasequences in the continental shelf, up to the present-day rim of the continental shelf, located at about 135 m of water depth, indicates a normal coastal progradation controlled by high siliciclastic supplies of the Sele river. The Sele prograding wedge is bounded at its base by a regional unconformity, marked by the downlap of seismic reflectors and at its top by another unconformity, indicated by the erosional truncation of the prograding reflectors. In slope areas, the stratigraphic relationships of paraconformity among the fourth-order parasequences supplied by the Sele river are evident from seismic stratigraphy.

Based on the geologic interpretation of multichannel seismic profiles, the tectonic setting of the Salerno Basin is controlled by two low-angle normal faults, such as the Capri-Salerno fault, WSW–ENE trending and bounding southward the Sorrento Peninsula and the Capo D'Orso fault, having an NE–SW trending [48]. Several high-angle normal faults, parallel to the Capri-Salerno faults, have been singled out near the shoreline of the Sorrento Peninsula [49; 50]. Some normal faults having a minor entity, parallel to the Capri-Sorrento fault, are evident from seismic data analysis. They are both antithetic and synthetic to the Capo D'Orso normal fault.

The sedimentary basins generated by lithospheric stretching are the sites of crustal extension as a consequence of passive rifting, assuming that the lithosphere has undergone extension with a uniform rate (i.e. uniform stretching) [51]. In a model of uniform stretching, the subsidence rates depend on the stretching rate and the lithospheric thickness [52]. The extension is often asymmetrical, since the faults propagate outward in a direction from the initial rifting of the upper continental crust. If the rifting process continues due to the critical values of stretching, a passive continental margin develops, characterized by the accumulation of thick sedimentary wedges deposited during the post-rifting phase.

The patterns of faults in the rift systems were originally considered to bound a symmetrical arrangement of horst and graben [53]. Regional surveys of seismic reflection [54] suggest, however, that the rifts are characterized by a set of half-graben, with a polarity of the throw reversing in correspondence to strike-slip faults, involving the rift system. Estimates of extension rates in half-graben systems may be obtained based on subsidence curves, calculated by removing the effects of sedimentary loads.

The activity of normal faults is responsible for local variation of subsidence rates or uplift rates, producing different depositional geometries in footwall and hanging wall blocks [55; 56]. The analysis of the seismic sequences deposited during the fault activity showing wedging depositional geometries offers the opportunity to investigate the tectonic movements along the fault.

Structural styles of extensional basins are often characterized by listric normal faults, diminishing their inclination proceeding downward, becoming parallel and/or subparallel in the detachment zone, located next to the transitional zone from upper to lower lithosphere. Since the listric faults produce a differential rotation of the strata between the hanging wall and the footwall, a set of imbricated block faults should show rotations progressively steeper in the main throw direction. Small anthitetic faults of compensation related to listric geometries often remove the exceeding uplift and the stratigraphic gap occurring in the central block [54; 57]. The extension in the half-graben basins is often associated with a syntectonic sedimentation or wedging, indicating tectonic downthrowing contemporaneous to sedimentation. The listric normal faults may connect with low-angle master faults, forming main levels of detachment; in this case, the type of shear on the fault must change from upper to lower levels [58].

The Salerno Basin is a Pleistocenic half-graben originated from the Capri-Sorrento Peninsula master fault, showing a throw in the order of 1.5 km, and is filled by marine clastic and epicontinental deposits. Strong reflectors, inclined toward SSE with apparent angles of 10°–15°, appear on seismic profiles, indicating the occurrence of levels of detachment in the carbonate multilayer. The extension in the Salerno Basin is controlled by groups of listric normal faults, joining on low-angle detachment plans. Another preferential detachment level is located at the top of Miocene flysch terranes. Bathymetry and high-resolution seismics of the Salerno Valley [47] indicate that the depositional processes of infilling prevail in the continental shelf located eastward, while the western sector of the half-graben is the site of processes of erosion and sedimentation still active in a canyon (the Salerno canyon) deep from 600 to 1000 m. The occurrences of wedging geometries in the sedimentary successions, of tectonic unconformities, and hummocky reflectors and chaotic facies at several stratigraphic levels indicate a strong synsedimentary tectonics and a strong uplift of the surrounding emerged areas during the Pleistocene. Anticlinalic structures with ENE–WSW axes suggest two rollover mechanisms resulting from the envelopment of listric faults and tectonic inversion of the half-graben, according to an N–S compression or a transpression along the master fault Capri-Sorrento Peninsula. Extensional processes seem to be still active in the Salerno Basin, being responsible for the present-day topography of the sea bottom and the foreland uplift along rift shoulders (Sorrento Peninsula).

A seismo-stratigraphic and morpho-bathymetric analysis on the continental margin of southern Campania in the elongment Sorrento Peninsula-Capri island [47] has been carried out based on multibeam bathymetric data and high-resolution multichannel seismics recorded by IAMC-CNR (Istituto per l'Ambiente Marino Costiero – Consiglio Nazionale delle Ricerche) of Naples, Italy, onboard of the R/V Urania (National Research Council of Italy). The acquisition of geophysical data, carried out in water depths ranging between 100 and 1000 m and calibrated on the integrated interpretation of bathymetric and seismic data in deep-sea depositional systems, has allowed the recognition of main morpho-structures cropping out at the sea bottom in slope and bathyal plain sectors and their correlation with the main seismic sequences. The geophysical data have furnished interesting evidence on deformation styles in the Salerno offshore, characterized by the occurrence of NNW–SSE-trending tectonic lineaments. The data have been recorded in an SEG-D format with a sample interval of 0.5 msec and a time window of 8 sec. The length of the seismic profile is about 15 km with a maximum coverage of 12 km for CDP. The processing of seismic reflection data has been divided into a preliminary phase, during which the editing and the immission of field geometry of data have been carried out, and into a phase of advanced processing, during which the noise removal and the sea-bottom multiples attenuation have been carried out. During the phase of advanced processing, the traces have been elaborated with different algorithms of two-dimensional filtering.

The velocity analysis has allowed for the plot of a preliminary seismic section, on which it has been possible to individuate which were more adequate processes to apply to these types of data. A spiking and predictive deconvolution has been applied to seismic sections, coupled with a time-variant filtering. The Salerno Valley represents a half-graben sedimentary basin, whose individuation has been controlled during the Early Pleistocene by the master fault Capri-Sorrento, bordering the southern margin of the Sorrento Peninsula with vertical throws in the order of 1.5 km, downthrowing Meso-Cenozoic carbonates under the Gulf of Salerno with a step of normal faults.

The geologic interpretation of the L6 seismic profile carried out according to seismo-stratigraphic criteria has allowed the individuation of an important unconformity, located between 1.7 and 1.8 sec (TWT; two way travel times), related with the top of the Meso-Cenozoic carbonate sequence, widely cropping out onshore in the structural high of the Sorrento Peninsula. The unconformity bounds upward the carbonate acoustic basement, which is strongly downthrown by normal faults and marks the base of Pleistocene filling of the Salerno Valley. The basin filling shows an average thickness of about 500 m. The strong synsedimentary tectonic activity, lasting up to the Late Pleistocene in correspondence to listric faults having an NNW–SSE trending, has controlled the triggering of submarine gravity instabilities, evident as chaotic acoustic facies interstratified in the stratigraphic succession.

New geological evidence derives from a digital terrain model (DTM) of the Salerno offshore, whose interpretation has evidenced a dense network of channels, eroding the southern slope of the Sorrento Peninsula and representing the seaward elongment of the onshore torrents [47].

Among the morpho-structures occurring in the Salerno offshore, particularly relevant are the morpho-structural highs having an NNW–SSE trending identified on the marine DTM (Fig. 3) [47]. The two highs are separated by an intra-slope basin, now filled by sediments, which was the site during the Pleistocene of an area of sedimentary transport known as the Salerno canyon [21; 22]. The multibeam bathymetry has evidenced the occurrence of intra-slope basins, NNW–SSE trending, narrow, and elongated, which characterize the whole margin of the Latium-Campania. The recent age of the extensional tectonics, which has originated from the morpho-structural highs, has prevented the accumulation of thick sedimentary drapes over the two morpho-structural highs.

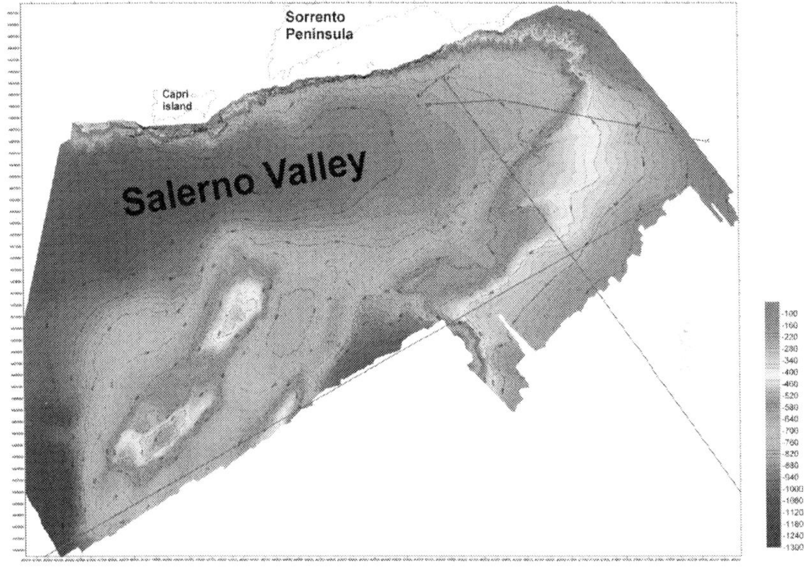

Figure 3. Digital terrain model (DTM) of the Salerno offshore (modified after Aiello et al., 2007).

3. Previous geological studies

3.1. General setting

The Cilento Promontory represents a structural high, interposed between the coastal depressions of the Sele Plain–Gulf of Salerno and the Gulf of Policastro, whose relief reaches 1700 m high. These reliefs are composed of thick successions of siliciclastic and carbonate sequences ("Flysch del Cilento" Auct.), dipping landward in the main carbonate reliefs of Southern Apennines ("Alburno-Cervati Unit" *Auct.*). Quaternary normal faults define the borders of the structure of the Cilento Promontory. Apart from the carbonate structures of Palinuro Cape and Bulgheria Mount and several isolated outcrops, the reliefs of the Cilento Promontory are formed by terrigenous rocks, which were accumulated in deep basins in a time interval ranging between the Late Mesozoic and the Late Miocene. The oldest formation pertain to the Northern Calabrian Units [41], representing the highest stratigraphic structural unit of this sector of southern Apennines. In the Cilento area, this is represented by a formation ranging in age between the Malm and the Oligocene, composed of dark shales, marls, and marly limestones, showing a deformation grade lower than the overlying units (Cilento Flysch). The Cilento Flysch includes, proceeding upward, the Pollica, San Mauro, and Monte Sacro Formations, with a total thickness of 1500 m [41].

The carbonate sequences exposed at the Palinuro Cape and in the carbonate massif of the Bulgheria Mount show paleo-sedimentary domains of slope–basin and show different facies with respect to the neritic limestones of carbonate platform cropping out in correspondence to the Alburno-Cervati massif [41]. During the Miocene, the Bulgheria Unit had been over-thrust by the Northern Calabride Unit and the overlying Cilento Flysch. Nonetheless, the Bulgheria terranes are exposed in outcrop due to the next uplift of the Bulgheria-Palinuro Cape area (Late Miocene–Early Pliocene) and the consequent erosion of the terrigenous units [59]. The uplift of the Bulgheria Mount has provoked the northward folding of carbonate sequences and their thrusting over the Calabrian Units with a south-verging reverse fault, well exposed along the southern border of the Cilento [41].

In the western sector of the Cilento Promontory, several morphological depressions have been filled by alluvial deposits, whose origin must be attributed to structural elements with an NNE–SWW trending in the Alento Plain and NW–SE trending in the Santa Maria di Castellabate and San Marco plains. Brancaccio et al. [5] have attributed the formation of these depressions to the Late–Middle Pleistocene; they include transgressive–regressive cycles referred to as glacio-eustatic sea-level oscillations of isotopic stages 9, 7, and 5 [60; 61], downthrown a few tens of meters with respect to the original heights, between the end of the Middle Pleistocene and the beginning of the Late Pleistocene.

The Cilento Promontory has been involved by a vertical uplift of 400 m during the Late Early Pleistocene and the Middle Pleistocene. Absolute estimates of the entity of tectonic uplift involving the Cilento Promontory have been obtained by the vertical distribution of Pleistocene marine terraces along the Cilento coasts. In the Northern Cilento, the oldest marine terraces (Middle Pleistocene) occur at maximum heights of 450 m, while the Emilian ones at heights of 350 m a.s.l. [62- 66].

Morphological elements in the coastal areas relative to paleostands of the sea level during the Late Pleistocene (isotopic stages 5e and 5c) have evidenced a substantial tectonic stability of this tract of shoreline from the Tyrrhenian to recent times [67]. The absence of relevant vertical movements during the last thousands of years is well testified by the height of Versilian beach deposits. These deposits occur in coastal fluvial depressions, incised during previous glacial regression for more than 2 km inner to the shoreline.

The present-day coastal cliffs are incised in the sandy and silty successions of the Pollica Formation. Next to the Licosa Cape promontory, this formation composes a wide terrace of marine abrasion located at heights of 4–10 m a.s.l., extending from the San Marco plain to the Ogliastro Marina Bay.

Northward of the carbonate blocks of Palinuro Cape and Bulgheria Mount, wide outcrops of a continental formation (Conglomerati di Centola) occur [68]. The lowest part of the formation, about 200 m thick, is represented by muddy, silty, and sandy deposits, with subordinate conglomeratic lenses, while the upper part is characterized by deltaic conglomerates with coarse-grained facies. The upper chronological boundary of the Centola conglomerates is furnished by Middle Pleistocene terraces of southern coastal slope of the Bulgheria Mount, which are strictly related to the terraces carving the Centola conglomerates at Palinuro.

Several structural stratigraphic studies have been carried out on the onshore and offshore areas of Agropoli. The most relevant ones are discussed here.

3.2. Stratigraphic and structural geology

Scandone et al. [69] have recognized a sedimentary succession in the Bulgheria Mount ranging in age from Trias to Miocene, whose characteristics let us to consider it as a transitional succession to open-sea sediments. It is worth noting that the differentiation with the Cilento carbonate sequence and the Sapri Mounts becomes marked starting from the upper part of the Early Liassic. General conditions of uniformity reestablish from the Miocene, since the calcarenites with *Miogypsinae* of the Bulgheria Mount show characteristics very similar to the ones of the Sapri Mountains, which are instead transgressive above Senonian or Paleocene limestones. The main structural pattern involving the area is constituted of a reverse fault located to the eastern margin of the massif located between San Severino and Scario. The extensional faults are mainly Apenninic (NW–SE) in direction and are younger than the reverse fault.

Ietto et al. [70] have carried out one of the first structural and stratigraphic studies on the Cilento Flysch in the area located between the Alento Valley and the Licosa Cape. These authors have distinguished for the first time the three main formations of San Venere, Pollica, and San Mauro, which then become an integrant part of next geological literature, being classified as shaley flysch, arenaceous flysch, and conglomeratic-marly-arenaceous flysch, respectively.

Guida et al. [68] have carried out a detailed geomorphologic study on the basin of Mingardo river (Cilento), with a particular regard to landslide phenomena and landslide hazard of the study areas. The applied themes of this research are based on geologic and morphologic surveys of the outcropping units and in particular on Pleistocene clastic deposits (Centola conglomerates). These deposits are composed of pebbles and blocks, highly rounded, immerged in an abundant sandy matrix. The source areas must be considered the morphological highs of the Cilento region, where the arenaceous and conglomeratic terms crop out. Several stratigraphic and structural units crop out in the basin, with terrigenous and carbonate successions. The morphologic trend is strictly related to the distribution of lithologies and geometries of geological bodies. In the geolithologic map, the lithostratigraphic sequences have been grouped into 13 lithological units with different lithotechnical characteristics. Moreover, the authors have elaborated some thematic maps relative to slope stability at the 1:10.000 and 1:25.000 scales. The morphological analysis elaborates on a morphometric map with five slope classes and a geomorphologic map, on which the most significant shapes of the relief are represented. Moreover, a detailed evaluation of the hazard to slide has been carried out, identifying the maximum hazard of slide deposits in a chaotic condition, while elevated conditions of instability individuate in terrigenous successions mainly composed of shales.

Critelli and Le Pera [71] have carried out a lithostratigraphic study on the Pollica Formation, a Miocene turbidite succession more than 500 m thick, representing the lowest stratigraphic portion of the Cilento Group [72]. The lithostratigraphic, sedimentologic, and compositional analyses have allowed in dividing the Pollica Formation into two distinct members, probably separated by an angular unconformity. Member A (Arenarie di Cannicchio), characterized by thin turbidites pelitic-arenaceous in a distal environment, is strongly deformed and folded. Member B (Arenarie di Pollica), Burdigalian in age, represents a turbidite system evolving from lobe facies to channel facies. The petrography of the arenitic terms of both the members records a marked compositional evolution, passing from lithic arkoses (A petrofacies) for member A and the lowest part of member B toward siliciclastic sandstones (B petrofacies) and volcanic–siliciclastic sandstones (C petrofacies). This change of depositional and physiographic characters of the basin contrasts with the change of the detrital modes. The latter ones suggest a strong restructuration of the source areas. The A petrofacies should have metamorphic terranes as a fundamental source area, while the petrofacies B and C should mark areas from more complex sources. Metamorphic rocks are progressively substituted by plutonic fragments, while locally large volcanic supplies are coeval with the sedimentation and extrabasinal carbonate clasts that are concentrated in single depositional events occur.

In the frame of a research program of systematic revision of the inner flysch of southern Apennines, Zuppetta et al. [73] have examined the terranes of the Albidona Formation in the area included between Sarconi and Trebisacce, which, at the Calabria-Lucania boundary, represents the highest formation of the Cilento Unit. In the study area these terranes unconformably overlie an ophiolitic nappe and the Crete Nere and Saraceno Formations, already deformed. This formation is characterized by arenaceous-pelitic facies, evolving upward towards arenaceous-conglomeratic facies. In this succession, slumpings and megaturbidites occur at several stratigraphic levels, testifying a strong tectonic activity. The age of the whole sampled succession obtained through the study of planktonic foraminifera ranges between the Late Oligocene and the Early Burdigalian. The obtained results give some consideration to the paleogeographic location of the sedimentary basin of the Albidona Formation, and precisely to its tectono-sedimentary evolution.

Russo et al. [74] have carried out a stratigraphic revision of the Cilento Flysch (Pollica and San Mauro Formations). Datations carried out through planktonic foraminifera have allowed to date the Langhian as the lowest part of the Pollica sandstones and the Serravallian-Early Tortonian as the highest part of the succession, represented by the San Mauro Formation. These data have allowed in interpreting this succession as a deposit of a piggyback basin, whose outer flank is composed of a frontal ramp, located in correspondence with the paleogeographic domains corresponding to the Bulgheria-Verbicaro Unit.

Zuppetta and Mazzoli [75] have carried out a structural analysis of the Cilento Unit, including the turbiditic and siliciclastic terranes, Langhian-Tortonian in age (Pollica and San Mauro Formations), recording a succession of pre-tectonic and tectonic events. The pre-tectonic events have originated from synsedimentary structures of slumping and syndiagenetic structures (sedimentary dykes). Early tectonic events have produced minor reverse fault conjugates with respect to the stratification, while the most important ones have produced a regional folded structure having an SW vergence. Late tectonic events have produced a wide folding of the carbonate terranes of the Alburno-Cervati Unit. Based on the paleotectonic unit proposed here, the Cilento Flysch represents the deformed filling of a Middle Miocene foredeep.

Critelli [76] has carried out a regional geological study in which the clastic successions of southern Apennines (among which the Cilento Flysch) have been put in relationship with the flexure of the lithosphere and the emplacement, in the nappes of the formation, of stratigraphic successions of the foreland basin of southern Apennines (Fig. 4) [76]. The clastic sediments of post-Oligocene sedimentary successions mainly derive from the allochtonous tectono-stratigraphic units of the northern Calabrian Arc. The stratigraphic and compositional relationships of clastic sediments of the sedimentary basins, interposed among the Calabrian microplate and the Adriatic microplate, implicate that the lithospheric flexuration of the Adriatic plate in conjunction with the tectonic accommodation of the Calabride crystalline nappes have been main control factors in the development of sedimentation of the foreland basin of Southern Apennines. The lithostatic load caused by the emplacement of the Calabrian nappes over the Adriatic plate, from Early Miocene to recent times, has provoked the flexural deformation of the Adriatic lithosphere, producing foredeep basins, forebulge areas, and back-bulge basins. The eastward migration of the lithosphere flexure happened through pulses as a response to the eastward transport of the Calabrian chain.

The stratigraphic sequences, key of the orogenic system of southern Italy, have been described by using new data and interpretations or through the reinterpretation of the data previously published (see the above mentioned corresponding figure) [76].

The denudation of the crustal terranes of the Calabrian Arc during the rapid uplift and erosion of the Middle–Late Oligocene since 10 my B.P. has produced abundant clastic sediments, which were accumulated in 1) an oceanic basin, Late Paleogenic in age (the Liguride Complex); 2) several Neogene foredeeps and depocenters in migration localized at the top of the foreland basins, migrating outward with the onset of the orogenic deformation; and 3) starting from the Late Tortonian, foreland basins and syntectonic back-arc foredeeps.

In particular, the Cilento Group (Cilento Flysch) [70] ranging in age between the Langhian and the Tortonian [72; 74; 77], with thickness ranging from 1200 and 2000 m, unconformably overlies the Liguride Complex, which is in turn unconformably overlain by the Gorgoglione Formation (Late Tortonian) and the Monte Sacro, Oriolo, and Serra Manganile Formations (Fig. 5) [76].

The Cilento Group consists of different turbiditic depositional systems [78]. More than the turbiditic siliciclastic strata, the Cilento Group includes several carbonate and clastic mega-strata (ranging in thickness from a few tens of meters and several hundreds of meters) and coarse-grained volcaniclastic debris flows and turbidites. The sandstones of the Cilento Group are lithic, volcano lithic, and rich in feldspars [79]. In the upper Cilento Group, the carbonate and clastic megastrata and the olistostromes record the main tectonic events on the active orogenic belt. The olistostromes are siliciclastic and include blocks of Calabride terranes and Liguride Complex [70; 80- 83; 78]. The clasts derived from the Liguride Unit appear only in the middle and upper part of the Cilento Group, suggesting initial signals of the emersion of the Liguride Complex. Clear signals of emersion and erosion of the Liguride Complex are recorded by the Piaggine sandstones, ranging in age between the Serravallian and the Tortonian) [84- 86]. The Piaggine sandstones are derived from the detrital content of the

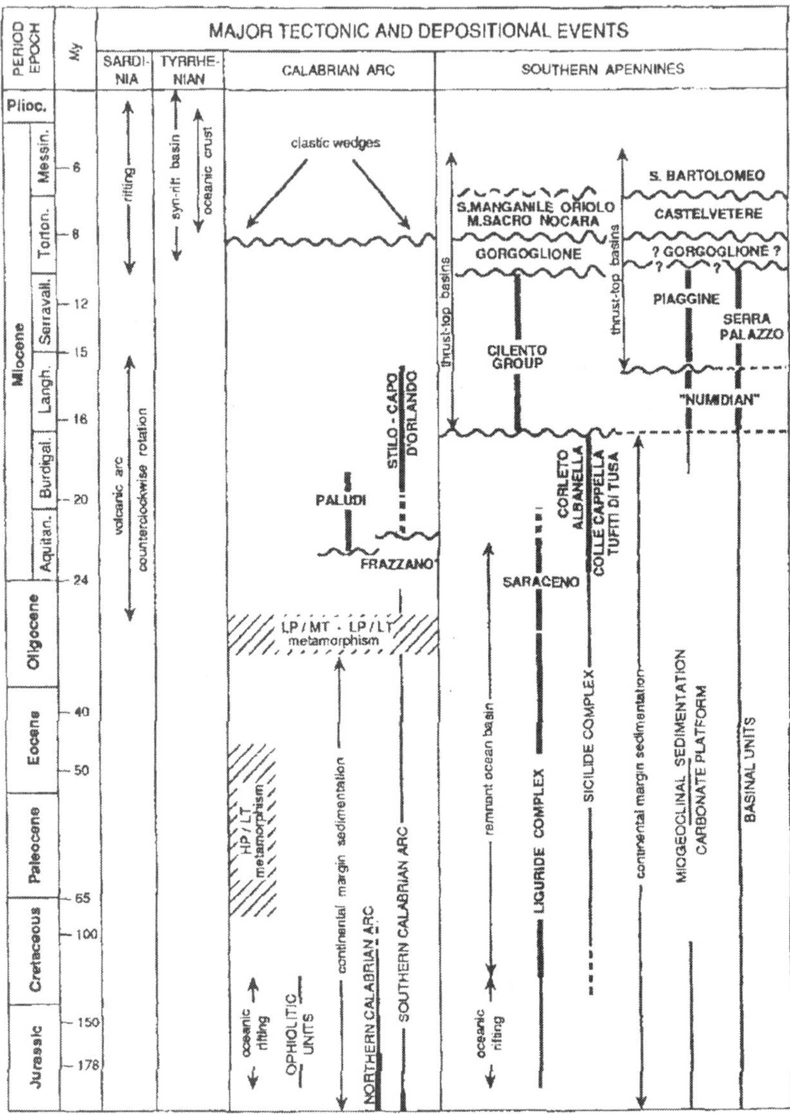

Figure 4. Stratigraphic diagram showing the main Meso-Cenozoic tectonic and depositional events of sedimentary sequences of the Cilento Flysch (modified after Critelli, 1999).

Liguride Complex suggesting that, at the Serravallian-Tortonian boundary, the Liguride Complex was in a subaerial location [76].

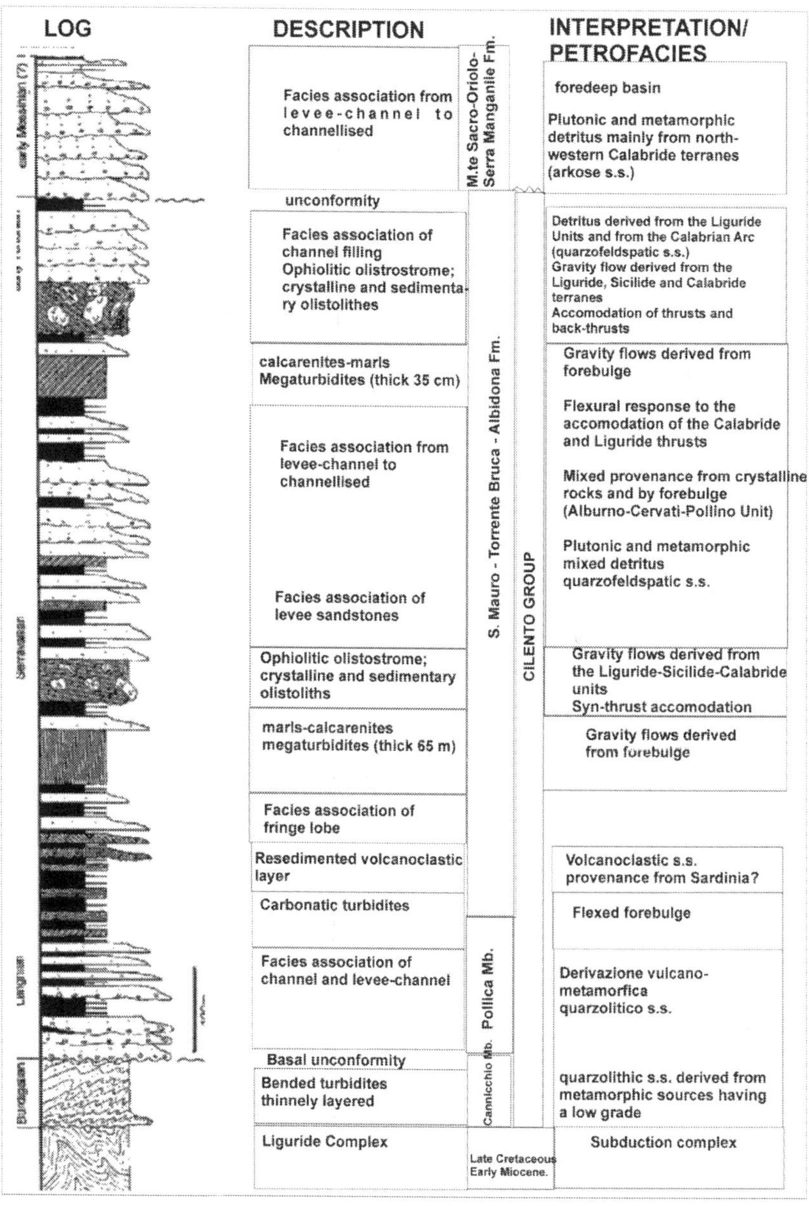

Figure 5. Stratigraphic column of the Cilento Group (modified after Critelli, 1999).

3.3. Subaqueous geology and Quaternary geology

Baggioni [87] has carried out a geomorphologic study of the Cilento coasts (southern Italy), evidencing that the present-day morphogenesis has been conditioned by the siliciclastic lithologies cropping out in the area, which can be easily eroded. This has provoked the formation of a platform of actual marine abrasion and has controlled the disappearance of the marine levels of the old Quaternary. Only a level attributed to Tyrrhenian 2 survived to the erosion involving the study area. The recent neotectonics is limited in entity, while the morphologic and stratigraphic evidence underlines the hypothesis of strong vertical movements during the Quaternary, conditioning the present-day trending of the shoreline. Some detailed morphologic sections on several localities of the Cilento shore have been produced.

Baggioni [88] has carried out a study on the evolution of the carbonate slopes of the Bulgheria Mount, evidencing that the northern slopes show different geomorphological aspects with respect to the southern ones; while the first ones are characterized by a main reverse fault, which puts in contact the flysch with the limestones, with a slope higher than 50°, the southern slopes are characterized by a flight of step of normal faults in carbonate rocks with slopes less than 40°.

Baggioni et al. [89] have carried out geomorphological studies in the surroundings of Marina di Camerota, where the occurrence of marine microfaunas has allowed differentiating the Late Pliocene terranes from the Calabrian to the post-Calabrian. Palynological analyses have allowed to put in relationship the evolution of the vegetation with that of the climate.

Ortolani et al. [90] have carried out a study of the Holocene geomorphologic evolution of the Velia coastal plain (Cilento) based on new archeological findings. New archeological studies carried out in the area of Velia (Cilento, Campania) have evidenced the occurrence, southward of Acropoli, of manufacts attributed to the VI century A.C. They are overlain by alluvial deposits including typical structures of liquefaction, interpreted as seismic. The archeological study and the stratigraphic analysis give further consideration to the recent evolution of the study area, particularly referring to the climatic events, which have conditioned several phases of aggradation and progradation of the old topographic surface and related paleo-shoreline of the alluvial coastal plain between the VI and IV century A.C. and since to recent epochs. Moreover, it can be evidenced that this alluvial period had an important influence on the Magna Grecia settlements, located in the alluvial coastal plains of southern Italy, and on social and economic developments of people.

Antonioli et al. [59] have carried out a geologic and geomorphologic study finalized to the recognition of the traces of old stands of sea level and the effects which they, interacting with the tectonic activity, have produced on the Quaternary evolution of the landscape. The data have been collected both in emerged and submerged sector, up to a water depth of 50 m through ARA dives carried out both on the coastal cliff and in many karstic cavities occurring in the carbonate promontory. In an emerged coastal belt, a succession of seven orders of marine terraces at water depths ranging between 180 and 2 m a.s.l., incised both in the Mesozoic carbonate rocks and in the flysch terranes cropping out northward of Palinuro Cape. The first five orders of terraces are represented by surfaces of marine abrasion, while the sixth and the seventh orders are represented by abrasion platforms and notches incised along the carbonate cliffs of Palinuro Cape. The traces of the beach lines found between 12–14 and 7–8 m should be coeval, if not older, of the Last Interglacial.

Russo [91] has singled out a fossiliferous level, Tyrrhenian in age in the Cala Bianca Bay (Marina di Camerota, Salerno). The Cala Bianca coastal cliff is characterized by a succession of dolomitic rocks, Upper Triassic in age, passing upward the dolomitic limestones of Early Triassic. In many places of the bay, clastic deposits crop out, both coarse-grained and fine-grained, related to continental levels sometimes including paleolithic remnants. Shells of Mollusks of marine coastal environment sampled in a sandy conglomeratic deposit have been dated back through isoleucine epimerization. The obtained values have allowed an attribution of these deposits to isotopic stage 5 and of the isotopic curve of the oxygen (125 ky B.P.).

Ascione and Romano [6] have carried out a detailed geomorphological study of the Bulgheria Mount through the identification of phases of uplift and subsidence, involving the area starting from the Early Pleistocene and the Middle Pleistocene. The end of vertical movements during the Late Pleistocene is demonstrated by the occurrence of Pleistocene paleobeaches, now uplifted at a water depth comparable with a highstand phase of 130 ky B.P., as recorded in stable areas of the Mediterranean region. The vertical movements of the area have been quantified through the study of marine terraces; the sum of these movements has produced a vertical uplift of the area of about 400 m starting from the Santernian, 150 of which were verified during the Middle Pleistocene. Along the Tyrrhenian coast of Southern Apenninic margin, other structural highs bounding the peri-Tyrrhenian depressions furnish evidence of subsidence episodes, which have interrupted the main trend of uplift of the area during the Quaternary. This, together with the distribution of Quaternary marine terraces and deposits, has evidenced that this sector of the Southern Tyrrhenian margin has undergone vertical movements of different entities.

3.4. Marine geology

The marine areas included in the geological map n. 502 "Agropoli" include a sector of continental shelf, developing with low gradients up to a water depth of 250 m. The topography of sea bottoms is particularly articulated and conditioned by the occurrence of an E–W-trending ridge along the 40°–14' parallel. This element represents the seaward prolongation of the Licosa Cape morpho-structural high and constitutes the acoustic basement unconformably overlain by Pleistocene prograding units of continental shelf. The acoustic basement shows an irregular topography, characteristic of a morpho-continental evolution and an acoustic facies with chaotic reflections, typical of flysch terranes.

The Pleistocene deposits are organized in prograding units separated among them by several unconformities; in particular, two surfaces of marine ingressions, probably pre-Tyrrhenian and Versilian, have been recognized. The surface of marine ingression, Versilian in age, crops out at the sea bottom and reworks the subaerial erosional surface formed during the Last Pleniglacial. Palinsest deposits have been detected on this surface at a water depth of 160 m, composed of coarse-grained organogenic sands with fragments of *Arctica islandica* with a prograding internal structure interpreted as beach deposits related to the last sea-level lowstand, downthrown at greater water depths by phenomena of margin subsidence [92].

All the marine areas in the sectors surrounding the Cilento Promontory seem to be involved by a recent tectonics, characterized by sub-vertical faults with an NE–SW trending, bounding small basins on the continental shelf and upper slope. They are elongated parallel to the shoreline and represent depocentral areas in the Late Pleistocene, conditioning the present-day physiography of the continental shelf.

The shelf break does not correspond, as it happens along the whole Tyrrhenian margin, to the last sea-level lowstand but is deeper, at 220 m. This underlines the importance of the structural elements in controlling the physiographic units in this sector.

Coppa et al. [45] have carried out a study of geomorphologic and faunistic elements of the Tyrrhenian continental margin between Punta Campanella and Punta degli Infreschi, characterized by three main morpho-structural units:

1. The continental shelf of the Gulf of Salerno, located in correspondence to the Sele basin. It is characterized by a gradient of 0.3°–0.8° and by a gradual shelf break, characterized by a thick Holocene sedimentation, underlain by the Wurmian erosional truncation, cutting older sediments.

2. The Cilento continental shelf located in correspondence to the structural high of the Cilento Promontory; this shows a variable slope, between 0.3° and 0.8° and its rim, located at 220 m of water depth, is abrupt. This is characterized by a scarce or absent Holocene sedimentation and by a wide diffusion. of relict morphologies. The continental shelf of the whole study sector is characterized by the abundance of microfossils and rests of plants.

3. Slopes and basins: the slope, with a gradient of 6°, is crossed by channels merging in basins; along this slope submarine slides are frequent.

Ferraro et al. [92] have carried out a study on the continental shelf between Licosa Cape and Palinuro Cape through subbottom and sparker profiles and sea bottom and subbottom samples in order to reconstruct the geological evolution of the continental shelf during the Late Pleistocene–Holocene. In its northern sector, the continental shelf is characterized by an offlap succession, which determines a seaward progradation of 10 km. Prograding units have been individuated, resulting progressively younger from land to sea. These sedimentary bodies refer to the general sea-level lowstand of the Late Pleistocene (isotopic stages 4, 3, and 2); this regression has happened in the frame of a fourth-order eustatic cycle. This area is characterized by normal faults, having an Apenninic trending, active since the beginning of the Late Pleistocene. These faults bound basin parallel to the shoreline, located both in platform and in slope.

The evolution of the continental shelf during the Late Pleistocene–Holocene can be schematized through the deposition of prograding units of the Late Pleistocene, referred to as isotopic stages 5, 4, and 3 [61], and then through the deposition, over a well-individuated ravinement surface, of sedimentary bodies characterized by middle-grained sands passing to silty shales, overlain by shales including a pumiceous level related to the A.D. 79 eruption of Vesuvius. These shales have been deposited during transgressive and highstand phases of the sea level, and they represent a thin drape over the erosional surface.

With the end of isotopic stage 5a, the sea level results in constant lowering since isotopic stage 2, when it occurs at a water depth of 120 m in the Mediterranean [93]. During this phase of forced regression of the sea level[94] a prograding wedge deposits, produced by several minor units in offlap, progressively more recent from land to sea. The outer beach is correlated with the phase of lowstand of isotopic stage 2. The acoustic basement cropping out at the sea bottom should stay in a subaerial environment from isotopic stage 4.

Trincardi and Field [95] have presented an interpretation of the stratigraphic architecture of the prograding deposits offshore the Licosa Cape promontory. The prograding wedges of outer shelf and shelf break have been deposited over an erosional surface located in continental shelf and at the shelf break during the last phase of eustatic sea-level lowstand. During the next phase of sea level rise, the erosion linked to the shoreline shifting provokes the formation of a wide ravinement surface, recognized starting from the isobaths of 120 m landward, which provokes the partial erosion of the forced regression prograding wedge. The thickest of them has been preserved. In fact, thick prograding wedges in continuity of sedimentation have been observed only at the platform margin, while on the continental shelf, thin morphological relics have been preserved, mainly in correspondence to intra-platform depressions or morphological steps.

Trincardi and Field [96] have carried out a detailed study of a wide submarine slide, involving the lowstand prograding deposits located westward to the Licosa Cape, based on the analysis of seismic profiles. The area of detachment is characterized by a wide mobilization surface located at its base, showing the deformation of the sediments above it *in situ* and which coincides with the sea bottom with a main downlap surface. The section involved by the slide is less than 20 m thick and has a local extension, but the deformation on the basal mobilization surface extends out of the slide area. Directly seaward with respect to the slide slope are the stratified muds, which do not show any evidence of deformation or movement. A part of the mobilized sediments has traveled toward the basin alongslope and then has accumulated to the base of a ridge parallel to the slope.

The Licosa slide follows a main pulse of coastal sedimentation along the lowstand shoreline. On many continental margins, the eustatic sea-level lowstand is considered as a main cause of sliding of poorly consolidated sediments deposited during previous highstand conditions. The Licosa slide confirms the role of control, also if indirect, played by the eustatic sea level fall on the formation of submarine slides. When the sea level lowers and reaches its lowstand position, the fluvial channels reach their maximum location of efficiency and a coastal clastic wedge is rapidly emplaced with relatively high gradients at the shelf margin.

4. Results

4.1. Morpho-bathymetry of the Agropoli continental shelf

The morpho-bathymetry of the Agropoli continental shelf, strongly controlled by onshore geology and geomorphology, has been studied through the construction of four geological maps, herein shown (Figs. 6, 7, 8 and 9).

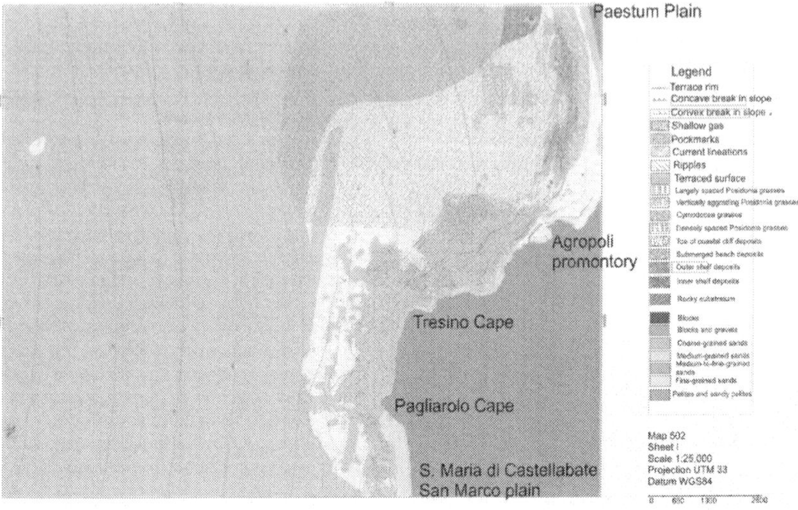

Figure 6. Marine geological map n. I NW, Agropoli geological map, scale 1:25.000 (modified after Aiello and Marsella, 2013).

The map n. I NW (scale 1:25.000) shows a continental shelf dipping with low gradients up to water depths of 95 m (Fig. 6). Its northeastern extremity, including the farthest southern sector of the Paestum Plain, is regarded as little ramps up to water depths of 35 m. Next to the Agropoli promontory, the convex isobaths up to 25 m are conditioned by the trending of the rocky shoreline and the outcrops of acoustic basement. The rocky promontory between the Tresino Cape and the Pagliarolo Cape (Fig. 6) controls the physiography of the submerged area. The isobaths are convex in correspondence to the two promontories up to 35 m of water depth shaping the trend of high coastlines. The physiography of the continental shelf is strongly conditioned by the geomorphologic setting of the surrounding emerged area; up to a water depth of 40 m the isobaths are concave and close, showing a high gradient of the continental shelf.

The map n. II SW (scale 1:25.000) is typified by the physiographic unit of the Licosa Cape high (Fig. 7), represented by an E–W trending ridge. The isobaths show an articulated E–W trending. The isobath of 15 m bounds a wide area showing remnants of terraced surfaces localized at several water depths. The southern sector of the structural high is characterized by pronounced concavities of the isobaths representing slide scars incised by drainage axes.

The map n. III SE (scale 1:25.000) is portrayed by the main physiographic unit of the Licosa Cape morpho-structural high (Fig. 8). Three main morpho-structural highs of the acoustic basement occur to the northwest. The N–S first high strikes at water depths of 120–145 m; the second E–W appears at water depths of 120–140 m; the third one, showing an elongated NNW–SSE trending, extends from 85 to 115 m of water depth.

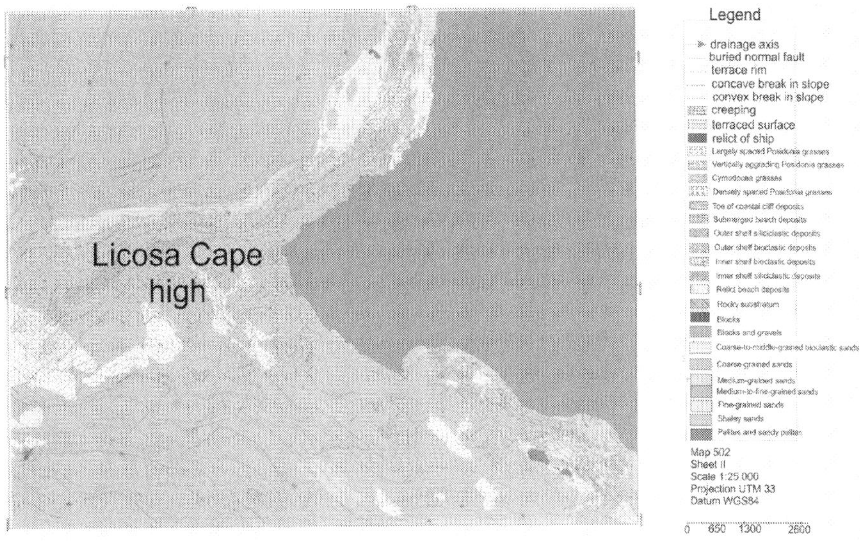

Figure 7. Marine geological map n. II SW, Agropoli geological map, scale 1:25.000 (modified after Aiello and Marsella, 2013).

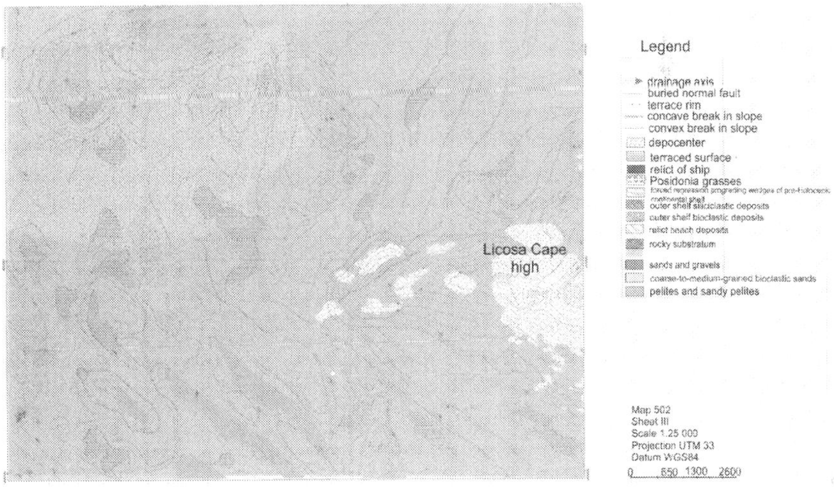

Figure 8. Marine geological map n. III SE, Agropoli geological map, scale 1:25.000 (modified after Aiello and Marsella, 2013).

Several bathymetric highs, showing an elongated shape and an NW–SE trending, are localized at water depths of 140–145 m. They are interpreted as relict morphologies corresponding to

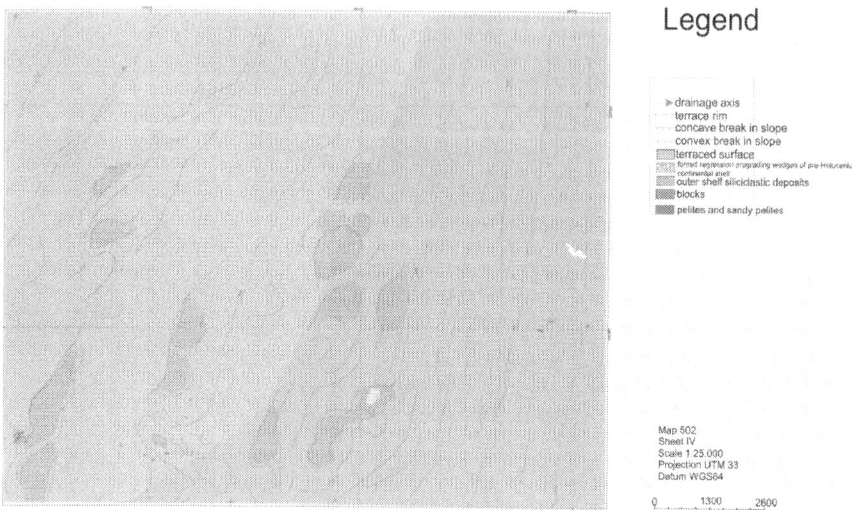

Figure 9. Marine geological map n. IV NE, Agropoli geological map, scale 1:25.000 (modified after Aiello and Marsella, 2013).

the last regressive phase and are represented by prograding sedimentary wedges. These sedimentary wedges signify portions of submerged beaches connected to fifth-order eustatic cycles, overlying the upper part of a wide Pleistocene prograding succession.

The map n. IV NE (scale 1:25.000) covers the continental shelf at water depths of 105–185 m (Fig. 9). Several convexities in correspondence to the isobaths of 135, 145, and 155 m probably correspond to remnants of terraced surfaces. The geological interpretation of seismic data has evidenced the outcrops at the sea bottom of a wide prograding succession dated to the Late Pleistocene and well evident on seismic profiles.

4.2. Geological units

The main geological units mapped in the geological map n. 502 Agropoli are described herein (Figs. 6–9). The Cenozoic substratum is composed of Cenozoic siliciclastic rocks, genetically related to the Cilento Flysch. The unit crops out in the inner shelf, particularly offshore the Licosa Cape morpho-structural high and is terraced at its top by polycyclic erosional surfaces and covered by wide beds of marine phanerogams. The littoral environment is characterized by the deposits of submerged beach and of toe of coastal cliff, incised in the arenaceous-silty successions of the Cilento Flysch.

The deposits of submerged beach are composed of gravels, sandy gravels, and coarse-grained sands with rounded to sub-rounded pebbles in a medium- to fine-grained sandy matrix; by coarse- to medium-grained sands, from rounded to sub-rounded, with pebbles and blocks, by medium- to fine-grained litho-bioclastic sands.

The deposits of toe of coastal cliff consist of siliciclastic heterometric blocks.

The inner shelf environment is characterized by innershelf deposits and bioclastic deposits. The inner shelf deposits consist of coarse-grained litho-bioclastic sands, middle-to fine-grained litho-bioclastic sands, and fine-grained pelitic sands.

The bioclastic deposits are made up of heterometric gravels, gravel sands, and bioclastic sands in a pelitic matrix. They represent the base of wide beds of marine phanerogams and overlie the top of outcrops of Cenozoic substratum.

The outer shelf environment is typified by clastic deposits and bioclastic deposits. The clastic deposits are distinguished by middle-fine grained sands localized at the top of wide outcrops of Cenozoic substratum, by pelites and sandy pelites. The bioclastic deposits are composed of bioclastic calcareous sands in a pelitic matrix, organized as sedimentary drapes located at the top of outcrops of Cenozoic substratum or of Late Pleistocene relic deposits (Figs. 6–9).

The lowstand system tract (LST) is distinguished by coarse-grained organogenic sands, grading upward into the middle-grained sands and pelitic drapes. They are relict littoral deposits organized as coastal prisms overlying the shelf margin progradations and representing portions of submerged beaches related to the last sea-level lowstand related to isotopic stage 2.

The LST deposits form NW–SE-trending coastal dunes, occurring in the southwestern sector of the area, at water depths ranging between 140 and 145 m (Figs. 6–9).

The Pleistocene relict marine units are represented by coarse- to fine-grained marine deposits, probably composed of well-sorted sands and gravels with bioclastic fragments and by medium-fine grained sands, with a pelitic coverage having a variable thickness but less than 2 m. They are localized in the NW and SW sectors of the Agropoli continental shelf and represent relics of beach and continental shelf environment interpreted as the remnants of older beach systems, related to isotopic stages 4 and 3 (Figs. 6–9).

Rocky units of the substratum

Cenozoic substratum

The Cenozoic substratum is composed of the siliciclastic rocks of the Cilento Flysch [2; 41; 77]. It crops out at the sea bottom next to the Licosa Cape morpho-structural high. Its top is marked by a polycyclic erosional surface involved by marine terraces. Wide grasses of marine phanerogams occur atthe top of the Licosa high up to a water depth of 30 m (Figs. 6–9).

Quaternary marine deposits

Littoral environment

Submerged beach deposits

The submerged beach deposits include gravels, sandy gravels, and coarse-grained sands, with rounded to sub-rounded pebbles in a middle-to fine-grained sandy matrix, coarse-to-middle-grained sands, from rounded to sub-rounded, with gravels, pebbles, and blocks.

Toe of coastal cliff deposits

The toe of coastal cliff deposits include both siliciclastic heterometric blocks, from metric to decametric and siliciclastic heterometric gravels, from centimetric to metric.

Inner shelf environment

Inner shelf deposits

The inner shelf deposits consist of litho-bioclastic coarse-grained sands, from well-sorted to poorly sorted, medium- to fine-grained litho-bioclastic sands, and fine-grained pelitic sands. The sandy fraction is composed of bioclastic fragments. Sandy ridges parallel to the shoreline have been interpreted on sidescan sonar profiles and reported on geological maps. Current lineaments parallel to the shoreline and perpendicular to the isobaths have also been identified.

Bioclastic deposits

The bioclastic deposits are made up of heterometric gravels, gravelly sands, and bioclastic sands, sometimes in a pelitic matrix. They are located at the top of outcrops of the acoustic basement.

Outer shelf environment

Outer shelf deposits

The outer shelf deposits are made up of medium- to fine-grained sands with lithoclasts and bioclasts and abundant rhyzomes of marine phanerogams, pelites and sandy pelites, rich in small Gastropods, Vermetids, and small mollusks. Current lineaments have been recognized based on sidescan sonar profiles.

Bioclastic deposits

The bioclastic deposits are made up of calcareous sands in an abundant pelitic matrix, overlying the top of outcrops of acoustic basement, correlated with the Cilento Flysch in the western Licosa structural high or at the top of palimpsest deposits.

Lowstand system tract

The LST deposits are composed of coarse-grained organogenic sands including abundant shell fragments and fragments of mollusks (*Arctica islandica*, cold host of the Pleistocene), fragments of Echinids and Bryozoans, grading upward to medium-grained sands and pelitic covers having a variable thickness, but less than 2 m. They form coastal prisms overlying the shelf margin progradations and represent portions of submerged beach linked to the last sea-level lowstand [60; 61]. The coastal dunes have an NW–SE trending and occur in the southwestern sector of the area at 140–145 m of water depth. Their age is Late Pleistocene–Holocene.

Pleistocene relict marine units

The Pleistocene relict marine units are made up of coarse- to-fine-grained marine deposits, of well-sorted sands and gravels with bioclastic fragments and of medium- to fine-grained sands with a pelitic coverage having a variable thickness, but less than 2 m. They are located in the

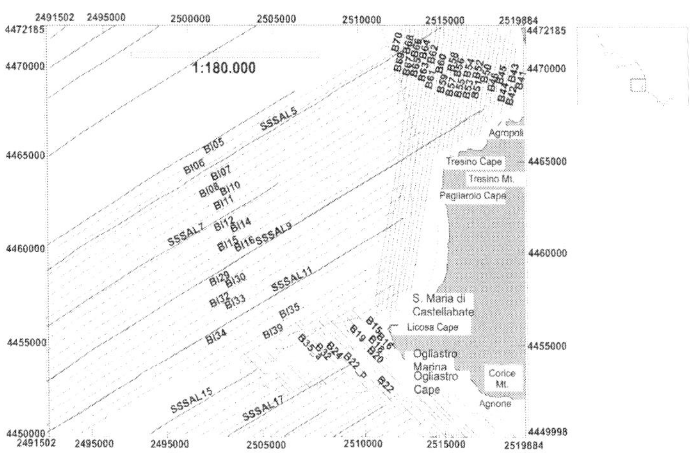

Figure 10. Seismic grid of the study area.

NW and SW sectors of the study area and represent palimpsests of beach and continental shelf environments.

The Pleistocene marine units, overlying the LST deposits, stand for the remnants of older beach systems, correlated to isotopic stages 4 and 3 [60; 61].

4.3. Seismo-stratigraphic interpretation

The seismo-stratigraphic interpretation of strictly spaced grid of subbottom Chirp profiles, located on the Campania continental shelf among the Solofrone river mouth and the Agnone town, has been carried out.

The seismic grid of study sections is located at water depths ranging between 10 and 150 m. Seismo-stratigraphic units separated by unconformities and/or paraconformities have been distinguished through the criteria of seismic stratigraphy [97; 98].

The acoustic basement has been mapped through the geological interpretation of Chirp profiles. It crops out at the sea bottom offshore the Licosa Cape, having an extension wider than the one previously singled out by maps reported in geological literature on the area [45; 92].

The study area, located between the Solofrone river mouth and the Agnone town, is distinguished in three sectors, characterized by different line trending and line spacing (Fig. 10).

A first grid of seismic sections has been recorded on the inner shelf bounded northward by the Solofrone river mouth and southward by the Licosa Cape (Fig. 10). It is NNE–SSW oriented, parallel to the 50 m isobath and located at water depths of 10–60 m, with a line spacing of 250 m.

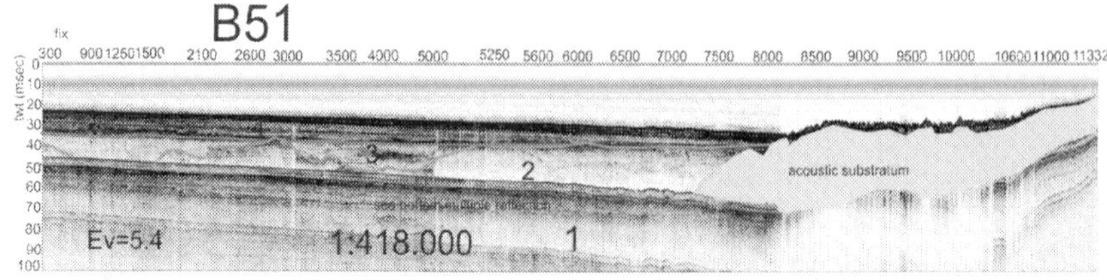

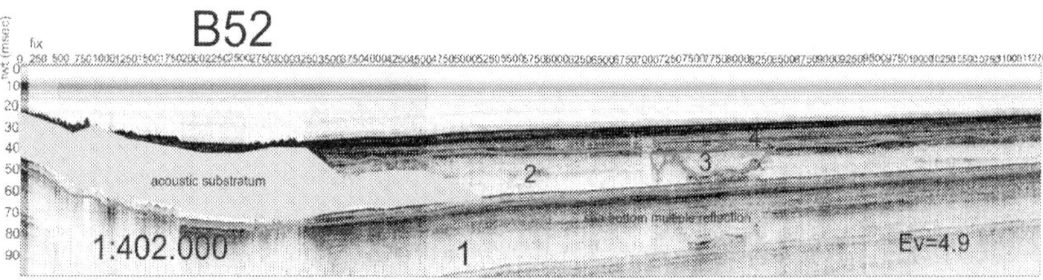

Figure 11. Seismic profiles B51 and B52 and corresponding geological interpretation (modified after Aiello and Marsella, 2013).

A second grid is located on the inner shelf bounded northward by the Licosa Cape and southward by the Agnone town (Fig. 10). It is NW–SE oriented, perpendicular to the 50 m isobath and located at water depths of 10–60 m, with a line spacing of 250 m.

A third set of NE–SW-oriented perpendicular sections covers the continental shelf bounded northward by the Solofrone river mouth and southward by the Agnone town, at water depths of 40–160 m. The line spacing is about 500 m.

Seismo-stratigraphic analysis of significant seismic sections, respectively, located in each of the abovementioned seismic grids, has been carried out. One main aim was the mapping of the rocky outcrops of the acoustic basement, while another aim was the mapping of marine terraces carving the rocky substratum. The measurement of the depth of marine terraces and their tentative correlation with the corresponding episodes of sea-level lowstand during the Late Quaternary have also been carried out [37].

The seismic profiles B51 and B52 and corresponding geologic interpretation are reported in Fig. 11. The seismic line B51, about 6-km long and located at water depths of 12–20 m, has recorded a stratigraphic succession about 65-m thick (100 msec TWT), an excellent stratigraphic record for the Chirp profiles.

The top of the acoustic basement (S unit on seismic profiles) is incised by marine terraces at water depths ranging between 18 and 21 m.

The first seismo-stratigraphic unit (Unit 1) is characterized by parallel seismic reflectors and shows an average thickness of 38 m (44 msec TWT). The unit is composed of marine sediments and bounded at its top by a paraconformity (Fig. 11).

The second seismo-stratigraphic unit (Unit 2) shows an acoustically transparent seismic facies and a wedge-shaped external geometry; its top is bounded by an erosional unconformity. According to the geological interpretation, the unit should be composed of coarse-grained sands, forming the filling of channels (Fig. 11).

The third seismo-stratigraphic unit (Unit 3) is characterized by discontinuous reflectors with high amplitude, alternating with acoustically transparent intervals and showing an average thickness of about 17 m (about 20 msec TWT).

Shallow gas pockets widely occur on seismic profiles and are in course of study with detail.

The seismic line B52, about 6 km long and located at shallow water depths of 16.5–19 m, has shown a stratigraphic architecture similar to that of the seismic line B51 (Fig. 11).

The first seismo-stratigraphic unit (Unit 1) is characterized by parallel seismic reflectors and shows an average thickness of 33 m (38 msec TWT). It is coeval with the acoustic basement (S unit; Fig. 11).

The second seismo-stratigraphic unit (Unit 2) is characterized by an acoustically transparent seismic facies and by a wedge-shaped external geometry, filling wide paleo-channels.

The third seismo-stratigraphic unit (Unit 3; Fig. 11) is characterized by an acoustic facies with discontinuous reflectors having a high amplitude, alternating to acoustically transparent intervals and showing an average thickness of 17 m (about 20 msec TWT). The unit is probably composed of coarse-grained sediments. Shallow gas pockets have been identified on the seismic profile and are in course of study through a detailed analysis.

The B15 seismic line, NW–SE trending, located at water depths of 5–42 m, has crossed wide outcrops of acoustic basement offshore the Licosa Cape (Cilento Promontory). The top of the acoustic basement represents a terraced surface having a low gradient; marine terraces have been identified as occurring at water depths of 31, 19, and 5 m (Fig. 12).

The B16 seismic line (Fig. 12), located at water depths of 12 and 47 m, has shown wide outcrops of rocky units of the acoustic basement. The stratigraphic architecture observed on seismic section is similar to the one recognized on the seismic section B15. The top of the acoustic basement is carved by marine terraces, occurring at water depths of 21, 14, and 11 m (Fig. 12).

The BL05 seismic profile is located at water depths between 112.5 and 42 m (Fig. 13). It has crossed about 82 m of sediment (97 msec TWT; Fig. 13). Nearshore, the S unit crops out at the sea bottom, delineating a wide terraced surface, degrading seaward with low gradients. At the margin of this surface, the S unit is downthrown under the recent sedimentary cover. The S unit is then uplifted forming a structural high extending for almost 1.4 km (Fig. 13).

The first seismo-stratigraphic (Unit 1) shows an average thickness of about 50 m (60 msec TWT). It overlies the acoustic basement and is limited at its top by an unconformity (BA reflector), characterized by the onlap on the morpho-structural high of the S unit (Fig. 13).

Lowstand deposits, characterized by prograding reflectors truncated upward by an erosional surface, have been interpreted (Fig. 13).

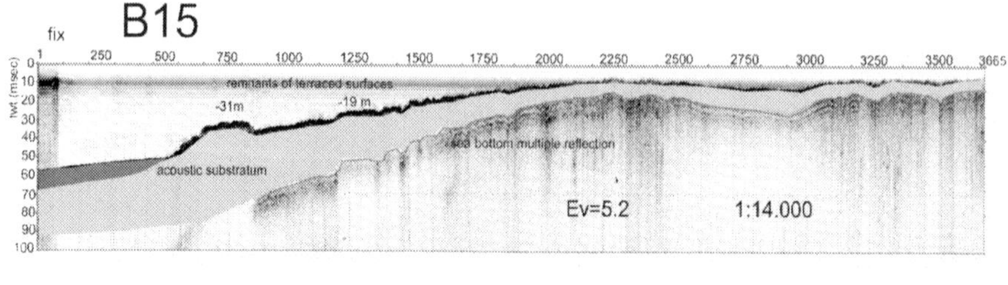

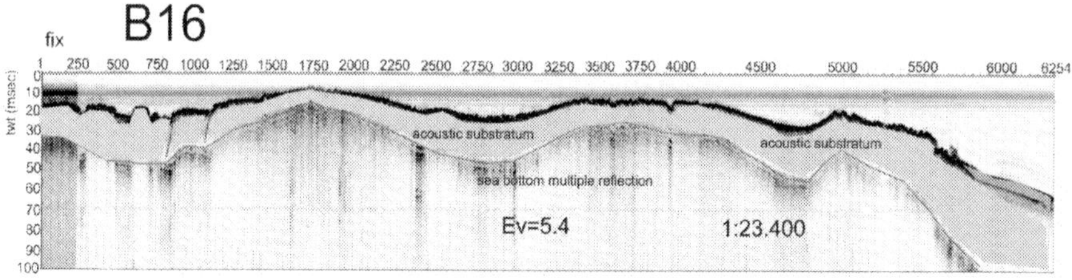

Figure 12. Seismic profiles BL15 and BL16 and corresponding geological interpretation (modified after Aiello and Marsella, 2013).

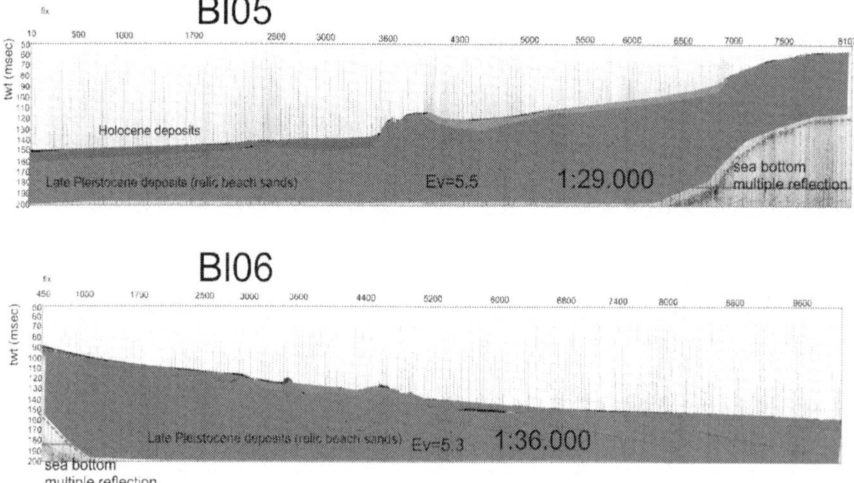

Figure 13. Seismic profiles BL05 and BL06 and corresponding geological interpretation (modified after Aiello and Marsella, 2013).

The second seismic sequence (Unit 2), having a wedge-shaped external geometry, is characterized by pockets of coarse-grained marine sediments, representing the filling of channels.

Marine terraces have been identified at 63 and 85 m of water depth (Fig. 13).

The BL06 seismic line, having a length of about 4 km, is located at water depths of 120 m. It has shown a stratigraphic architecture similar to the ones in the BL05 seismic line (Fig. 13).

5. Conclusions

The geological evolution of coastal and marine environments offshore the Cilento Promontory (submerged beach, inner shelf, outer shelf, lowstand system tract, Pleistocene relict marine units, and rocky units of the substratum) has been discussed through the identification of corresponding deposits and their cartographic representation. The seismic stratigraphy of the Agropoli continental shelf (southern Campania) has been highlighted through the geological interpretation of a densely spaced grid of Chirp seismic profiles [37].

The geological interpretation of Chirp profiles located in the morpho-structural high of the Licosa Cape has been largely documented in the acoustic basement (S unit), cropping out at the sea bottom near shore and dipping seaward below the Quaternary marine deposits, forming the recent sedimentary cover. Remnants of terraced surfaces located at several water depths have supported the morpho-evolution of the acoustic basement during the Late Quaternary.

Author details

Gemma Aiello* and Ennio Marsella

*Address all correspondence to: gemma.aiello@iamc.cnr.it

Istituto per l'Ambiente Marino Costiero (IAMC), Consiglio Nazionale delle Ricerche (CNR), Calata Porta di Massa, Porto di Napoli, Napoli, Italy

References

[1] D'Argenio B., Pescatore T., and Scandone P.(1973) Schema geologico dell'Appennino meridionale (Campania e Lucania). Accademia Nazionale dei Lincei, Quaderno 183, Atti del Convegno "Moderne Vedute sulla Geologia dell'Appennino", Roma.

[2] Bigi G., Cosentino D., Parotto M., Sartori R., and Scandone P. (1992) Structural model of Italy. Monografie Progetto Finalizzato Geodinamica, CNR, Roma.

[3] Catalano R., Fabbri A., Argnani A., Bortoluzzi G., Correggiari A., Gamberi F., Ligi M., Marani M., Penitenti D., Roveri M., Trincardi F. (1996) Linee guida al rilevamento delle aree marine del Servizio Geologico Nazionale. APAT, Bozza n. 1.

[4] Fabbri A., Argnani A., Bortoluzzi G., Correggiari A., Gamberi F., Ligi M., Marani M., Penitenti D., Roveri M., and Trincardi F. (2002) Carta geologica dei mari italiani alla scala 1: 250.000. Guida al rilevamento. Presidenza del Consiglio dei Ministri, Dipartimento per i Servizi Tecnici Nazionali, Servizio Geologico, Quaderni serie III, 8, pp. 1–93.

[5] Brancaccio L., Cinque A., Romano P., Rosskopf C., Russo F., and Santangelo N. (1995) L'evoluzione delle pianure costiere della Campania: geomorfologia e neotettonica. Memorie della Società Geografica Italiana, 53, 313–336.

[6] Ascione A. and Romano P. (1999) Vertical movements on the eastern margin of the Tyrrhenian extensional basin. New data from Mt. Bulgheria (Southern Apennines, Italy). Tectonophysics, 315, 337–356.

[7] Violante C., Tonielli R., De Lauro M., Bellonia A., Di Martino G., and Innangi S. (2005) Indagini acustiche e caratteri ambientali nell'area marina di Punta Licosa (Parco Nazionale del Cilento, Italia meridionale). Poster GEOSED (Associazione Italiana per la Geologia del Sedimentario).

[8] Marghany M. (2013a) Three-dimensional coastal deformation from Insar Envisat satellite data. In: Murgante B., Misra S., Carlini M., Torre C.M., Nguyen H.-Q., Taniar D., Apduhan B.O., and Gervasi O. (Eds.), Computational Science and Its Applications – ICCSA 2013, vol. 7972, pp. 599–610, Springer-Verlag Berlin Heidelberg.

[9] Marghany M. (2013b) DInSAR technique for three-dimensional coastal spit simulation from Radarsat-1 fine mode data. Acta Geophysica, 61 (2), 478–493.

[10] Marghany M. (2014a) Modelling wave refraction pattern using AIRSAR and POL-SAR C-band data. In: Murgante B., Misra S., Carlini M., Torre C.M., Nguyen H.-Q., Taniar D., Apduhan B.O., and Gervasi O. (Eds.), ICCSA 2014, Guimaraes, Portugal, June 30–July 3, 2014, Proceedings, Part VI, pp. 518–531.

[11] Marghany M. (2014b) A three-dimensional of coastline deformation using the sorting reliability algorithm of ENVISAT interferometric synthetic aperture radar. In: Marghany M. (Ed.), Advanced Geoscience Remote Sensing, pp. 105–121. INTECH Publisher, Croatia.

[12] Marghany M. (2014c) Simulation of tsunami impact on sea surface salinity along Banda Aceh coastal waters, Indonesia. In: Marghany M. (Ed.), Advance Geoscience Remote Sensing, pp. 229–251. INTECH Publisher, Croatia.

[13] Marghany M. (2014d) Simulation of three dimensional of coastal erosion using differential interferometry synthetic aperture radar. Global Nest Journal, 16 (1), 80–86.

[14] Ortolani F. and Aprile F. (1978) Nuovi dati sulla struttura profonda della Piana Campana a sud-est del fiume Volturno. Bollettino della Società Geologica Italiana, 97, 591–608.

[15] Fabbri A., Gallignani P., and Zitellini N. (1981) Geologic evolution of the peri-tyrrhenian sedimentary basins. In: Wezel F.C. (Ed.), Sedimentary Basins of the Mediterranean Margins, pp. 101–126. Technoprint, Bologna.

[16] Bartole R. (1984) Tectonic structure of the Latian-Campanian shelf (Tyrrhenian sea). Bollettino di Oceanologia Teorica e Applicata, 2, 197–230.

[17] Bartole R., Savelli D., Tramontana M., and Wezel F.C. (1983) Structural and sedimentary features in the Tyrrhenian margin off Campania, Southern Italy. Marine Geology, 55, 163–180.

[18] Finetti I. and Morelli C. (1974) Esplorazione sismica a riflessione dei Golfi di Napoli e Pozzuoli. Bollettino di Geofisica Teorica Applicata, 16 (62/63), 175–222.

[19] Trincardi F. and Zitellini N. (1987) The rifting of the Tyrrhenian basin. Geomarine Letters, 7, 1–6.

[20] Mariani M. and Prato R. (1988) I bacini neogenici costieri del margine tirrenico: approccio sismico-stratigrafico. Memorie della Società Geologica Italiana, 41, 519–531.

[21] Aiello G., Budillon F., de Alteriis G., Di Razza O., De Lauro M., Marsella E., Pelosi N., Pepe F., Sacchi M., and Tonielli R. (1997a) Seismic exploration of the perityrrhenian basins in the Latium-Campania offshore. Abstract Int. Cong. International Lithosphere Program Task Force: Origin of Sedimentary Basins, Torre Normanna (Palermo, Italy), June 1997.

[22] Aiello G., Budillon F., de Alteriis G., Ferranti L., Marsella E., Pappone G., Sacchi M. (1997b) Late Neogene tectonics and basin evolution of Southern Italy Tyrrhenian margin. Abstract Int. Cong. International Lithosphere Program Task Force: Origin of Sedimentary Basins, Torre Normanna (Palermo, Italy), June 1997.

[23] Bertotti G., Marsella E., Pepe F., Pelosi N., and Tonielli R. (1999) SISTER99: a seismic campaign to investigate the kinematics of South Tyrrhenian extensional regions. Giornale di Geologia, 61, 25–36, Bologna, Italy.

[24] Chiocci F.L., Martorelli E., Sposato A., Aiello G., Baraza J., and Budillon, F. (1998a) Prime immagini TOBI del Tirreno centro-meridionale. Geologica Romana, 34, 207–222.

[25] Chiocci F.L., Tommasi P., Aiello G., Baraza J., Bosman A., Cristofalo G., De Lauro M., Ercilla G., Estrada F., Farran M., Martorelli E., Ortolani U., Sarli R., Senatore M.R., Tonielli R., Jacobs C.L., Matthews D., Rouse I., and Wallace R.F. (1998b) Submarine debris avalanche off the southern flank of Ischia volcanic island, Gulf of Naples. Second International Symposium of the Geotechnics of Hard Soils – Soft Rocks, Napoli, 12–14 October 1998.

[26] Aiello G., Marsella E., and Sacchi M. (2000) Quaternary structural evolution of Terracina and Gaeta basins. Rendiconti Lincei, 11, 41–58.

[27] Aiello G., Budillon F., Cristofalo G., D'Argenio B., de Alteriis G., De Lauro M., Ferraro L., Marsella E., Pelosi N., Sacchi M., and Tonielli R. (2001) Marine geology and morpho-bathymetry in the Bay of Naples (South-Eastern Tyrrhenian sea, Italy). In: Faranda F.M., Guglielmo L., and Spezie G. (Eds.), Structures and Processes of the Mediterranean Ecosystems, Special Volume, pp. 1–8. Springer-Verlag, Milano, Italy.

[28] Aiello G., Angelino A., D'Argenio B., de Alteriis G., De Lauro M., Ferraro L., Marsella E., Pelosi N., Sacchi M., and Tonielli R. (2004) Carta magnetica del Golfo di Napoli (Tirreno meridionale). Bollettino della Società Geologica Italiana, 123, 333–342.

[29] Aiello G., Angelino A., D'Argenio B., Marsella E., Pelosi N., Ruggieri S., and Siniscalchi A. (2005) Buried volcanic structures in the Gulf of Naples (Southern Tyrrhenian sea, Italy) resulting from high resolution magnetic survey and seismic profiling. Annals of Geophysics, 48, 1–15.

[30] Aiello G., Cicchella A.G., Di Fiore V., and Marsella E. (2011a) New seismo-stratigraphic data of the Volturno Basin (northern Campania, Tyrrhenian margin, southern Italy): implications for tectono-stratigraphy of the Campania and Latium sedimentary basins. Annals of Geophysics, 54 (3), 265–283.

[31] Aiello G., Marsella E., Cicchella A.G., and Di Fiore V. (2011b) New insights on morpho-structures and seismic stratigraphy of the Campania continental margin based on deep multichannel profiles. Rendiconti Lincei, 22 (4), 349–373.

[32] Aiello G., Marsella E., and Di Fiore V. (2012a) New seismo-stratigraphic and marine magnetic data of the Gulf of Pozzuoli (Naples Bay, Tyrrhenian sea, Italy): inferences for the tectonic and magmatic events of the Phlegrean Fields volcanic complex (Campania). Marine Geophysical Researches, 33 (2), 93–125.

[33] Aiello G., Giordano L., Marsella E., and Passaro S. (2012b) Seismic stratigraphy and marine magnetics of the Naples Bay (Southern Tyrrhenian sea, Italy): the onset of new technologies in marine data acquisition, processing and interpretation. In: Elitok O. (Ed.), Stratigraphic Analysis of Layered Deposits, Chapter 2, pp. 21–60. Intech Science Publishers, Croatia.

[34] Aiello G., Marsella E., and Passaro S. (2012c) Stratigraphic and structural setting of the Ischia volcanic complex (Naples Bay, southern Italy) revealed by submarine seismic reflection data. Rendiconti Lincei, 23, 387–408.

[35] Secomandi M., Paoletti V., Rapolla A., Aiello G., Marsella E., Ruggieri S., and D'Argenio B. (2003) Analysis of the magnetic anomaly field of the volcanic district of the Bay of Naples, Italy. Marine Geophysical Researches, 24, 207–211.

[36] D'Argenio B., Aiello G., de Alteriis G., Milia A., Sacchi M., Tonielli R., Budillon F., Chiocci F.L., Conforti A., De Lauro M., D'Isanto C., Esposito E., Ferraro L., Insinga D., Iorio M., Marsella E., Molisso F., Morra V., Passaro S., Pelosi N., Porfido S., Raspini A., Ruggieri S., Terranova C., Vilardo G., and Violante C. (2004) Digital Elevation Model of the Naples Bay and adjacent areas, Eastern Tyrrhenian sea. Atlante di Car-

tografia Geologica scala 1:50.000 (progetto CARG), Servizio Geologico d'Italia (APAT), 32nd International Congress "Firenze 2004", Editore De Agostini, Italy.

[37] Aiello G. and Marsella E. (2013) The contribution of marine geology to the knowledge of marine coastal environment off the Campania region (Southern Italy): the geological map n. 502 "Agropoli". Marine Geophysical Researches, 34, 89–113.

[38] Aiello G. and Marsella E. (2014) The Southern Ischia canyon system: examples of deep sea depositional systems on the continental slope off Campania. Rendiconti Online della Società Geologica Italiana, 32, 28–37.

[39] Aiello G. and Marsella E. (2015) Interactions between late quaternary volcanic and sedimentary processes in the Naples Bay, southern Tyrrhenian sea. Italian Journal of Geosciences, 134 (2), 367–382.

[40] Sacchi M., Infuso S., and Marsella E. (1994) Late Pliocene-Early Pleistocene compressional tectonics in the offshore of Campania. Bollettino di Geofisica Teorica Applicata, 36, 141–144.

[41] Bonardi G., Amore F.O., Ciampo G., De Capoa P., Miconnet P., and Perrone V. (1992) Il Complesso Liguride Auct.: stato delle conoscenze e problemi aperti sull'evoluzione pre-appenninica ed i suoi rapporti con l'Arco Calabro. Memorie della Società Geologica Italiana, 41, 17–35.

[42] Capaldi G., Civetta L., Di Girolamo P., Lanzara R., Orsi G., and Scarpati C. (1987) Volcanological and geochemical constraints on the genesis of Yellow Tuffs in the Neapolitan-Phlegrean area. Rend. Accademia Scienze Fisiche e Matematiche Special Issue, 25–40.

[43] Ortolani F. and Torre M. (1981) Guida all'escursione nell'area interessata dal terremoto del 23-11-80. Rendiconti della Società Geologica Italiana, 4, 173–214.

[44] Agip (1977) Temperature sotterranee. Inventario dei dati raccolti dall'Agip durante la ricerca e la produzione di idrocarburi in Italia. Elli Brugnora Segrate, Milano, pp. 1–1390.

[45] Coppa M.G., Madonna M., Pescatore T.S., Putignano M.L., Russo B., Senatore M.R., and Verrengia A.(1988) Elementi geomorfologici e faunistici del margine continentale tirrenico tra Punta Campanella e Punta degli Infreschi (Golfo di Salerno). Memorie della Società Geologica Italiana, 41, 541–546.

[46] D'Argenio B., Ortolani F., and Pescatore T. (1986) Geology of Southern Apennines. Geologia Applicata e Idrogeologia, 6, 135–161.

[47] Aiello G., Di Fiore V., Marsella E., and D'Isanto C. (2007) Stratigrafia sismica e morfobatimetria della Valle di Salerno. Atti 26° Convegno Nazionale GNGTS, Roma, Dicembre 2007, Extended Abstract, pp. 495–498.

[48] Aiello G., Marsella E., Di Fiore V., and D'Isanto C. (2009) Stratigraphic and structural styles of half-graben offshore basins in Southern Italy: multichannel seismic and

Multibeam morpho-bathymetric evidences on the Salerno Valley (Southern Campania continental margin, Italy). Quaderni di Geofisica, 77, 1–31.

[49] Capotorti F. and Tozzi M. (1991) Tettonica trascorrente nella Penisola Sorrentina. Memorie della Società Geologica Italiana, 47, 235–249.

[50] Milia A. and Torrente M.M. (1997) Evoluzione tettonica della Penisola Sorrentina (margine peritirrenico campano). Bollettino della Società Geologica Italiana, 116, 487–502.

[51] McKenzie D. (1978) Some remarks on the development of sedimentary basins. Earth and Planetary Science Letters, 40, 25–32.

[52] Allen P.A. and Allen J.R. (1990) Basin analysis. Blackwell Scientific Publications, London (England), 451 pp.

[53] Liggett M.A. and Ehrenspeck H.E. (1974) Pahrnagat shear systems, Lincoln County, Nevada. Argus Exploration Report, NASA-CR, E74-10206, 10 pp.

[54] Bally A.W., Bernoulli D., Davis G.A., and Montadert L. (1981) Listric normal faults. Oceanologica Acta, pp. 87–102.

[55] Gawthorpe R.L., Fraser A.J., and Collier R.E.L. (1994) Sequence stratigraphy in active extensional basins: implications for the interpretations of ancient basin fills. Marine and Petroleum Geology, 11 (6), 642–658.

[56] Collier E.L. and Gawthorpe R.L. (1995) Neotectonics, drainage and sedimentation in central Greece: insights into coastal reservoir geometries in syn-rift sequences. In: Lambiase J.J. (Ed.), Hydrocarbon Habitat in Rift Basins, pp. 165–181. Geological Society, London, Special Publication, no. 80.

[57] Ellis P.G. and McClay K.R. (1988) Listric extensional fault systems – results of analogue model experiments. Basin Research, 1, 55–70.

[58] Bally A.W. and Oldow J.S. (1984) Plate tectonics, structural styles and the evolution of sedimentary basins. Unpublished Course Notes, 238 pp.

[59] Antonioli F., Donadio C., and Ferranti L. (1994) Guida all'escursione. Note scientifiche Geosub, 94, Palinuro, De Frede, Napoli.

[60] Shackleton N.J. and Opdyke N.D. (1973) Oxygen isotope and paleomagnetic stratigraphy of equatorial pacific core V28-238: oxygen isotope temperature and ice volume on a 10 year scale. Quaternary Research, 3, 39–55.

[61] Martinson D.G., Pisias N.G., Hays J.D., Imbrie J., Moore T.C., and Shackleton J. (1987) Age dating and the orbital theory of the ice ages: development of a high resolution 0 to 300.000 year chronostratigraphy. Quaternary Research, 27, 1–29.

[62] Ciampo G. (1976) Ostracodi pleistocenici di Cala Bianca (Marina di Camerota, Salerno). Bollettino della Società Paleontologica Italiana, 15, 3-23.

[63] Ciampo G., Perrone V., and De Pascale B. (1984) Revisione stratigrafica delle Formazioni di Pollica e di S. Mauro (Flysch del Cilento – Appennino meridionale). Bollettino della Società Geologica Italiana, 103, 333–339.

[64] Baggioni M., Suc J.P., and Vernet J.L. (1981) Le Plio-Pleistocene du Camerota (Italie meridionale): Geomorphologie and paleoflores. Geobios, 14 (2), 229–237.

[65] Lippmann-Provansal M. (1987) L'Apennin Campanien meridionel (Italie). Etude geomorphologique. These de Doctorat d'Etat en Geographie Physique, Univ. D'Aix, Marseille.

[66] Borrelli A., Ciampo G., De Falco M., Guida D., and Guida M. (1988) La morfogenesi del Monte Bulgheria (Campania) durante il Pleistocene inferiore e medio. Memorie della Società Geologica Italiana, 41, 667–672.

[67] Romano P. (1992) La distribuzione dei depositi marini pleistocenici lungo le coste della Campania. Stato delle conoscenze e prospettive di ricerca. Studi Geologici Camerti, Volume Speciale, 1992/1, 265–269.

[68] Guida M., Guida D., Iaccarino G., Metcalf G., Vallario A., Vecchio V., and Zicari G. (1979) Il Bacino del Mingardo (Cilento): Evoluzione geomorfologica, fenomeni franosi e rischio a franare. Geologia Applicata e Idrogeologia, 14 (2), 119–198.

[69] Scandone P., Sgrosso I., and Bruno F. (1963) Appunti di geologia sul Monte Bulgheria (Salerno). Bollettino Società Naturalisti in Napoli, 1972, 19–27.

[70] Ietto A., Pescatore T., and Cocco E. (1965) Il Flysch Mesozoico-Terziario del Cilento occidentale. Bollettino Società Naturalisti in Napoli, 74, 396–402.

[71] Critelli S. and Le Pera E. (1990) Litostratigrafia e composizione della Formazione di Pollica (Gruppo del Cilento, Appennino meridionale). Bollettino della Società Geologica Italiana, 109, 511–536.

[72] Amore F.O., Bonardi G., Ciampo G., De Capoa P., Perrone V., and Sgrosso I. (1988) Relazioni tra flysch interni e domini appenninici: reinterpretazione delle formazioni di Pollica, San Mauro e Albidona e l'evoluzione infra-medio-miocenica delle zone esterne sud-appenniniche. Memorie della Società Geologica Italiana, 41, 285–297.

[73] Zuppetta A., Russo M., Turco E., and Gallo G. (1984) Età e significato della Formazione di Albidona in Appennino meridionale. GSA Bulletin, 109, 698–708.

[74] Russo M., Zuppetta A., and Guida A. (1995) Alcune precisazioni stratigrafiche sul Flysch del Cilento (Appennino meridionale). Bollettino della Società Geologica Italiana, 114, 353–359.

[75] Zuppetta A. and Mazzoli S. (1995) Analisi strutturale ed evoluzione paleotettonica dell'Unità del Cilento nell'Appennino Campano. Studi Geologici Camerti, 13, 103–114.

[76] Critelli S. (1999) The interplay of lithospheric flexure and thrust accommodation in forming stratigraphic sequences in the southern Apennines foreland basin system, Italy. Rendiconti Lincei, 10, 257–326.

[77] Zuppetta A. and Mazzoli S. (1997) Deformation history of a synorogenic sedimentary wedge, northern Cilento area, southern Apennines fold and thrust belt, Italy. GSA Bulletin, 109, 698–708.

[78] Valente A. (1993) Studi sedimentologici sulla successione miocenica di Monte Sacro (Flysch del Cilento). PhD Thesis, University of Naples "Federico II", 170 pp.

[79] Critelli S. and Le Pera E. (1994) Detrital modes and provenance of Miocene sandstones and sandstones of Southern Apennines thrust-top-basins. Journal of Sedimentary Research, A64, 824–835.

[80] Cocco E. and Pescatore T. (1968) Scivolamenti gravitativi (olistostromi) nel Flysch del Cilento (Campania). Bollettino Società Naturalisti in Napoli, 77, 51–91.

[81] Carrara S. and Serva L. (1982) I ciottoli contenuti nei flysch cretacico-paleogenici e miocenici e nei depositi post-tortoniani dell'Appennino meridionale. Loro significato paleotettonico. Bollettino della Società Geologica Italiana, 101, 441–496.

[82] Di Girolamo P., Morra V., and Perrone V. (1992) Ophiolitic olistoliths in Middle Miocene turbidites (Cilento Group) at Mt Centaurino (southern Apennines, Italy). Ofioliti, 17, 199–217.

[83] Valente A. (1991) Caratteri sedimentologici di una successione torbiditica nel Cilento orientale (Appennino meridionale). Memorie della Società Geologica Italiana, 47, 191–196.

[84] Sgrosso I. (1981) Il Significato delle Calciruditi di Piaggine nell'ambito degli eventi del Miocene inferiore nell'Appennino Campano-Lucano. Bollettino della Società Geologica Italiana, 117, 679–724.

[85] Sgrosso I. (1998) Possibile evoluzione cinematica miocenica nell'orogene centro-sud appenninico. Bollettino della Società Geologica Italiana, 117, 679–724.

[86] Castellano M.C., Putignano M.L., and Sgrosso I. (1997) Sedimentology and stratigraphy of the Piaggine sandstones (Cilento, Southern Apennines, Italy). Giornale di Geologia, 59, 273–287.

[87] Baggioni M. (1973) Le bordure du la Plain du Sele. Etude geomorphologique. Mediterranée, 3.

[88] Baggioni M. (1976) L'evolution quaternaire des versants calcaires dans le massif du Mont Bulgheria (Italie meridionale). CEGERM, V, 57–60.

[89] Baggioni M. (1981) Neotectonique et geomorphologie dans l'Apennin campanien (Italie meridionale). Revue de Geologie Dinamique et de Geographie Physique Paris, 23 (1), 41-54.

[90] Ortolani F., Pagliuca S., and Toccaceli R.M. (1991) Osservazioni sull'evoluzione geo-morfologica olocenica della Piana costiera di Velia (Cilento, Campania) sulla base di nuovi rinvenimenti archeologici. Geografia Fisica Dinamica Quaternaria, 14, 163–169.

[91] Russo F. (1994) Segnalazione di un livello fossilifero riferibile al Tirreniano a Cala Bianca (Marina di Camerota, Salerno). Memorie Descrittive della Carta Geologica d'Italia, LII, 395–398.

[92] Ferraro L., Pescatore T., Russo B., Senatore M.R., Vecchione C., Coppa M.G., and Di Tuoro A. (1997) Studi di geologia marina sul margine tirrenico: la piattaforma conti-nentale tra Punta Licosa e Capo Palinuro (Tirreno meridionale). Bollettino della Soci-età Geologica Italiana, 116, 473–485.

[93] Bonifay E. (1975) L'Ere Quaternaire: definition, limites and subdivision sur la base de la chronologie Méditerranee. Bulletin de la Societe Geologique de France, 17, 380–393.

[94] Posamentier H.W., James D.P., Allen G.P., and Tesson M. (1992) Forced regressions in a sequence stratigraphic framework: concepts, examples and exploration signifi-cance. AAPG Bulletin, 76, 1687–1709.

[95] Trincardi F. and Field M.E. (1991) Geometry, lateral variation and preservation of downlapping regressive shelf deposits: Eastern Tyrrhenian sea margin, Italy. Journal of Sedimentary Petrology, 61, 775–790.

[96] Trincardi F. and Field M.E. (1992) Collapse and flow of lowstand shelf-margin depos-its: an example from the Eastern Tyrrhenian, Italy. Marine Geology, 105, 77–94.

[97] Mitchum R.M., Vail P.R., and Thompson S. (1977) Seismic stratigraphy and global changes of sea level, part 2: the depositional sequence as a basic unit for stratigraphic analysis. In: Payton C.E. (Ed.), Seismic Stratigraphy – Applications to Hydrocarbon Exploration, vol. 26, pp. 53–62. AAPG Memoir, Tulsa (USA).

[98] Vail P.R., Mitchum R.M., Todd R.G., Widmier J.M., Thomson S., Sangree J.B., Bubb J.N., and Hatlelid W.G. (1977) Seismic stratigraphy and global changes of sea level. In: Payton C.E. (Ed.), Seismic Stratigraphy – Applications to Hydrocarbon Explora-tion, pp. 49–212. AAPG Memoire, no. 26, American Association of Petroleum Geolo-gists, Tulsa, OK.

Engineering Tools for the Estimation of Dredging-Induced Sediment Resuspension and Coastal Environmental Management

Iolanda Lisi, Marcello Di Risio, Paolo De Girolamo and Massimo Gabellini

Additional information is available at the end of the chapter

http://dx.doi.org/10.5772/61979

Abstract

In recent years, increasing attention has been paid to environmental impacts that may result from resuspension, sedimentation and increase in concentration of chemicals during dredging activities. Dredging dislodges and resuspends bottom sediments that are not captured by dredge-head movements. Resuspended sediments are advected far from the dredging site as a dredging plume and the increase in the suspended solid concentration (SSC) can strongly differ, in time and space, depending on site and operational conditions. Well-established international guidelines often include numerical modelling applications to support environmental studies related to dredging activities. Despite the attention that has been focused on this issue, there is a lack of verified predictive techniques of plume dynamics at progressive distances from the different dredging sources, as a function of the employed dredging techniques and work programs, i.e., spatial and temporal variation of resuspension source. This chapter illustrates predictive techniques to estimate the SSC arising from dredges with different mechanisms of sediment release and to assess the spatial and temporal variability of the resulting plume in estuarine and coastal areas. Predictive tools are aimed to support technical choices during planning and operational phases and to better plan the location and frequency of environmental monitoring activities during dredging execution.

Keywords: Dredging, resuspension source, plume modelling, SSC assessment, dredging management

1. Introduction

Dredging activities are widely used in estuarine and coastal areas to maintain or improve the designed depth of navigation channels or basins (i.e. ports and harbours), for the creation or

the improvement of facilities, for beach nourishment and to carefully remove and relocate contaminated materials (i.e. remedial or environmental dredging). These activities generally involve processes of removing materials from the bottom and relocating them elsewhere. Some of the sediments removed from the bottom are not captured by the dredge-head movement as part of the dredging operation, and thus the fine-grained fraction of resuspended sediments is dispersed in the water column ([1]). Furthermore, in the case of environmental dredging, some of the contaminated sediment may be left on the bottom as *"no dredging operation can remove every particle of contaminated sediment"*([2]). In recent years, increasing attention has been paid to the environmental impact due to the dredging activities and to reduce any physical, chemical and biological changes related to the sediment resuspension and pollutants (if any) dispersion (e.g. [2–5]). Four issues relevant to environmental dredging (the so-called four R's) were recognized ([2,6]): sediment resuspension, contaminants release, residual contaminated sediment produced by and/or remaining after dredging and environmental risk. We focus on the first R herein, i.e. the resuspension of dredged sediments. Indeed, the increase in the suspended solid concentration (hereinafter referred to as SSC) and the subsequent resettling of sediments transported as a dredging plume can bring adverse impacts on water quality (i.e. pollutants and nutrient dispersion, dissolved oxygen sags), on aquatic ecosystem (i.e. limiting of the photosynthetic process, smothering on the gills of bivalves or fish, inability to detect predators) and on human health (e.g. [7–9]). The assessment of environmental effects and risk exposure to the turbidity extent of dredging plumes always requires site-specific evaluations and should take into account the natural variability of local background turbidity ([10,11]). Hence, the magnitude of resuspension due to dredging should be placed in context with other sediment resuspension events or sources (i.e. river flow, the overflow from a barge, vessel traffic, etc.) and carefully managed. Establishing quantifiable tolerance limits for increase in turbidity and chemicals concentration (if any) is thus complex, while it is critical for the assessment of environmental impacts that may result from the dispersion and resettling processes of resuspended sediments. Hence, both monitoring and modelling data should be preliminary stated in order to provide control parameters useful for subsequent monitoring and management plan during the execution of dredging activities ([12,13]). Meaningful criteria to limit environmental impacts related to the extension of dredging plumes require the estimation of the SSC at the dredging source, and thus the definition of its source strength (i.e. the mass of resuspended sediment per unit time at the dredging point) and geometry ([14]). The resuspended rate (or sediment loss rate) close to dredging sources and the spatial and temporal variability of resulting plumes can significantly vary based on site and operational parameters, such as: sediment properties (i.e. bulk density, grain size, mineralogy), environmental (i.e. dredging water depth, thermoaline stratification, currents) and operational (i.e. dredging type, production rate, thickness of dredge cuts, velocity of dredging cycle) conditions (e.g. [1,6,14–17]). The resuspended sediment can be released from dredges by a wide range of mechanisms and at different levels in the water column and the resulting plumes are complex in terms of spatial distribution and time evolution. Understanding how different dredging types can operate is thus essential to provide standard methods for assessing increase in SSC and other adverse effects that may result from resuspension, sedimentation and increase in concentration of chemicals (e.g. [2,3,18,19]). Usually, three phases are identified for the dredging plume development at different distances from the dredging location ([2,3]): the dredging zone; the near field zone; and the far field zone (or passive plume). The starting phase of the plume dynamic is strongly dependent on the temporal and spatial variation of the source strength (expressed as concentration by weight) and geometry at the dredging zone. The dredging zone is usually small (approximately few meters around the moving dredge-head equipment). Within this zone, the grain size distribution and the mass flux of SSC entering into the near field are relatively uniform and mainly affected by selected operational conditions. Subsequently, SSC are subjected to differential settling during the plume development (i.e. the coarser particles settle faster than the finer particles) and only the finer fraction moves out from the near field to the far field zone. The plume behaviour becomes reasonably predictable and quantifiable at the edge of the dredging zone. Generally, the plume passes from the dredging zone into the near field over a timescale of seconds and a distance of few meters, while the far field transition occurs (over a time scale ranging from minutes to hours) within few hundreds of meters far from the dredging location ([2,3,6]). Within the far field zone, the plume dynamic is mainly driven by environmental conditions in which dredging takes place and resulting dispersion and settling processes are of the same order of magnitude. Consequently, oceanographic features, hydrodynamic conditions and background SSC (or turbidity) levels are all important factors in determining the plume extension. Depending on the plume dynamic in the far field, there can be significant spatial and temporal variations of the SSC distribution and of the related environmental effects ([13]). Tighter control in the form of strict regulations, proper enforcement on monitoring and mitigating measures can help to prevent, at least to minimize, the adverse environmental impacts ([10]). However, relying both on mathematical modelling and on field measurements, appropriate management and monitoring measures can be designed before the dredging execution and the dredging plan may be optimized to achieve the environmental objectives while maintaining desired production rates ([20]). Well-established international guidelines and past researches aimed at supporting environmental studies for projects that involve the handling of sediments are now available. Most of these include the use of numerical modelling as a valuable tool ([13,21–26]) to predict the area interested by SSC increase. Different models are currently used to investigate advection and dispersion phenomena of any substances (i.e. sediment and pollutants) and their effects on the water quality and ecological processes (i.e. larval dispersion, nutrient level) with respect to different environmental windows (e.g. storm or low-frequency flood event and seasonal or annual scenarios, i.e. [27–31]). In particular, source models are especially developed for estimating the source strength and the SSC moving out the dredging zone for different site and operational conditions (e.g. [14–17,32–34]). Then, near field and far field numerical models are commonly used for quantitatively estimating the planar and vertical extension of the plume dynamics close to (near field models) and far from the resuspension sources (far field models, e.g. [35–40]). As already pointed out, depending on the variability of site and operational conditions, the resuspended rate at dredging sources and the subsequent overall sediment transport can significantly vary (in terms of magnitude, duration and location of the induced SSC) during the execution of dredging projects. Thus, to be truly effective as a dredging project management tool, models should be capable of running starting from different dredging sources (i.e. continuous or time varying sources) also imposed as initial conditions. This allows the evaluation of a number of alternative dredging scenarios so that those with the least probabilities of detrimental impacts with respect to different environmental site conditions can be determined. Despite the attention that has been focused on this issue, there is a lack of verified predictive tools of dredging plumes usable for environmental and management purposes ([41]). In this contest, the development and the application of models to estimate the sediment loss rate from different hydraulic and mechanical dredges is one of the most important efforts to address this wide-ranging and complex problem (e.g. [14,15,17]).

This chapter aims at providing a general insight about the problem of sediment plumes arising from dredging activities. In particular, the available predictive techniques aimed to estimate the near field SSC induced by different dredging sources are detailed. Then, a novel one-dimensional analytical model is proposed to provide a fast order of magnitude of the plume dynamic when time-varying sources and time-varying currents have to be considered. Moreover, an integrated and replicable numerical suite, combining a series of numerical tools, is proposed to support the design phase of a dredging project and environmental studies in coastal and estuarine areas (especially in semi-enclosed basin). The proposed integrated numerical approach, within which empirical formulations based on field and experimental observations are used as forcing terms of mathematical models, is applied to the Augusta Harbour (Eastern coast of Sicily Island – Italy) case study. Results, obtained for different hypothetical dredging activities, show hydrodynamics and induced SSC dispersion features and thus the ability to support technical choices during planning and operational phases.

2. Dredging techniques and mechanisms of sediment resuspension

"Dredging" is a general term covering a wide variety of activities. A review of dredging applications reveals three main groups of dredging works: (1) capital dredging involves the creation of new or improved facilities (i.e. harbour basins, deeper navigation channels, area of reclaimed land) for industrial or residential purposes; (2) maintenance dredging concerns the removal of siltation from the bottom of navigation channels and ports in order to maintain the designed depth and (3) environmental or remedial dredging requires the careful removal of the contaminated material, and it is often linked to the further treatment, reuse or relocation of such sediments. With regards to the working principles for processes of removing sediments, dredges may be categorized into two broad categories: mechanical and hydraulic dredges (i.e. [42,43]). The selection of the most appropriate equipment and method to perform dredging depends on different environmental factors, e.g. volume, physical and geotechnical features of the sediment to be removed, the presence of contaminants at different depths of the bottom layers, the dredging depth and the distance to the disposal area. SSC levels of the resuspended sediment along the water column depend on the complex interaction between these environmental factors and different mechanisms of sediment resuspension from specific dredging techniques. The most commonly used mechanical and hydraulic dredges are described in the following, by pointing out the main sediments resuspension mechanisms that may occur during the excavation of material for different dredging equipment. The description of dredges presented in this section is mainly adapted from [42] and [43]. Mechanical dredges make use of mechanical excavation equipment for cutting the bottom and raising material. The material is picked up by mechanical excavators mounted on a pontoon kept in fixed position into the seabed and placed in disposal areas. When the disposal area is too far from the dredge site, the sediments is transported by a barge. These dredges may be used to excavate most types of materials but for the most cohesive consolidated sediments and solid rock. The most common mechanical dredges are the grab- (or clamshell) dredge (GD), basically a conventional cable crane, and backhoe- (or dipper) dredge (BHD), basically a conventional hydraulic excavator (Figure 1). These dredges types move by using anchors, spud carriage systems or by a tug for significant distances, and are typically used in both capital and maintenance projects. They are typically used in areas where hydraulic dredges cannot work (particularly around docks and piers or within other restricted areas), or where the disposal area is too far from the dredging site. The excavating system has to be repositioned at every cycle. The dredging process can be controlled accurately (e.g. [44]), since it is used extensively for removing relatively small volumes of material (i.e. a few tens or hundreds of thousands of cubic meters) particularly within other restricted areas. Production rates are low if compared to the hydraulic dredges and it is highly dependent on the operator's skill. Up to 500 m^3/hr can be achieved with the largest backhoe dredges.

Figure 1. Mechanical dredges (U.S. Army Corps of Engineers – Downloaded from www.flickr.com)

The dredge begins the digging operation by dropping the bucket in an open position from a point above the water. The bucket penetrates into the bottom material and the material is sheared from the bottom and contained in the bucket compartment. The sediment is removed nearly at its in situ density. Conventional mechanical dredges resuspend sediment when the bucket hits the seabed, when the hoisting turbulence washes away part of the load and because of spillage as the bucket is lifted or lowered through the water column. Moreover, significant amounts of suspended sediment can be released when its contents are loaded into a barge. For mechanical dredges, sediment resuspension concentrations can be significantly affected by different operative parameters, such as bucket overfilling, over-penetration, bucket speed when contacting bottom and bucket speed when lifting bucket off bottom. Hence operator controls can be utilized in order to reduce sediments resuspension. Loss of loose and fine sediments is generally influenced by the fit and condition of the bucket, the hoisting speed and the properties of the sediment. Then, variations of the jaw grab dredge design have been developed in recent years to minimize loss of sediment and allow better precision. In particular, special watertight buckets, collectively referred to as "environmental" buckets, are used in contaminated sediments to minimize the turbidity generated during dredging operations. Generally, the "environmental" buckets move along a horizontal profiling by removing thin layers of sediments in order to achieve high accuracy. The horizontal profiling allows the filling of the bucket with low water content. Furthermore, "environmental" buckets can have other features such as one-way vents in the top of the dredge to reduce downward pressure during deployment and rubber seals to prevent loss of sediments deployed in order to further reduce sediment resuspension.

Hydraulic dredges pump the dislodged sediment as a mixture of water and sediment. The dilution of excavated materials, typically with 5–30% of solids content, can vary significantly depending upon the soil type and the attainable layer thickness. The loosened sediment is pumped via a pipeline to the desired location (disposal area), or uncommonly loaded into a barge, making use of centrifugal pumps for the transport processes (raising and horizontal transport) installed on the floating pontoon. Hence, hydraulics dredges operate on an almost continuous dredging cycle and they are suitable for maintaining harbours, canals and outlet channels, when the removed material has to be pumped ashore (e.g. for reclamation) or for long distances to upland disposal areas. Depending on the size of the suction dredge and the characteristics of the excavated material, the output rate can vary widely, from 50 to 5,000 m³/hr. The most common types of hydraulic dredges used in coastal water are stationary- (SD) and cutter- (CSD) suction dredges based on the breaking up system of the mass of sediment (Figure 2).

Figure 2. Hydraulic dredges (U.S. Army Corps of Engineers – Downloaded from www.flickr.com)

Stationary suction dredger (SD) is the simplest type of hydraulic dredge. The suction pipe is lowered into the sea bottom and seabed material is sucked up by suction action of the dredge pump which is often mounted on the suction ladder. Only relatively loosely packed granular material or silt can be dredged with this equipment. Water is added to the removed sediments for transportation purposes. The application of water jet near the suction pipe can improve excavation efficiency. Generally, the amount of suspended sediment close to the suction mouth depends on the difference between jet-flow and suction-flow. The suction dredge has the disadvantages that considerable resuspension is possible due to relatively uncontrolled process of applying suction. Furthermore, accurate dredging is not possible, due to the relatively uncontrolled production process by suction. Hence, the suction dredge is less suitable for selective dredging. Improved precision can be achieved only by using specific operating precautions (in the order of 1.0 m up to 10 cm), e.g. when the suction entrance is modified by installing a wide and flat head, such as in the so-called Dustpan Dredge. Cutter suction dredge (CSD) is an efficient and versatile sediment mover commonly used in both large and small dredging projects. It makes use of different types of rotating cutterhead (or breaking up system) mounted at the end of a suction pipeline to dislodge a wide range of seabed materials, including clay, silt, sand, gravel, stiffer cohesive sediments and soft rock (such as softer types of basalt and limestone). During operation, the cutterhead dredge swings from side to side alternately using the port and starboard spuds as a pivot. Good dredging location accuracy may be achieved because the movement of the dredging-head is controlled from a fixed point (the working spud). Cables attached to anchors on each side of the dredge allow the control of lateral movements. Sediment resuspension induced by hydraulic dredges shows relatively simple physics. The main sediment resuspension is due to the disturbance of the seabed around the dredge-head, because not all dislodged sediments are entrained into the suction pipe line (e.g. [34]). The transport of the dredged material from the bottom to the surface is fully enclosed into the suction pipe, then, unless there is leakage of pipelines, there should be no sediment losses. Significant sediment release can also occur during the loading phases into the dredge or into barges, due to overflow or to possible leakage from the barge of the pumped sediments as a highly diluted mixture. Compared with most other hydraulic dredges, the rate of sediment resuspension from cutter suction dredges is expected to be relatively constant during the main part of the swing (especially when homogeneous materials must be removed). As the dredge approaches the end of the swing, it slows down and the sediment resuspension rate may be reduced (e.g. [3]). Different operating conditions such as: cutterhead rotation, swing speed, pump rate, bank height, water depth, shape of the cutter-head, thickness and cut direction (particularly if no over-depth is allowed) can also affect the resuspended sediment that leaves the dredging zone (the turbulent zone around the ladder and the cutter-head action). In particular, low production rates and shallow cuts for hydraulic dredges can increase resuspension. However, the high production rate may increase the resuspension when currents and wave energy are high because more sediment can be dislodged and captured by the dredge head. According to [1], the increase of SSC induced by dredging operations does not only depend on the employed dredging techniques, either mechanical or hydraulic, but can also be related to the geotechnical property and the water content of the dredged sediment (i.e. liquidity, density, porosity or void ratio, plasticity and nature of a fine-grained soil). In particular, very soupy sediments resuspend more easily and effects related to SSC are larger for sediments with higher liquidity. Generally, silts are more liquid than clays at the same water content, while the liquidity increases with a decrease in the plasticity index of the sediment. Moreover, an increase in obstacles to normal dredging, such as debris, cobbles, boulders, bedrock and rock outcroppings, may result in an increase in resuspension. Of these obstacles, debris poses the greatest problem to resuspension when mechanical dredges are used. Indeed, debris can prevent closure or seal of the clamshell, causing significant leakage or loss of dredged material to the water column. Debris can also disrupt the capture of sediment by cutter head dredges and increase dispersion of the dis-lodged sediments. Then, the separate removal operation of impediments can resuspend nearly as much sediment as the dredging operation as well as increasing the liquidity of the material for subsequent dredging. As already outlined for mechanical dredges, specific operating precautions can now successfully be used to minimize the sediment resuspension due to "conventional" dredges and a series of newer "environmental" dredges are specifically designed to careful remove contaminated materials with respect to reductions in resuspension (and release), and operational efficiency for removal and transportation (e.g. [12,42]). Resuspension can be minimized using specific operating precautions, by proper equipment selection, modifying equipment and operation techniques or by containment of resuspended sediment such as silt curtains or screens. Equipment selection is sometimes limited by availability, but operational controls can be considered for a wide range of equipment types. The reader is referred also to [12,42,46] for more details on the commonly used "environmental" dredges types and environmental operating precautions.

3. Resuspension dredging source formulations

The assessment of environmental impacts related to the exposure of sediment plume resulting from dredging activities requires the estimation of the near field re-suspended sediment concentrations (SSC, kg/m^3) moving out the dredging zone (i.e. dredging source). Few conceptual models to predict the resuspended sediment mass rate at the dredging point, and thus its source strength and geometry, as a function of different site (i.e. sediment properties, water depth, currents) and dredge-operating (i.e. dredge type, dredge-head dimension) parameters were developed by e.g. [14–16,32–34,47]. Considerable resuspended sediment field data were collected in the vicinity of various hydraulic and mechanical dredging projects to develop these models. However, since the used data sets cover a relatively narrow set of site and operational conditions, the proposed formulations have limited predictive values and require specific adjustments to serve as predictive tools of SSC for different dredging projects. Typical resuspension rates have been reported in [45] and [15]. As highlighted by [19], the resuspended source induced by dredging may be defined by the estimation of the SSC (mg/l) moving out the dredging zone, by the sediment loss rate (kg/s) in the water column, or by the total mass of resuspended sediment expressed as a fraction of the dredging production rate (kg/m^3). The concept of the sediment flux across the sources boundaries can also be used (e.g. [18]). The available conceptual methods for estimating the sediment resuspension induced by different (i.e. hydraulic and mechanical) dredging sources are reported in the following. Such formulae are divided in "basic formulations", which are based on the use of tabular data, and "empirical formulations" containing a set of dimensionless parameters related to operating and site characteristics.

3.1. Basic formulations

Ref. [16] introduced the turbidity generation unit (TGU) concept. It corresponds to the unit of suspended sediment or turbidity (kg/m^3) generated if a unit of quantity of bottom sediment will be dredged. In Nakai's formulation ([16]), the TGU (kg/m^3) is expressed as a function of the total quantity of resuspended sediments generated by dredging (W_0, kg) and the total quantity of dredged materials (Q_s, in m^3):

$$TGU = \frac{K\,W_0}{Q_s} \tag{1}$$

where the coefficient K is given by the ratio (R_{74}/R_0) of the fraction (by weight) of sediments smaller than 74 μm (R_{74}) and the fraction of sediments smaller than the particles whose critical resuspension velocity equals to the local current velocity (R_0). The variable W_0 strongly depends upon the specific dredging operation, thus equation (1) can be rewritten as follows:

$$TGU = KC\rho \tag{2}$$

where C is a coefficient depending upon dredge type (and thus sediment characteristics) and ρ is the specific weight of dredged sediments (kg/m^3). Equations (1) and (2) give the turbidity generated by a unit of dredged sediment. It may be observed that Nakai's formulation assumes uniform vertical and horizontal sediment distributions in the water column and a constant and unidirectional velocity. The W_0 was estimated by field data of suspended sediment collected down-drift (30–50 m) the dredging area for different dredging projects. After these measurements, Nakai used a simple relationship to infer the total mass of turbidity (or suspended sediments) arising from the dredge:

$$W_0 = AuC_r \tag{3}$$

where C_r (Kg/m^3) is the measured net SSC generated by dredging; A is the cross section of the turbid area (m^2), obtained by multiplying its width (m) and the water depth (m); and u is the mean tidal current velocity (m/s). Nakai rewrote equation (1) in the appropriate form in order to define the dredging source in term of resuspended rate (or sediment loss rate), redefining W_0 as the turbidity generation rate (kg/s) and Q_s as the sediment removal rate (m^3/s).

$$W_0 = \frac{TGU\,Q_s}{K} = \frac{R_0\,Q_s}{R_{74}}TGU \tag{4}$$

In particular, equation (4) allows estimating the total quantity of the turbidity generated by dredging (W_0) when TGU values, sediment removal rate (Q_s) and characteristics of dredged sediments (R_{74} e R_0) are known. It could be useful to note that the term $Q_s\,R_0$ represents the removed sediment mass per unit time (m³/s) with a sediment settling velocity sufficiently low to stay in suspension for a long time. Then, it is quite clear that the implementation of Nakai's formulation as predictive tool requires the knowledge of TGU values, depending upon the dredging technique. TGU values estimated for different site and operational conditions are reported in [16], along with values of critical resuspension velocities estimated for different sediment size and type of dredged sediments. The Nakai's TGU values were integrated by [17] with their "suspension parameter" (similar to TGU definition) on the basis of further SSC field measurements. Although the TGU method is now commonly used for open clamshell, cutterhead and hopper dredges, it has significant application limitations. Indeed, there are a relative small number of SSC data (i.e. collected for different sediments and equipment) used to derive the TGU values and information on key sites and operational descriptors related to dredging method and techniques are lacking. Moreover, some inconsistencies may occur because TGU formulation is derived only for unidirectional currents and it does not include the SSC due to the overflow during the loading phases (especially for hydraulic dredges). Inconsistencies can also be derived by the basic assumption used to infer TGU values that all particles > 5 μm resettle in the vicinity of the dredge. In particular, recent SSC measurements showed that a considerable fraction of particles ranging from 20 μm up to 30 μm can leave the near field zone ([1]). Thus, according to [1], a considerable overestimation of the multiplier (R_{74}/R_0), especially when sediments with low clay fractions are removed, can lead to an overestimation of the TGU values. In particular, [15] stated that inconsistencies of the TGU method can be related to the R_{74} parameter that could lead to an erroneous estimation of resuspended sediments. Indeed, an increase in the sediment fraction smaller than 74 μm (and thus a reduction of the ratio $1/R_{74}$ in equation 4) leads to a non-physical decrease in the estimated resuspended loss rate. Nevertheless, TGU method can be considered inconsistent only when comparable changes of R_0 and R_{74} parameters simultaneously occur. To overcome such application limitations, [34] and [15] introduced the resuspension factor (R). It is defined as the mass of resuspended sediments relative to the total mass of dredged sediments (expressed as percentage):

$$g = R\left(\frac{R_{74}V_sC_s}{360}\right) \tag{5}$$

where g is the predicted rate of resuspended sediment leaving the near field zone for a specific dredging operation (in g/s), V_s is the volumetric rate of in situ removal sediment (m³/h), C_s is the in situ sediment concentration (kg/m³) and R_{74} is the fraction of sediments smaller than 74 μm leaving the near field, or smaller than the largest sediment size subject to far field transport.

A series of values of R values are reported in [15,34] for different dredging operations covering a variety of conditions. However, since resuspension factor model has been developed on the basis of a limited number of data set, resulting ranges of estimated R values have shown a high variability for the different examined dredging techniques.

3.2. Parametric formulations

Empirical formulations, developed for estimating resuspended sediment and its source strength, relate non-dimensional variable groups of dredge operating parameters to SSC observed for different operational (dredge type, dredge-head dimension, controllable aspects of dredge operation, etc.) and site conditions (sediment type and grain size, water depth and local hydrodynamics conditions, etc.). These formulations solve some of the application limitations of the TGU and resuspension factor models, based on tabular approach. Empirical models to estimate the SSC of the released sediment moving out from the dredging zone into the far field, and its associated source strength and geometry, were developed by [14,24,33,47] only for clamshell (open and enclosed) and cutterhead dredges. Ref. [33] proposed a relationship between the net concentrations of resuspended sediments C_r (kg/m^3) induced by cutterhead dredges and two dimensionless groups related to "kinematic" operating parameters:

$$C_r = 0.15 \left(\frac{V_s}{V_i} \right)^{2.869} \left(\frac{V_t}{V_i} \right)^{1.027} \tag{6}$$

where V_s is the swing velocity at the tip of the cutter (m/s), V_i is the suction intake velocity at the cutter blades, V_t is the tangential velocity of the cutter blades at the top of the rotation relative to the surrounding water (m/s). It has to be stressed that the tangential velocity V_t is relative to the surrounding water, then it depends on the swing velocity and on the direction of the swing (i.e. either starboard-to-board or board-to-starboard, $V_t = | V_s \pm V_c |$, being V_c the tangential velocity of the rotation of the cutter blades). The relationship (6) was obtained on the basis of Calumet Harbour dredging project ([48]) during which only one cut thickness was used, being roughly equal to the cutter diameter (i.e. mud-line full penetration as defined by [14]). Then, the cut thickness parameter was not considered in the analysis and any dimensionless group related to the used cut thickness. Ref. [14] confirmed the results of [33] and performed a similar regression analysis, including more SSC data from other dredging sites (i.e. Savannah River and James River in addition to Calumet Harbor) by including further dredge sizes, operating parameters and sediment characteristic (e.g. [15]). Indeed, in addition to operating kinematic parameters (i.e. V_i, V_t, V_s), [14] also included geometric operating parameters (e.g. length of cutterhead, L, diameter of the cutterhead, D_{ch}, and the degree of cutterhead penetration, D) and the sediment mean grain size (d). Ref. [14] proposed to use the same approach of [33] by considering the influence on the first coefficient (i.e. 1.150 in eq. 6) of the cut thickness. In particular, in the Collin's regression analysis ([14]), the net SSC (C_r, in kg/m^3) is estimated as:

$$\frac{C_r}{\varrho \cdot 10^{-6}} = F_F F_D \left(\frac{V_s}{V_i}\right)^{2.848} \left(\frac{V_t}{V_i}\right)^{1.022} \tag{7}$$

where the SSC concentration is related to the density of water above mud-line (ρ, g/m³), to the same cinematic parameters (V_s, V_i and V_t) proposed by [33] and where F_F and F_D are dimensionless parameters aimed at describing the influence of cut thickness upon SSC concentration. Indeed, F_F (full cut parameter) accounts for the cutter size (characteristic size of cutterhead, L) and the median grain diameters (d):

$$log_{10}\left(F_F\right) = 10^{\left[10^{-4}\left(\frac{L}{d}\right)/13.32\right]^{-7.04}} - 2.05 \tag{8}$$

F_D is the dimensionless parameter (non-full-cut parameter) that accounts for the thickness of the cut relative to the cutter diameter (D_{ch}):

$$F_D = 1 + 1.9039\left(D-1\right)^2 + 0.4116\left(D-1\right)^7 \tag{9}$$

where dimensionless parameter D ($=d_f/D_f$) is the ratio of the actual cut depth with respect to the mud-line (d_f) and the full cut depth (D_f), hence representing buried ($D>1$), partial ($D<1$) and full ($D=1$) cutting. Then, F_D is equal to 1 for full cut dredging, while F_D is greater than 1 for both partial and buried cutting. The relationship providing full cut parameter (F_F) was developed as a function of dimensionless sediment grain size alone as other parameters, such as water depth, did not changed for full cut dredging project used to perform the regression analysis. It has to be stressed that, on one hand, the full cut parameter does not take into account the cutting mode and, on the other hand, the non-full-cut parameter (F_D) takes into account the differences between types of cutting modes.

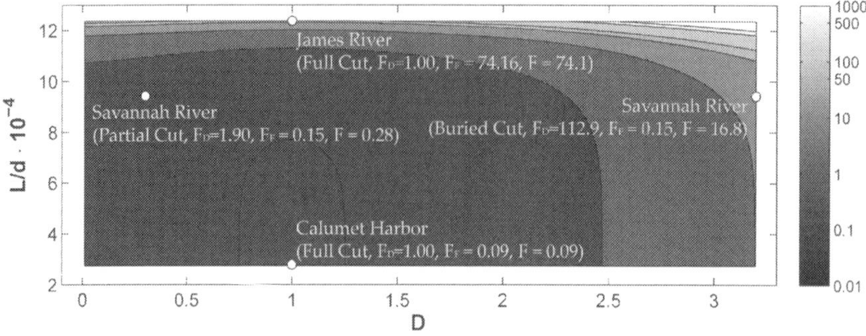

Figure 3. Variation of F ($=F_D F_F$, see equations 8 and 9) as a function of cut parameter (D) and dimensionless grain size (L/d). White circles refer to field data used by [14].

It could be noted that also the non-full-cut parameter may depend upon the dimensionless sediment diameter. Nevertheless, non-full-cut dredging was performed during only one project (i.e. Savannah River), then L/d parameter was constant and such dependence cannot be identified, even if it exists ([14]). Figure 3 shows the estimated values of F ($=F_D\,F_r$) as a function of dimensionless dredged grain diameter (L/d) and of dimensionless cut parameter (D) along with the points indicating the field data used by [14] to define relationships (7), (8) and (9). It could be noted that, obviously, the lower the grain size, the higher the SSC concentration. As far as the buried cut is concerned, it could be observed that the SSC concentration rapidly increases as the sediment above the cutter head tends to fall and slough and the suction pipe may not be able to remove it, then an increase in the resuspended sediment concentration occurs ([14]). When partial cut is considered, the SSC concentration slightly increases due to the increased exposition of the cutter head blades to wash-off phenomena, if compared to the full cut configuration ([14]).

Ref. [34] analyzed field data collected during cutterhead dredging project performed in the James River, Savannah River, Calumet River and the Acushnet River, for a total of 106 observations, and proposed two different empirical models aimed at predicting the rate of sediment resuspended by the dredge (g) as a percentage of the total dredged sediment. These models are based on the hypothesis that the majority of the resuspended sediment arising from the dredge-head is due to the wash off phenomena occurring at the cutter blades that pass through the water after they have passed through the sediments. Hence, the sediment particles that adhere to the blades are washed into the water column and, if the sediment grain size is small enough, are dispersed to the far field. Ref. [34] proposed a first "dimensional" model by using stepwise linear regression aimed at catching empirical relationships between the involved operating parameters, such as the swing velocity, the angular velocity of the cutterhead, the surface of the blades exposed to wash-off, etc., and the rate of resuspended sediments:

$$g_D = \frac{\left|V_s \pm \beta\pi d_c\right|^{1.864}}{27.4\,C_s V_s t_c L^{15.143}}\left(\frac{A_E}{d_c}\right)^{14.143} \tag{10}$$

where g (%) is the predicted rate of resuspended sediment and transported away from the dredging site as a percentage of the sediment mass dredged, A_E is the total surface area of cutter blades exposed to washing ([32]), V_s is the swing velocity at the tip of the cutter (m/s); C_s in situ sediment concentration (kg/m³), t_c is the thickness of cut (m), L is the length of the cutterhead (m), d_c the cutter diameter and β the cutter rotational speed (rps). Moreover, [34] proposed a "non-dimensional" model based on the definition of a series of dimensionless variable groups. Even if conceptually similar, [34] found a different relationship:

$$g_{ND} = \frac{\left(L d_c\right)^{1.966}\left|V_s \pm \beta\pi d_c\right|^{1.966}\left(V_s A_E\right)^{1.804}}{1.099\,Q^{3.770}} \tag{11}$$

where Q is the volumetric dredge production (m³/s). Ref. [34] did not clearly indicate which of the two models is superior to the other by using the standard statistical parameter used to evaluate the models' skill. However, they found that the non-dimensional model given by equation (11) is better if compared to dimensional model given by equation (10). Furthermore, [34] stated that the models application outside the validity range may lead to very high, i.e. conservative, estimates.

As far as mechanical dredging is concerned (i.e. open and enclosed bucket), [14] has developed an empirical model to eimate the dredging-induced sediment resuspension rates as a function of bucket size, sediment size (related to the settling velocity) and operational techniques (i.e. dredging depth and cycle time). Indeed, [48] identified bucket impact, its penetration into the bottom and withdrawal as the main contributors to sediment resuspension. Ref. [14] indicated a further resuspension source when the bucket exits from the water and swung to the point of bucket opening. Resuspended sediment concentration (C_r, in g/m³) at the dredging source are extrapolated by a linear regression analysis, obtained for measurements of SSC at various depths and distances from a number of dredging sites (i.e. St. John River, Black Rock Harbor e Calumet River):

$$\frac{C_r}{\varrho \cdot 10^{-6}} = 0.00235 \left(\frac{b}{\omega_s T} \right)^{3.033} \tag{12}$$

relating the SSC to the characteristic bucket size (b, m), to the sediment fall velocity (ω_s, m/s) and to the duration of the dredging cycle (T, s). Relationship (12) may be rearranged in order to highlight the influence of each parameter:

$$\frac{C_r}{\varrho \cdot 10^{-6}} = 0.00235 \left(\frac{b}{h} \frac{h}{\omega_s T} \right)^{3.033} \tag{13}$$

where h is the water depth. It is quite clear that the bigger the bucket size (b) compared to the water depth (h), the greater the resuspended sediment concentration. Moreover, the bigger the cycle duration (T) compared to the time spent by the sediment to settle (h / ω_s), the lower the resuspended sediment concentration.

3.3. Resuspended source strength

The use of mathematical modelling, either analytical or numerical, often needs the definition of the resuspension source at a point, along a line or over a surface ([14]). The strength of the resuspended sediment source, i.e. the mass (or weight) flux of sediments (e.g. [18,19]) introduced in the near field from the dredging zone, can be expressed in term of the product of SSC and velocities distributed over the source surface boundaries (e.g. [14]). Hence, the definition of sediment source strength to be used within mathematical modelling requires a description of the geometry of the boundaries source and the local velocity at those boundaries ([14]), while the concentration may be estimated by the relationships given in the previous section. In the following, the source strength models proposed by [14] are summarized. As far as cutterhead suction dredges are concerned, the resuspended sediment source volume from the bottom is assumed approximately equal to the geometry of a semi-ellipsoid with its minor axis and major axis equal to the maximum radius (D_{ch}) and length (L_{ch}) of the dredge-head, respectively. Nevertheless, as the cutterhead swings during the dredging activities, the actual volume over which the SSC (C_r) typically occupies is a volume larger than the defined semi-ellipsoidal source volume. Thus, the length and the vertical extent of this source volume can be estimated respectively as $(1+K'_{ch})L_{ch}$ and $(1+K_{ch})D_{ch}$, defining the cutterhead length (K'_{ch}) and diameter (K_{ch}) size factors, usually lower than 1.5. If $(1+K_{ch})D_{ch}$ exceeds the water depth, the vertical extension will be limited by the water depth. Hence, cutterhead mean source strength can be expressed as:

$$q_{ch} = C_r V_t \left(1 + K_{ch}\right)\left(1 + K'_{ch}\right) D_{ch} L_{ch} \tag{14}$$

where q_{ch} represents the sediment loss rate (g/s), i.e. the rate at which the mass of sediment is introduced into the near field with a given mean concentration (C_r, mg/l given by equation 7), V_t is the tangential velocity of the cutter blades (m/s) at the top of the cutterhead, for both overcutting $(V_t=V_c+V_s)$ and undercutting $(V_t=V_c-V_s)$ conditions. It has to be stressed that the velocity at the top of the cutter is a representative velocity at all points of the resuspension source. Recently, [47] proposed to use the net velocity estimated along the whole washing region.

For a bucket dredge, the resuspended sediment source volume can be defined as the apparent bucket footprint multiplied by the water depth (h) at the point of the dredging location ([14]). Thus, the dredging source can be idealized as a cylindrical geometry. This geometry is only approximate because SSC can occupy a volume larger than the idealized cylindrical source volume for the turbulent mixing near the dredge-head. Hence, the source strength must be computed taking into account the increase in effective bucket size due to turbulent mixing by means a size factor (K_{cb}). Through this schematization of the sediment loss rate and the available field data, the function representative of the mean source strength (q_{db}) for typical cycle of the bucket motion is defined by:

$$q_{bd} = 2b^2 \left(\frac{h}{T}\right)\left(1 + K_{cb}\right) \frac{C_r}{f_u + 2f_0 + f_d} \tag{15}$$

where C_r is given by equation (12) and K_{cb} is a size factor (equal to 1÷2). The terms f are the fractions of the cycle time needed to perform operations, i.e. f_u is the fraction of the dredging cycle time over which the bucket is rising in the water, f_d is the fraction of the cycle time over which the bucket is descending in the water, f_0 is the fraction of the cycle time for which the bucket is out of the water. It could be noted that the fraction of the cycle time during which the bucket rests on or is dragged along the bottom is considered as almost negligible. The term $2h/T$ (i.e. the ratio of twice the water depth and cycle duration) is the estimation of the representative velocity at the source boundary.

4. A simplified engineering tool to estimate dredging plume transport

Simplified models may be used as economical and effective tools to provide a first order of magnitude of the far field dispersion of resuspended sediments (e.g. [4]). Typically, only steady-state simplified analytical models are available in the literature (e.g. [49–52]). Recently, [4] have proposed an analytical model aimed at describing the temporal evolution of dredging-induced resuspension in the far field forced by a continuous source term. Although simplified, such models allow carrying out a fast worst case assessment that may reveal the need of performing more complex and computational costly two- and/or three-dimensional numerical modelling. In the followings, a novel analytical model proposed by [26] is described. The model is one-dimensional, forced by whatever source term, i.e. it may vary in space and time, and the velocity field may vary in time:

$$\frac{\partial C}{\partial t} + U \frac{\partial C}{\partial x} - K \frac{\partial^2 C}{\partial x^2} = q(x,t) \tag{16}$$

where C is the concentration of resuspended sediments (i.e. depth-averaged SSC), U is the current field (constant in space), K is the diffusion coefficient, x is the spatial coordinate and t is the elapsed time. The source term (q) may vary both in time and space in order to describe the dredging resuspension source that moves in the domain and change over time as a function of operational technique and work program. Thus, plume is vertically well mixed (as assumed by [4]) and the diffusion/dispersion along the direction perpendicular to the current U is neglected. Furthermore, any sediment settling is accounted for, even if the sediment may flocculate (e.g. [50]). Hence, the model is intended to provide a worst case scenario. Moreover, the resuspension source term is constant over the entire water column, then aimed at describing mechanical dredging or shallow water hydraulic dredging (i.e. $h < 2.5 D_{ch}$). Applying Fourier Transforms, [26] solved equation (16) for different dredging scenarios. The solution reads:

$$C(x,t) = \sum_{i}^{N} \frac{C_0}{2\sqrt{k\pi}} \int_{t_i \le t}^{t_i + T_{di} \le t} \left[H(t - t_i) - H(t - t_i - T_{di}) \right] f(x_i, t - \tau) d\tau \tag{17}$$

where C_0 is the dredging induced SSC, H is the Heaviside step function, f is a function describing the dredging activities (hereinafter referred to as "dredging function"), N is the number of dredging cycle ($N=1$ for continuous hydraulic dredging, $N>1$ for mechanical dredging), T_{di} and x_i are the i-th dredging cycle duration and location, respectively. If constant velocity U_0 is considered, the dredging function may be expressed as follows:

$$f(x,t) = \frac{1}{\sqrt{t}} exp \left[-\frac{(x - U_0 t)^2}{4kt} \right] \tag{18}$$

If sinusoidal velocity $U=U_0 cos\omega t$ (ω the angular frequency, rad/s) is considered, the dredging function reads as follows:

$$f\left(x,t-\tau\right)=\frac{1}{\sqrt{t-\tau}}\,exp\left\{-\frac{\left[x-U_0\,/\,\omega\left(sin\omega t-sin\omega\tau\right)\right]^2}{4k\left(t-\tau\right)}\right\}\tag{19}$$

It could be noted that the dredging function expressed by equation (19) becomes equation (18) if angular frequency tends to be infinity (i.e. constant velocity). The integral solution (17) may be computed by means of standard numerical techniques. By way of example, Figure 4 shows the results obtained for mechanical dredging activities moving along the x axis. Each dredging cycle has a duration of 60 s and the angular frequency ω was kept constant to 0.05 rad/s (a quite high value of the angular frequency was used in order to highlight the model capability in catching the role of time variability of the environment current). In the figure, dimensionless concentration is shown (i.e. $C^*=C/C_0$). Example results' inspection reveals how the model is capable to catch the main features of plume dispersion with high concentrations close to the dredging points and decreasing concentrations in the surroundings due to the oscillating current action.

5. Numerical modelling to estimate the SSC dredging plume

5.1. Integrated approach

When preliminary evaluation suggests using a more sophisticated modelling approach, integrated numerical suite may be applied in the design phase of a dredging project. In particular, a comprehensive modelling approach was proposed by [26] to support technical choices and environmental studies related to the development of dredging plume in estuarine and coastal (or semi-enclosed basin) areas.

The integrated numerical system is inspired by standard numerical suites worldwide used to oil spill modelling, with exception of oil weathering process description (e.g. [53,61]). Basically, the proposed system is made of three modules, hereinafter referred to as hydrodynamic module (HM), source module (SM) and transport module (TM): TM computes SSC distribution in the far field as a function of the resuspended sediment source (occurring during the process of removing sediment from the bottom with different mechanical and hydraulic dredges types) estimated by means of SM, and the flow field variability computed by HM. The implementation of the integrated numerical system includes the collection of input data describing the oceanographic features and environmental conditions (i.e. wind fields, tidal oscillations, etc.), resuspension dredging source and other factors affecting the plume dispersion (i.e. turbulent diffusion coefficients). Indeed, three-dimensional hydrodynamic and sediment transport models are highly recommended as the best practice to assess environmental impacts related to the SSC and pollutants (if any) dispersion due to dredging activities. These types of models

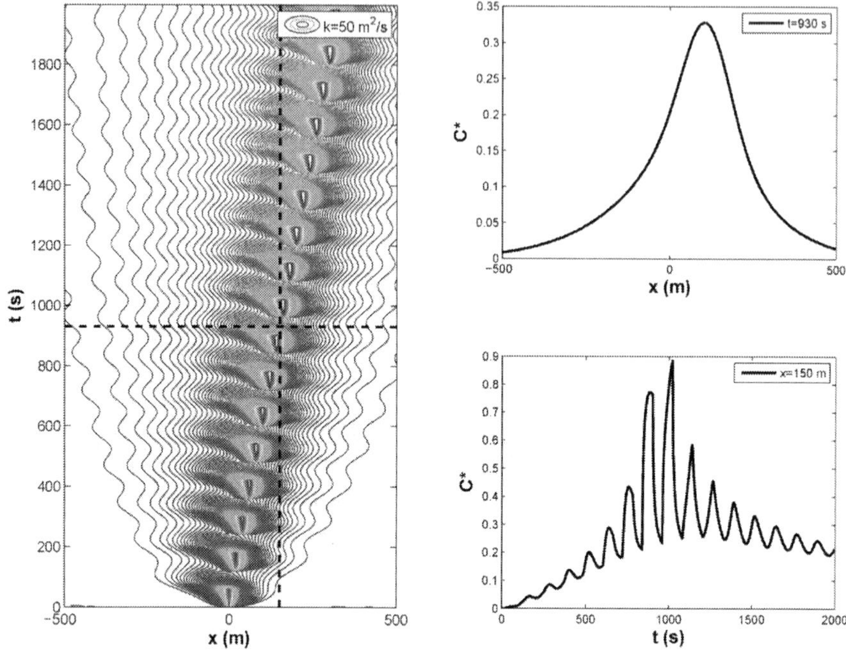

Figure 4. SSC as a function of: space (x) and time (t, left panel); space (x) for fixed time (t = 930 s, right upper panel); time (t) for fixed location (x=150 m, right lower panel). Numerical parameter: U_0 = 1 m/s, T_{di}=60 s, ω = 0.05 rad/s, k = 50 m²/s)

provide a realistic estimation of the planar and vertical extension of the plume dynamic, both close (near field zone) and far from the resuspension sources (far field zone, e.g. [2,37,55–57]), for different environmental (i.e. environmental forcing, sediment properties, water stratification, currents) and operational (i.e. dredge types, dredging water depth, production rate, thickness of cuts, velocity of dredging cycle) conditions. Then, the implementation of three-dimensional models allows to take into account the three-dimensional features of dredging sources (i.e. of the sediment release mechanisms during the excavation of bottom sediments) that affect the SSC distribution in the water column moving out from the immediate vicinity of the dredge (dredging zone) into the far field. The model domain should be suitable for environmental studies and thus discretized in horizontal and vertical directions with a resolution appropriate to reproduce the complexity of fluid flow and physical processes related to the dredged-induced resuspended sediment (advection, dispersion and sediment settling). Accordingly, a proper vertical discretization should be used to simulate the stratification. In detail, the simulation of dredging plume requires the definition of the dredging source, in term of its source strength and geometry, and thus of the sediment loss rate leaving the immediate vicinity of the dredge (near field area). The proper implementation of the source model (SM), based on empirical formulations (see the previous section), requires the knowledge of key components of the dredging programme (such as the dredge methodology, dredge schedule, sediment resuspension rate) in order to estimate source strength. These key components may not be well defined in the design phase of a dredging project, and this will typically be reflected in the accuracy of the predictions and in the choice of mitigation measures ([20]). Initial assumptions could require adjustments that are often unknown in the planning phases of a dredging project as they are usually based on the best available information. It could be noticed that some of the dredging parameters may be varied at the design stage to optimize the dredging activities in order to minimize the environmental impacts. The TM aims at estimating the spatial distribution and the time variation of SSC in the far field. Actually, Lagrangian and Eulerian models may be used. The Lagrangian model (i.e. random walk model, e.g. [11,25,36,58–59]) solves the trajectories equations of passive "label particles" released into the computational domain (each labelled with a representative concentration) including a diffusive term aimed at describing the turbulent fluctuations. The Eulerian approach is based on the numerical resolution of the standard advection–diffusion equation forced by the current velocities flow field and by the concentration source.

In order to identify the best timing for dredging periods, time-series from numerical results should be long enough to take into account the most representative meteomarine conditions and their seasonal variations. The use of field data and remote-sensed data (such as aerial photography) are also highly recommended to validate numerical results, in term of SSC and coastal bathymetry changes, during planning and operational phases of dredging activities [60–61].

It has to be highlighted that the implementation of specific data analysis procedures to synthesize numerical results is required to compare environmental effects related to different dredging projects, and thus to support decision and risk assessment. The reader is referred to [27] that provided standard methods and statistical parameters to relate the severity of environmental effects with magnitude, duration and frequency of the exposure to high predicted SSC levels.

5.2. Augusta Harbour case study: Hydrodynamics and SSC dispersion features

The application of the proposed approach to Augusta Harbour case study (Eastern coast of Sicily Island – Italy) is briefly described in the following. The harbour was formed from the closure of most of the natural Augusta bay (early 1950s) through the construction of three breakwaters (Figure 5). The main objective of the study is to evaluate the plume dispersion resulting from hypothetical dredging activities performed within the harbour. Model results are expected to allow catching key features of dredging plumes arising from different mechanical and hydraulic dredging sources, as a function of stratified circulation patterns, as described in the following.

The sheltered area of this harbour, roughly equal to 24 km^2, extends about 7.0 km alongshore and 3.5 km cross-shore with average water depth equal to about 15.0 m. The harbour is connected to the open sea by the eastern inlet (about 450 m wide and 40 m deep) and by the southern inlet (about 300 m wide and 20 m deep) and through the Ponte Rivellino channel

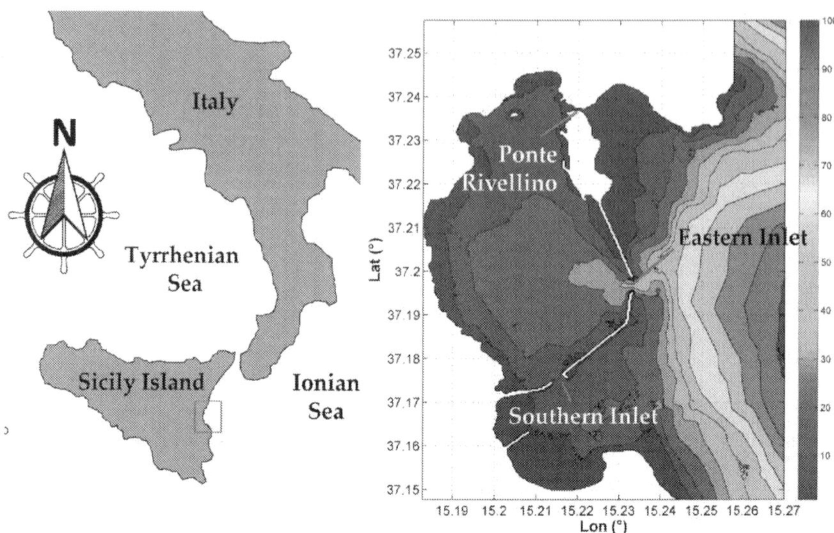

Figure 5. Location, bathymetry and key features of Augusta Harbour

(with extremely limited water depth that does not exceed 1.0 m). The small rivers flowing within the Harbour (i.e. Mulinello, Marcellino and Cantera rivers) are characterized by seasonal and discontinuous freshwater discharges. Physical characterization results showed a mainly sandy-muddy and muddy seabed. Only areas close to the western coastal zone and breakwaters are characterized by the outcropping of bedrock and coarser sediments. The harbour may be considered one of the most important Italian harbours as petrochemical pole and for bunker operations, ship repairs, ship maintenance and goods loading and unloading. Marine sediments have been highly polluted by heavy metals, especially in the southern area ([27,62,63]). Past researches payed attention to the hydrodynamic characterization of the area (e.g. [21]) in order to estimate the flushing features ([63]) related to potential environmental impacts induced by dredging activities ([27]). It is commonly accepted in the literature that tide- and wave-driven currents play a minor role in the dispersion of the plume in the Augusta Harbour [21,26]. Indeed, due to the bathymetry (Figure 5) with high water depths, the main circulation is driven by the wind field, then it is suggested to use a three-dimensional numerical modelling, as firstly done by [26] and [21]. The three-dimensional Regional Ocean Modelling System (ROMS, e.g. [66]) has been used to estimate the circulation pattern within the harbour forced by wind fields and tidal oscillations. ROMS is a finite difference three-dimensional, free-surface, terrain-following numerical model solving the Reynolds-averaged Navier-Stokes equations, using both the hydrostatic and Boussinesq hypothesis (the reader is referred to [64] for details). The estimated current velocities and resuspension sources are then used to assess the plume dynamic. Hydrodynamic and transport models are used within the framework of decoupled approach, as hydrodynamic effect of bottom variations (due to dredging activities) upon the circulation may be neglected (e.g. [27]). The computational domain covers the whole harbour area and an offshore area needed to avoid boundaries effects on numerical results (Figure 5). A sensitivity analysis was performed to ensure an appropriate horizontal and vertical resolution, based on the bathymetry data ([60]), and on the main forces acting on the harbour. A finite difference grid with uniform spatial resolution of 40 m covering an area of 9.8 km × 12.2 km (about 75,000 computational points) was implemented. In the vertical direction, 10 layers in the sigma coordinates were used (i.e. terrain following coordinates, e.g. [66]). Numerical simulations were performed in order to analyze frequent conditions. As already anticipated, waves forcing is not considered at all since preliminary numerical simulations showed that waves-induced circulation is negligible if compared to tidal- and wind-induced ones ([21, 27, 65]). Moreover, the freshwater inputs from rivers were neglected because of their short duration and local hydrodynamic effects. Wind parameters and tidal forcing (i.e. wind speed and direction and tidal harmonic constants) are based on the analysis (e.g. [65]) of the measurements collected by the Catania Station, belonging to the Italian Mareographic Network ([62]), deployed 35 km north from Augusta Harbour.

In the following, only the numerical results for wind speed equal to 5.2 m/s coming from 35°N are presented (the reader may refer to [21,26,65] to a comprehensive description of other scenarios). This scenario, characterized by small wind speed with an exceedance frequency of about 100 days per years, is selected to simulate wind-induced circulation that frequently occurs. It has to be stressed that tide-driven circulation is much less intense if compared to the wind-driven circulation. The analysis of hydrodynamic numerical results for the considered scenario (Figure 6) reveals that current intensities within the harbour reach higher values near the surface where the circulation is primarily forced by the wind action. At the bottom, higher values of current speed are generally reached close to the breakwaters and in the central part of the harbour. Moreover, the whole berthed area is interested by a significant spatial variability of the velocity field distribution. It could be observed that the current field (speed and direction) in the vertical direction varies as a function of the harbour morphology and bathymetry and of forcing conditions. At the free surface, the circulation is mainly directed as the forcing wind field. On the other hand, especially in the central part of the harbour, a strong stratification of the circulation along the water column occurs, with current direction close to the seabed opposite to wind direction. This reveals the importance of using 3D numerical model for assessing any environmental effects related to the SSC dispersion in marine and coastal water characterized by vertical flows variability, as highlighted by [21, 27]. Examples of dredging plume dynamics obtained by means of Lagrangian approach are shown by Figure 7, as representative of both mechanical and hydraulic dredges performed at selected hypothetical dredging-areas (DP1, DP2, DP3 and DP4, see Figure 7). In the figure, dark symbols in the right panels indicate the fate of resuspended particles close to the bottom (i.e. by hydraulic dredges), while light symbols indicate the fate of resuspended particles close to the free surface (i.e. by mechanical dredges). It could be noted that, coherently to the current stratification produced by wind action, the dredging plume close to the bottom can escape from the enclosed basin to the open sea through a deep current characterized by a direction opposite to the surface one. Instead, at the surface layer, suspended sediments can remain confined inside the harbour depending on climatic conditions. It is worth to stress again that the simulated spatial current

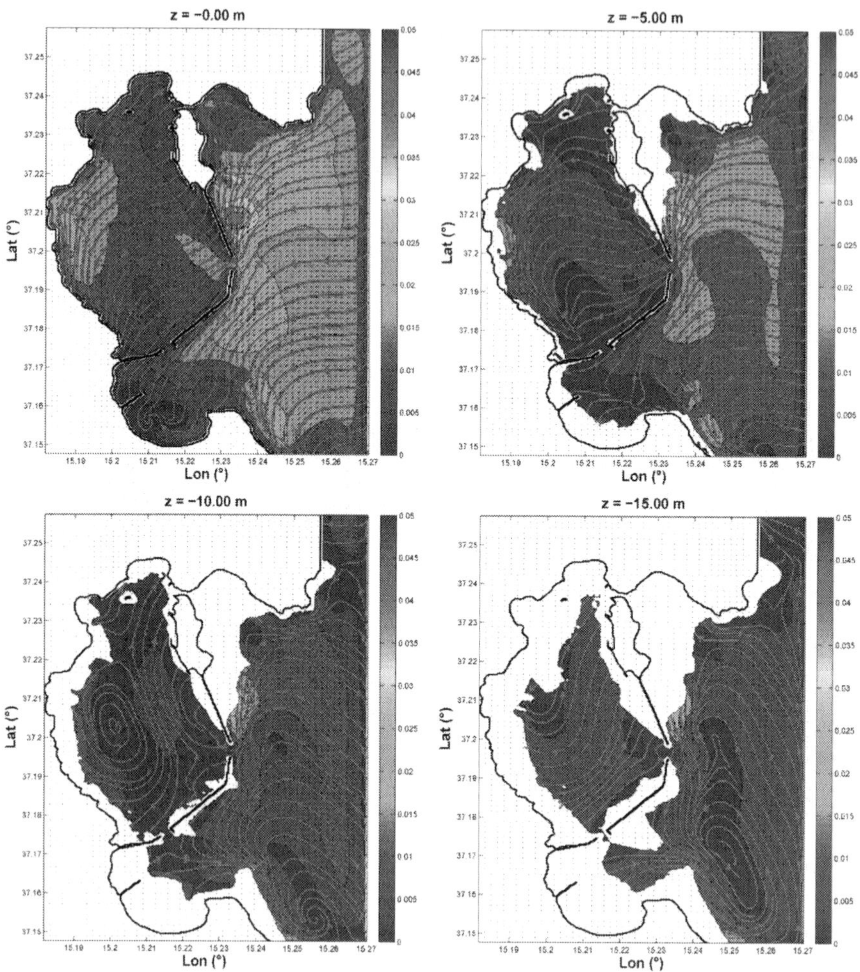

Figure 6. Hydrodynamic numerical results for a frequent wind condition (speed: 5.2 m/s; dir.: 35°N). Symbol z is the vertical coordinate directed upward starting at the still free surface. Red arrows indicate streamlines, colour scale refers to current intensities and black arrows to the local current direction.

speed distribution and variability highlight the importance of using three-dimensional numerical modelling to ensure an effective environmental management of dredging activities.

Figure 7 shows greater extensions of mechanical dredging plumes when activities are performed at DP3 and DP4 work areas if compared to DP1 and DP2 ones. In these cases, significant levels of SSC generally occur along the whole water column. While, for hydraulic dredges, characterized by a spill layer located near the bottom, greater spatial confinement as well as a higher intensity of SSC is generally observed in the immediate vicinity of the dredge. Generally, hydraulic dredges seem to generate considerable values of SSC at the surface layer only where hypothetical work areas are characterized by significant current intensities (e.g. DP2 and DP4, Figure 7).

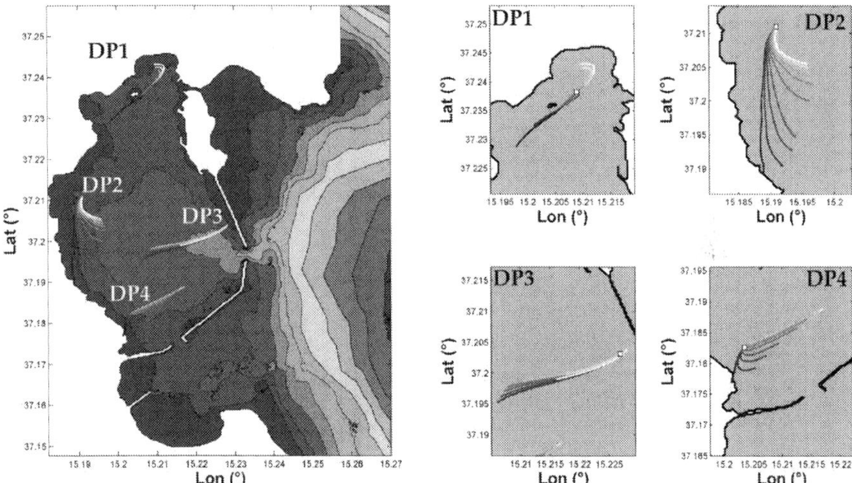

Figure 7. Dredging plume modelled by Lagrangian mode for DP1, DP2, DP3 and DP4 hypothetical dredging points. Colour scale in the left panel refers to the bathymetry. In the right panels, black, grey and white symbols refer to the resuspension of particles for bottom, intermediate and surface layers, respectively.

6. Concluding remarks

This chapter describes the available predictive techniques to estimate different (mechanical and hydraulic) dredging resuspension sources, occurring during the process of removing sediment from the bottom, and the SSC variability related to the resulting plume dynamic. Dredging techniques most commonly used in coastal water along with potential resuspension mechanisms are reviewed. Then, predictive tools aimed at estimating the resuspension source strength and geometry based on field data are detailed. The resuspension source characterization is the first step needed to estimate the dredging plume development and related effects on water quality, aquatic ecosystem and human health. Both simplified and sophisticated tools aimed at estimating the plume dispersion are described in some detail. In particular, a novel one-dimensional analytical model is proposed to provide a fast order of magnitude of the plume dynamics when time-varying sources and time-varying currents have to be considered. The results obtained by the simplified model may be used to decide if more complex numerical modelling is needed. In such a case, an integrated and replicable numerical approach, combining a series of numerical tools, is proposed to support technical choices and environmental studies during planning and operational phases of dredging projects in estuarine and coastal (or semi-enclosed basin) areas. The application of the proposed integrated numerical approach to Augusta Harbour (Sicily Island, Italy) case study is briefly described to show typical results achievable in the design stage of the four hypothetical (mechanical and hydraulic) dredging projects.

Author details

Iolanda Lisi[1*], Marcello Di Risio[2], Paolo De Girolamo[3] and Massimo Gabellini[1]

*Address all correspondence to: iolanda.lisi@isprambiente.it

1 ISPRA - Institute for Environmental Protection and Research, Rome, Italy

2 University of L'Aquila, Department of Civil, Construction-Architectural and Environmental Engineering (DICEAA), Environmental and Maritime Hydraulic Laboratory (LIAM), L'Aquila, Italy

3 "Sapienza" University of Rome, Department of Civil, Constructional and Environmental Engineering (DICEA), Rome, Italy

References

[1] Palermo M.R., Schroeder P.R., Estes T.J., Francingues, N.R. Technical Guidelines for Environmental Dredging of Contaminated ERDC/EL TR-08-29US Army Corps of Engineers, Engineer Research and Development Center. 2008.

[2] Bridges T.S., Ells S., Hayes D., Nadeau S., Palermo M.R., Patmont C., Schroeder P. The Four Rs of environmental dredging: Resuspension, Release, Residual, and Risk. ERDC/EL TR-08-4. US Army Corps of Engineers. 2008.

[3] HR Wallingford Ltd & Dredging Research Ltd. Protocol for the Field Measurement of Sediment Release from dredges. Produced for VBKO TASS, Issue 1, p. 83. 2003.

[4] Shao D., Purnama A., Sun T. Modeling the Temporal Evolution of Dredging-Induced Turbidity in the Far Field. J Wat Port Coastal Oc Eng 2015. 141(5), 04015001.

[5] Wilber D.H., Clark D.G. Biological Effects of Suspended Sediments: A Review of Suspended Sediment Impacts on Fish and Shellfish with Relation to Dredging Activities in Estuaries. N Am J Fish Manage 2001. 21(4): 855–875.

[6] Bridges T. S., Gustavson K.E., Schroeder P., Ells S. J., Hayes D., Nadeau S.C., Patmont C. Dredging processes and remedy effectiveness: Relationship to the 4 Rs of environmental dredging. Integr Environ Assess Manage 2010. 6(4): 619–630.

[7] Manap N., Voulvoulis N. Risk-based decision-making framework for the selection of sediment dredging option. Sci Total Environ 2014. 496: 607–623.

[8] PIANC, Permanent International Association of Navigation Congresses. Environmental risk assessment of dredging and disposal operations. Report of Working Group 10 of the Environmental Commission. p. 40. 2006.

[9] Suedel B.C., Clarke J.U., Wilkens J., Lutz C.H., Clarke D.G. The effects of a simulated suspended sediment plume on Eastern Oyster (Crassostrea virginica) Survival, Growth, and Condition. Estuar Coast 2014. 38(2): 578–589.

[10] Erftemeijer P.L., Lewis R.R.R. Environmental impacts of dredging on seagrasses: a review. Mar Pollut Bull 2006. 52(12): 1553–1572.

[11] Sofonia J.J., Unsworth R.K. Development of water quality thresholds during dredging for the protection of benthic primary producer habitats. J Environ Monitor 2010. 12(1): 159–163.

[12] Palermo M.R., Averett D.E. Environmental dredging – a state of the art review. 2nd International Symposium on Contaminated Sediments: Characterization, Evaluation, Mitigation/Restoration, Management Strategy Performance, Quebec City, Canada: 12–17. 2003.

[13] PIANC, Permanent International Association of Navigation Congresses, Dredging and port construction around coral reefs, PIANC EnviCom, Report 108. ISBN: 978-2-87223-177-5. 2010.

[14] Collins M. Dredging-Induced Near-Field Resuspended Sediment Concentration and Source Strengths. US Army Corps of Engineers, Waterways Experiment Station, p. 232. 1995.

[15] Hayes D., Wu P.Y. Simple Approach to TSS Source Strength Estimates. In: Proceedings of Western Dredging Association WEDA XXI, Houston. 2001.

[16] Nakai O. Turbidity Generated by Dredging Projects. Proceedings of the 3rd U.S./Japan Experts Meeting, US Army Engineer Water Resources Support Center.1978.

[17] Pennekamp J.G.S., Eskamp R.J.C., Rosenbrand W.F., Mullie A., Wessel G.L., Arts T., Decibel I.K.Turbidity caused by dredging; viewed in perspective. Terra et Aqua 1996. 64: 10–17.

[18] Becker J., van Eekelen E., van Wiechen J., ve Lange W., Damsma T., Smolders T., van Koningsveld M. Estimating source terms for far field dredge plume modelling. J Environ Manage 2014. 149: 282–293.

[19] John S.A., Challinor S.L., Simpson M., Burt T.N., Spearman, J. Scoping the assessment of sediment plumes arising from dredging. CIRIA Report C547, p. 192. 2000.

[20] Savioli J.C., Magalhaes M., Pedersen C., Van Rijmenant J., Oliver M.A., Fen C.J., Rocha C. Dredging - how can we manage it to minimise impacts. Proceedings of 7th International Conference on Asian and Pacific Coasts, Bali, Indonesia, p.6. 2013.

[21] De Marchis M., Freni G., Napoli E. Three-dimensional numerical simulations on wind- and tide-induced currents: The case of Augusta Harbour (Italy). Comput Geosci 2014. 72: 65–75.

[22] Edwards K.P., Hare J.A., Werner F.E., Blanton B.O. Lagrangian circulation on the Southeast US Continental Shelf: Implication for larval dispersal and retention. Contin Shelf Res 2006 26: 1375–1394.

[23] EPA, Environmental Protection Agency. Environmental Assessment Guideline for Marine dredging proposals, Environmental protection Authority, Western Australia, p. 36. 2011.

[24] Hayes D., Je, C.H. DREDGE Module User's Guide. Department of Civil and Environmental Engineering, University of Utah. 2000.

[25] Jouon A., Douillet P., Ouillon S., Fraunie P. Calculations of hydrodynamic time parameters in a semi-opened coastal zone using a 3D hydrodynamic model, Contin Shelf Res 2006. 26: 1395–1415.

[26] Lisi I. Development of an integrated system to the estimation and the environmental management of resuspension induced by dredging activities. [thesis, in Italian]. Univ. of L'Aquila, Italy. 2012.

[27] Feola A., Lisi I., Venti F., Salmeri A., Pedroncini A., Gabellini M., Romano E. Platform of integrated tools to support environmental studies and management of dredging activities. J Environ Manage 2015. In press.

[28] IMDC, International Marine & Dredging Consultants. Environmental Impact Assessment windmill farm Rentel, Numeric modelling of dredging plume dispersion. Prepared for Rentel NV, version 2.0. 2012.

[29] Jiang J., Fissel D.B. Modeling Sediment Disposal in Inshore Waterways of British Columbia, Canada. Estuarine and Coastal Modeling. ASCE. 2011.

[30] Jiang J. Investigation of Key Parameters for 3-D Dredging Plume Model Validation. J Ship Oc Eng 2014. 4: 129–-139.

[31] Liu J.T., Chao S., Hsu R.T. Numerical modeling study of sediment dispersal by a river plume. Contin Shelf Res 2002. 22(11–13): 1745–1773.

[32] Crockett T.R. Modeling near-field sediment resuspension in cutter [thesis]. Lincoln, NEUniv. of Nebraska. 1993.

[33] Hayes D. Development of a Near Field Source Strength Model to Predict Sediment Resuspension from Cutter Suction Dredges. M.S. Thesis, Mississippi State University. 1986.

[34] Hayes D., Crockett T.R., Ward T.J., Averett D. Sediment resuspension during Cutterhead dredging operation. J Wat Port Coastal Oc Eng 2000. 126 (3):153–161

[35] Bell R.G., Reeve G. Sediment plume dispersion modelling: Comparison of a larger dredger and the New Era NIWA, Report HAM2010–119, Port Otago Ltd, p. 57. 2010.

[36] Bilgili A., Proehl J., Lyncl D., Smith K., Swift M.R. Estuary/Ocean exchange and tidal mixing in a Gulf of Maine Estuary: A Lagrangian modeling study. Estuar Coast Shelf Sci 2005. 65(4): 607–624.

[37] Chao X., Shankar N. J., Fatt C.H. A three-dimensional multi-level turbulence model for tidal motion. Oc Eng 1999. 26(11): 1023–1038.

[38] Kim K., Je C. Development of a framework of automated water quality parameter optimization and its application. Environ Geol 2006. 49(3): 405–412.

[39] Shankar N.J., Cheong H.F., Sankaranarayanan S. Multilevel finite-difference model for three-dimensional hydrodynamic circulation. Oc Eng 1997. 24(9): 785–916.

[40] Zhang Q.Y., Gin K.Y.H. Three-dimensional numerical simulation for tidal motion in Singapore's coastal waters. Coastal Eng 2000. 39(1): 71–92.

[41] SKM, Sinclair Knight Merz Pty Ltd, Improved dredge material management for the Great Barrier Reef Region, Great Barrier Reef Marine Park Authority, Townsville. 2013.

[42] USACE, US Army Corps of Engineers. A review of environmentally improved techniques for dredging contaminated mud. TECHNICAL NOTE TN 3/2003. Fill Management Division GEO, p. 29. 2003.

[43] Eisma D., Dredging in Coastal Water. Published by Taylor & Francis plc., London, UK, ISBN: 978-0-415-39111-5, p. 244. 2006.

[44] Otten M.T., Webb R. Dredging Systems Used to Meet Remediation Goals - Evaluation of Three Successful Case Histories. In: Proc. of West. Dredg. Ass. WEDA. 2008.

[45] Anchor Environmental C.A. L.P. Literature review of effects of resuspended sediments due to dredging operations. Los Angeles Contaminated Sediments Task, p. 140. 2003.

[46] IADC, International Association of Dredging Companies. Environmental Aspects of Dredging – Guide 4: Machines, Methods and Mitigation. International Association of Dredging Companies (IADC), The Netherlands, p. 80. 1998.

[47] Henriksen J., Randall R., Socolofsky S. Near-field resuspension model for a cutter suction dredge. J Wat Port Coastal Oc Eng 2011. 138(3): 181–191.

[48] Hayes D.F., McLellan T.N., Truitt C.L. Demonstrations of innovative and conventional dredging equipment at Calumet Harbor, Illinois. Miscellaneous Paper EL-88-1, U.S. Army Engineer Waterways Experiment Station, Vicksburg, MS. 1988.

[49] Je C. H., Hayes D.F. Development of a two-dimensional analytical model for predicting toxic sediment plunges due to environmental dredging operations. J Environ Sci Health A 2004. 39(8): 1935–1947.

[50] Je C., Hayes D.F., Kim K. Simulation of resuspended sediments resulting from dredging operations by a numerical flocculent transport model. Chemosphere 2007. 70:187–195.

[51] Kuo A. Y., Welch C. S., Lukens R. J. Dredge induced turbidity plume model. J Wat Port Coastal Oc Eng 1985. 111(3): 476–494.

[52] Kuo A. Y., Hayes D. F. Model for turbidity plume induced by bucket dredge. J Wat Port Coastal Oc Eng 1991. 117(6): 610–623.

[53] De Dominicis M., Pinardi N., Zodiatis G., Lardner R. MEDSLIK-II, a Lagrangian marine oil spill model for short-term forecasting – part I: theory. Geosci Model Dev 2013. 6: 1851–1869.

[54] Zafirakou A., Palantzas G., Samaras A., Koutitas C. Oil Spill Modeling Aiming at the Protection of Ports and Coastal Areas. Environ Processe 1–13. 2015.

[55] Davies A.M., Xing J., Huthnance J.M., Hall P., Thomsen L. Models of near-bed dynamics 714 and sediment movement at the Iberian margin. Prog Oceanogr 2002. 52(2–4): 373–397.

[56] Gessler D., Hall B., Spasojevic M., Holly F., Pourtaheri H., Raphelt N. Application of 3D Mobile Bed, Hydrodynamic Model. J Hydraulic Eng 1999. 125(7): 737–749.

[57] Kim C.S., Lim H.S. Sediment dispersal and deposition due to sand mining in the coastal waters of Korea. Contin Shelf Res 2009. 29(1):194–204.

[58] Hunter J.R.; Craig P.D., Phillips H.E. On the use of random walk models to with spatially variable diffusivity. J Comput Phys 1993. 106(2): 366–376.

[59] Visser A.J. Using random walk models to simulate the vertical distribution of particles in a turbulent water column. Marine Ecol Progr Ser 1997. 158: 275–281.

[60] Marghany M., Hashim M., Cracknell A.P. Simulation of shoreline change using AIRSAR and POLSAR C-band data. Environ Earth Sci 2011. 64(4), 1177–1189.

[61] Marghany M., Hashim M., Cracknell A.P. 3-D reconstruction of coastal bathymetry from AIRSAR/POLSAR data. Chin J Ocean Limnol 2009. 27(1), 117–123.

[62] ISPRA, Italian Institute for Environmental Protection and Research. Progetto preliminare di bonifica della rada di Augusta inclusa nel Sito di bonifica di Interesse Nazionale di Priolo: fase I e II. Tech. Report, in Italian, p. 253. 2008.

[63] Romano E., Bergamin L., Celia Magno M., Ausili A. Sediment characterization of the highly impacted Augusta harbour (Sicily, Italy): modern benthic foraminifera in relation to grain-size and sediment geochemistry. Environ Sci Process Impact 2013. 15(5): 930–946.

[64] Hedstrom, K.S., Draft technical manual for a coupled sea-ice/ocean circulation model (version 4). 2012.

[65] Lisi I., Taramelli A., Di Risio M., Cappucci S., Gabellini M. Flushing efficiency of Augusta Harbour (Italy). J Coastal Res 2009. SI 56: 841–845.

[66] Haidvogel D. B., Arango H. G., Hedstrom K., Beckmann A., Malanotte-Rizzoli P., Shchepetkin A. F. Model evaluation experiments in the North Atlantic Basin: Simulations in nonlinear terrain-following coordinates. Dynam Atmos Ocean 2000. 32: 239–281.

[67] Pawlowicz R., Beardsley B., Lentz S. Classical tidal harmonic analysis including error estimates in matlab using t tide. Comput Geosci 2002. 28(8): 929–937.

Neo-tectonic Movement in the Pearl River Delta

Neo-tectonic Movement in the Pearl River Delta (PRD) Region of China and Its Effects on the Coastal Sedimentary Environment

Yuanzhi Zhang, Zhaojun Huang, Yu Li, Qing Yu, Bing Wang and Tingchen Jiang

Additional information is available at the end of the chapter

http://dx.doi.org/10.5772/61981

Abstract

This chapter presents the Late Quaternary neo-tectonic movement in the Pearl River Delta (PRD) region of China and its effects on the coastal sedimentary environment. Taking the Xilin Hill fault as the example, we have explained the reason of the PRD formation, inferred its interactions with local/regional climate and sea-level changes, and analyzed the evolution of the PRD formation and its controlling factors of the coastal sedimentary environment.

Since Late Quaternary, the intense change of coastal sedimentary environment in the PRD region has resulted in accumulating the thick sedimentary layers with the different causes and properties which were influenced by the fluctuation of sea levels, the change of ancient climate, and human activities. Based on the interpretation of high resolution of these sediments, such as the period, sedimentary facies, and discontinuous facies, we can infer the evolution of the ancient climate of regional Quaternary environment in the background of global change to know more details about the formation of the PRD. It is noted that the neo-tectonic movement and the fault activities played a critical role in controlling the coastal geological environment. This is helpful to further investigate the evolution of the PRD coastal sedimentary environment, the history of the inversion region, and the sedimentary facies model of the interaction between land and sea. More investigations will be conducted in the near future.

Keywords: Neo-tectonic movement, Pearl River Delta, Quaternary geology, coastal sedimentary environment, sea-level change

1. Introduction

It is an important question for discussion of coastal geomorphology [1] to study the coastal sedimentary environment and the impact from the Quaternary neo-tectonic movement in river estuarine–coastal regions, such as the Pearl River basin. Pearl River Delta (PRD) is a dynamically evolving landform that occurs at the mouth of Pearl River systems where the sediments are deposited as they are moving to South China Sea (SCS). There are several controlling factors accounting for the formation of the PRD, including neo-tectonic movements, sea-level changes, fluvial processes, and human activities (e.g., [29, 30]).

The PRD Quaternary environment is an important topic in response to global climate change, as it has a complicated environment, in which the PRD is located in the special position of the transition of land–sea, sensitive to the climate change and sea-level rising [29].

A number of previous studies investigated the PRD region and made a great achievement, including the deposition, new tectonic movement, fault activities, sea-level changes, and ancient environment evolution. For example, the relationship between several activities of the fracture structure and river terraces by the causes of the block tectonic geomorphology in the PRD was first proposed and discussed in the early 1980s (e.g., [12, 30, 31]). Later on, the amplitude and rate of fault block tectonic movement were reported by applying the results of the dating of Quaternary sediment chronology since Quaternary period (e.g., [2]). However, it was addressed to analyze the characteristics of the fault block tectonic movement and the dynamic tectonic stress field since Late Tertiary (e.g., [6, 33]). In addition, the crustal stability of several districts in the PRD region was analyzed by using geochemical testing results (e.g., [28, 34]).

From the detailed information of sediments, the environmental history of the sedimentary sequences in Late Quaternary can be reconstructed to understand the evolution of the ancient climate of regional Quaternary environment in the background of global change and to infer the changes of global climate and sea-level rising (e.g., [29]). In a word, the Quaternary geology of the PRD region is of great significance.

In regard to the causes of the formation and development of the PRD since Late Quaternary, it is still being an academic controversy. Some researchers believe that the development of the PRD was mainly influenced by rising sea level, or caused by Lushan glacial and postglacial climate changes. In comparison, the others believe that the climate fluctuation did not change too much during Lushan glacial and postglacial, as the amplitude of fluctuation is only a few meters to tens of meters since 30 ka B.P. (B.P. means before present where "present" represents 1950) in the north of South China Sea. However, a number of SCS transgressions since the postglacial period had provided the necessary condition for the development of the PRD (e.g., [13, 16, 18, 19, 26]). Nevertheless, there are some issues to be further investigated. For example, there are two kinds of disparate opinions about the age ascription of the lower cycle. One is that the lower cycle may be dating back to about 30,000–40,000 years, while the other is that it may belong to the period of about 100,000 years more or less.

In recent 20 years, advanced techniques such as interferometric synthetic aperture radar (InSAR) and differential synthetic aperture interferometry (DInSAR) have been widely applied to construct digital elevation models of the earth's topography and to image centimeter-scale displacements associated with crustal deformation and the flow of ice sheets. As a result, coastal geomorphology can also benefit from these techniques [23]. In particular, such techniques have made a great progress in the application of coastal geomorphology to interferometrically measure centimeter-scale motions associated with unstable slopes, land subsidence, fluctuating soil moisture, and water levels [20].

In this chapter, we first review the neo-tectonic movement in the PRD based on the theory of Quaternary geology, then analyze the reason of the PRD formation and its controlling factors, which integrated the results of the deposition, sea-level changes, fracture, and fault block, and finally discuss the effects on the PRD coastal sedimentary environment.

2. Study area

The Pearl River Delta region is located at 21°20′–23°30′N and 112°40′–114°50′E, Guangdong Province, South China, which was formed in the past 9000 years [16, 17]. It is one of the fastest developing regions in China and experiencing a rapid rate of economic growth. As it is shown in Figure 1, Pearl River flows through the PRD and then into the South China Sea. In recent years, it is one of the most densely urbanized regions and one of the main hubs of China's economic growth.

The length of the Pearl River is 2214 km, and the drainage system of the Pearl River covers about 425,700 km² (e.g., [17]). But the area of the receiving basin is 9750 km², which includes the deltaic plains, the estuary, and rocky islands but excludes the area of Hong Kong [11].

The Pearl River consists of three major sub-rivers, namely the East River (or Dongjiang), the North River (or Beijiang), and the West River (or Xijiang), which flow into the Pearl River estuary (PRE) before inflowing the South China Sea. The flat lands of the delta are crisscrossed by a network of tributaries and distributaries of the Pearl River system. The PRD is actually two alluvial deltas, separated by the core branch of the Pearl River. The North River and the West River flow into the South China Sea and Pearl River in the west, while the East River only flows into the Pearl River proper in the east [17].

The PRD belongs to the East Asian monsoon region and the climate is low subtropical. The temperature of the delta area is relatively high, with the average temperature of the whole year varying from 21.5 to 22.5 °C. The highest temperature is 38 °C during the past years, but the minimum temperature occurs every 10 years, low to 0 °C. In addition, the rainfall is abundant, with the average annual rainfall of 1600 mm or more. However, during the dry seasons, the distribution of rainfall is not even. The main modern natural vegetation includes evergreen broad-leaved forests, coniferous forests, grass slopes, mangroves, etc. [11].

Due to the uplift of Tibetan Plateau, the PRD was formed during Tertiary and Quaternary. Its active fault systems developed as a result of the land subsidence of the receiving basin during

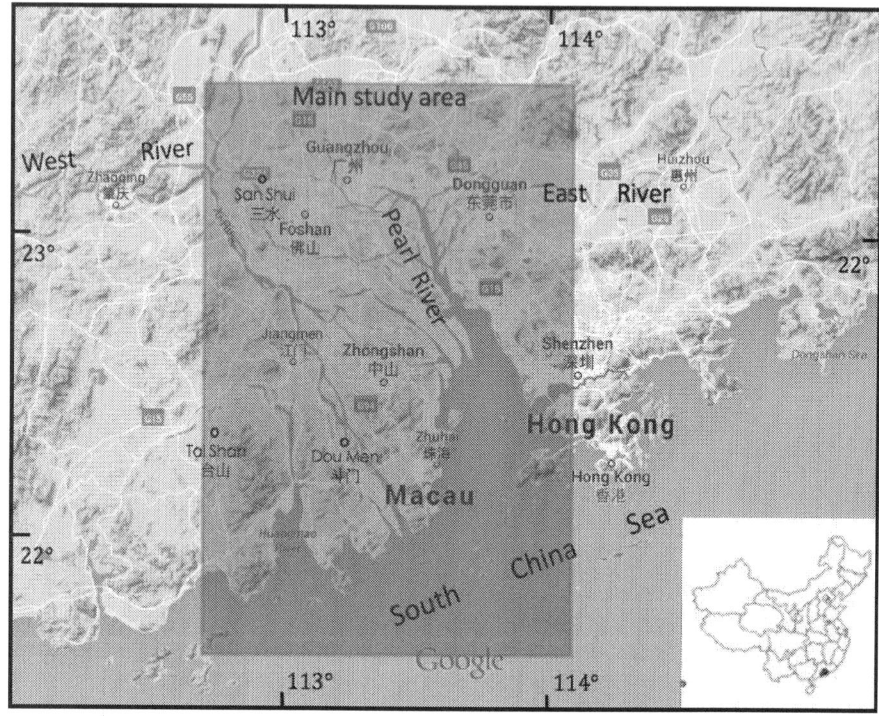

Figure 1. The study area of the PRD region (courtesy of Google Maps as the background of the PRD region).

Late Quaternary. A number of valleys were disconnected by rocky islands and several hills. In the lower reaches of the East River, there is a shallow paleo-valley with the depth of 30 m. Meanwhile, two paleo-valleys with the depth from 30 to 40 m are found in the North River and the West River. When the three paleo-valleys crash into the southeastern basin, they are merged. At that time, another separate paleo-valley was formed in the southwestern basin [16, 17].

The PRD geomorphological characteristics with more than 160 island bumps on the plain are represented by several geomorphic types, such as hills, platforms, and monadnocks. Its area accounts for about one fifth of the whole delta area. In recent times, there are hundreds of islands distributed in the waters of the Pearl River estuary, with the altitudes varying from 100 to 300 m [13, 14, 15].

As the deposition rate of the delta estuary and coastal waters was relatively high, a great of geological information was recorded in deep layer sediments. The sedimentary information is important for analyzing the reason of the PRD formation and controlling factors, which integrated the results of the deposition, fracture, and fault block, and discussing the effects on the PRD coastal sedimentary environment.

3. Methodology

The methodology of this research includes traditional geological survey and applications of advanced techniques. Hence, we applied stratigraphic and sedimentary analysis, carbon dating, and InSAR and DInSAR techniques in the PRD investigation. By interpreting the high resolution of sediments, the detailed information, such as the periods, sedimentary conditions, sedimentary facies, and discontinuous facies, and sequential relationships between the strata can be obtained [2, 29].

Radiocarbon 14 (14C) dating is a common means for chronological divisions. The dating principle is that after the carbonaceous materials died, the carbon in the body could stop exchange with the outside world. Then the content of 14C is reduced by the rate of decay. On comparing the residual content of radiocarbon 14 samples with that of present similar samples, we could deduce the epoch from the time after exchanging 14C. As the half-life period of 14C is 5568 years, it will be reduced by about 1000 times after 10 half-life periods [2, 9, 29]. So the modern radioactive detector almost cannot work. Considering these conditions, the 14C dating method is suitable for carbonaceous materials dating back to 300–50,000 years. As the precision is about 1–2%, the older the sample, the greater the error is [25].

4. Results and discussion

4.1. Structure characteristic and faults distribution

PRD is located within the Eurasian Plate, adjacent to the oceanic crust of South China Sea. It is subjected to the extrusion in different directions of Philippine Sea Plate and Indian Ocean Plate. The PRD tectonic stress field is mainly affected by the pushing of the South China near south-north. The PRD fault depression originated in the Cretaceous and continued to the inherited activities in the basin of Quaternary (e.g., [7, 12, 30, 32]).

As a result of a series of crustal movements, the fault structure and fault basin were formed, namely fold: syncline structure and anticline structure. The PRD basal strata are old, which is mainly composed of metamorphic rocks of Sinian and Paleozoic eras. The metamorphic rocks of Sinian are distributed in northern and eastern parts of the delta, whereas those of Paleozoic are found in the periphery of the delta. Granites intruded when the Yanshanian movement occurred from Jurassic to Cretaceous. More important is the foundation of the topographic outline of the ancient gulf. With the invasion of granites and fault activities, the fault block mountain and fault basin developed rapidly (e.g., [30]). After the erosion and eroding, fault blocks were turned into hilly and mountainous lands. The fault basin was developed by accumulating the materials of mountain erosion. Consequently, there existed the hilly land and marine strata composed of volcanic rocks with the volcanic activities and marine trans-gression. Later on, the delta basin was uplifted to lands and endured erosion and incision during the Middle Tertiary period when the crust changed rapidly by the Himalayan move-ment [7]. After that, the materials from erosion and eroding were carried to the place which is far away from the PRD, more than 200 km. During the Late Tertiary period, several delta basins were rifted by the fault block movements. When it came to Middle–Late Pleistocene of Quaternary, the delta experienced a depositional phase. At that time, the plain areas were subsided, but fringe areas were uplifted [12]. Under the combined effects of fault block subsidence and sea-level changes since the Quaternary, the depression within the rifts usually formed two sets of delta-accumulated strata, with the large range of distribution and thickness of sediments [3, 32].

The PRD has three groups of main striking fault systems, namely northwest (NW), nearly east-west (EW), and northeast (NE) faults. By the analysis of the characteristics of seismic risks in the PRD, it is suggested that these three groups jointly cut these fault blocks into seven sub-fault blocks [3, 12, 30, 33] as shown in Figure 2. Although the formation times of the three group faults are different, EW and NE faults were formed earlier than the NW faults in the Yansha-nian movement. Delta basal unit was also cut into several fault blocks with different sizes and velocities as the interaction of the three groups of these faults [78]. In Figure 2, the fault blocks (I–VII) and faults (1–13) are summarized and described as [12, 30, 31, 33]: I Dongjiang Delta fault block, II Tan Hoi fault block, III Wuguishan fault block, IV Shiqiao–Guangzhou fault block, V Xijiang–Beijiang Delta fault block, VI Pearl River estuary fault blocks, and VII Near-shore SCS (South China Sea) fault block, while the faults are named as:

1. Sanshui–Luofushan Fault,

2. Guangzhou–Conghua Fault,

3. Dongguan–Houjie Fault,

4. Shiqiao-Xinhui Fault,

5. North foothills of Wuguishan Fault,

6. South foothills of Wuguishan Fault,

7. Shenzhen–Zhuhai Fault,

8. Near-shore shallow water of PRE Fault,

9. Xijiang Fault,

10. Baini–Shawan Fault,

11. Nangang–Humen Fault,

12. Hualong–Huangge Fault, and

13. Yamen Fault.

As shown in Figure 2, among the fault blocks mentioned above, the fault blocks from I to IV were mainly controlled by northeast faults. Fault blocks (I–IV) were developed into the Dongguan basin and the Xinhui basin, which were formed before Quaternary. The delta sedimentary facies were filled with the two faults by about 20–30 m deep. During the Yan-shanian movement, crystalline rocks were broken along the northeast trending and invaded

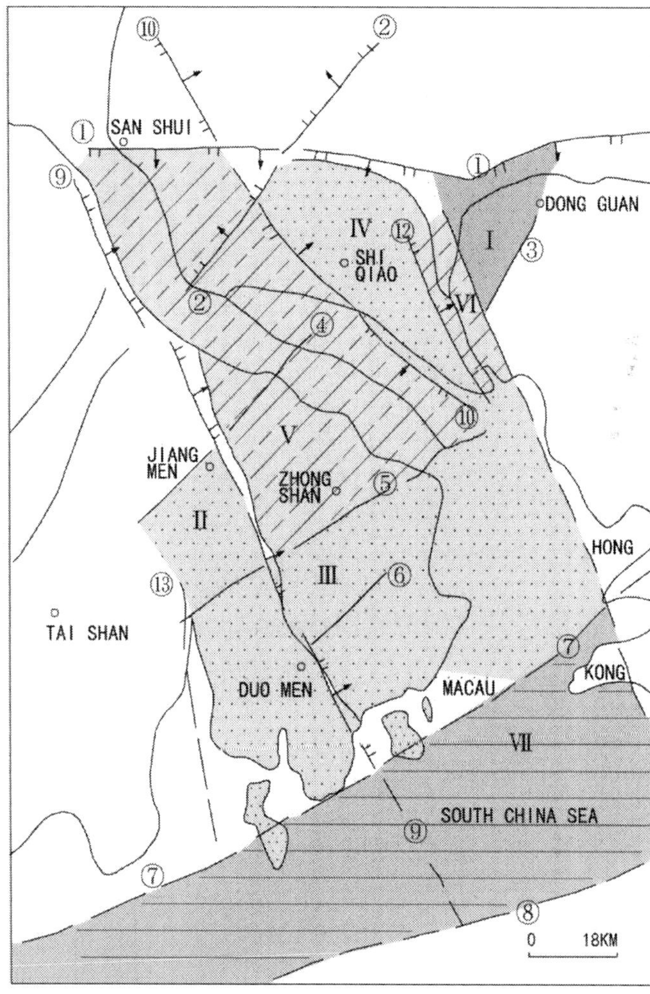

Figure 2. The schematic diagram of the fault blocks in the Pearl River Delta (summarized and courtesy of [7, 12, 30, 32, 33].

into the raised rear part of Wuguishan. According to the study of radioisotope dating, three group faults moved vertically at the rate of less than 1 mm/year since Late-Pleistocene, and thus the main faults had not obvious signs of active tectonic dislocations [4, 24].

The north of striking fault block IV is principally controlled by Sanshui–Luofushan fault, while Baini–Shawan and Hualong–Huangge faults are in control of the south of this fault block. Among them, it is said that Sanshui–Luofushan and Baini–Shawan faults had obvious active tectonic dislocations during Early-Pleistocene, but there was no dislocation in Late-Pleistocene.

Until Late-Pleistocene, the Pearl River's development was operated by Sanshui–Luofushan and Hualong–Huangge faults [6, 32, 33].

Fault blocks V and VI are controlled by the NW faults. The fault blocks are subjected to the incision of the NE faults. Among the two sets of faults, the former has the feature of normal faults, while the latter has the feature of reverse faults. The tectonic geomorphology was well developed, including the northwest trough and valley, as well as the northeast ridge. The thickness of Quaternary sediments varies from about 40 to 50 m, while some parts may be up to 60 to 70 m. However, there is a huge difference in south and north. Usually, the thickness of the northern PRD is less than 40 m, even that the weathering crust in large areas gave a push up by several meters to about 100 m from Early Quaternary. At that time, the southern weathering crust was buried with several meters. It is clear that that the activity of faults in north is much weaker than that in south (e.g., [32]).Fault block VII is controlled by the northeast faults and is also cut by the northwest faults. The thickness of Quaternary trough and valley can be up to more than 50 m. According to the estimation of radioisotope dating, the sedimentation rate of fault block VII is up to 3 mm/a in middle–late Pleistocene. The activities with high sedimentation rates and fracture rates lasted for a long time during Late Quaternary. It is evident that the activity of this fault block is stronger than the other six. Moreover, the fault is dating back to 30,000–40,000 years, indicating that its formation period is young, and thus the activity of fault block VII is relatively new (e.g., [32, 33]).

Due to the different activities of the fault zones since Quaternary, the intensity of the seven fault blocks is also different. According to the regulations of South China coastal seismic belt seismicity, the various seismic risk regions of PRD can be divided into three grade areas, that is, the activity of South China Sea fault is intense, while Xijiang fault, Beijiang fault, and estuary fault are posterior and also weak [6, 33].

4.2. Neo-tectonic movements

To a certain extent, tectonic movements affect the distribution of sediments, which is the result of the tectonic movements. As PRD is located within the continental plates of Guangzhou in South China, the average thickness of crust is about 28–30 km. According to the speculation, it was caused by the upburst of upper mantle with a high density. As a result, the neo-tectonic movement has inherited the tectonic framework since Yanshanian movement. Several borders of the PRD region are controlled by NE tectonic movements. On the other hand, the activity of NW tectonic movements dominated the trend of sedimentary geomorphology. Tectonic lines of the above two groups jointly manipulated the pattern of tectonic geomorphology in that the direction of NW is as the long axis and the direction of NE is as the short axis (e.g., [2, 6]).

It is acknowledged that in the tectonic uplift areas, the higher the landform is, the older the stratum is. However, in the tectonic subsidence areas, the lower the sedimentary layer is, the older the stratum is, and each new layer is laid down horizontally over older ones in a process [2, 6, 21] as shown in Figure 3.

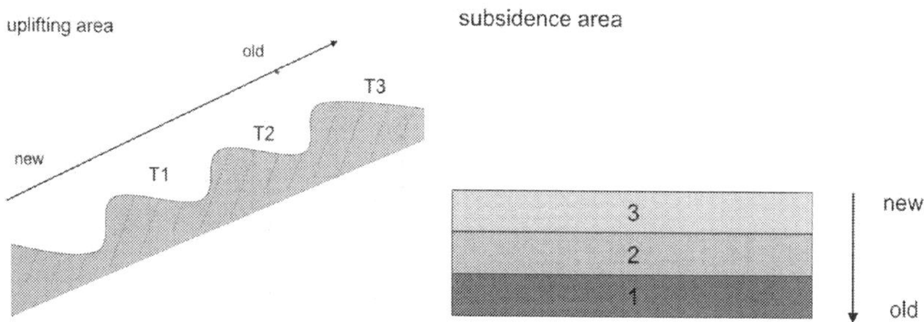

Figure 3. Ages sketch map of sediments in the uplifting and subsidence areas.

There is no doubt that the upper cycle of two sedimentary cycles in PRD belongs to Holocene, and the lower one is dating back to Late-Pleistocene [35, 36]. By analysis of the PRD sediments, it is suggested that there has existed two progressive cycles since Late-Pleistocene. However, there is no consensus about the epoch ascription of the lower cycle. If it is dating back to about 30,000–40,000 years, it should be middle–late Late-Pleistocene (i.e., MIS3). But it should belong to early Late-Pleistocene, if it is dating back to about more than 100,000 years (i.e., MIS5).

According to the previous investigation, the sea level of MIS3 was about 50 m lower than the current level. However, the average thickness of the PRD sediments in Quaternary is only about 30 m. Therefore, in theory, it is impossible that there was deposition only during MIS3. It is inferred that the lower cycle may be dating back to about 120,000 years, belonging to MIS5 (e.g., [22, 29]).

Although the PRD region is located in the continental margin of the north of South China Sea, compared with other continental margins, there is no volcanic activity but weak seismic activity. Consequently, the crust is relatively stable. But the region is influenced by the expansion of South China Sea. The faults developed in the coast of South China, especially the basement faults with the NE and near EW directions. As the PRD is cut by the faults and formed several fault blocks with different movement directions and velocity, the neo-tectonic movement of the PRD is characterized by the fault activities and differentiated movements of these fault blocks [6, 30, 33].

In addition, the tectonic movement has a great significance in the development of PRD. The delta was cut into a series of fault blocks by the basement faults which developed or reactivated during Quaternary. These fault blocks were dominated by subsidence in the vertical movement, moving at different directions and rates, which provided accommodation for Quaternary sediments. As a result, the evolution process of PRD was mainly controlled by the faults since a lot of evidence has been found from the distribution of Quaternary sediments, drill holes, and geologic profiles (e.g., [7, 33]).

According to the different distribution of fluvial gravels developed since the first marine transgression in Late Quaternary, it is obvious that the paleo-drainage patterns have been greatly changed as a result of fault activities (e.g., [12]). In short, during the PRD formation and evolution, neo-tectonic movements not only influenced its sediments but also controlled the river channel changes.

4.3. A case study of neo-tectonic movements at Xilin Hill

According to the previous studies, it was found that the Late Quaternary layers were uplifted, tilted, and faulted (e.g., [33]). These layers are located at the Xilin Hill, between Guangzhou and Foshan, in the north of PRD. On the south of the Xilin Hill, the Holocene black carbonaceous mud of the depth about 2 m was intercalated with sands. It is found that it contains Chinese cypress old tree, covering on vermiculated laterite weathering granite [8, 33]. This implies that the strata contain not only the information of faulting but also the records of neo-tectonic movement, sea-level changes, and paleo-environmental changes in the north of the PRD region.

Xilin Hill is located in the east of Guangzhou–Conghua fault and the west of Baini–Shawan fault. The bottom of Xilin Hill is covered with the granites but the top is covered by the residues of red soil and Quaternary sedimentary soil. Meanwhile, the Quaternary is distributed in the slope southeast of Xilin Hill, in which the middle layer of soil layer deposition is greater than 3.5 m. Among them, the fourth layer was dating back to 40,000 years.

Xilin Hill fault with NE direction is referred to the cross section on the western mountain of Xilin Hill, which directly cuts the Quaternary sediments. It is found that the fault not only cuts Quaternary stratum but also that two sets of sedimentary layers of Quaternary are an angular unconformity (e.g., [33]). The fault strikes 30° and trends NW with the inclination of 72°. That is, a normal fault with the maximum distance of 53 cm.

According to the preliminary study, the strata can be divided into four parts from old to new as layers A, B, C, and D (e.g., [32, 33]) as given below:

Layer A: The grayish white sands are filled with the lower part of the layer. It goes up to the gray and black clay. This layer contains humus. The phenomenon of soil gasification is obvious. In the top of the layer, the limonite shells were developed intermittently. The depth of limonite shells varies from 1 to 2 mm. The whole depth of this layer is 50 cm or more. Moreover, the granites were vermiculated and undergo laterization.

Layer B: The grayish sand and brownish clay are alternated with the total thickness of about 4 m. The coarse sands and fine gravels are surrounded by black clays with the depth of 60 cm. The gray clay and coarse sand and fine gravels are alternated. The top is covered with the silty clays. Meanwhile, the top of orange-red coarse sands is covered with peat layers. There are some brick-red middle coarse sands in this layer which are not weathered. The thickness of each sub-layer varies from 30 to 40 cm.

Layer C: This layer is covered with poor bedded light orange coarse sands and fine gravels. The maximum diameter of the gravel is 0.5 m. However, the bedding is not developed in this area. The slope sediments are better developed.

Layer D: the fourth layer is modern sediments, dating back to 40,000 years.

The thickness of A, B, and C is more than 10 m. Layer A and layer B were tilted south-ward. However, the inclination angle is different. The inclination angle of the former one is about 30° but the latter one is about 4°. Although layer B is covered over layer A, the two layers show an angle unconformity (e.g., [8, 32, 33]). It is evident that a long hiatus occur-red between the two layers. Additionally, there are modern gullies and slope sediments, which embedded layers A and B. These three layers make up the pedestal terrace with the height about 22 m. As in Figure 4, there exist cutting and superimposed relationship between layers B, C, and D [33].

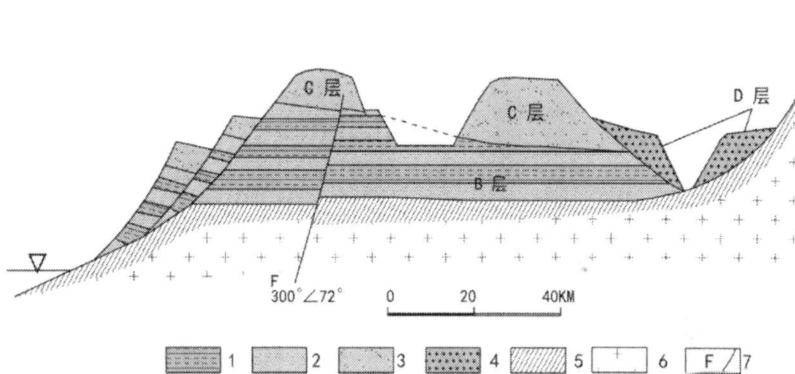

Figure 4. Section of the Late Quaternary pedestal terrace at the Xilin Hill in the north PRD (summarized and courtesy of [8, 33].

Figure 4 also shows the slope sediments of Holocene (layer D). The pedestal terrace is com-posed of the granites since Late Yanshanian period, making up of layers B and C. It is clear that layers A, B, and C are formed as the pedestal terrace in the elevation of 7 m. The height of the sediments of the pedestal terrace is about 22 m.

As seen in Figure 4, layer B is cut by a fault. The strike of fault is 30°, the inclination angle is 72°, and the tendency is northwest. The side structure suggested that there was some dis-placement as a result of dextrorotation, which is the dextral normal fault, as the upper side of the fault coasted down with several centimeters [8, 32, 33] as shown in Figure 5.

Here, the Late Quaternary pedestal terrace at Xilin Hill shown as in Figure 4 can be summarized and described as 1, clay layer; 2, sand layer; 3, sand and gravel layer; 4, gravel layer; 5, vermiculated laterite weathering crust; 6, granites of Late Yanshanian period; and 7, faults.

As shown in Figure 5, it is obvious that the layer B is cut by a fault. The fault is the eastern branch of the Guangzhou–Conghua fault, in which the upper plate slides down. When the fault incised upward to the layer C, it was bifurcated, with decreasing amplitude and occurrence changes. The period of the fault activities may be dating back to the Late-Pleistocene. According to the compared ages of vermiculated laterite weathering crust and other adjacent similar river terraces, the age of layer B should be dating back to the late Middle-Pleistocene to the early Late-Pleistocene [8, 32].

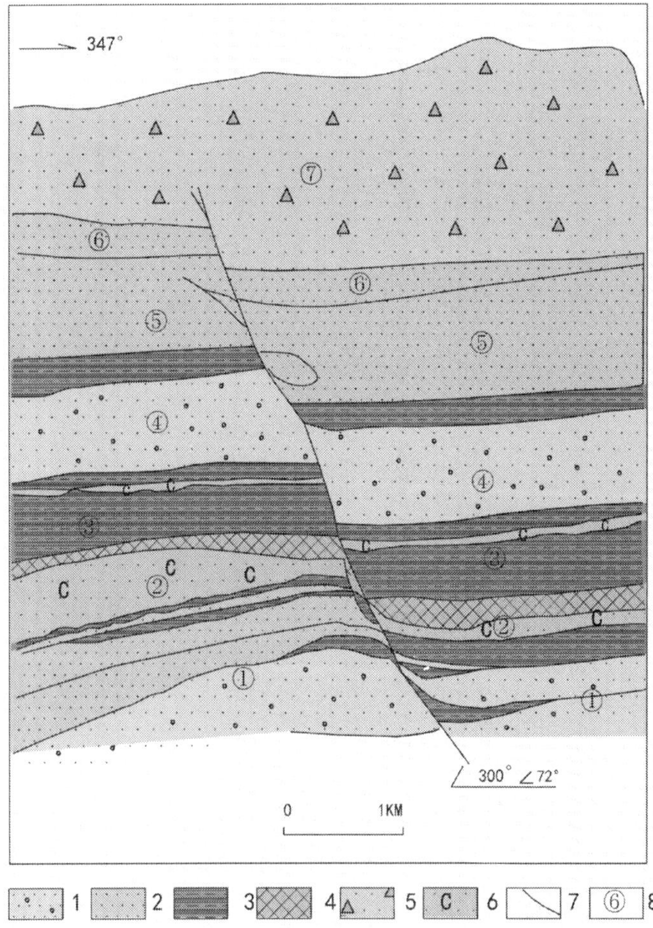

Figure 5. Sketch map of Pleistocene offset fault at the Xilin Hill in the northern PRD (summarized and courtesy of [8, 32].

There existed a fault with the direction nearly EW in the south of Xilin Hill, forming the obvious boundary between the hilly in the north and the PRD plain in the south. The peat mud layer of Holocene is found in the south of this fault. Gray thin layer of sands is covered with the black carbonaceous mud, which contains a large number of cork and Chinese cypress ancient trees and covers on the vermiculated laterite weathering granite. In the lower place, vermiculated laterite weathering crust appears [32, 33].

Here, the Late Quaternary pedestal terrace at Xilin Hill as shown in Figure 5 can be summarized and described as 1, sand gravel; 2-coarse sand; 3, clay; 4, silty clay; 5, slope sediments of sand and gravel; 6, carbonaceous mud; 7, fault; and 8, number of layers (1–6: layer B and 7: layer C).

The fault of Xilin Hill is within the range of maximum value of gravity anomaly in the PRD. The fracture dislocation and fault movement caused by the fault are connected by the deep faults. In general, there occurred transgressions twice in PRD since Pleistocene.

As the PRD is a fault block delta, the evolution of this delta accounts for several factors, including fault block movement and discrepant movement between the fault blocks and fault activities. Based on the development of vermiculated laterite weathering crust below the layer A, it is thought that the PRD region had been subjected to a long weathering before the deposition of Quaternary. Moreover, layers B and D may correspond to the first and the second transgressions, respectively. The entire evolutionary process is recorded perfectly from layer A to layer D [32]. Therefore, three sedimentary layers above the pedestal terrace not only contain the information of fault activities but also reflect the significant relationship among the changes of sedimentary facies, the migration of ancient rivers, and neo-tectonic movements. The important information mentioned above provides the basis for studying the formation of the PRD. In addition, it plays an important role in civil engineering construction, urban planning, and geo-hazard prevention in the further study.

4.4. The PRD sedimentary environment

Sedimentary facies are the sediment bodies combined with a certain characteristic of lithology, texture, structure, and paleontology signs, reflecting the depositional environment at that time [5, 10, 13, 19, 26]. In general, the sedimentary facies are a comprehensive reflection of the sediments and their sedimentary environment. The sedimentary environment not only controls the texture, structure, and compositions of sediments, but also affects the distribution and combination of fossils of plants.

Thus, the study of sedimentary facies is the foundation of recovering the sedimentary environment, and is also an important base for stratigraphic divisions. The sequence of the PRD sedimentary facies is divided into four major environments (e.g., [19]), namely delta plain, delta front, tidal estuary, and shallow-sea shelf (Figure 6).

As shown in Figure 6, the PRD sedimentary environment can be summarized and described as follows:

1. I delta plain

Delta plain (I) is the part of land, which is the district between the trend line of dry seasons and modern shoreline. There are several platforms and more than 160 monadnocks in the Pearl River Delta plain. The river network developed well and branches of water channels were interwoven into the network. There are more than a hundred of water channels with the length

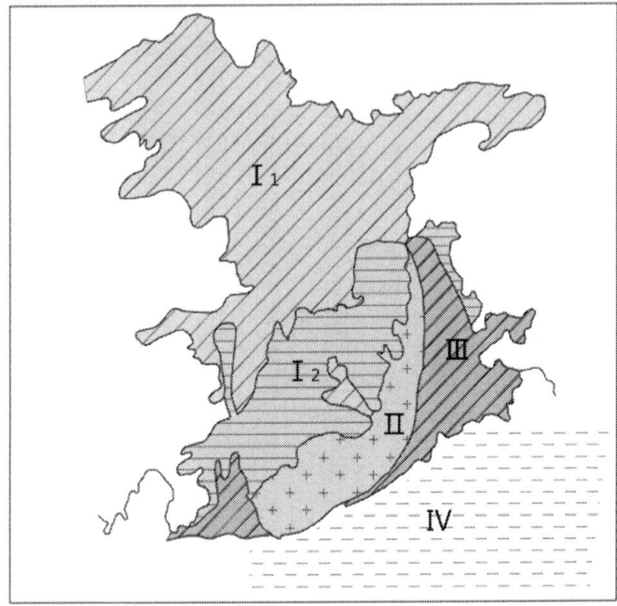

Figure 6. Sketch maps of recent sedimentary environments of the PRD (I delta plain, II delta front, III tidal estuary, IV shallow-sea shelf) (courtesy of [10, 19].

of about 1700 km. Meanwhile, the delta plain can also be divided into two parts: the upper delta plain (I_2) and the lower delta plain (I_1). The upper delta includes distributary channel, distributary depression, and floodplain, while the lower delta includes distributary channel and floodplain.

2.   II delta front

Delta front (II) is usually formed as a result of the interaction of sea and river, including estuary dam, natural levee, front slope, distributary gulf, and shallow ford. In this area, therefore, the activity of the deposition is the most, and further on, the river sediments shall be once more superposed and distributed to cause more complicated types of sediments and sedimentary structures.

3.   III tidal estuary

Tidal estuary (III) is mainly influenced by the tidal action. The related depositions are tidal channel, tidal dam, tidal shallow ford, lower tidal plateau, and upper tidal plain.

4.   IV shallow-sea shelf

Shallow-sea shelf (IV) is distributed in the 30 m depth of estuary, which are rich in organisms of shallow sea and mollusks. In addition, the content of organic matter is very high (about

1.5%). The sediment is gray silty clay in the northern part. But due to the wave erosion, these sediments are being coarsened to be mixed with sand, silt, clay, and/or fine sand [10, 19].

5. Conclusions

In this chapter, we present the Late Quaternary neo-tectonic movement in the Pearl River Delta region of China and its effects on the coastal sedimentary environment. By the example of the Xilin Hill fault, we gave the reason of the PRD formation, inferred its interactions with local/ regional climate and sea-level changes, and discussed the evolution of the PRD formation and its controlling factors of the coastal sedimentary environment [8, 10, 19, 32]. Thus, some conclusions can be summarized as below:

1. According to the analysis of sedimentary facies and carbon dating, the PRD evolution process since Late-Pleistocene (about 40 ka B.P.) can be divided into three stages called as pre-deltaic (40–32.5 ka B.P.), paleo-deltaic (32.5–7.5 ka B.P.), and neo-deltaic (7.5 ka B.P. to present).

2. Quaternary sedimentary layers of the PRD region in the vertical stacking sequence are such a positive sequence that the lower is coarse-grained, while the upper and middle is fine-grained.

3. Based on the study of the modern sedimentary environment, the Pearl River runoff and estuarine current interact to promote the formation and development of the delta.

4. Due to the different activities of the fault zones since Quaternary, the intensity of the seven faults is different from each other. The activity of South China Sea fault is the greatest, while Xijiang fault, Beijiang fault, and estuary fault are posterior and also weak (e.g., [32]).

5. The neo-tectonic movement plays a significant role during the evolution of the PRD. It not only controlled the changes of river channels but also influenced the sediments of delta. The two stages of gravel stratum with fluvial facies are represented by the location of the ancient valley before the two transgressions. However, the difference of distribution of gravel stratum with fluvial facies could reflect that the changes of river channels between the first transgression and the second one as a result of the fault activities.

6. The Xilin Hill fault revealed many important and basic issues in the changes of Late Quaternary environments, neo-tectonics, and fault activities (e.g., [32]).

7. There are several controlling factors for the PRD formation, including sedimentary evolution, the Quaternary ice age, sea-level changes, and human activities.

In the near future, more investigations will be conducted for the further study of the Late Quaternary neo-tectonic movement in the PRD region of China and assess its effects on the coastal sedimentary environment.

Acknowledgements

This research is jointly supported by the "2015 Jiangsu Program for Innovation Research and Entrepreneurship Groups, China", the National Science Foundation of China (41271434), and the Priority Academic Program Development of Jiangsu Higher Education Institutions (PAPD).

Author details

Yuanzhi Zhang[1,2*], Zhaojun Huang[2], Yu Li[2], Qing Yu[2], Bing Wang[2] and Tingchen Jiang[3]

*Address all correspondence to: yuanzhizhang@cuhk.edu.hk

1 School of Marine Sciences & Jiangsu Research Center for Ocean Survey Technology, Nanjing University of Information Science and Technology, Nanjing, Jiangsu Province, China

2 Center for Housing Innovations, the Chinese University of Hong Kong, Shatin, Hong Kong, China

3 School of Geodesy & Geomatics Engineering, Huaihai Institute of Technology, Lianyungang, Jiangsu Province, China

References

[1] Bird, E., 2008. *Coastal Geomorphology: An Introduction*, 2nd Ed., ISBN-9780470517291, Wiley, New York, pp. 436.

[2] Chen, W., Zhang, H., Zhang, F., 1991. Chronological study on neo-tectonic movement in the Zhujiang (Pearl River) Delta area, *Seismology and Geology*, 13(2), 213–219.

[3] Chen, W., 1998. Quaternary geology and seismic risk assessment in the Pearl River Delta and its vicinity area, *South China Journal of Seismology*, 18(2), 58–62.

[4] Chen, W., Zhao, H., Chang, Y., Lu, B., 2001. The rate of late Quaternary vertical motion of the Zhujiang (Pearl River) Delta, *Seismology and Geology*, 23(4), 527–536.

[5] Chen, G., Zhang, K., Chen, F., 1994. Sedimentary paleo-geography of late Pleistocene in the Pearl River Delta, *Quaternary Research*, 1, 67–74 (in Chinese).

[6] Chen, G., Zhang, K., Chen, H., He, X., 1995. Research on latest reactivity of faults in the Pearl River Delta, *South China Journal of Seismology*. 15(3), 16–21 (in Chinese).

[7] Chen, G., Zhang, K., Li, L., 2002. Development of the Pearl River Delta in SE China and its relations to reactivation of basement faults, *Journal of Geosciences of China*, 14(1), 17–24 (in Chinese).

[8] Dong, H., Huang, C., Zeng, M., Chen, W., 2012. Feature and formation mechanism of the Quaternary fault plane at Xilingang, Foshan City, *Seismology and Geology*, 34(2), 313–324 (in Chinese).

[9] Driese, S.G., Li, Z.H., Horn, S.P., 2005. Late Pleistocene and Holocene climate and geomorphic histories as interpreted from a 23,000 ^{14}C yr B.P. paleosol and floodplain soils, southeastern West Virginia, USA, *Quaternary Research*, 63(2), 136–149.

[10] Gao, F., 2006, Environmental evolution of the Pearl River Delta in Late Quaternary, Master Thesis, Zhongshan University, pp. 91.

[11] Huang, Z., Zong, Y., 1982. Grain size parameters and sedimentary facies of the Quaternary sequences in the Pearl River Delta, China, *Tropical Geography*, 2, 82–90 (in Chinese).

[12] Huang, Y., Xia, F., Chen, G., 1983. The fault controlling function in the formation of the Pearl River Delta, *Journal of Oceanography*, 5(3), 316–327 (in Chinese).

[13] Huang, Z., Li, P., Zhang, Z., Li, K., 1985. Characteristics of the Quaternary deposits in the Pearl River Delta, *Geological Review*, 31(2), 159–164 (in Chinese).

[14] Huang, Z., Li, P., Zhang, Z., Zong, Y., 1986. Sea-level change since the Late Pleistocene along the south coast of China. In: Qin, Y., Zhao, S. (Eds.), *China Sea Level Changes*. China Ocean Press, Beijing, pp. 178–194 (in Chinese).

[15] Huang, Z., Li, P., Zhang, Z., Li, K., 1987. The geomorphological evolution of the Pearl River Delta. In: Gardiner, V. (Ed.), *International Geomorphology 1986 Part I*. Wiley & Sons Ltd, New York, pp. 989–997.

[16] Li, P., Huang, Z., Zhang, Z., Li, K., 1984. The Quaternary strata of the Pearl River Delta, Geographic Sciences, 2, 1–6 (in Chinese).

[17] Li, P., Qiao, P., Zheng, H., Fang, G., Huang, G., 1990. *The Environmental Evolution of the Pearl River Delta in the Last 10,000 Years* (in Chinese). China Ocean Press, Beijing, pp. 154.

[18] Long, Y., Huo, C., 1990. Sedimentary characteristics of late Quaternary in the Pearl River Delta, *Marine Sciences*, 4, 7–13 (in Chinese).

[19] Ma, D., Xu, M., Zhou, Q., Zhang, G., Lan, X., 1988. Sedimentary facies sequences of the Pearl River Delta, *Marine Geology & Quaternary Geology*, 8(1), 43–53.

[20] Marghany, M., Sabu, Z., Hashim, M., 2010. Mapping coastal geomorphology changes using synthetic aperture radar data, *International Journal of Physical Science*, 5(12), 1890–1896.

[21] Pedoja, K., Shen, J. W., Kershaw, S. et al., 2008. Coastal Quaternary morphologies on the northern coast of the South China Sea, China, and their implications for current tectonic models: a review and preliminary study, *Marine Geology*, 255, 103–117.

[22] Siddall, M., Rohling, E.J., Thompson, W.G.., Waelbroeck, C., 2008. Marine Isotope stage 3 sea level fluctuations: data synthesis and new outlook, *Reviews of Geophysics*, 46, RG4003, 1–29.

[23] Smith, L.C., 2002. Emerging applications of interferometric synthetic aperture radar (InSAR) in geomorphology and hydrology, *Annals of the Association of American Geographers*, 92(3), 385–398.

[24] Song, F., Wang, Y., Li, C., Chen, W., Huang, R, Zhao, H., 2003. Late Quaternary vertical dislocation rate on several faults in the Pearl River Delta area, *Seismology and Geology*, 25 (2), 203–210.

[25] Stuiver, M., Reimer, P.J., Braziunas, T.F., 1998. High-precision radiocarbon age calibration for terrestrial and marine samples, *Radiocarbon*, 40, 1127–1151.

[26] Wang, J., Cao, L., Wang, X., Yang, X., Yang, J., Su, Z., 2009. Evolution of sedimentary facies and paleo-environment during the late Quaternary in Wanqingsha area of the Pearl

[27] River Delta, *Marine Geology & Quaternary Geology*, 29(6), 35–41.

[28] Xu, N., Jiang, Y., Jia, J., Wang, J., Liu, H., 2005. Crustal stable zones and its characteristics in the Pearl River Delta, *Geological Hazards & Environmental Protection*, 16, 395–399.

[29] Yim, W.W.-S., 1999. Radiocarbon dating and the reconstruction of late Quaternary sea-level changes, *Quaternary International*, 55, 77–91.

[30] Zhang, H., 1980. Fault-block delta, *Journal of Geography*, 1, 18–30 (in Chinese).

[31] Zhang, H., 1983. Fault function in the formation and development of the Pearl River Delta, *Journal of Oceanography*, 2, 202–221 (in Chinese).

[32] Zhang, K., Chen, G., Zhuang, W., Peng, Z., Hou, W. 2009. New evidences for late Quaternary tectonic movement in the northern Pearl River Delta, *South China Journal of Seismology*, 29, 22–26 (in Chinese).

[33] Zhang, X., 2009. Studies of Guangzhou-Conghua fault zone and its activities, Master Thesis, Zhongshan University, pp. 86.

[34] Zhong, J., Zhan, W., 1996. Analysis for neo-tectonic movement and its crustal stability in the Pearl River Delta, *South China Journal of Seismology*, 16, 57–63 (in Chinese).

[35] Zong, Y., Lloyd, J.M., Leng, M.J., Yim, W.W.-S., Huang, G., 2006. Reconstruction of Holocene monsoon history from the Pearl River estuary, southern China, using diatoms and carbon isotope ratios, *The Holocene*, 16, 1–13.

[36] Zong, Y., Huang, G., Switzer, A.D., Yu, F., Yim, W.W.-S., 2009. An evolutionary model for the Holocene formation of the Pearl River delta, China, *The Holocene*, 19, 129–142.

Different Aspects of the Coastal Pollution

Review of Mercury Circulation Changes in the Coastal Zone of Southern Baltic Sea

Magdalena Bełdowska

Additional information is available at the end of the chapter

http://dx.doi.org/10.5772/61991

Abstract

Despite its undoubted usability, mercury (Hg) is the most toxic metal and one of the most toxic elements. The problem of mercury toxicity was only widely explored in the second half of the 20th century, following cases of fatal poisonings as a result of the consumption of contaminated fish and grains preserved with mercury compounds. According to HEL-COM reports, Hg emission in the Baltic region at the beginning of the 21st century was lower than during the 1980s. In addition to mercury transformation, climate warming, particularly in the autumn-winter season, is another factor contributing to the changes in mercury circulation, especially in the area of land-sea contact. The increase in rainfall, particularly in the summer, is of particular importance for the marine environment. This is related to an increased inflow of Hg with wet precipitation, but the warm season is also favourable for intensive growth of sea organisms and, consequently, a faster accumulation of chemical substances, including toxic ones. As a result, the concentration of mercury in organism biomass increases.

Keywords: mercury, aerosols, precipitation, Hg input, macrophyta, coastal zone

1. Introduction

One of the coastal pollution sources is an oil spill, such as the Gulf of Mexico disaster [1]. However, coastal land use has a large impact on the quality of seawater. One coastal chemical pollution is mercury (Hg). Despite its undoubted usability, mercury (Hg) is the most toxic metal and one of the most toxic elements. It is as yet unknown whether there are any beneficial functions performed by mercury in live organisms. It is, however, proven to have neurotoxic, nephrotoxic, immunotoxic, mutagenic, embryotoxic, and allergising properties [2, 3]. It is one of the factors causing a number of disorders, such as Alzheimer's disease, Parkinson's disease, autism, amyotrophic lateral sclerosis, or lupus [2, 4–7]. The main route of mercury introduction to the

human system is the consumption of fish and seafood. Owing to the fact that mercury is capable of penetrating quickly through the placenta (foetus barrier), some countries have introduced recommendations against the consumption of predatory fish by pregnant women [3].

The problem of mercury toxicity was only widely explored in the second half of the 20th century, following cases of fatal poisonings as a result of the consumption of contaminated fish [8] and grains preserved with mercury compounds [9]. As a consequence, Hg emission and deposition has been monitored in many regions of the world. In the Baltic region, the role has been performed by the Baltic Marine Environment Protection Commission, also known as the Helsinki Commission or HELCOM. The important pathway of Hg transport is the atmosphere, but in coastal zones, rivers represent the main source of the metal. According to HELCOM reports [10, 11], Hg emission in the Baltic region at the beginning of the 21st century was about 20–30% lower than during the 1980s. Climate warming, particularly in the autumn-winter season [12], is another factor contributing to the decline in Hg emission from the main source, the burning of fossil fuels, into the coastal zone of the southern Baltic. On the other hand, in the seaside air rich in halides, there are conducive conditions for a quick transformation of gaseous mercury to the aerosol form, which causes Hg to be removed faster to the ground. In the zone where marine and terrestrial air masses meet, this process is aided by high humidity. This results in larger aggregates being formed, which drop out of the atmosphere faster [13–15]. During warm autumn and winter months phytoplankton blooms can occur [16], leading to the bioaccumulation of mercury from atmospheric deposition, the inflow of which is always higher in the heating season as opposed to the warm part of the year [17]. In such cases, mercury compounds do not drop to the bottom of the water basin but are accumulated by phytoplankton and thus included in the trophic chain [18]. Climate changes in the coastal zone of the southern Baltic point to the shortening of the icing period [19], which affects the circulation of mercury between near-bottom water, sediments, and benthic organisms [20]. The reason for this is the prolongation of the macrophytobenthic vegetative period and zoobenthic activity, which can, in turn, increase the absorption of pollutants from near-bottom water and porewater and intensify the mixing of surface sediments. In recent years, the intensification of extreme natural phenomena has also been observed [12]. More frequent precipitation and floods can contribute to the reemission and remobilisation of mercury transported into the sea from land. The above observations led to the formulation of the following research goals: (I) to determine the factors that influence mercury concentration in aerosols, precipitation, and rivers in the southern Baltic region; (II) to determine the factors that influence the degree of Hg inflow via atmospheric precipitation and rivers into the southern Baltic; (III) to assess the role of macrophytobenthos as a carrier of mercury, transported to the sea contemporarily as well as in the past, into the trophic chain; and (IV) to determine the tendencies in the changes of Hg circulation in the marine environment under the influence of climate changes occurring in the southern Baltic region.

2. Materials and methods

The research was conducted at the coastal zone of Gulf of Gdańsk (southern Baltic Sea) (Figure 1) during 2005-2013. The solid samples were analysed by means of an advanced mercury analyser AMA 254 thermal desorption analyser. The detection limit for solid materials was 0.005 ngg^{-1}. Water samples for mercury analysis were oxidised by the addition of BrCl and pre-reduced with hydroxylamine hydrochloride solution 1 h prior to analysis by CVAFS (TEKRAN 2600), according to US EPA method 1631 [21]. Quality control procedures for water samples included blanks and water spiked with mercury nitrate in the range of 0.5–25 ng/L and produced adequate precision (1% RSD) and recovery (98%–99%). The detection limit was as low as 0.05 ng/L. More details about sampling stations, collections, and chemical analysis are in publications [22–28].

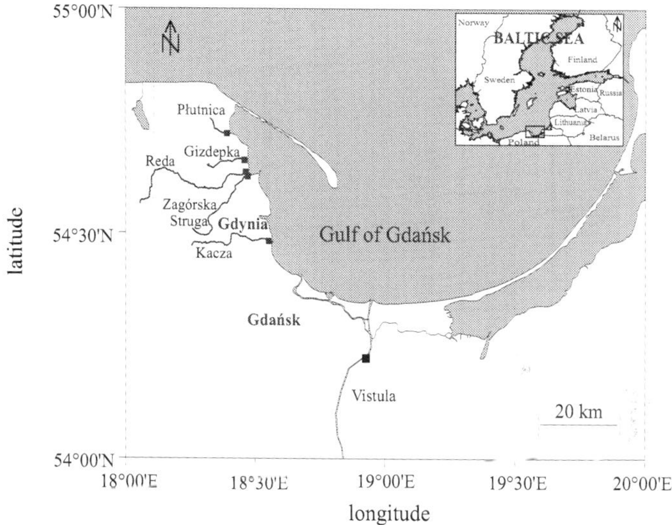

Figure 1. Map of the study area.

3. Results

The research carried out during 2005–2013 has shown that, in the Polish coastal zone, mercury concentrations both in aerosols and in rains [23, 24] were comparable to the values obtained at other stations within the Baltic [29, 30, 32, 33]. The mean annual Hg concentration in aerosols (Hgp) (Figure 2) was 1% of total Hg concentration in the air, which is typical of unpolluted areas [22, 31].

In the six studied rivers (Vistula, Reda, Zagórska Struga, Płutnica, Kacza River, and Gizdepka) terminating in the Baltic in the region of the Polish coast, the mean annual Hg concentrations were similar to the value of 5 ng Hg dm^{-3}, which is considered to be the global mean concentration in watercourses. Annual medians were lower than 7 ng Hg dm^{-3} (Figure 2) [24, 25].

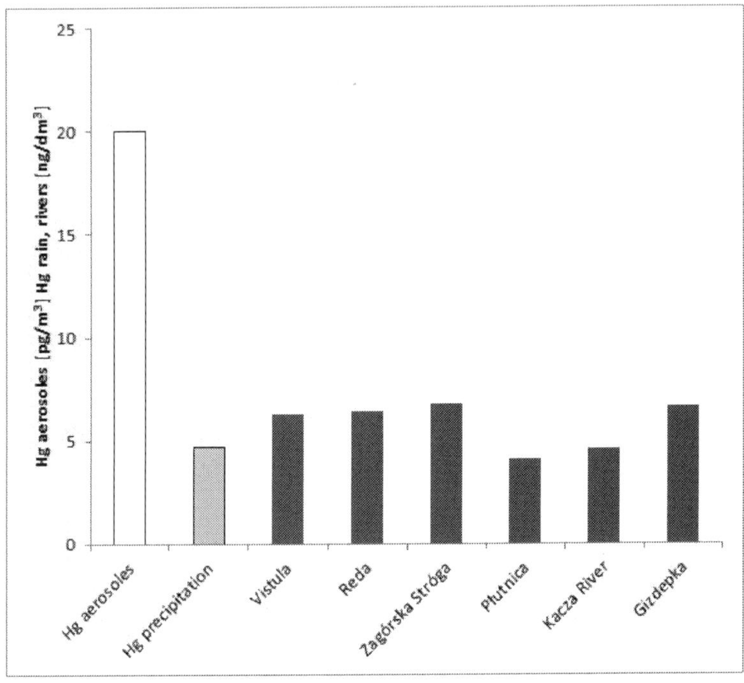

Figure 2. Hg median concentration in aerosols (pg g⁻¹); precipitation (ng dm⁻³) and in rivers (ng dm⁻³).

4. Discussion

4.1. Changes of Hg concentration in the atmosphere and rivers of the southern Baltic region

Hg concentrations in the air and rivers varied in the particular months of the year [22–25, 28]. In the region of the Polish coastal zone of the Baltic, at the station in Gdynia, the Hg concentration in large aerosols was higher than in small particles (on average, 93% of the total Hg in aerosols) [22]. The reverse was observed in air masses that had been transported from distant polar and subpolar regions high above the ground, particularly from over volcanoes in Iceland. A significant source of Hg in the southern Baltic region is the combustion of fossil fuels. Therefore, Hg concentrations in aerosols, both large and small, were higher in terms of statistical significance in the heating season than in the non-heating season (Figure 3). Moreover, in the heating season, Hg concentration in both large and small aerosols was lower in the maritime air masses than in continental air masses (Figure 3). This is related to the intensive combustion of fossil fuels for heating purposes in this season. The research showed a significant role of individual home furnaces, which are relatively numerous in and around the Tricity, in the formation of mercury concentrations in aerosols and rains [22, 23]. Hg concentration in rains in both seasons (heating and non-heating) was not found to be different in terms of statistical significance [23]. In the coastal zone of the sea, a place where the polluted continental air masses meet the cleaner, humid and halogen-rich marine air masses, an important factor in forming the volume of mercury concentration, both in aerosols and in rains, was found to be that of Hg transformation. In such conditions, gaseous mercury Hg(0) can be oxidised to Hg(II) and adsorbed on condensation nuclei, which, in consequence, can lead to an increase of Hg concentration in rains and in large aerosols in marine sea masses, particularly in the warm season [22, 23]. These are significant processes, particularly in the warm season, when an increased emission of gaseous mercury from the sea into the air is observed [34]. As a consequence, in the non-heating season, Hg concentration in large aerosols was higher than in the maritime air masses compared to continental air masses [22], and Hg concentration in rains was the highest in summer (Figure 3) [23]. Additionally, in summer, mercury in dry air evaporated off the aerosol surface, which led to a drop in Hg(p) concentration in terrestrial air masses [22].

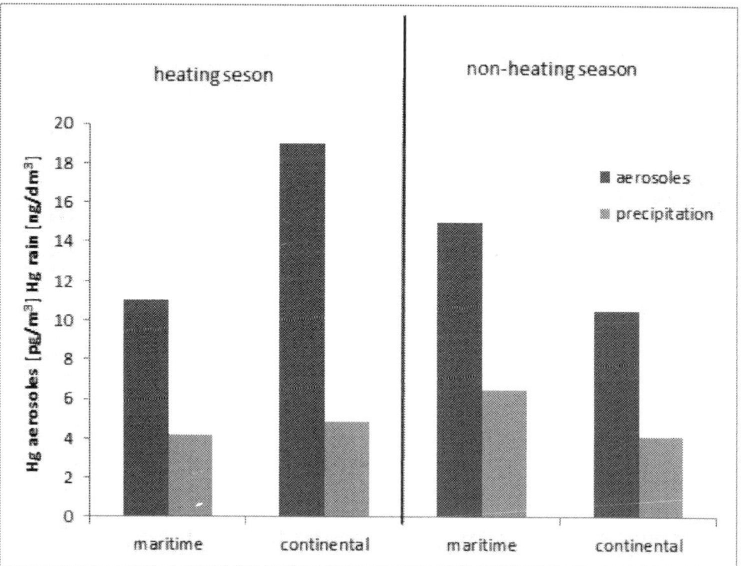

Figure 3. Median Hg concentrations in aerosols (pg/m³) and in precipitation (mg/dm³) in maritime and continental air masses during the heating and non-heating seasons.

In rivers flowing into the Baltic along the Polish coast, Hg concentration was much less directly dependent on meteorological parameters than was the case with aerosols and rains. The level of mercury concentration in rivers was indirectly related to the volume and height of precipitation [24]. Depending on the type of catchment and the length and height of wet precipitation, Hg concentration was either diluted in a river or increased as a result of the metal being washed off land. An increase in mercury concentration was observed in sections of rivers flowing through urbanised areas. The washing off of pollutants from street and pavement surfaces, pitches, or car parks as a Hg source into rivers is also confirmed by increased metal concentrations in water from storm sewers collected both during and after rain. Hg remobilisation from agricultural areas also increased Hg concentration in rivers compared to the source section. In the 20th century, mercury compounds were commonly used in agriculture as fungicides, which is why such areas are currently a potential source of this metal into rivers and the sea [24].

4.2. Factors influencing Hg inflow into the southern Baltic

The inflow of mercury into the southern Baltic changed over the course of a year [25]. The size of Hg load introduced with dry precipitation mainly depended on air temperature, particularly in the heating season. In that period, dry deposition of Hg depended mainly on Hg concentration in aerosols in both marine and terrestrial air masses. Wet Hg deposition, on the other hand, increased with the height of precipitation. In the non-heating season, the size of Hg load introduced in this way depended on both the height of precipitation and the Hg concentration in rains and, analogously, that introduced with dry precipitation: on the concentration of Hg in aerosols and the deposition velocity [25].

Taking into account the average wind speed at the sampling station and the trajectories of air masses, the areas were defined from which mercury was brought to coastal zone: $Vw < 1$ m s^{-1} local sources; $1 \leq Vw < 3$ m s^{-1} regional sources; and $Vw \geq$ m s^{-1} remote sources [25]. There is a possible tendency of increasing Hg inflow from the atmosphere, which may reveal itself in the following conditions: in the heating season, the inflow of atmospheric Hg will increase with the number of cases where continental air masses from regional and distant sources, with the predominance of coal burning in home furnaces, flow over the coastal zone of the southern Baltic and with the increasing incidence of rains from marine air masses from distant sources (Figure 4) [25].

The atmospheric Hg load will also increase as air temperature drops and the amount of burnt coal increases, as well as with a rise in the intensity of gaseous mercury transformation into the particulate form [22, 25]. Taking into account the non-heating season, it was estimated that the more frequent are the inflows of air masses from regional sources (both marine and terrestrial) and rains from continental regional air masses, the greater the load of mercury coming from the atmosphere will be (Figure 4) [23, 25].

Hg inflow via rivers also fluctuated in the particular months over the course of the study period [25]. A larger load of mercury was introduced in the cold season, when the metal penetrated into rivers directly from increased atmospheric deposition and with the washing out of pollutants from land — those deposited there both contemporarily and in the past. In addition, meltwater also contributed to the transportation of Hg from land into rivers and the sea. Rains were also a significant factor forming the inflow of Hg via rivers. Changes occurred in various directions depending on the type of catchment: intensive rains were either conducive to the washing out of Hg from land or contributed to the thinning down of the metal concentration in the river.

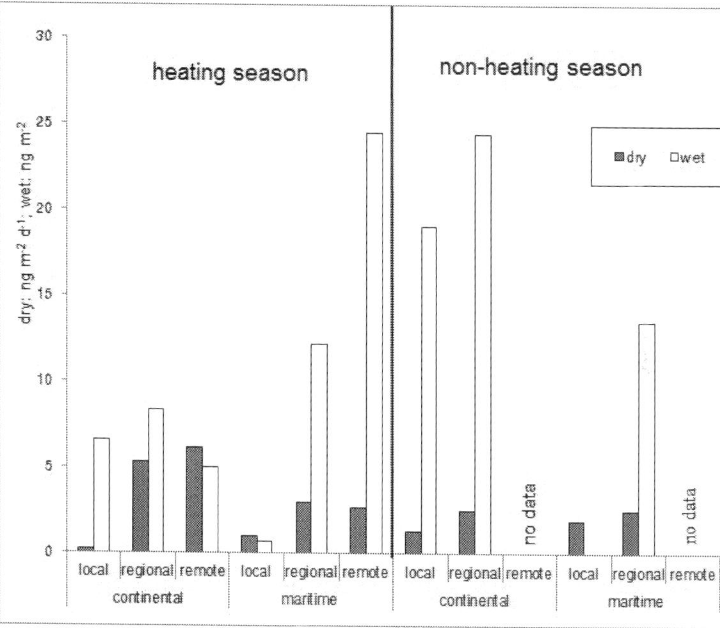

Figure 4. Atmospheric deposition (dry and wet) of Hg during the heating and non-heating seasons in maritime and continental air masses form local, regional, and remote sources.

4.3. Macrophytobenthos as a carrier of Hg into the trophic chain in the coastal zone of the southern Baltic

Mercury that reaches the sea is either included in the trophic chain through accumulation by phytoplankton or phytobenthos or becomes sedimented in the seabed, from where it can be reintroduced to the cycle. A good indicator of the size and changes of metal concentrations in water is macrophytobenthos [35, 36]. It follows from the research on Hg concentration in macrophytobenthos, which has been carried out since 2006, that over a long time period the changes in concentrations in the particular seasons are statistically insignificant [26]. This is related to the disappearance of stark differences in air and water temperatures between particular seasons (the warming of the autumn-winter-spring season, lack of icing, and rainy summer). It has, however, been observed that there are differences in Hg concentration in macrophytobenthos in the particular regions of the Polish Baltic coast [26]. Benthic flora in the coastal zone of the open sea was considerably less polluted with mercury than the material from the Pomerania Bay or the Gulf of Gdansk. The urbanisation of the coastal zone has contributed to the increase in Hg concentration in macroalgae. Surface run-off and the inflow of rainwater (e.g. from storm sewers), to the coastal zone of the gulf, have led to an increase in mercury concentration in macroalgae there compared to regions located far away from those sources. On the other hand, the inflow of pollutants from river catchment areas contributed to an increase in the concentration of the metal in marine vascular plants. The suspension carried via rivers into the sea, depending on the type of the outlet, was deposited close to the shore to a smaller (e.g. Vistula outlet) or greater degree (e.g. Oder outlet), becoming an Hg source into porewater and, consequently, into vascular plants. The capability of vascular plants to accumulate chemical substances from porewater also contributed to the absorption of mercury that had been deposited in sediments in the past. As a consequence, *Potamogeton pectinatus* (a vascular plant) had a 60% higher bioconcentration coefficient than *Furcellaria lumbricalis* (a macroalga) [26].

4.4. Possible influence of climate changes on the circulation of Hg in the coastal zone of the southern Baltic

In recent years, an intensification of extreme natural phenomena has been observed [12, 37]. Intensive rains and floods cause increased the washing off of chemical substances from land, which, in consequence, contributes to the inclusion in the river course of mercury deposited for decades. This is a highly significant problem, as calculations have shown that currently about 50% of mercury deposited into the bottom sediments of the Gdansk Basin becomes reintroduced to the water column. If more mercury flows into sediments (e.g. with surface run-off as a result of intensive rains), the streams of mercury emission from sediments may increase by a few to more than 10% above the difference resulting from the change in the inflow [38]. Thus, the bottom sediment will increase its share as a source of Hg into the marine ecosystem.

During the largest flood (2010) on the River Vistula, within a relatively short time (31 days), 1.2 tonnes of Hg penetrated into the Baltic, amounting to 75% of the annual load in 2010 [27]. Such a large mass of mercury introduced was caused not only by the large volume of water (12 km^3, which amounted to 21% of the yearly flow of the Vistula) but also by a several-fold increase in the Hg concentration compared to the time before the flood (Figure 5). During the culmination of the first flood wave, the concentration of total mercury (Hg tot) in the Vistula was over 200 ng dm^{-3}, whereas the Hg concentration median from before the flood was 6.3 ng dm^{-3}. That was caused by the washing out of pollutants, deposited over years, from the soil, as well as from cemeteries, landfills, farmyards, waste depositories, etc. The high water level, persisting for a long time, apart from washing of Hg out of land, contributed to the deposition of suspension (enriched with mercury) close to and in the riverbed. This is of considerable significance as, under the influence of another rise in the water level, either due to a flood, flooding, or seasonal meltdown, it is very likely that mercury, in the form of methylmercury (the most toxic form), will be released from the bed and transported into the sea [27, 39, 40]. Such a situation was observed in July and September 2010 [25].

The mercury introduced into the coastal zone of the sea during the flood in 2010 was adsorbed by phytoplankton, contributing to a fourfold increase in its concentration in algae in the period between two flood waves reaching the Gdansk Basin, compared to the period before and after the flood [27]. Mercury bound to the fine sediment fraction, particulate suspended matter, and in its dissolved form was transported by the strong flood current from the river mouth far into the southern Baltic. As a consequence, the concentration of Hg in the sediments of the southern

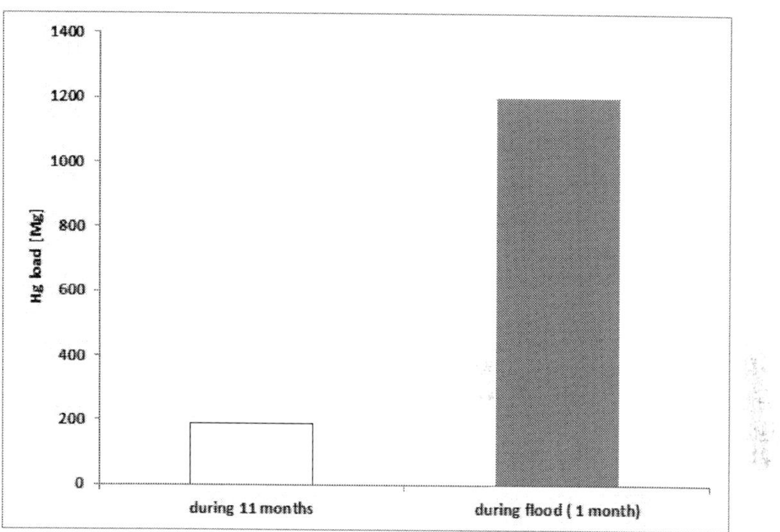

Figure 5. Hg load introduced by Vistula into Baltic seawater during 1 month of flood and during the rest months.

Baltic following the flood increased by as much as 500% compared to the period before the flood and, in relation to the fine fraction content, by as much as 1500%. This state persisted for nearly 2 years [27].

The mercury concentrations measured during the flood both in the Vistula and in the Gulf of Gdansk exceeded the safe values for water organisms. Hg concentrations in herring measured in 2011 by Woroń and Danowska [41] were the highest and were found to be 67% higher than in the period 1998–2010, while, in mussels, they were 60% higher than in that period [27].

According to HELCOM reports, mercury inflow into the Baltic has decreased since the 1990s, thanks to industrial modernisation [11]. Moreover, climate warming [12] has been contributing to the decrease in the proportion of the main Hg source in the Polish coastal zone (i.e. fossil fuel burning), which reduces the inflow of mercury with atmospheric deposition and rivers [25]. This is of particular importance in regions where emission from individual home furnaces is predominant. The inflow of Hg with dry deposition during winter, when subzero air temperatures were predominant, was eight times higher than in warm winter, abounding in temperatures above zero [25]. On the other hand, a warm winter, as opposed to a frosty one, is conducive to faster mercury transformation in the zone where the marine and terrestrial air masses meet, thus shortening the time of Hg residence and its transportation over long distances [22]. As a result of the lengthening of the warm season (winter starting later and ending earlier) [12], phytoplankton blooms are observed, particularly in the coastal zone, as early as the end of January or the beginning of February [16, 42, 43, 44]. In this situation, Hg from atmospheric deposition, the inflow of which by this route is always greater in the heating season than in the warm season [22, 25], does not drop to the bottom of the water basin but is accumulated by phytoplankton. As a consequence, Hg concentration in the phytoplankton of the Puck Bay was found to be higher in the heating season than in the non-heating season [28]. In this way, as a result of the warming of the winter season, taking into account the persistence of raised alga biomass, a higher Hg load is included in the trophic chain over the course of a year. During an anomal warm autumn (according to the rating introduced by the Institute of Meteorology and Water — the State Research Institute [37]), raised Hg concentrations were also observed in peryfitone covering stones [28]. Its mean concentration in that season was three times higher than during a thermally normal winter.

As a result of the warming of the winter season in the southern Baltic region, there are increasingly many winters with no icing even in the coastal zone of the gulfs, or with icing lasting much shorter [19], particularly compared to the years 1946–1991, when the average duration of the icing season in the inner Puck Bay was 90 days [45, 46]. The lack of icing is conducive to the thriving of macroalgae, which consequently prolongs the period of metal accumulation by fauna. On the one hand, the reduction of Hg emission into the environment has led to a drop in the concentration of the metal in macrophytobenthos of the Polish coastal zone of the southern Baltic [26]. On the other hand, the prolongation of the vegetative season and the improvement of environmental quality have led to an increasingly intense growth and coverage of the seabed by marine fauna. That has brought about the inclusion into the trophic chain of mercury deposited contemporarily as well as in the past. Seagrass beds are places where animal organisms thrive; they are also the consumers of phytobenthos, directly accumulating mercury. Thriving phytobenthos contributes to the faster inclusion of Hg into the trophic chain, and this is of particular importance in areas where the seagrass beds of *Zostera marina* are regrowing, as that is where the highest Hg concentrations were measured. In this way, while macrophytobenthos can be said to clear water and sediments, it also transfers Hg from sediments, which has relatively low bioaccessibility for higher organisms, to the higher trophic levels [26, 28].

High phytobenthos and zoobenthos biomass persisting during a warm autumn [37] caused the Hg mass contained in benthic organisms in that season to be five times higher than during a thermally normal winter [28, 37]. Owing to that, despite the decreasing Hg emission, the warming of the climate in the southern Baltic region is conducive to the prolongation of the period within which mercury is included in the food chain. Between 1999 and 2014, there were 12 instances in which autumn was observed to be above the thermal norm [37].

Climate changes are also observed in the summer. Increase of temperature led to enhanced emission/re-emission of gaseous Hg from seawater [34, 47]. In warm, marine air masses, there is intensive transformation of gaseous mercury into the particulate form, and rains occurring at that time lead to an increase in the Hg load introduced to the sea with wet precipitation [23]. In July and August, there is a higher incidence of intensive rains, reaching as much as 220% of the mean total precipitation from the period between 1971 and 2000 [37]. This leads to an increase of about 12% to the annual inflow of atmospheric Hg [28]. Intensive rains, particularly after a dry spell, were also favourable for the washing out of Hg from the catchment area, which also boosted the load of Hg introduced to the sea with rivers [24, 25]. This is of great significance, as at such a time phytoplankton biomass in the sea is normally high, presenting conditions conducive to the inclusion of a higher Hg load to the trophic chain, in comparison to a summer without anomalously intensive rains.

5. Conclusion

The presented phenomena and processes are especially significant for marine organisms that thrive in the coastal zone of bays. This is of particular importance, as these regions are attractive for tourists and consumers of fish, which are caught in these parts along with occasional seafood.

Owing to the high toxicity of mercury, the Baltic countries are obliged to reduce its emission into the environment. The main source of Hg is the burning of fossil fuels, which is why an increase in its inflow to the Baltic may be observed in the heating season compared to the non-heating season. Many restrictions concern the energy sector. The conducted studies showed that not only industrial coal combustion but also regional individual home furnaces have an important influence on the deposition of this metal. However, along with the adopted methods of mercury removal from fumes, an important factor influencing the emission of this metal is the warming of the cold season: late autumn-winter, early spring. Anomalously high air temperatures translate to less heating needs, which in turn reduces coal burning. On the other hand, the warming of the winter season prolongs the vegetative season of marine organisms, extending the period within which Hg is included in the food chain. As a consequence, despite the decreasing mercury emission into the natural environment, a larger load of the metal may be included in the marine trophic chain over the course of a year.

The decrease in Hg inflow to the Baltic is not the only result of climate changes. More intense rainfalls also lead to the washing out of Hg from both the atmosphere and from land. The conducted studies showed that the remobilisation and discharge of Hg occurs in most catchments of rivers flowing into the Gulf of Gdansk. This, to a large extent, is the effect of urbanisation. Tarmac-ing or pouring concrete over roads and pavements limits the retention of Hg in the soil. Moreover, intensive rains and floods wash out the metal from agricultural areas where it was widely used in the past as a fungicide. This is why the type of catchment ought to be taken into consideration while drawing up regulations concerning the reduction of Hg emission by particular states. The present studies were conducted in the drainage area of the Gulf of Gdansk, but analogous processes occur in other regions on the same latitude. The choice of suitable methods of drainage area management may significantly regulate the discharge of Hg into the sea.

Chemical substances that reached the sea with the extreme Vistula flood in 2010 were both deposited in the coastal zone and being carried far out into the Baltic, contributing to an increase in their concentrations in surface sediments. The research carried out after the flood indicated the need to monitor this process in sediments and, more importantly, in commer-cially attractive fish at least 2 years after the flood. Moreover, during a flood, a large load of polluted suspension drops to the riverbed or on the flooded areas, from where it can be washed off and transported to the sea during the next intensive floods or meltdowns.

The increase in rainfall, particularly in the summer, is of particular importance for the marine environment. This is related to an increased inflow of Hg with wet precipitation, but the warm season is also favourable for intensive growth of sea organisms and, consequently, a faster accumulation of chemical substances, including toxic ones. As a result, the concentration of mercury in organism biomass increases.

The conclusions described above should be taken into account while preparing regulations concerning Hg emission by particular countries. Studies have shown that an increase in mercury concentration in marine organisms is not always linked to an increased inflow of anthropogenic Hg, directly influenced by humans, but may also be related to climate changes occurring in a given area, the type of catchment, as well as to the predominance of either marine or terrestrial air masses.

The conducted studies show that even if Hg emission remains the same, its inflow into the Baltic will change as conditions, such as air temperature, number and intensity of rains, wind speed, and type of atmospheric circulation, change.

Author details

Magdalena Bełdowska*

Address all correspondence to: m.beldowska@ug.edu.pl

Institute of Oceanography, University of Gdansk, Gdynia, Poland

References

[1] Cheng Y., Li X., Xu Q., Garcia-Pineda O., Andersen O.B., Pichel W.G. SAR observation and model tracking of an oil spill event in coastal waters. Mar. Pollut. Bull. 2011;62(350–363). DOI: 10.1016/j.marpolbul.2010.10.005

[2] Zahir F., Rizwi S.J., Haq S.K., Khan R.H. Low dose mercury toxicity and human health. Environ. Toxicol. Pharmacol. 2005;20:351–360.

[3] Bose-O'Reilly S., McCarty K.M., Steckling N., Lettermeier B. Mercury exposure and children health. Curr. Prob. Pediatr. Adolesc. Health Care. 2010;40:186–215.

[4] Wermuth L., Koldkjur, O.G., Bjerregaard P. Parkinson's disease among Inuit in Greenland: Mercury and lead as risk factors. J. Neurol. Sci. 2005;238(1):375. DOI: 10.1016/S0022-510X(05)81449-2

[5] Johnson F.O., Atchison W.D. The role of environmental mercury, lead and pesticide exposure in development of amyotrophic lateral sclerosis. NeuroToxicology. 2009;30(5):761–765.

[6]   Salerno J.P. System lupus erythematosis and mercury toxicity. Personalised Med. Univ. 2015;1(1):81–83.

[7]   Yassa H.A. Autism: A form of lead and mercury toxicity. Environ. Toxicol. Pharmacol. 2014;38:1016–1024.

[8]   Takeuchi T., Kambara T., Morikawa N., Matsumoto H., Shiraishi Y., Ito O. Pathological observations of the Minamata disease. Acta Pathol. Jpn. 1959;9:769–776.

[9]   Rustam H., Hamdi T. Methyl mercury poisoning in Iraq. Brain. 1974;97:499–510.

[10] HELCOM. The fourth Baltic Sea pollution load compilation (PLC-4). Baltic Sea Environ. Proc. 2004;93

[11] HELCOM. Towards a tool for quantifying anthropogenic pressures and potential impacts on the Baltic Sea marine environment: A background document on the method, data and testing of the Baltic Sea pressure and impact indices. Baltic Sea Environ. Proc. 2010;125

[12] HELCOM. Climate change in the Baltic Sea area: HELCOM thematic assessment in 2013. Baltic Sea Environ. Proc. 2013;137

[13] Sommar J., Gårdfeldt K., Strömberg D., Fenga X. A kinetic study of the gas phase reaction between the hydroxyl radical and atomic mercury. Atmos. Environ. 1997;35:3049–3054.

[14] Hedgecock I.M., Pirrone N. Mercury and photochemistry in the marine boundary layer-modelling studies suggest the in situ production of reactive gas phase mercury. Atmos. Environ. 2001;35:3055–3062.

[15] Chand D., Jaffe D., Prestbo E., Swartzendruber P.C., Hafner W., Weiss-Penzias P., et al. Reactive and particulate mercury in the Asian marine boundary layer. Atmos. Environ. 2008;42:7988–7996.

[16] Wasmund N., Uhlig S. Phytoplankton trends in the Baltic Sea. J. Mar. Sci. 2003;60:177–186.

[17] Bełdowska M., Falkowska L., Siudek P., Gajecka A., Lewandowska A., Rybka A., Zgrundo A. Atmospheric mercury over the coastal zone of the Gulf of Gdańsk. Oceanol. Hydrobiol. Stud. 2007;36(3):9–18.

[18] Bełdowska M., Jędruch A., Bełdowski J., Szubska M., Kobos J., Mudrak-Cegiołka S., Graca B., Zgrundo A., Ziółkowska M., Kiełczewska J., Lewandowska E., Wasowska K., Falkowska L. Rtęć w strefie brzegowej Zatoki Puckiej (południowy Bałtyk). In: Falkowska L, editor. Rtęć w środowisku Identyfikacja zagrożeń dla zdrowia człowieka. Gdańsk: Wydawnictwo Uniwersytetu Gdańskiego; 2013. p. 97–103.

[19] Kożuchowski K. Contemporary climate warming in Poland. Papers Global Change. 2009;16:41–53.

[20] Bełdowska M., Jędruch A., Bełdowski J., Szubska M. Mercury concentration in the sediments as a function of changing climate in coastal zone of southern Baltic Sea — preliminary results. E3S Web Conf. 2013;1. DOI: 10.1051/e3sconf/2013016002

[21] USEPA (US Environmental Protection Agency). Method 1631, Revision E: Mercury in water by oxidation, purge and trap, and cold vapor atomic fluorescence spectrometry. US Environmental Protection Agency, Office of Water 4303. 2002;EPA-821-R-02-019:1-46.

[22] Bełdowska M., Saniewska D., Falkowska L., Lewandowska A. Mercury in particulate matter over the Polish zone of the southern Baltic Sea. Atmos. Environ. 2012;46:397–404. DOI: 10.1016/j.atmosenv.2011.09.046

[23] Saniewska D., Bełdowska M., Bełdowski J., Falkowska L. Mercury in precipitation at an urbanised coastal zone of the Baltic Sea (Poland). AMBIO. 2014;43(7):871–877. DOI: 10.1007/s13280-014-0494-y

[24] Saniewska D., Bełdowska M., Bełdowski J., Saniewski M., Szubska M., Romanowski A., Falkowska L. The impact of land use and season on the riverine transport of mercury into the marine coastal zone. Environ. Monitor. Assess. 2014;186(11):7593–7604. DOI: 10.1007/s10661-014-3950-z

[25] Bełdowska M., Saniewska D., Falkowska L. Factors influencing variability of mercury input to the southern Baltic Sea. Mar. Poll. Bull. 2014;86:283–290. DOI: 10.1016/j.marpolbul.2014.07.004

[26] Bełdowska M., Jędruch A., Słupkowska J., Saniewska D., Saniewski M. Macrophyta as a vector of contemporary and historical mercury from the marine environment to the trophic web. Environ. Sci. Pollut. Res. 2015;22(7):5228–5240. DOI: 10.1007/s11356-014-4003-4

[27] Saniewska D., Bełdowska M., Bełdowski, J., Jędruch A., Saniewski M., Falkowska L. Mercury loads into the sea associated with extreme flooding. Environ. Pollut. 2014;191:93–100. DOI: 10.1016/j.envpol.2014.04.003

[28] Bełdowska M. The influence of weather anomalies on mercury cycling in the marine coastal zone of the southern Baltic — future perspective. Water Air Soil Pollut. 2015;226:2248. DOI: 10.1007/s11270-014-2248-7

[29] Wängberg I., Munthea J., Pirroneb N, Iverfeldta Å., Bahlmanc E., et al. Atmospheric mercury distribution in northern Europe and in the Mediterranean region. Atmos. Environ. 2001;35(17):3019–3025.

[30] Wängberg I., Munthe J., Ebinghaus R., Gardfeidt K., Iverfeldt A., Sommar J. Distribution of TPM in northern Europe. Sci. Total Environ. 2003;304:53–59.

[31] Lamborg C.H., Fitzgerald W.F., Vandal G.M., Rolfhus R.R. Atmospheric mercury in northern Wisconsin: sources and species. Water Air Soil Pollut. 1995;80:198.

[32] Bartnicki J., Gusev A., Aas W., Valiyaveetil S. Atmospheric Supply of Nitrogen, Lead, Cadmium, Mercury and Dioxins/Furans to the Baltic Sea in 2008. EMEP Centers Joint Report for HELCOM EMEP/MSC-W Technical Report. 2010;2

[33] Bartnicki J., Gusev A., Aas W., Valiyaveetil S. Atmospheric Supply of Nitrogen, Lead, Cadmium, Mercury and Dioxins/Furans to the Baltic Sea in 2009. EMEP Centers Joint Report for HELCOM EMEP/MSC-W Technical Report. 2011;2

[34] Marks R., Bełdowska M. Air-sea exchange of mercury vapour over the Gulf of Gdańsk and southern Baltic Sea. J. Mar. Syst. 2001;27:315–324.

[35] Filipovic-Trajkovic R., Ilic S.Z., Sunic L., Andjelkovic S. The potential of different plant species for heavy metals accumulation and distribution. J. Food Agric. Environ. 2012;10:959–964.

[36] Stankovic S., Kalaba P., Stankovic A.P. Biota as toxic metal indicators. Environ. Chem. Lett. 2014;12:63–84. DOI: 10.1007/s10311-013-0430-6

[37] IMGW PIB (Institute of Meteorology and Water Management National Research Institute). Polish Climate Monitoring Bulletin [Internet]. 2015 [Updated: 2015]. Available from: http://www.imgw.pl/extcont/biuletyn_monitoringu

[38] Bełdowski J., Pempkowiak J., Miotk M. Mercury fluxes through the sediment water interface and bioavailability of mercury in southern Baltic Sea sediments. Oceanologia. 2009;51:263–285.

[39] Jackson T.A. Biological and environmental control of mercury accumulation by fish in lakes and reservoirs of northern Manitoba. Can. J. Fish. Aquat. Sci. 1991;48(12): 2449–2470. DOI: 10.1139/f91-287

[40] Heaven S., Ilyushchenko M.A., Kamberov I.M., Politikov M.I., Tanton T.W., Ullrich S.M., Yanine E.P. Mercury in the River Nura and its floodplain, Central Kazakhstan: II. Floodplain soils and riverbank silt deposits. Sci. Total Environ. 2000;260:45–55. DOI: 10.1016/S0048-9697(00)00566-0

[41] Woroń J., Danowska B. Metale ciężkie. In: Zalewska T., Jakusik E., Łysiak-Pastuszak E., Krzymiński W, editors. Bałtyk Południowy w 2010 roku. Charakterystyka wybranych elementów środowiska. Warszawa: Wydawnictwo IMGW-PIB; 2012. p. 158–160.

[42] Łysiak-Pastuszak E. Oxygen and nutrients in the southern Baltic Sea. Ocean. Stud. 1996;1–2:41–76.

[43] Witek B., Pliński M. The first recorded bloom of *Prorocentrum minimum* (Pavillard) Schiller in the coastal zone of the Gulf of Gdańsk. Oceanologia. 2000;42(1):29–36.

[44] Błaszczyk A., Kobos J., Hohlfeld N., Toruńska A., Hebel A., Mazur-Marzec H. First report of saxitoxin production by *Alexandrium ostenfeldii* in Puck Bay (southern Baltic

Sea). In: 32nd International Conference of Polish Phycologists. Do thermophilic species invasion threaten us? Book Abstr. 2013; 2013; p. 12.

[45] Girjatowicz J.P. Lodowe warunki Zatoki Gdańskiej. Przegląd Geograficzny. 1988;1–2:93–112.

[46] Szefler K. Zlodzenie. In: Korzeniewski K, editor. Zatoka Pucka. Gdańsk: Fundacja Rozwoju Uniwerystetu Gdańskiego; 1993.

[47] Bełdowska M., Zawalich K., Falkowska L., Siudek P., Magulski R. Total gaseous mercury in the area of southern Baltic and in the coastal zone of the Gulf of Gdansk, during spring and autumn. Environ. Protect. Eng. 2008;4

Computatonal Analysis of System and Design Parameters of Electrodeposition for Marine Applications

Gabriel A. Farotade, Patricia A. I. Popoola and Olawale M. Popoola

Additional information is available at the end of the chapter

http://dx.doi.org/10.5772/62334

Abstract

The financial prudence of the global world is shaken due to the vigor induced by corrosion as the degradation of essential assets, namely, oil and gas platforms and marine machineries, appears as a red-flagged situation. The exacerbation created by chemical degradation of these assets is as a result of the presence of saltwater, and the highly dominating activity of salt in the atmosphere poses a critical influence in selecting the mode of corrosion prevention to be integrated. As the hunger for longer term service and cost effectiveness of protection increases, studies are clustered in the search of new corrosion-resistant coatings while adhering to the increasing stringent environmental code of practice. Porosity is a key factor which is considered in the development of corrosion-resistant coatings, stimulating localized forms of corrosion, such as galvanic and crevice corrosion. Nevertheless, researches have been implemented to coin this challenge by acquiring full understanding of effective parameters of salt bath conditions, their interactions, and influence on the degree of uniform electrodeposition. A significant contribution is presented in the light of reviewing the possibility of computational analysis of system and design parameters in optimizing the deposition rate and quality.

Keywords: Corrosion, Porosity, Electrodeposition, System and design parameters, Designs of Experiments, Marine environments

1. Introduction

Corrosion has been a menace, for decades, to global development, especially in oil and gas assets, mostly in coastal and marine environments. Maintenance is a better option than replacement as this prevents loss of time and cuts cost. In order to improve the quality and minimize the time taken for the process of maintenance with the consideration of the fleeting environmental regulations, new anticorrosion coatings are being developed, giving cogni-

zance to improved service life, better protective performance, and improved mechanical properties in aggressive marine conditions. These coatings have a minimum thickness that gives tolerance for turbulence effect of seawater in order to withstand degradation. As a result of fluid turbulence on asset regions close to sea level, there is excessive deposition of salt content on the coating surface due to saltwater splashes; corrosion is of high level of proneness if pores are present in the coatings. Localized chemical attacks take place, resulting in catastrophic failure of oil transport vessels during service.

Porosity is an important issue to consider in examining the life expectancy and performance of coated structures as it may exist due to dissolution of prime elements from the coating into the environment. For instance, the effect of the dissolution may be the formation of galvanic cell between substrate and electroless nickel coating on steel immersed in an acidic environment; hydrogen evolution takes place, increasing the pore dimension and speeding up localized corrosion of the substrate. Porosity is a critical factor to be considered in the design of coatings as many researchers have implemented and provided various methods to curb its existence.

Firstly, electrodeposition offers a variety of appropriate techniques for the fabrication of tailored metallic and composite coatings, which include powder metallurgy, nitriding, metal spraying, and vacuum deposition. It is, simply, the use of current flow to mitigate the desired cation of a material in a solution for coating of the material to be attained as a uniform, thin film on a conductive metal substrate. It is a successful way of dealing with the issue of porosity in coatings because this technique is compatible with the deposition of a wide scope of materials, such as nickel, zinc, cadmium, aluminum, and their respective composites in severe marine environments. Secondly, electrodeposition offers the flexibility of controlled processing of these corrosion-resistant coatings. However, expertise is highly appreciated in the full knowledge of the manipulation of the system and design parameters employed for optimum deposition characteristics. Computational analysis of the effect of these parameters and their interactions on the deposition quality can be achieved based on various designs of experiments (DoE). These DoE offer the possibility of understanding the cause-and-effect of the parameters on the final product and finding the interactions between variables and running effective experiments in a cost-effective manner. They also provide the opportunity of building mathematical models to support the correlation between product properties and significant processing parameters, establishing a platform for future consideration.

2. Marine corrosion

Corrosion occurs fastest in the marine environment as a result of the high concentration of salt content in the atmosphere. The contained salt increases the susceptibility of marine assets to degradation.

2.1. Salinity

In relation to chemical perspective, seawater is a solution of salts dissolved in water. Typically, the salinity of an ocean can be measured at a value between 3% and 4% [1]. The concentration of [Cl–] ions in water determines the degree of salinity [2]. Evaporation of water can probably pose as a determinant of the occurring differences in local salinity, which, in turn, increases the degree of salinity of seawater [3]. When water exchange is restricted in a certain region of the sea, a free water inflow into the sea will lead to a drop in the salinity of seawater [4, 5].

2.2. Temperature

Another factor that affects the salinity of seawater is temperature; as the temperature decreases, concentration of dissolved salt decreases, that is, there is a drop in solubility [6, 7]. This drop in solubility reduces the conductivity of the seawater, increases its electrolyte resistivity [8], and encourages the dissolution of oxygen. With an increase in the oxygen dissolution, corrosion rate becomes greater [2].

2.3. Geographical oceans

Influence of corrosion rates on marine equipment, assets, ships, and oil platforms can be influenced by geographical factors. The combined effect of salinity and temperature can be seen in various oceans, which eventually lead to variation in corrosion rates. Oceans situated at regions close to the North and the South poles are prone to ice formation due to low temperature. Brine is trapped within crystals, thereby depleting the level of salinity. Oceans close to the equator possess higher degree of salinity as this can have a higher influence on the corrosion rate of machinery.

3. Corrosion

Corrosion is referred to as the degradation of a material's properties and performance as a result of the material's reactivity with its environment. Most of the time, the reaction turns out to be inexorable. This enables the rapid consideration of research and development, seeking for better paths of eradicating its effect. Polymers and ceramics undergo degradation in the form of aging and selective dissolution, respectively. Corrosion is majorly associated with metallic structures, especially structures that are in the vicinity of marine regions. Ships, oil platforms, and vessel assets are the major victims of corrosion based on the highly induced attack by seawater. With increasing industrialization, the cost of battling corrosion has consistently been rising as the years go by. Based on the thermodynamic phase for corrosion, it is of little wonder that the attributed cost with corrosion has skyrocketed. Many studies conducted for more than three decades have revealed that the yearly expenditure on corrosion control is close to 3.1% of the Gross National Product (GNP) of the country. This works out to a value above $275 billion annually [9]. In 2001, the United States experienced an uprising in its cost of corrosion control to a shocker of over $600 billion, representing about 5% of the GNP of the country. This could be interpreted in such a way that for every family of four, the net annual expenditure on fighting corrosion is $8,000. In addition, corrosion seems to be the nation's most significant cause of financial calamity; it costs more than natural disasters and automobile accidents [10]. Figure 1 shows the result of the study on corrosion cost distribution of the United States carried out by the United States Federal Highway Authority in collaboration with the National Association of Corrosion Engineers (NACE) in 2001.

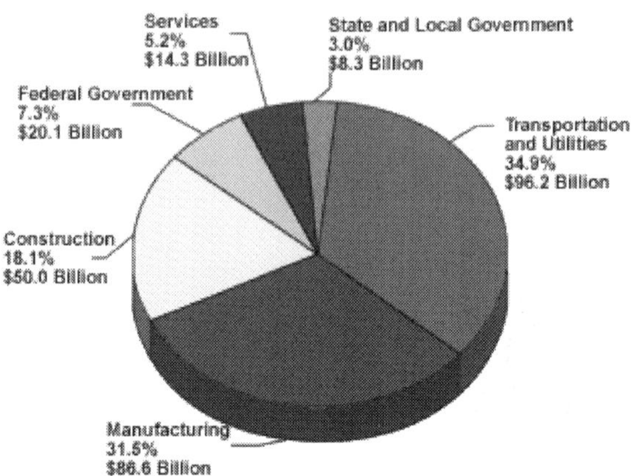

Figure 1. Corrosion Cost Distribution [11].

4. Chemistry of corrosion

Corrosion is defined as the deterioration of a component that occurs due to its interactive reactivity with the environment. When a metal comes in contact with a liquid, an electrochemical reaction is set up involving a transfer of electrons across the metal/liquid boundary. Conservation of energy is observed as metals return to their original state; the energy required for corrosion to occur is the energy responsible for converting the ores to metals. This energy responsible for the conversion is a free energy known as Gibb's free energy. When an atom leaves the crystal structure of a metal, it exceeds the interatomic forces binding the atoms together in the crystal matrix. The surface atom experiences lower attractive forces compared with the internal atoms, a significant increase in temperature can enable the outer atom to leave the structure, leaving its electrons behind [12]. This is why corrosion is rapid with increased temperature in a corrosive medium. The equation below illustrates the electron transfer of a metallic atom leaving the surface of the metal, an oxidation reaction:

$$M_{(s)} \rightarrow M^{n+}_{(aq)} + ne^{-} \tag{1}$$

where 'M' symbolizes the metal and 'n' is the number of electrons left by the metal atom on the bulk metal.

As the atom leaves the surface, it immediately appears as an ion in the liquid. With subsequent dissolution of the metal atoms to form ions, the weight of the bulk material reduces, creating a rough metal surface. The metal ion reacts with dissolved oxygen present in the liquid to form the metal oxide. Most times, the corrosive medium appears stagnant, thereby leaving the oxide product on the corroded surface; in the case of iron, the oxide formed is known as 'rust'. The oxidation and reduction mechanism is shown in Figure 2. Even though it appears that corrosion takes a generic routine, there are different forms of corrosion as a result of the localization, nature of corrosive medium, and chemistry of the metal itself. Corrosion is independent of distinct electrodes, as it may occur at micro scale. Numerous anodic and cathodic sites may appear on a single surface where oxidation and reduction reactions take place [13].

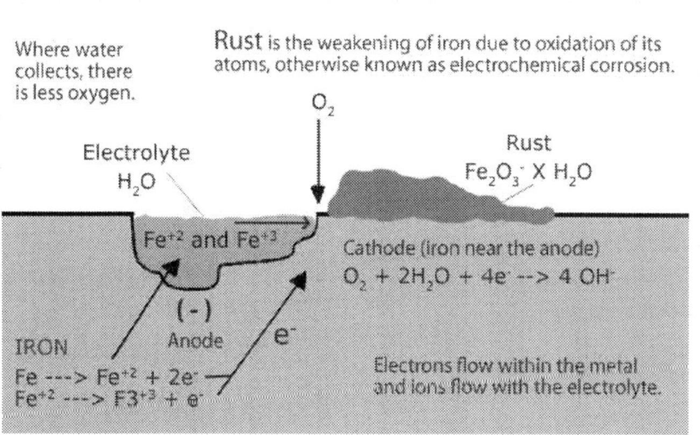

Figure 2. Corrosion Mechanism of Iron to Produce Rust [14].

5. Forms of corrosion

Useful information of the history of corrosion of a failed material can be carefully acquired if adequate observation is made. Based on the appearance of the corroded component, there are eight different forms of corrosion.

5.1. Uniform or general corrosion

Uniform or general corrosion is the most common form of corrosion. It is frequently attributed with electrochemical reactions, which prevails in a homogeneous manner across the surface of the component that is exposed to the atmosphere. As corrosion proceeds, the component reduces in dimension as failure becomes imminent. This form of corrosion is known to be the most disastrous form of corrosion, which may lead to a massive deterioration of metal. As its name appears, it affects a wide span region unlike the other forms of corrosion that are localized in nature. Therefore, its effect may be predicted experimentally [15]. It can be controlled by cathodic protection or by applying coatings and paints and also by introducing dimensional allowance (Figure 3).

Figure 3. (a) Schematic Representation of Uniform Corrosion and (b) a Weathering Steel [16].

The degradation of coating structures for surface protection usually results in uniform corrosion. Characteristics observed on such surfaces are muting of shiny metallic surfaces, discoloration of iron-based materials, and rough impressions by corrosive media. Surface degradation is the result of a negatively altered coating system and consideration should be given to repair; otherwise, it may lead to worse situations of corrosion attacks [16].

5.2. Crevice corrosion

Crevice corrosion is a localized form of corrosion attack. It exists with metals and alloys, which have unstable passive films in the presence of chloride and hydrogen ions. It is common with nonferrous metals, such as aluminum, zinc, titanium, and copper, but not with stainless steels. This form of corrosion attack occurs at crevices, joints, riveted regions, flanges, or isolated regions beneath where debris deposits can be found. Sand and dust deposits with scales and products from degradation can act as preferred sites where corrosive electrolyte can be reintroduced with high level of challenge [17]. For crevice corrosion to take place, the opening must be spacious for accommodating and retaining the electrolyte. Though there are no specific dimensions for crevices, this form of corrosion can be present where there are gaps of few micrometers. The presence of high level of chromium can reduce the possibility of crevice corrosion [18]. Also, considering flanges and joints, compromise must be made in terms of the opening because extremely narrow openings may lead to galvanic or cell corrosion, whereas a larger opening will allow accumulation of electrolyte that initiate crevice corrosion (Figure 4).

5.3. Intergranular corrosion

This form of corrosion is observed at the microlevel of metals situated at grain boundaries. Rapid attack of neighboring grain boundaries will eventually lead to cracking of the grain structures. Sometimes, elements and phases may be precipitated out of the grains and migrate to the grain boundaries as a result of processes, such as aging; these migrated precipitates can

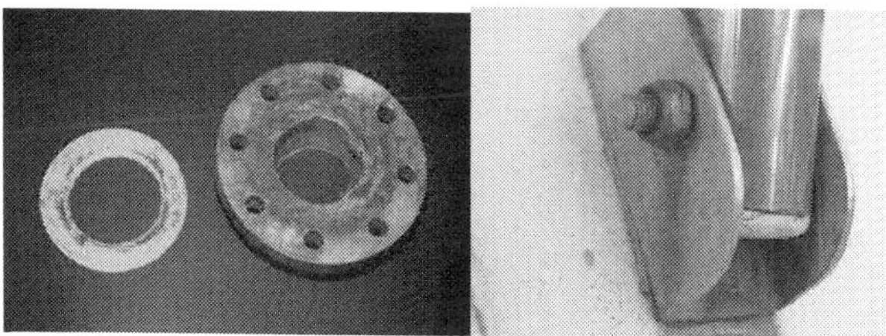

Figure 4. (a) Crevice Corrosion on a Steel Flange [17] and (b) Typical View of Crevice Corrosion.

be a mismatch with the grain boundary composition, in this manner, causing a depletion of intergranular bonds. Also, adjacent grains may lose significant elements such as chromium to chromium-rich grain boundary precipitates, leaving those regions deficient of chromium and susceptible to corrosion in a corrosive medium (Figure 5).

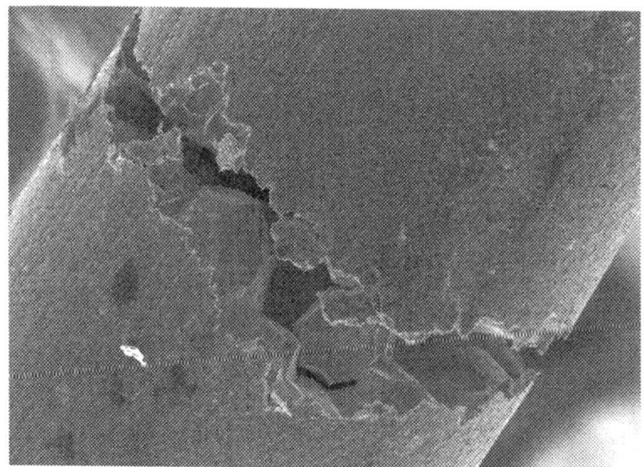

Figure 5. Intergranular Corrosion [19].

5.4. Pitting corrosion

Pitting corrosion has been one of the most expensive forms of corrosion because of its unpredicted manner of attack. It is a form of confined degradation which yields effects in the form of pits and holes. This form of corrosion may exist in stainless steels whether in acidic or neutral chlorides. An example of such media is the seawater. Pitting corrosion concentrates on the weakest link, which is a region where passive layer might not be tough enough. The mechanism is very rapid as a damaged surface experiences successive deterioration within a short period of time (Figure 6).

Pitting is dependent on temperature, the chloride concentration, and pH value. A low-corrosive environment will aid a natural repair of a steel grade pit in the presence of oxygen. Natural seawater causes more harm in terms of pitting than any other chloride-contained corrosive media as a result of its biological interaction. This requires the prevention of microorganism growth by applying biocides in seawater piping systems [20].

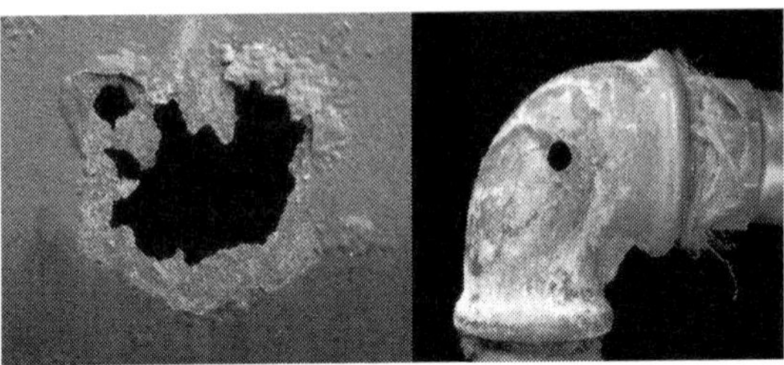

Figure 6. Pitting Corrosion [21].

5.5. Erosion corrosion

Erosion corrosion is a rapid rate of corrosion action on metals and alloys as a result of relative movement of corrosive fluid to the metal of alloy surface. With rattling flow of the fluid, the inner surface of a pipe or a tube can be washed resulting in erosion and then a leakage. This form of corrosion attack can occur due to bad internal design of component, for instance, a protruding inclusion which was meant to be trimmed can perturb a laminar flow of the fluid. Subsequent washing of a localized region on the inner surface by high-velocity fluid flow can initiate erosion corrosion. Pitting may result from a collaboration of erosion and corrosion. For erosion corrosion to be mitigated, pipe diameters should be increased while turbulent flow will be eliminated. Also, highly affected regions, such as elbows and bends, should be designed with thicker materials [16] (Figure 7).

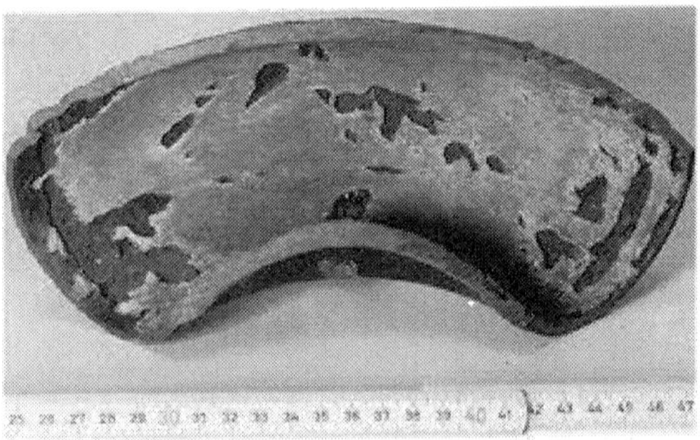

Figure 7. Erosion Corrosion on Inner Surface of an Elbow Pipe [22].

The other three forms of corrosion are hydrogen embrittlement, stress corrosion cracking, and galvanic corrosion. Hydrogen embrittlement can be experienced in acidic media where the hydrogen ions act on the surface of the metal in contact; stress corrosion cracking occurs with materials having a combination of stress-induced conditions under corrosive environment; and galvanic corrosion takes place between two dissimilar metals in contact.

It can be seen that necessary ways of protecting metals from the actions of corrosive media, mostly in the marine environment, should be looked into. This will drastically reduce the massive amount of expenditure on corrosion control, especially in industries such as marine, defense, and oil and gas sectors.

There are two major forms of corrosion that dominate the morphology of degradation: galvanic corrosion and electrolytic corrosion.

6. Surface engineering

Surface improvement of engineering components is expedient in order to protect these components from premature failure and effect of corrosion during service [23]. Surface engineering can be described as a broad set of technologies that focuses on the design and modification of surface characteristics of a component. Surface engineering can be seen in two major ways for improved surface characteristics and bulk material, that is, surface coating and surface modification, whereas surface shape design is a secondary type of surface engineering. Surface coating processes entail deposition of layers in the form of molten or semi molten material on the surface of a metal (substrate). Part of the prime reasons for surface coatings is to enhance the surface behavior rather than altering the bulk composition of the material. There are various techniques of applying surface coatings, which include chemical and physical vapor deposition (C & PVD), thermal and plasma spraying, cladding, and electrodeposition [24]. With surface modification, processes like surface hardening and alloying performed either by flame, laser, electron beam, induction, or elevated energy treatments (ion implementation, carburizing, and nitriding) are considered. Surface modification processes are employed to regulate surface tribological properties. Sometimes, there can be a combination of surface coating and surface modification, as in the case of laser cladding [25]. The major challenge of these techniques is the presence of porosity as this coating defect lays a strong platform for localized forms of corrosion, even a general form of corrosion.

7. Electrodeposition

Electrodeposition can also be referred to as electroplating since electroplating occurs during the process of electrodeposition. This is a process that makes use of electric current to mitigate cations from a preferred solution and migrate them as a thin-film coat on a conductive plate immersed in an electrolyte. Deposition of coatings can be achieved to obtain an improved surface with an excellent protective coating via electroplating technique [23]. The deposition of the material (the cations receive electrons at the cathode and are reduced to their metal atoms) as coatings on the conductive plate can be influenced by various factors, that is, the nature of cathode, electrolyte (salt bath), electric current, and so on (Figure 8).

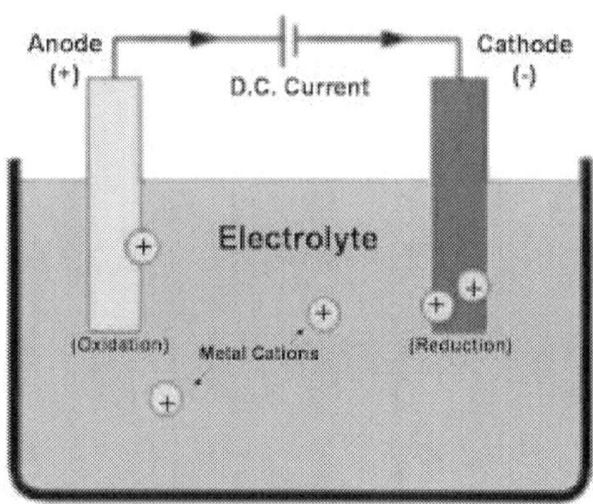

Figure 8. Electrodeposition Setup.

Porosity is among the major causes of discontinuities in electrodeposited coatings. The other causes are elevated internal stresses, corrosion, and wear of the plated surface. Porosity may be deleterious as its presence can expose coated surfaces to corrosive media, deplete mechanical properties, and affect density and diffusion properties in a negative manner. Pores that are created based on increase in temperature during service can lower the adhesion ability of a deposited layer. The presence of pores in metal coatings that readily produce thin passive film may be given less consideration compared with noble metals. For instance, if zinc is coated on a steel substrate, the zinc will be sacrificial by cathodically protecting the steel substrate. Furthermore, the presence of pores can allow the degraded products to accumulate on the substrate surface, leading to a localized form of corrosion, such as pitting. A material that contains pores or voids formed during electrodeposition is prone to deterioration of properties. The mechanical feature of a pore-contained coating is identified by a massive reduction in toughness and strength with increasing pore volume as ductility is also decreased.

7.1. Electroless plating

Electroless plating can be defined as a method of application of protective coatings: a chemical reduction–oxidation reaction that takes place as metallic ions are automatically aided in reduction. This can be achieved in an aqueous solution, which includes a reducing agent while deposition is made without flow of electrons [26]. Deposition of coating and its quality in electroless plating can be influenced by various variables that include the bath concentration, bath volume, acidity or basicity of the bath, bath history, temperature, and bath additives [27]. Electroless plating has better yield and quality with better surface profile compared with conventional electrolytic deposition; this is due to the flexibility of manipulating process parameters to obtain tailored coating dimensions. Figure 9 illustrates this benefit.

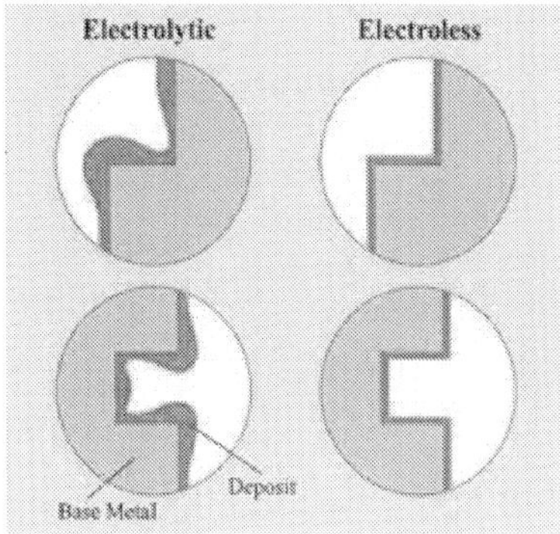

Figure 9. Schematic Illustration of Advantage of Electroless Plating [28].

8. Process parameters

For optimized system of uniform deposition and elimination of porosity in deposited coatings, the knowledge of the influence of deposition parameters is essential. Current density, time, nature of electrolyte, voltage, salt bath composition, temperature, stirring rate, pH value, and substrate type are the prime parameters that influence the deposition rate. Adequate control of the relationship between these parameters will yield an optimized deposition. Current density is inversely proportional to time while, theoretically, conductivity of the substrate varies inversely as the time [29]. This shows that a mathematical relationship can be achieved for various electrodeposition setup depending on the aforementioned parameters.

8.1. Current density

The current density is the rate of movement of a certain quantity of electricity per unit volume within the electrolyte. It is observed that the larger the current density, the more nucleation sites are widely distributed across the cathode [30]. Also, the current density favors the formation of larger grain sizes, with a lower strength compared with smaller grain sizes. Therefore, large current density produces more nuclei sites, whereas small current density will support small grain growth; adjusting the current density will tailor the quality and the properties of the coating to a better performance [31].

8.2. Deposition time

The time taken for deposition of uniform coating can be affected by other variables, for instance, a larger current density will reduce the time of deposition or an electrolyte with higher molecular mass will slow down the deposition process and increase the deposition time. The stability of an electroless bath strongly influences the time of coating deposition. Higher stability of bath may lead to increase in time of deposition, and thus a steady renewal and replacement will be achieved with respect to unit area [32].

8.3. Substrate type

Properties of the substrate, such as the conductivity, surface profile, and chemical composition, can determine the substrate's affinity for rapid coating deposition [33]. High electrical conductivity will ensure faster movement of electrons to the cathode (substrate) surface where reduction will occur. The more the supply of electrons, the faster the rate of metal deposition.

8.4. Bath composition (Electrolyte)

The bath composition is dependent on the molecular mass of the additives and also their viscosities. Bath age, bath formulation, bath temperature, and stability are integrated in the net influence on coating deposition rate. Usually, a rise in the bath temperature has a way of inducing an increase in metal solubility and stability in the electrolyte, resulting to a surge in transport number and conductivity of the specie, though the viscosity of the electrolyte is reduced [34]. Another bath parameter that influences deposition rate is the stirring rate; a higher stirring rate will lead to increase in bath agitation and decreased time to attain adsorption on the cathode surface, hence deposition rate is increased.

9. Computational analysis of process parameters

Predictive models, simulations, and theories have been developed over the years to subdue the hardship of material development. Response surface model, artificial neural networks, and Taguchi models are forms of predictive models which have been used to achieve optimal deposition. These models have also eliminated the use of trial and error methods in experiments. The high cost of resources, time, and labor has been taken care of by the advent of DoE. DoEs make experiments easier by linking data compilation with experimental procedures to analysis of results and prediction of optimized output and performance.

Generally, several experiments are often needed to be carried out considering different plating parameters and plating functionality in order to achieve a model. An artificial neural network model was developed for the deposition of copper and then compared with regression model. Validation was performed by conducting another set of experiment with the same conditions and then comparison was made. The validation showed a percentage error, which was less than 1.0 [34].

10. The future

Studies are ever increasing on the development of new and effective techniques on electrodeposition. Oil platform assets, ship vessels, and marine components will be able to function at optimized performance for longer duration without the cost and labor of repair and maintenance. Coatings that will be able to be highly unresponsive to severe marine environments and adhere strongly to underlying substrate are becoming more achievable irrespective of harsh cascading of saltwater splashes on ship hulls, propeller panels, and rig supports. Consequently, the global cost on corrosion control will diminish to an appreciable extent, though it will still be a serious discussion ever-present in the world of engineering and technology.

Author details

Gabriel A. Farotade[1], Patricia A. I. Popoola[1*] and Olawale M. Popoola[2]

*Address all correspondence to: popoolaapi@tut.ac.za

1 Department of Chemical, Metallurgical and Materials Engineering, Tshwane University of Technology, Pretoria, South Africa

2 Centre for Energy and Electrical Power, Department of Electrical Engineering, Tshwane University of Technology, Pretoria, South Africa

References

[1] Williams P.D., et al. (2010). *The Role of Mean Ocean Salinity in Climate*. Dynamics of Atmosphere and Oceans, volume 49, pp. 108-123.

[2] Zakowski k., et al. (2014). *Influence of Water Salinity on Corrosion Risk – The Case of the Southern Baltic Sea Coast*. Environment Monitoring Assessment, volume 186, pp. 4871-4879.

[3] Da-Allada C.Y., et al. (2013). *Seasonal Mixed-layer Salinity Balance in the Tropical Atlantic Ocean: Mean State and Seasonal Cycle*. Journal of Geophysical Research: Oceans, volume 118, no.1, pp. 332-345.

[4] Zhang Y. and Yan D. (2012). *Seasonal Variability of Salinity Budget and Water Exchange in the Northern Indian Ocean from HYCOM Assimilation*. Chinese Journal of Oceanology and Limnology, volume 30, no.6, pp. 1082-1092.

[5] Rhode J. and Winsor P. (2002). *On the Influence of the Freshwater Supply on the Baltic Sea Mean Salinity*. Tellus Series: A-Dynamic Meteorology and Oceanography, volume 55, no.5, pp. 455-456.

[6] Helber R.W., et al. (2012). *Temperature versus Salinity Gradients below the Ocean Mixed Layer*. Journal of Geophysical Research: Oceans. Doi: 10.1029/s2011JC007382.

[7] Bingham F.M., Foltz G.R., and McPhaden M.J. (2010). *Seasonal Cycles of Surface Layer Salinity in the Pacific Ocean*. Ocean Science, volume 6, pp. 775-787.

[8] Sharqawy M.H. (2013). *New Correlations for Seawater and Purewater Thermal Conductivity at different Temperatures and Salinities*. Desalination, volume 313, pp. 97-104.

[9] Shaw B.A. and Kelly R.G. (2005). *What Is Corrosion?* The Electrochemical Society Interface.

[10] Bushman J.B. *Financial Impact of Corrosion on the Economy of the United States*. Bushman and Associates, Inc., Corrosion Consultants, Ohio, USA.

[11] Ahluwalia H (2003). *Combating Plate Corrosion: Improving Corrosion Resistance Through Welding, Fabrication Methods*. The Fabricator, 2003.

[12] Singh R. (2013). *Pipeline Integrity Handbook: Risk Management and Evaluation*. Gulf Professional Publishing, Houston, TX, pp. 320.

[13] Natarajan K.A. NPTEL Web Course, Lecture 1: *Corrosion: Introduction, Definitions and Types*, Advances in Corrosion Engineering.

[14] Hock Tools. (2001). *The Perfect Edge: The Ultimate guide to Sharpening for Wood Workers*. https://hocktools.wodpress.com/2011/10/24/rust-part-1/

[15] Fontana M.G. and Mitra M.S. (1953). *The Eight Forms of Corrosion and Corrective Measures*, In Symposium on Industrial Failure of Engineering Metals & Alloys, Feb. 5-7, 1953, NML, Jamshedpur.

[16] National Association of Corrosion Engineers (NACE) Website. www.nace.org/corrosion-central/corrosion-101/.

[17] The Multimedia Corrosion Guide. *Crevice Corrosion*. www.cdcorrosion.com/mode_corosion/corrosion_crevice.htm.

[18] British Stainless Steel Association. *Principles and Prevention of Crevice Corrosion*.

[19] Lewis O. (2014). *Forms of Corrosion: Intergranular Corrosion and Stress Corrosion*. https://degradtioneng.wordpress.com/2014/10/23/forms-of-corrosion

[20] Sandvik Materials Technology. www.smt.sandvik.com/en/materials-center/corrosion/wet-corrosion/pitting/

[21] Merus Website. http://www.merusonline.com/in-general/pitting.

[22] Wear Mechanisms Tutorials. http://sirius.mtm.kuleuven.be/Research/corr-o-scope/hcindex1/tutorial3.htm

[23] Popoola A.P.I., Fayomi O.S.I., and Popoola O.M. (2012). *Electrochemical and Mechanical Properties of Mild Steel Electro-Plated with Zn-Al*. International Journal of Electrochemical Science, volume 7, pp. 4898-4917.

[24] Grainger S. and Blunt J. (1998). *Engineering Coatings—Design and Application*, Woodhead Publishing, pp. 10-41.

[25] Kennedy D.M., Xue Y., and Mihaylova E. (2005). *Current and Future Applications of Surface Engineering*. The Engineers Journal (Technical), volume 59, pp. 287-292.

[26] Zaki A. (2006). *Introduction to Corrosion, Principles of Corrosion Engineering and Corrosion Control*. Butterworth-Heinemann, Oxford.

[27] Sudugar J., Lian J., and Sha W. (2013). *Electroless Nickel, Alloy, Composite and Nano Coatings – A Critical Review*. Journal of Alloys and Compounds, volume 571, pp. 183-204.

[28] Classic Plating Inc. *Electroless Nickel Plating.* http://www.classicplating.com/En.html.

[29] Tuaweri T.J., Adigio E.M., and Jombo P.P. (2013). *A Study of Process Parameters for Zinc Electrodeposition from a Sulphate Bath*. International Journal of Engineering Science Invention, volume 2, issue 8, ISSN: 2319-6734, pp. 17-24.

[30] Sagane F., Ikeda K., Okita K., Sano H., Sakaebe H., and Iriyama Y. (2013). *Effect of Current Densities on the Lithium Plating Morphology at a Lithium Phosphorus Oxynitride Glass Electrolyte/Copper Thin Film Interface*. Journal of Power Sources, volume 233, issue 34, pp. 34-42.

[31] Sano H., Sakaebe H., Senoh H., and Matsumoto H. (2014). *Effect of Current Density on Morphology of Lithium Electrodeposited in Ionic Liquid-Based Electrolytes*. Journal of the Electrochemical Society, volume 161, issue no. 9, pp. A1236-A1240.

[32] Bulasara V.K., Thakuria H., Uppaluri R., and Purkait M.K. (2011). *Combinational Performance Characteristics of Agitated Nickel Hypophosphate Electroless Plating Baths*. Journal of Materials Processing Technology, volume 211, issue no. 9, pp. 1488-1499.

[33] Raiessi R., Saatchi A., Golozar M.A., and Szpunar J.A. (2005). *Effect of Surface Preparation on Zinc Electrodeposited Texture*. Surface and Coating Technology, volume 197, pp. 229-237.

[34] Subramanian K., Periasamy V.M., Pushpavanam M., and Ramasamy K. (2009). *Predictive Modeling of Copper in Electro-Deposition of Bronze Using Regression and Neural Networks*. Portugaliae Electrochimica Acta, volume 27, issue no. 1, pp. 47-55.

Chapter 7

Fabrication and Properties of Zinc Composite Coatings for Mitigation of Corrosion in Coastal and Marine Zone

Patricia A.I. Popoola, Nicholus Malatji and Ojo Sunday Fayomi

Additional information is available at the end of the chapter

http://dx.doi.org/10.5772/62205

Abstract

Deterioration of metals and alloys during service due to corrosion and wear phenomena shortens materials' life span and structural integrity particularly in aggressive environments such as coastal and marine. This degradation also limits the use of these materials in most industrial applications. Therefore, the improvement of the quality of these materials in order to combat these challenges in industry remains critical. Surface modification techniques are employed to enhance materials' properties to enable better performance and to extend their applications in demanding environments. Electrodeposition has been a useful method developed to improve the corrosion and mechanical properties of materials. In the present contribution, ample knowledge about electrodeposition of Zn composite/nanocomposite coatings and their characteristics are reviewed to address coastal and marine degradation of metals and alloys.

Keywords: Corrosion Degradation, Mitigation, Nanocomposite, Electrodeposition, Tribology

1. Introduction

Metallic materials and structures (mainly steel) exposed to marine and coastal environments are constantly at risk of corrosion, which compromises their structural integrity and life span. Deterioration of materials due to corrosion can also have other serious implications such as economic costs, safety and environmental issues. The corrosion of these materials is induced by atmospheric pollutants mainly chloride ions. According to [1] the atmospheric salinity is dependent mainly on the distance from the sea shore (concentration of chloride salts is high near the shore and reduces with increasing distance from the shore) and wind. The salinity in marine and coastal atmosphere is caused by wind which stirs up and entrains sea water. The generated marine aerosol droplets are transported few hundred of meters inland by the same

mechanism. Other factors that affect salinity include latitude, ruggedness of the coastline and undulation of the land surface. A high level of salinity is associated with accelerated corrosion rates and formation of unstable corrosion products [2]. Therefore, metallic structures found in marine and coastal environments have higher corrosion rates than those found inland (since chloride concentration is high close to the sea). Despite the magnitude of corrosion in the respective environments, prevention and combating corrosion remain critical to improve the service life of materials. Figures 1 and 2 show surface degradation induced by marine corrosion and the effect of atmosphere on the corrosion of low carbon steel.

Figure 1. Accelerated Low Water Corrosion Observed on Steel Piles above Water during Spring Tide [2]

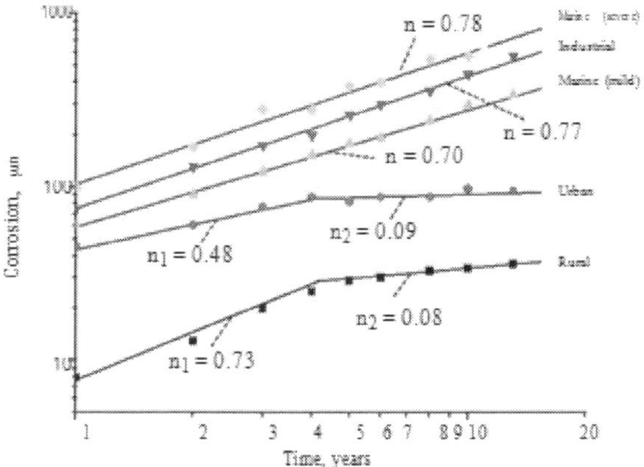

Figure 2. Effect of Atmosphere on the Corrosion of Mild Steel

Surface modification and enhancement of properties of the base metal by applying protective coatings has played a significant role in the fight against corrosion. Coatings offer protective barrier between the corrosive environment and the substrate [3]. There are various types of protective coatings that range from nickel, copper, zinc, aluminium, etc., but this chapter focuses only on zinc coatings. Zinc coatings have been successfully used over the past decades to mitigate the corrosion deterioration of mild steel due to their good anti-corrosion properties, ease of fabrication and affordability. Zinc is more active than iron or steel and, thus, offers sacrificial protection to the base metal [4]. Several industries employ extensively this metal for protecting steel used in manufacturing of components, parts or structures against chemical or electrochemical degradation in rural, urban, industrial, coastal and marine atmospheric conditions [5]. However, the exposure of these coatings to aggressive conditions such as industrial, coastal and marine environments induces formation of white rust [6]. Basic reactions that take place during the corrosion of zinc are shown in Equations 1 and 2.

Anodic reaction:

$$Zn \rightarrow Zn^{2+} + 2e^-$$

(1)

Cathodic reaction:

$$O_2 + 2H_2O + 4e^- \rightarrow 4OH^-$$

(2)

Zinc oxides and hydroxides are commonly the main constituents in the corrosion products which form as a result of corrosion and tend to act as a passivation layer preventing further damage. However, the chemistry of the corrosion products depends on the level and nature of pollutants in the environment. The constituents of the corrosion products in various atmospheres are presented in Table 1. The presence of these atmospheric pollutants (chlorides, carbonates and sulphates) also promotes dissolution of the protective film and thus exposing the fresh metal for further deterioration. Other factors contributing to the degradation of zinc include time of exposure and wetness, pH, stability of corrosion products, relative humidity and erosion [6-8].

To make zinc-based coatings to be strong and durable during application in harsh conditions, metals that are more chemically stable than the matrix are incorporated. Several metals such as aluminium, nickel, copper, cobalt, magnesium and manganese are added to fabricate zinc alloy coatings. These additions promote the evolution of uniform microstructure with limited surface defects and formation of stable and adherent passivation film that can withstand further attack. Many researchers have studied these coatings and found out that the coatings possessed not only excellent corrosion when compared to their traditional counterparts but also good mechanical properties [9-11]. Therefore, the incorporation of these elements into the coatings also extends their applications. The good performance of the coatings has made them to be substitute of cadmium sacrificial coatings and reduce the usage of organic chelating agents and chrome passivation that are not environmentally friendly [12-13]

Atmosphere	Test Site	Phases	Sulphate (mg/m^2)	Chloride (mg/m^2)
Rural	El Escorial	ZnO (zincite)	267	143
Urban	Madrid	ZnO (zincite)	609	106
Industrial	Bilbao	ZnO (zincite) $Zn_4SO_4(OH)_6.H_2O$ (zinc hydroxysulphate)	1057	194
Marine	Barcelona	ZnO (zincite) $ZnCO_3$ (smithsonite) $Zn_5(OH)_8Cl_2. H_2O$ (simonkolleite) $NaZn_4Cl(OH)_6SO_4.6H_2O$ (sodium zinc chlorohydroxysulphate)	2592	5663
Marine	Alicante	ZnO (zincite) $ZnCO_3$ (smithsonite) $Zn_5(OH)_8Cl_2. H_2O$ (simonkolleite) $NaZn_4Cl(OH)_6SO_4.6H_2O$ (sodium zinc chlorohydroxysulphate)	3949	15261

Table 1. Phases present and soluble salt contents from the corrosion products [6]

The ever changing and demanding industry calls for further advancement in the quality of the zinc coatings to cater for the new or improved applications. Exploration, investigation and development of new materials therefore remain critical to finding viable solutions for counteracting the change and demand. Metal oxides, carbides, nitride and so on have been codeposited with Zn to fabricate high corrosion, thermal and wear-resistant composite coatings. The incorporation of these particles into a metal matrix can be accomplished in micro, semi-micro and nano sizes. The composite dispersed particles in micro and semi micro size produced porous structured coatings which have limited corrosion control application. In search for more advanced materials that can cater for these limitations, nano-sized particles were found to have a potential that can eradicate the drawbacks associated with micro-sized particles [14].

Inclusion of nano-sized materials into a Zn matrix to produce nanocomposite coatings offers novel properties that are not exhibited by traditional coatings. The exotic properties such as excellent corrosion resistance, high temperature oxidation resistance, excellent self-lubricating and tribological properties possessed by these materials enable their extensive use for several applications [15]. The incorporation of the second phase particles into the matrix enable excellent performance of the resultant deposits in demanding environments such as coastal and marine. Nanoparticles of metal oxides, carbides, nitrides, borides, etc., have been used as bath additives to improve the properties of Zn matrix. The nanoparticles include Al_2O_3, TiO_2, CeO_2, SiO_2, SiC, Si_3N_4, TiN, TiB_2, etc.

Electrodeposition is one of the most important surface engineering techniques that is used for fabrication of Zn composite/nanocomposite coatings. This technique exhibits several advantages over the other deposition techniques. The advantages include low cost, versatility, simplicity of operation, high production rates and few size and shape limitations. The quality/ performance of the nanocomposite coatings produced by this process depend on a number of operating parameters that include particle concentration in the bath, pH, temperature, current and bath agitation speed. These parameters need to be optimized to achieve the desirable coatings quality [16].

Fabrication and study of the properties of zinc composite/nanocomposite coatings using electrodeposition technique have not been thoroughly reviewed in literature. Therefore, this chapter focuses on providing ample information and in-depth analysis of the electrodeposition process (mechanism of incorporation of particles into a metal matrix and optimization of process parameters) and properties (morphological and structural characteristics, hardness, corrosion and wear) of the resultant deposits. This will help zinc electroplaters and corrosion engineers to understand zinc composite/nanocomposite electrodeposition to enable them to produce high quality coatings with improved functional properties.

2. Fabrication of zinc alloy/composite coatings

Application of coatings on metal surfaces that are prone to corrode is one of the most versatile methods that are employed to protect the metal from degradation. A coating acts as a physical barrier between the corrosive environment and metal, thus providing protection to the substrate from chemical deterioration. In tribological applications, the coating also offers load bearing capabilities. To ensure effective coating performance, the coating must be compatible with the substrate, have good adhesion and possess low porosity. Incorporation of reinforcement nanomaterials into the primary matrix to form composite coating enables production of coatings with advanced properties and good performance during application.

There are several techniques available for fabrication of composite coatings. These include, but not limited to, hot-dipping, thermal spraying, sol gel, vapour deposition, laser cladding, and electro- and electroless plating. The coatings resulting from these techniques exhibit different properties and find use in various applications.

2.1. Hot dipping

Hot-dip galvanizing is a deposition technique for coating iron or steel substrate with zinc by immersing the material of interest in a zinc molten bath [5]. This technique is a batch process and is mainly used on heavy structural steel sections. Galvanizing can be found in almost every major application and industry where iron or mild steel are employed. The surface of the steel requires preparation prior to deposition, and the preparation process includes the following steps:

1. Oil and grease are cleaned from the steel mostly with warm caustic soda to ensure proper chemical interaction between the substrate and molten zinc.

2. The steel is then pickled in dilute solution of hydrochloric or sulphuric acid to remove rust and scale.

3. The steel is immersed in a fluxing bath containing ammonium chloride.

4. The fluxed large sections of steel are then directly dipped into the zinc molten bath for coating.

5. The hot-dip galvanized steel can sometimes be further treated by centrifuging method to reduce the excess zinc on the steel.

Coatings on small components, wires, tubes and sheets can be applied using continuous galvanizing process. The coatings produced by continuous galvanizing process are thinner than that of the batch hot dipping [17]. The bonding of the coating produced by hot dipping is promoted by the interaction of the molten zinc and iron in the steel to form a series of Zn–Fe alloy [18]. The simplicity of galvanizing process enables wide range of applications in protection of steel more than other deposition techniques. The automotive industry is one of the industries that mostly depended on this process. The galvanized sheets are used for the assembling of the body of the car. Other automotive parts are also galvanized to protect them from corrosion attack [19].

2.2. Thermal spraying

Thermal spraying involves the melting and propulsion of metal particles onto a desired substrate using a spraying gun. The material (feedstock) which is to be applied as a coating is fed through the nozzle in a form of powder, wire or rod. It is fed with a stream of gas or air and is melted by a suitable gas-oxygen mixture. The melted material is propelled towards the surface to be coated and forms a consolidated coating on impact [17]. The three main variables that govern this process include the temperature of the flame, the velocity of the particles that are sprayed onto the work piece to fabricate a coating and the nature of the material used to form the coating. There are several types of thermal spraying, which include flame or torch, electric arc and plasma spraying. Thermal sprayed coatings can be made to have good adhesion, controlled thickness and can be applied to existing structures. Surface preparation is required to achieve good adhesion. Coatings fabricated by this technique possess strong bond with the substrate due to the interaction between the coating material and the steel. However, the fabrication of coatings using this technique requires high temperatures and the interaction between the substrate and the coating material makes it difficult to refurbish them when they are worn out.

2.3. Laser cladding

Laser cladding is another technique that is used to fabricate composite coatings on metal surfaces. This high energy technique employs laser beam to create a melt pool on the surface of the metal substrate in which the reinforcement powder is injected simultaneously. Coatings formed using laser cladding technique are dense and are employed in applications where good wear, corrosion and other surface-related properties are required. Laser cladding also has the capability to control composition and properties within a fabricated structure by either pre-blending or combining different elemental powders using multiple feeder systems at the laser focal zone, which enables tailoring of properties suited for specific engineering applications [20]. The use of laser in coatings technology is costly due to the high energy demand but still finds a variety of use in applications where durability and strength are required.

2.4. Electrodeposition

Electrodeposition is a technique that is used to apply adherent metallic coating on metal substrate by passing current between two electrodes immersed in an aqueous electrolyte containing ions of the metal of interest causing the metal ions to be deposited on the cathode. The anode is caused to be dissolved to replenish the metal ions in solution by flowing of current and make the cathode to be covered with the deposited metal [21]. Figure 3 shows a schematic diagram of a typical electrolytic cell used for the electrodeposition.

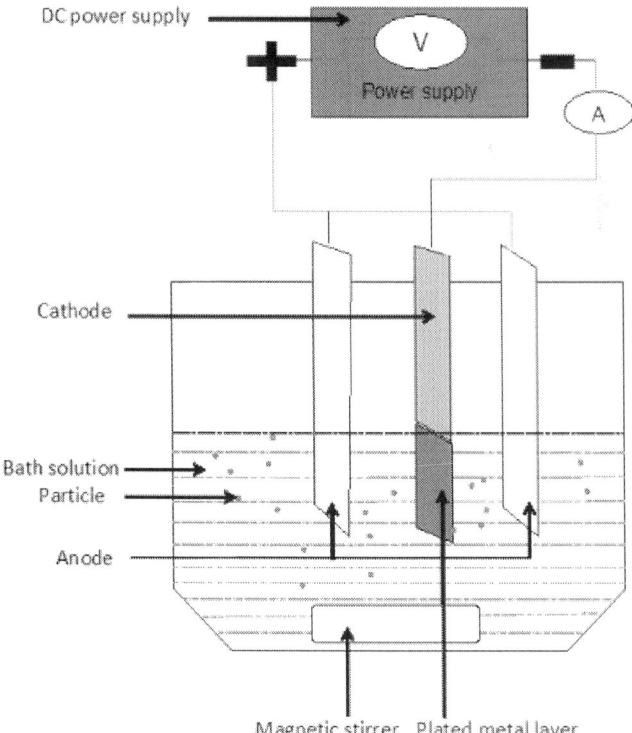

Figure 3. Typical Electrolytic Cell for Electrodeposition of Metals

The reactions that take place during plating process can be depicted as follows:

Cathode reactions:

$$M^{n+} + ne^- \rightarrow M^\circ \tag{3}$$

$$2H^+ + 2e^- \rightarrow H_2 \tag{4}$$

Anode reactions:

$$M^\circ \rightarrow M^{n+} + ne^- \tag{5}$$

$$4OH^- \rightarrow 2H_2O + O_2 + 2e^- \tag{6}$$

The quality of coatings produced by this technique depends on a number of operating parameters that include the bath composition, throwing power, temperature and bath agitation. Coatings exhibiting different properties such as microstructure, corrosion resistance, adhesion, thickness, hardness, thermal stability and wear resistance can be produced by manipulating these parameters. The flexibility of the electrodeposition enables the technique to find wide variety of applications in mitigation and control of corrosion mainly for small components. However, application of these coatings on large metal structures is beyond the capabilities of this process [21].

Electrodeposited coatings were primarily used for corrosion resistance and decoration (metal finishing). However, industrial technological advancement demand coatings that exhibit functional properties such as wear resistance, electrical properties, magnetic properties, solderability and thermal stability [22]. Zn electrodeposits are constantly improved to cater for this demand and electrodeposition technique makes coatings possible and simple at a lower cost compared to other deposition techniques.

2.5. Electroless plating

Electroless plating is a deposition technique used for coating metal bases by auto-catalytic or chemical reduction of metal ions without the application of external current. The internal current required for deposition is provided by the oxidation of reducing agents present in the solution. The quality of the coatings produced by this process makes it to gain more advantage over electrodeposition technique. The coatings exhibit excellent corrosion and wear resistance properties. The desired quality can be attained by controlling operating parameters such as temperature, pH and the composition of the bath. The bath constituents involved in electroless plating include metal ions source, reducing agent, buffering agent, complexing agent and wetting agent [23]. One of the most important bath constituent is a reducing agent since it supplies electrons to reduce metal ions to enable deposition. There are several available reducing agents, which include sodium hypophosphates, amino boranes, sodium borohydrides and hydrazines. Amongst them, sodium phosphates are commonly used due to lower costs and better corrosion resistance of the resulting coating. The reactions involved during electroless plating are as follows:

$$H_2PO_2^- + H_2O \rightarrow H_2PO_3^- + H^+ + 2e^- \tag{7}$$

$$2H_2PO_2^- + H + e \rightarrow P + H_2PO_3^{2-} + H_2O + H_2 \tag{8}$$

$$M\left(NH_3\right)_n^{2+} + 2e^- \rightarrow M + nNH_3 \tag{9}$$

$$2H^+ + 2e^- \rightarrow H_2 \tag{10}$$

Electroless nickel coatings has assumed the greatest commercial importance among electroless coating. These coatings possess good surface properties, and incorporation of second phase particles enhances the characteristics of the resultant deposits. The second phase particles codeposited with the nickel include silicon carbide, aluminium oxide, carbon nanotubes, etc. The dispersion of these particles requires good bath stability, agitation, particle size, particle concentration and surfactants. This plating technique is complex, requires many additives, operates at elevated temperatures, contain phosphorus or borine and applicable for mostly nickel coatings.

3. Composite electrodeposition

3.1. Mechanism of particle incorporation

Researchers have attempted to investigate the mechanism of codeposition to understand the way particles are incorporated into metal matrix over the past decades. Many models have been developed which gives insight into the mechanism of codeposition of particles on the cathode. Most models that have been developed involve one or more of the following processes: (i) migration of the charged particles to the cathode by electrophoresis, (ii) adsorption of the particles on the surface of the electrode by van der Waals forces and (iii) mechanical incorporation of the particles into the layer [24]. In 1972, Guglielmi developed a two-step adsorption model that incorporates both adsorption and electrophoresis to describe the mechanism of particle codeposition. In the first step, the particles are covered with metal ions which cause them to be weakly adsorbed on the cathode. The second step includes the strong adsorption of the particles by Coulomb force which result in subsequent entrapment of the particles within the metal matrix [25]. The model has been accepted by many researchers; however, it excludes the particle characteristics and mass transfer during electrodeposition process. This model has been improved by other authors to eliminate the drawback associated with it. [26] used three modes of current density (low, intermediate and high) to differentiate the reduction of adsorbed ions on the particles. The model was developed to improve Guglielmi's model. It involves three steps, namely: (i) forced convective-diffusion of particles to the surface of the cathode, (ii) loose adsorption of particles on the cathode and (iii) irreversible incorporation of particles by reduction of adsorbed ions. Bercot et al. [27] improved Guglielmi's model by incorporating the corrective factor (third-order polynomial equation) to cater for its drawbacks. One model that has been greatly accepted by scholars is Kurozaki's model, which includes the movement of the particles to the cathode by mechanical agitation. This model involves three-step processes for incorporation of solid particles in to a metal matrix, which include:

1. Uniformly distributed particles are conveyed to the electric double layer by mechanical agitation.

2. Charged particles are transported to the cathode surface by electrophoresis.

3. Particles are adsorbed at the surface of the cathode due to the Coulomb force between the particles and adsorbed anions.

Figure 4 demonstrates the five consecutive steps involved in codeposition of particles into metal matrixes. The regions include formation of ionic cloud around the particles, convective movement towards the cathode (convection layer, typical length less <1 mm), diffusion through a concentration boundary layer (diffusion layer, typical dimensions of hundreds of μm), electrical double layer (typical dimensions of nm) followed by adsorption and entrapment of particles [28]. This model shows the step by step progression of particle codeposition from the electrolytic bulk solution until they are incorporated into the metal matrix. An electroactive cloud that surrounds the particles that are added into the electrolyte is created, which transport the covered particles to the hydrodynamic boundary layer through convection.

The particles migrate through this layer and are taken to the cathode through diffusion. The deposited and incorporated inert particles are embedded into the matrix by discharged metal ions after the ionic cloud is wholly or partially reduced [24]. The manner of incorporation of particles on the metal matrix depends mainly on the electrodeposition process parameters. Some of the most important parameters include the speed at which the bath is stirred, the applied current density and electrolyte composition. Bath agitation serve as a medium that assists particles to be transported to the cathode; while applied current density and electrolyte composition are responsible for the formation of ionic cloud around the introduced particles. There are three possibilities for particles to be incorporated into a metal matrix: (i) coatings that are just covered by adsorbed particles on the surface, (ii) coatings containing entrapped particles and (iii) coatings containing particles truly embedded uniformly into the metal matrix [29].

These three possibilities for explaining the manner of particle incorporation into the matrix is clearly demonstrated in Figure 5.

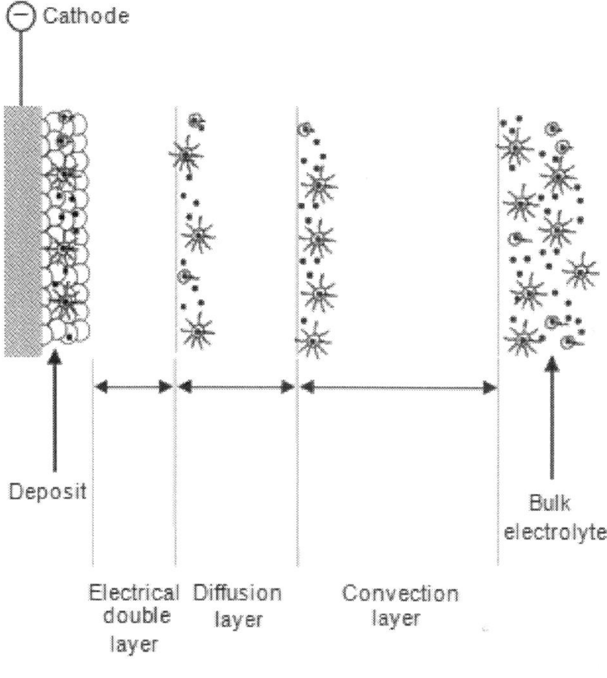

Figure 4. Codeposition Mechanism of Particles into a Metal Deposit [24]

The first type of coating that comprises many adsorbed particles on the surface is shown in Figure 5(a). These type of coatings exhibit inferior chemical, mechanical and thermal stability properties due to the unoccupied crevices, pores and microholes on the surface of the coatings. The localized inclusion of particles in the metal matrix as shown in Figure 5(b) is caused by agglomeration of particles in the solution and rapid deposition rate occurring as an effect of unoptimized process parameters. The poor performance of these coatings due to inhomogenous distribution of particles that causes surface defects does not allow them to be used for technological applications. The third coatings with truly homogenous and embedded particles within the matrix are corrosion and wear resistant possessing excellent microhardness [29]

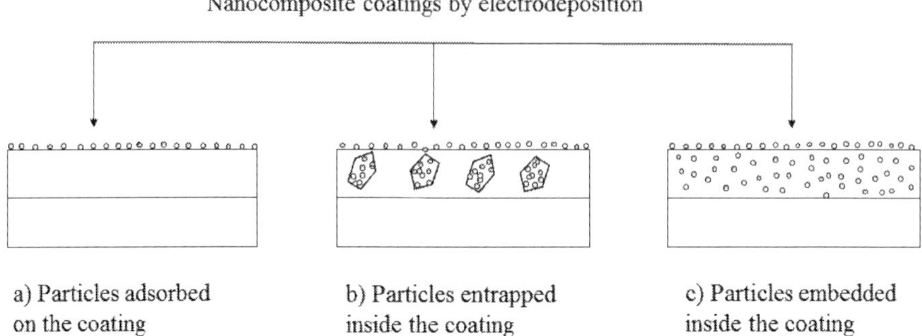

Figure 5. Schematic Representation of Different Codeposition Possibilities

3.2. Effect of operating parameters

Incorporation of composite nanoparticles into metal coatings depends mainly on the experimental parameters. The amount and nature of incorporation of the particles in the metal matrix determine the quality and success of the fabrication of the composite coatings. Several common process parameters and their effect on the properties of the coatings are discussed below.

3.2.1. Applied current density

Current is one of the important experimental parameters that require optimization in electrodeposition since the success of the process depends solely on current. The process uses electrical current to accelerate the dissolution and flowing of electrons to be deposited at the cathode. Studies have shown that at lower current densities, harder and brighter coatings with increased concentration of incorporated particles are formed. Current density of 0.8 A/cm^2 yielded microhardness value of 400 Hk, and increasing the current density to 1.6 A/cm^2 resulted in a decrease of 340 Hk. The mechanism of codeposition can be explained by the fact that at lower current densities, more time for achieving certain thickness is required compared to higher current densities. Due to this requirement, availability of particles at the cathode is raised and the possibility of the particles to be uniformly embedded in the coatings is increased. This results in formation of good quality composite coatings that can withstand trying environments [30]. [31] reported no significant effect on the microstructure and uniform distribution of Al$_2$O$_3$ nanoparticles in the nickel matrix due to the variation of current density from 0.01 to 0.03 A/cm^2. However, the concentration of the particles in the coatings was found to vary depending on the experimental condition (current density) the deposit was fabricated from. At low current density, increased concentrations of Al$_2$O$_3$ nanoparticles were present.

[32] studied the influence of applied current on the microstructure of electrodeposited ceria coatings for enhanced oxidation resistance of nickel-based alloy. The results showed that high applied current density lead to formation of crack network on the microstructure of the coatings. The development of the crack network was attributed to the rising of internal stresses

caused by rapid deposition kinetics [32]. Increased current densities enhance the deposition kinetics and promote rapid codeposition, which forms coatings that exhibit no improved properties. Increased voltage gives results that are in contrast with findings obtained with variation of applied current density. Weight gain, brightness, fineness and adhesion were improved when applied voltage was increased from 0.6 to 0.8 V [33] Other reported work showed that cathode current efficiency was achieved at low current densities when the influence of process parameters on cathode current efficiency were studied on electrodeposition of Zn–SiO$_2$ composite coatings [34].

3.2.2. Particle characteristics and concentration in the bath

The size, shape, type and concentration of the particles used for the fabrication of composite coatings have significant influence on the manner of incorporation mechanism of codepostion and properties of the subsequent coatings. The incorporation of metal oxide, carbides, nitrides, borides and so on has been reported to exhibit excellent surface properties by several authors [24, 35-37]. However, it has been found that micron- and submicron-sized particles incorporated into the coatings leave surface defects such as microholes, pores and crevices on the metal matrix, which then serve as a weak site for material degradation when exposed to a corrosive medium [38]. Findings have shown that when the size becomes smaller there are high chances for more of the particles to be codeposited into the matrix. Solution where 300 nm Al$_2$O$_3$ particles were used resulted in low amount of particles present in the nickel coatings compared to solutions where a size of 50 nm was used [38]. Nanoparticles possess excellent and exotic properties that are not contained in the bulk material even when they are of the same kind [15]. Incorporation of metal oxide, carbides and nitride nanoparticles yielded high microhardness, better corrosion resistance, good thermal stability and wear resistance [39, 36, 40].

The concentration of the particles in the electrolytic bath has significant effect on the properties of the coatings, and different particles reach optimum concentration in the solution at different bath loading levels. ZrO$_2$ nanoparticles were found to yield best microhardness values when the bath loading was at 5 g/L during their incorporation into a zinc matrix. The results showed a linear relationship between the particle concentration in the bath and their contents in the coating [36]. This behaviour is in harmony with Guglielmi's two-step adsorption model. A linear relationship between current efficiency and bath particle loading was observed by [34]. The increasing concentration of silica particles in the bath resulted in reduction of the current efficiency. The decrease might have been caused by solution resistance due to the increasing concentration of SiO$_2$ particles in the solution. This contrasting result suggests that every particle has its own peculiar behaviour when added into bath solution and hence unpredictable. The work reported by [41] showed that the optimum particle loading for Al$_2$O$_3$ incorporated into Zn–Ni matrix is 5 g/L and deviation from this value resulted in poor electrochemical behaviour of the coatings. The reduced corrosion inhibition of the particles beyond the optimal concentration has been ascribed to agglomeration of the particles in the bath, leading to lack of availability of particles to be deposited on the cathode. Another reason is the development of poor quality coatings caused by the generation of defects in the coatings due to rapid codeposition. Nano-Cr$_2$O$_3$ was effective at 50 g/L and increment to 75 g/L resulted in no significant effect on the microhardness of nickel-based coatings [30]. Good anti-corrosive properties, better thermal stability and microhardness have been achieved by adjustment of bath particle of SiO_2 to 5 g/L in a Zn–Ni matrix.

3.2.3. Bath temperature

Temperature does not play a major role in electrodeposition as it does in electroless plating, which depends mainly on the temperature and pH of the bath. However, increasing of temperature is known to enhance reaction rates in chemical processes. High temperatures increase the supply of ions to the cathode and the rate of nucleation of particles at the cathode [42]. This behaviour of the particle due to increases in the temperature has negative impact on the quality of the coatings. [12] reported the increase of the content of cobalt in zinc coatings due to the increment of temperature. However, raising the temperature was also found to reduce the polarization potentials and increase the amount of zinc in the deposit. So it is important for optimum temperature to be established to avoid all the drawbacks that are associated with increases in deposition temperature.

3.2.4. Bath agitation

The agitation of the bath during electrodeposition serves two functions: keeping the nanoparticles in suspension and the transportation of the particles to the cathode for codeposition. This actually explains that the incorporation of the particles in the metal matrix depends on the agitation speed of the bath during electrodeposition. Increasing of agitation speed improves the chance of the availability and adsorption of the particles by entrapment into the matrix. However, speeds beyond the optimal levels remove the particles from the cathode before they can be incorporated. The study of the effect of agitation speed on the microhardness revealed an increase in microhardness to maximum value of 395 Hk at 400 rev/min. Further increase to 600 rev/min had no effect on the microhardness of the coatings [30]. Agitation speeds of 900 and 1200 rev/min gave good microstructures when SiC particles were incorporated into a Ni matrix. Deviation from these values resulted in undesired appearance of microstructure, and it was concluded that the agitation must neither be too low or high but optimum [43].

Figure 6 demonstrates the effect of rotation speed on the incorporation of polytetrafluoroethylene (PTFE) particles in nickel coatings as studied by [27]. Different optimum rotation speeds were established, depending on the particle loading which is 500 rev/min at 10 g/dm^3, 600 rev/min at 20 g/dm^3 and 700 rev/min at 30 g/dm^3. Similar results were obtained when two baths were investigated for the effect of agitation speed on the current efficiency. The bath that contained 26 g/L SiO_2 showed higher current efficiency compared with the bath loaded with 13 g/L. The increase in current efficiency was attributed to enhanced mass transfer due to agitation action [34]. These results showed that achieving optimum rotation speed depends on the particle concentration of the solution.

Figure 7 shows methods that are used in industry and laboratories to keep particles in suspension and transport them to the cathode during electroplating. These include physical dispersion of particles by bath agitation (mechanical stirrers, reciprocating plates, pumping,

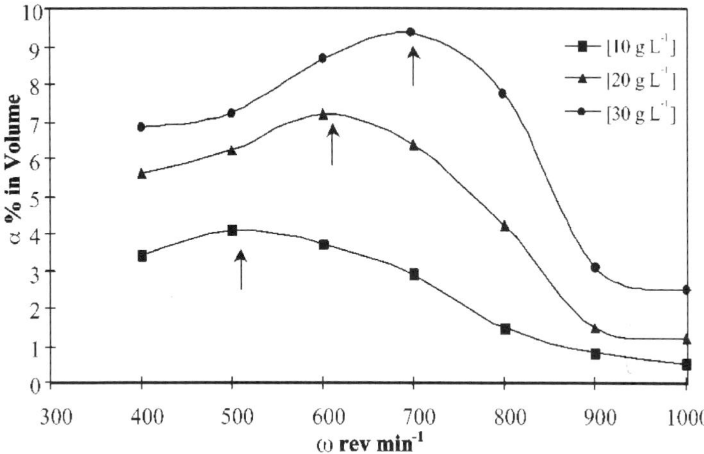

Figure 6. Effect of Stirring Speed on the Amount of PTFE Particles Embedded into Ni matrix, ● 10 g/L, ○ 20 g/L and ▼ 30 g/L of PTFE particles [27]

rotating disk or cylinder electrodes and parallel plate channel flow) and chemical dispersion by surfactants (cationic surfactants or combination of cationic and anionic surfactants).

3.2.5. Electrolytic pH

Bath electrolytic pH has an influence on the behaviour of the bath since reactions can only occur under certain pH conditions. [45] reported an increase in deposition rate due to increasing electrolyte pH in electroless plating of Ni–Fe–P alloys. However, pH values higher than 9 was found to cause solution instability due to precipitation of ferrous hydroxide during plating. Hypophosphite oxidation influences the rate of deposition, and this reaction is dependent on the pH of the solution. Electrokinetic measurements showed ZrO_2 nanoparticles were positively charged in pH less than 4 which aided the particles to be transported to the cathode. The addition of the particles in solutions of pH 3.5 and 2.5 did not cause any new process, but some kind of surface blockage occurred on the electrode favouring reduction reactions [36]. The reduction reactions that could possibly occur are shown in Equations 11 and 12 below:

$$H_3O^{3+} + 2e^- \rightarrow H_2 + OH^- \tag{11}$$

$$2H_2O + 2e^- \rightarrow H_2 + OH^- \tag{12}$$

The formation of inhibiting species such as hydrogen and nickel hydroxide depends on the pH of the bath and other process parameters. Hydroxide inhibiting species forms at high pH,

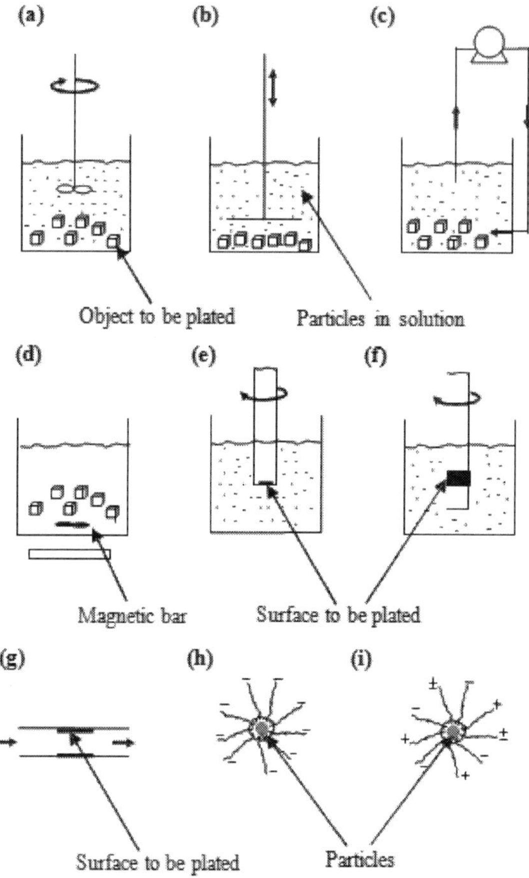

Figure 7. Bath Agitation Strategies: (a) Overhead Blade Stirrer; (b) Reciprocating Plate Plunger; (c) Pumped Electrolyte; (d) Magnetic Stirrer; (e) Rotating Disk; (f) Rotating Cylinder Electrode; (g) Flow between Parallel Plates; (h) Cationic Surfactant and (i) Combination of Ionic and Non-Ionic Surfactants [44]

and the evolution of hydrogen occurs at low pH as depicted by Equations 5 and 6. The occurrence of this species affects the degree of texture of the coatings, and it can be easily controlled by adjusting the pH. The preferential orientation of Ni–Cu films of (100) at pH between 3.3 and 2.6 was changed to (111) by changing the electrolytic pH to 2 [46]. [47] obtained insufficient zinc coating in the low current density region at low pH between 2.5 and 3.5. The increasing of the pH to 4 yielded satisfactory results with a bright and mirror finish deposit. Further increasing of the pH above 4.5 gave negative results and the deposit became dull. The activity of bath additives depends also on the electrolytic pH, and hence it is important to optimize the pH when best results are sought.

3.2.6. Deposition time

The thickness and the amount of incorporated nanoparticles in the composite coatings depend on the deposition time. [34] studied the effect of deposition time on the current efficiency of electrodeposited Zn–SiO$_2$ coatings. The current efficiency was found to increase with deposition time; the optimum time established was 180 seconds. Beyond this point, the current efficiency started to reduce which might be due to agglomeration of the particles in the bath. When these agglomerated particles are codeposited, blocking on the cathode due to the reduction of active surface area for deposition and a subsequent decrease in current efficiency occur [34]. Three distinct regions have been observed on the coating: a relatively compact underlayer of zinc, a discontinuous middle layer of dendritic trees of zinc with vertical silica in between and a topcoat of predominantly dense silica. The development and build up of these layers occur as a function of pH and time. The first layer of zinc forms with the good current efficiency due to the availability of the active surface area. The thickness of the coatings increases with time and pH also increases, causing agglomeration of particles in solution. The codeposition of the agglomerated particles reduce the active surface area and localized deposition, leading to the development of zinc dendritic 'trees' struggling to build up through the thick mass of silica top coat. The increase in the thickness of the coatings depends not only on the time but also on the bath particle loading. [48] obtained similar results when investigating the effect of time on the weight gain of zinc deposits.

[32] studied the influence of deposition time on the deposit growth, current density and coating thickness. Figure 8 shows the relationship between deposition time, deposit weight gain, thickness and the microstructures obtained from different deposition times. The results showed that deposit growth and thickness was not affected by the change in current density but increased with electrodeposition time. The deposit weight gain and thickness was constant up to 30 minutes of plating process. However, the coatings produced beyond 10 minutes of electrodeposition revealed a growing number of cracks [32].

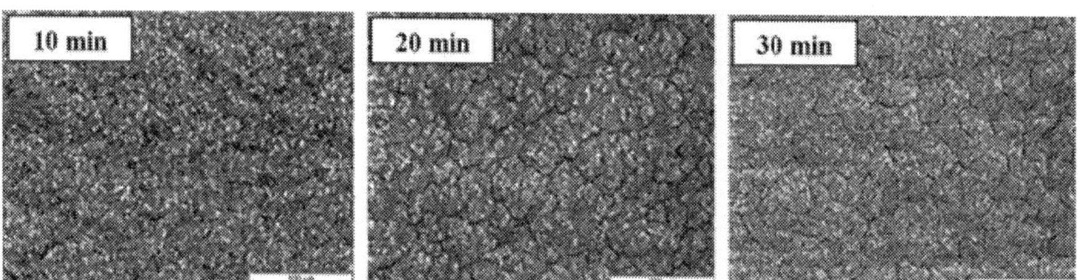

Figure 8. Surface Morphology of Electrodeposited Films with Increasing Time at Room Temperature in 0.1 mol/L Ce(NO$_3$)$_3$.6H$_2$O for an Applied Density of −1 mA/cm^2 [32].

3.2.7. Chemical additives

Chemical additives play a major role in electrodeposition. These chemical agents (surfactants and stabilizers) enhance the brightness, induce grain refinement and improve deposit properties. The commonly used additives for electrodeposition of zinc include gelatin, dextrin, thiourea, glycin and glycerol [49]. Others that have been recently reported include CTAB (Cetyltriammonium Bromide), SDS (Sodium Dodecyl Sulphate), TEA and MEA [4, 36, 50]. Their addition during plating has a significant effect on the resultant deposits. They alter crystallographic orientation of the matrix and perturb grain growth, leading to microstructures with small-sized grains. High concentrations of the surfactant in the bath have been reported to positively affect the microstructural properties of the produced deposits [49]. The added additives also increased the content of the reinforcement particles and also control their distribution in the matrix.

4. Properties of electrodeposited composite coatings

Zinc matrix nanocomposite coatings find various applications in industry due to the unique properties that they exhibit over the traditional metal or alloy coatings. There are several composite coatings that have been produced using electrodeposition, but the ones that have received most commercial use include zinc and zinc-nickel nanocomposites. These novel coatings have been developed to meet the industrial demand and perform better under demanding environments at a reduced thickness than traditional coatings [29]. Zinc and zinc-nickel composite coatings exhibit enhanced properties such as self-healing properties, high hardness, good thermal stability, corrosion and wear resistance.

4.1. Zn matrix composites

Electrodeposition of zinc composite coatings is motivated by the development of novel coatings with enhanced surface properties for combating corrosion phenomena. Due to the ability of Zn to sacrificially protect steel from corrosion, it makes the metal to be an attractive matrix for fabrication of advanced composite coatings for demanding application. Several inert second phase particulates have been incorporated into the matrix, and improvement in the surface characteristics of the base coatings was achieved. The choice of the particles to be incorporated and formulation are guided by the properties expected to be achieved and potential applications where the coatings can be used. Single or double particle loading to fabricate binary and ternary composite coatings has been reported in literature.

4.1.1. Composition, morphological and structural properties

The incorporation of 5 and 10 g/L of sub-micron Al_2O_3 into Zn matrix were found to follow the morphological orientation of Zn coating but with partial homogeneous structures [24]. The particles were also found to have no chemical interaction with the matrix, and the particle content in the coating was subject to the concentration of the particle in the plating bath. However, [51] reported formation of Zn_xSn_x phases when SnO_2 was codeposited with Zn. The chemical interaction observed in this work may be due to the change in operating parameters and bath formulations. The solubility of particles depends on various factors, and temperature and pH play a major role. Therefore, the difference obtained by the authors may be due to the increment of operating temperature to 40 °C. The addition of 7 g/L of the reinforcement particles yielded microstructures with no surface pores. The grain size was also found to be not dependent on the content of particles in the bath. [49] observed a correlation in bath particle loading and content of incorporated particles in the coating. Increasing particle concentration in the electrolyte solution improved the codeposition of the particles (bath solutions with high particle concentration yielded coatings with high particle content). This result is supported by Guglielmi's two-step adsorption model [25]. The model suggests that increasing the particle content in the bath increases their availability to the cathode thus enhancing codeposition.

4.1.2. Corrosion resistance

The main challenge that is faced by steels in marine and coastal environments is corrosion degradation, and enhanced corrosion resistance is one of the key properties that the fabricated coatings must exhibit. The chloride in the sea water accelerates material loss in this atmosphere. $Zn–TiO_2$ composite coatings exposed corrosive atmosphere of 3.65% NaCl proved to possess better corrosion resistance than mild steel as studied by [35]. The embodiment of the particles brought a positive potential shift of 0.38 V and decreased the current density from 0.07 to 0.0000211 A/cm^2. Similar results were obtained by the authors when SnO_2 particles were incorporated into the matrix. The inclusion of the second phase particles into the metallic matrix induces formation of stable film for further protection against corrosion. The passive film acts a physical barrier separating the fresh matrix from the corrosive media. The distribution of particles in the matrix also has a strong bearing on the corrosion resistance. Samples with uniform distributed particles have better anti-corrosive properties than those with agglomerated particles. Therefore, it is mandatory for the production of the composite coatings to be optimized to produce coatings with better particle dispersion and enhanced microstructure.

4.1.3. Other properties

4.1.3.1. Hardness

Incorporation of submicron particles into metal matrix is known mainly for their excellent improvement of mechanical properties. The fabrication of Zn composite coatings from baths solution containing 5 and 10 g/L of submicron alumina particles drastically increased the hardness of Zn from 1.77 to 2.55 and 2.37 GPa, respectively [24]. The irregular morphological characteristics exhibited by samples fabricated from 10 g/L might have been responsible for reduction in hardness. Incorporation of second phase particles into the matrix impedes the growth of grains leading to the formation of microstructures with small-sized crystals [36]. Therefore, uneven distribution of particles in the matrix promotes formation of microstructures with chemical and physical inhomogeneity. Multiple loading of particles has also been found to improve the matrix by [52]. The incorporation of TiC–TiB mixed ceramic particles into Zn matrix have positive synergetic effect on the mechanical properties. The ternary composite exhibited high hardness values as compared to Zn–TiC and Zn–TiB. The synergetic properties of the mixed reinforcement particles improve the properties of the individual ones.

4.1.3.2. Wear resistance

Wear resistance is one of extractive properties that coatings should possess for durability and increased life span during service. However, zinc coatings are known for their poor performance when they are subject to wear environment. The development of composite coatings also aims at improving this property to extend the applications of the coatings. The inclusion of reinforcement particles hinder plastic deformation due to the dispersive strengthening effect exhibited by the particles [51]. The authors also reported that the alteration of microstructure and strengthening of the coatings by formation of compact and crack-free coating add to the minimization of friction due to the presence of embedded particles The particles also form a protective barrier between the matrix and the wearing medium, thus reducing direct contact and friction coefficient [53].

4.2. Zn nanocomposite coatings

4.2.1. Morphological and structural properties

The inclusion of nanoparticles in a metal matrix has been reported to induce grain refinement and modify the structural characteristics of the matrix. During codeposition, second phase particles are adsorbed on the growing crystal and retard its further growth. The adsorbed particles increase the number of nucleation sites, which results in reduction of the grain size [36]. The morphology of composite coatings depends mainly on the manner of particle dispersion and content of particles in the metal matrix [4]. Morphological and structural analysis results obtained by [54] revealed a reduction of crystal size of Zn matrix due to the incorporation of TiO_2 nanoparticles. The nanocomposite coating followed the same morphological pattern with the matrix, but with smaller and smooth grains. The deposits also showed minimal surface defects such as pores and microholes. $Zn–TiO_2$ nanocomposites produced from baths containing the highest particle concentration (16 g/L) have been reported to affect the morphology of the plain Zn coatings [37]. The nanocomposite coatings exhibited surface morphology different from the matrix. High particle content and orientation of particles (agglomeration) in the coating causes an evolution of a new structure. The introduction of SiC were found to cause deposition overpotential leading to the formation of coatings with grains perpendicularly oriented to the substrate as compared to the parallel hexagonal plates showed by Zn matrix [55]. The incorporation of the particles onto the cathode was also reported to perturb horizontal grain growth in the coatings. Therefore, the morphological characteristics of the coatings depend solely on the bath parameters and formulation.

4.2.2. Corrosion resistance

Addition of second phase particles to Zn matrix promotes the formation of fine microstructure with good corrosion resistance. The particles perturb the growth of Zn crystals and induce the formation of new nucleation sites resulting in small-sized crystals and a microstructure with minimal defects [13]. Uniform dispersion of the nanoparticulates in the matrix fills the microholes, pores, crevices, gaps and other surface defects of the coatings leading to the formation of compact microstructures. These defects can serve as active sites for corrosion if present. The particles have also been reported to act as physical barriers between corrosive media and the matrix, thus minimizing the occurrence of defect corrosion [55]. Codeposition of black carbon nano-sized particles with Zn reduced the current density when the coatings were tested in 3.5% sodium chloride solution [56]. The metal dissolution of the composites was also found to occur at high anodic potentials at a steady rate as compared to the Zn coating. [36] reported zirconia nanoparticles to influence the kinetics of the anodic and cathodic electrochemical reactions. Therefore, the particles can promote the perturbation of anodic reaction and favour the formation of a stable passivation layer. Therefore, the presence of second phase particles in the matrix has positive effect on the corrosion resistance of the coatings. However, optimum particle loading has to be established to obtain coatings with enhanced anti-corrosive properties. Agglomeration of particles in the coating was found to occur when Al_2O_3 and SiO_2 particle contents in the electrolyte solution were beyond optimal concentrations [57]. The authors established optimal particle loading for excellent corrosion resistance is 5 g/L for both Al_2O_3 and SiO_2. Beyond these optimal conditions, the corrosion resistance of the coatings was compromised. Incorporation of non-uniform agglomerates in the matrix promotes formation of surface defects and causes chemical heterogeneities in the coating. This allows penetration of the corrosive media into the matrix in weak sites, leading to the degradation of the corrosion resistance of the entire system.

4.2.3. Other properties

4.2.3.1. Hardness

Microhardness of a metal matrix mainly depends on the microstructural properties of that particular matrix. When force or load is exerted on a material, the dislocation movements are induced which lead to the weakening of the material's mechanical properties [58]. Incorporation of nanoparticles into a matrix perturbs dislocation movement and grain boundary sliding, thus improving the hardness of the composite coatings. Grain refinement and uniform incorporation of reinforced particles result in improvement of hardness. [56] obtained an increase in microhardness from 72 to 90 HV when 2 g/L of nano-sized carbon particles were incorporated into Zn matrix. The composite coating also showed improved morphology with small-sized crystals. Another factor that affects microhardness of a Zn matrix is the content of the reinforcement material added into the coating. The increase of bath particle content increases the chances of the particles to be incorporated into the metal matrix [54-55].

4.2.3.2. Wear resistance

The coatings' resistance to wear degradation is one of the important properties sought for metal matrix composites to exhibit. Many industrial applications where these coatings are employed require them to possess excellent wear resistance. The addition of reinforcement into a metal matrix promotes grain reduction and hinders plastic deformation of the material [59]. This is attributed to particles' grain refining and dispersive strengthening characteristics exhibited by these additives. The reinforcement particles also possess self-lubricating properties and hence reduce friction, thus improving the wear resistance of the matrix. The incorporation of nano-sized TiO_2 into Zn matrix carried out by [54] reduced the wear loss of Zn from 11 to 5 mg. The results obtained by [35] showed an improvement in wear resistance and reduction of friction coefficient of the Zn matrix due to the addition of TiO_2 particles.

4.3. Zn alloy nanocomposite coatings

There are several zinc alloy nanocomposite coatings ranging from Zn-Ni, Zn-Co, Zn-Fe, etc. to matrix-based composites. Zn-Ni matrix nanocomposites have received much attention as compared to the others. Nickel is chemically stable and has better mechanical properties than Zn, and its incorporation into Zn matrix notably improves the corrosion and wear resistance of the coatings.

4.3.1. Composition, morphological and structural properties

Uniformly distributed nano-sized Al_2O_3 incorporated into Zn-Ni matrix improves the compaction of grains in the coatings and minimizes surface defects [60]. The crystallite size of Zn-Ni matrix was reduced from 40.9 to 20.68 nm. The crystallite size was found to be dependent on the particle content of nano-Al_2O_3 present in the solution. The authors also reported the increment of nickel content in the coatings due to the addition of the reinforcement particles. The work conducted by [41] confirms these findings. Zn_5Ni_{21} phase was increased as the result of incorporation of the nanoparticles into the matrix. [39] also showed similar results when SiO_2 nanoparticles were added. The nickel content was increased from 6.3 to 12.3 wt.% while zinc was decreased from 93.7 to 87.7 wt.%. Inclusion of particles into the matrix has significant influence on the surface morphology of the resulted deposits. Al_2O_3 solutions changed hemispherical nodules of Zn-Ni to cauliflower-like surface morphology [61].

4.3.2. Corrosion resistance

The automobile industry extensively uses Zn-Ni coatings for corrosion protection of steel. These coatings are chemically stable than their zinc counterparts. Addition of second phase nanoparticles into Zn-Ni enhances anti-corrosive and mechanical properties. Al_2O_3 nanoparticles have been incorporated into Zn-Ni matrix from alkaline solutions by [41] The coatings fabricated from baths containing 5g/L Al_2O_3 possessed excellent corrosion resistance when tested in 0.2g/L Na_2SO_4 electrolyte. The corrosion current was drastically reduced, and the polarization resistance increased. These findings are similar to those obtained by [60] where the current density was decreased from 0.19 to 0.092 $\mu A/cm^2$. The results obtained by the authors show that the particles impart inhibitory effect to the corrosion of the matrix. Al_2O_3 particles exhibit low level of electronic conductivity and can disturb corrosion current when they present in the coating. SiO_2 nanoparticles codeposited with Zn-Ni matrix have reported to protect the matrix from corrosion in 3% NaCl solution [39]. Electronegative Zn rich η phase has been found to improve the protective life of the coatings.

4.3.3. Other properties

Other properties that Zn-Ni–based nanocomposite coatings possess include high hardness and good thermal stability. [60] obtained improvement of microhardness of 640 HV when they incorporated Al_2O_3 nanoparticles into Zn-Ni matrix with the aid of ultrasound for bath agitation. $Zn–Ni–SiO_2$ nanocomposite coatings produced from bath containing 5g/L SiO_2 increased the microhardness value of the matrix from 140 to 396 HV [39]. The microhardness values of the composite remained higher than that of the matrix even when the samples were subjected to annealing temperatures of 200 °C for a duration of 24 hours (shown in Figure 9) Second phase reinforcement particles have pinning effect on the grain growth of the coatings due to elevated temperatures. The particles reduce surface oxidation and relaxation of internal stressed.

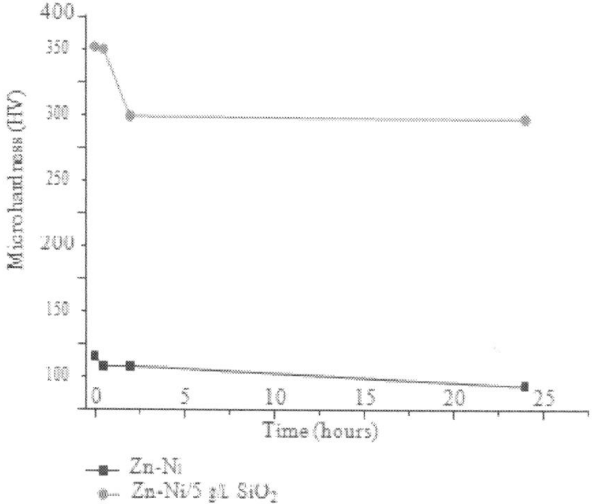

Figure 9. Microhardness evolution of Zn-Ni and $Zn-Ni-SiO_2$ nanocomposite coatings subjected to heat treatment conditions of 200 °C for 0.5, 2 and 24 hours time range.

5. Conclusions

From the current review, it can be concluded that electrodeposited Zn composite coatings have high potential to serve as protective films for protecting components and pieces that are made of mild steel from corrosion in coastal and marine environments. Their improved surface properties such as corrosion resistance, high hardness, low friction and wear resistance due to the incorporation of second phase particles enable them to find various applications. The advantage of inclusion of inert ceramic particles in metal deposits is that harder, and chemically stable coatings can be produced at reduced thickness. However, the quality of electrodeposited coatings is not dependent on the presence of reinforcement particles only but on a number of process parameters (pH, plating current, temperature, deposition time, stirring speed and bath composition); hence, optimization of the process is required to produce desired and high quality coatings. More work on further improvement of electrodeposited zinc coatings is required to enable them to serve as alternative or replacements to other coatings produced by high cost and energy intensive surface modification processes.

Author details

Patricia A.I. Popoola*, Nicholus Malatji and Ojo Sunday Fayomi

*Address all correspondence to: popoolaapi@tut.ac.za

Department of Chemical, Metallurgical and Materials Engineering, Tshwane University of Technology, Pretoria, South Africa

References

[1] Alcantara, J., Chico, B., Diaz, I., de la Fuente, D. & Morcillo, M. 2015. Airborne Chloride Deposit and its Effect on Marine Atmospheric Corrosion of Mild Steel. *Corrosion Science*, 97, 74 – 88.

[2] James, M.N. & Hattingh, D.G. 2015. Case Studies in Marine Concentrated Corrosion. *Engineering Failure Analysis*, 47, 1 – 15.

[3] Popoola, A.P.I., Olorunniwo, O.E. & Ige, O.O. 2014. Developments in Corrosion Protection.Intech.

[4] Ranganatha, S. Venkatesha, T.V. Vathsala, K. & Punith Kumar, M.K. 2012. Electrochemical Studies on Zn/nano-CeO$_2$ Electrodeposited Composite Coatings. *Surface & Coatings Technology*, 208, 64–72.

[5] Roberge, P.R. 2008. *Corrosion Engineering, Principles and Practice*. New York, NY: McGraw Hill.

[6] De La Fuente, D., Castano, J.G. & Morcillo, M. 2007. Long-term Atmostpheric Corrosion of Zinc. *Corrosion Science*, 49, 1420 – 1436.

[7] Azmat, N.S., Ralston, K.D., Muddle, B.C. & Cole, L.S. 2011. Corrosion of Zn under Acidified Droplets. *Corrosion Science*, 53, 1604 – 1615.

[8] Veleva, L., Acosta, M. & Meraz, E. 2009. Atmospheric Corrosion of Zinc Induced by Runoff. *Corrosion Science*, 51, 2055 – 2062.

[9] Sylla, D., Creus, J., Savall, C., Roggy, O., Gadouleau, M. & Refait, P.H. 2003. Electrodeposition of Zn-Mn Alloys on Steel From Acidic Zn-Mn Chloride Solutions. *Thin Solid Films*, 424, 171-178.

[10] Hosking, N.C., Strom, M.A., Shipway, P.H. & Rudd, C.D. 2007. Corrosion Resistance of Zinc-Magnesium Coated Steel. *Corrosion Science*, 49, 3669–3695.

[11] Popoola, A.P.I & Fayomi, O.S. 2011b. Performance Evaluation of Zinc Deposited Mild Steel in Chloride Medium. *International Journal of Electrochemical Science*, 6, 3254–3263.

[12] Lodhi, Z.F., Mol, J.M.C., Hovestad, A., Terryn, H. & DE Wit J.H.W. 2007. Electrodeposition of Zn-Co and Zn-Co-Fe Alloys from Acidic Chloride Electrolytes. *Surface & Coatings Technology*, 202, 84–90.

[13] Punith Kumar, M.K & Srivastava, C. 2013. Morphological and Electrochemical Characterization of Electrodeposited Zn-Ag Nanoparticle Composite Coatings. *Material Characterization*, 85, 82–91.

[14] Benea, L., Bonora, P.L., Borello, A. & Martelli, S. 2001. Wear Corrosion Properties of Nanostructured Sic-Nickel Composite Coatings obtained by Electroplating. *Wear*, 249, 995-1003.

[15] Khan, W.S. & Asmatulu, R. 2013. *Nanotechnology Emerging Trends, Markets, and Concerns. Nanotechnology Safety*. Amsterdam: Elsevier.

[16] Tjong, S.C. & Chen, H. 2004. Nanocrystalline Materials and Coatings. *Materials Science Engineering*, 45, 1–88.

[17] Chandler, K.A. & Bayliss D.A. 1985. *Corrosion Protection of Steel Structures*. England: Elsevier Applied Science Publishers Ltd.

[18] Revie, R.W. & Uhlig, H.H. 2008. *Corrosion and Corrosion Control: An Introduction to Corrosion Science and Engineering*. New York: John Wiley and Sons, Inc.

[19] Park, H. 1997. The Role of Texture and Morphology in Optimizing the Corrosion Resistance of Zinc-based Electrogalvanized Coatings. D.Phil. Thesis. Montreal. McGill University.

[20] Palcic, I., Balazic, M., Milfelner, M. & Buchmeister, B. 2009. Potential of Laser Engineered Net Shaping (LENS) Technology. *Materials and Manufacturing Processes*, 24(7–8), 750–753.

[21] Mudali, U.K. & RAJ, B. 2008. Corrosion Science and Technology: Mechanisms, Mitigation and Monitoring. New Delhi: Narosa Publishing House.

[22] Nemes, P.L., Lekka, M., Fedrizzi, L. & Muresan, L.M. 2014. Influence of the Electrodeposition Current Regime on the Corrosion Resistance of Zn–CeO_2 Nanocomposite Coatings. *Surface Coatings and Technology*, 252, 102 – 107.

[23] Sudagar, J., Lian, J. & Sha, W. 2013. Electroless Nickel, Alloy, Composite and Nano Coatings – A Critical Review. *Journal of Alloy and Compounds*, 571, 183–204.

[24] Sancakoglu, O., Culha, O., Toparli, M., Agaday, B., & Celik, E. 2011. Co-deposited Zn-submicron Sized Al_2O_3 Composite Coatings: Production, Characterization and Micromechanical Properties. *Materials and Design*, 32, 4054–4061.

[25] Guglielmi, N. (1972). Kinetics of the Deposition of Inert Particles from Electrolytic Baths. *Journal of the Electrochemical Society*, 119, 1009–1012.

[26] Hwang, B.J. & Hwang, C.S. 1993. Mechanism of Codeposition of Silicon Carbide with Electrolytic Cobalt. *Journal of Electrochemical Society*, 140, 979–984.

[27] Bercot, P., Pena-Munoz, E. & Pagetti. 2002. Electrolytic Composite Ni-PTFE Coatings: An Adaptation of Guglielmi's Model for the Phenomena of Incorporation. *Surface Coatings Technology*, 157, 282.

[28] Roos, J.R., Celis, J.P., Fransaer, J. & Buelens, C. 1990. The Development of Composite Plating for Advanced Materials. *Journal of Materials*, 42, 60–73.

[29] Khan, T.R., Erbe, A., Auinger, M., Marlow, F. & Rohwerder, M. 2011. Electrodeposition of Zinc-silica Composite Coatings: Challenges in Incorporating Functionalized Silica Particles into a Zinc Matrix. *Science and Technology of Advanced Materials*, 12(5), 1–9.

[30] Srivastava, M. Balaraju, J.N., Ravishankar, B. & Rajam, K.S. 2010. Improvement in the Properties of Nickel by Nano-Cr_2O_3 Incorporation. *Surface & Coatings Technology*, 205, 66–75.

[31] Saha, R.K. & Khan, T.I. 2010. Effect of Applied Current on the Electrodeposited Ni–Al_2O_3 Composite Coatings. *Surface Coatings and Technology*, 205, 890-895.

[32] Bouchaud, B., Balmain, J., Bonnet, G. & Pedraza, F. 2013. Optimizing Structural and Compositional Properties of Electrodeposited Ceria Coatings for Enhanced Oxidation Resistance of a Nickel-based Superalloy. *Applied Surface Science*, 268, 218–224.

[33] Popoola, A.P.I. & Fayomi, O.S.I. 2011a. Effect of Some Process Variables on Zinc Coated Low Carbon Steel Substrates. *Scientific Research and Essays*, 6(20), 4264–4272.

[34] Tuaweri, T.J., Adigio, E.M. & JOMBO, P.P. 2013. Influence of Process Parameters on the Cathode Current Efficiency of Zn/SiO_2 Electrodeposition. *International Journal of Mechanical Engineering and Applications*, 1(5), 93–99.

[35] Fayomi, O.S.I., Popoola, A.P.I. & Loto, C.A. 2014. Tribo-Mechanical Investigation and Anti-Corrosion Properties of $Zn-TiO_2$ Thin Film Composite Coatings from Electrolytic Chloride Bath. *International Journal of Electrochemical Science*, 9, 3885 – 3903.

[36] Vathsala, K. & Venkatesha, T.V. 2011. $Zn-ZrO_2$ Nanocomposite Coatings: Electrodeposition and Evaluation of Corrosion Resistance. *Applied Surface Science*, 257, 8929–8936.

[37] Fustes, J., Gomes, A. & Da Silva Pereira, M.I. 2008. Electrodeposition of Zn-TiO$_2$ Nanocomposite Films-effect of Bath Composition. *Solid State Electrochemistry*, 12, 1435–1443.

[38] Low, C.T.J., Wills, R.G.A. & Walsh, F.C. 2006. Electrodeposition of Composite Coatings Containing Nanoparticles in a Metal Deposit. *Surface & Coatings Technology*, 201, 371–383.

[39] Hammami, O., Dhouibi, L., Bercot, P., Rezrazi, E. & Triki, E. 2011. Study of Zn-Ni Alloy Coatings Modified by Nano-SiO$_2$ Particles Incorporation. *International Journal of Corrosion*, 12, 1–8.

[40] Lee, C.K. 2012. Wear and Corrosion Behavior of Electrodeposited Nickel–Carbon Nanotube Composite Coatings on Ti–6Al–4V Alloy in Hanks' Solution. *Tribology International*, 55, 7–14.

[41] Blejan, D. & Muresan, L.M. 2013. Corrosion Behavior of Zn-Ni-Al$_2$O$_3$ Nanocomposite Coatings Obtained by Electrodeposition from Alkaline Electrolytes. *Materials and Corrosion*, 64, 433–438.

[42] Muralidhara, H.B., Arthoba Naik, Y., Balasubramanyam, J., YOGESH KUMAR, K., H. & VEENA, M.S. 2012. Nanocrystalline Zinc Coating on Steel Substrate using Condensation Product Of Glycyl-Glycine (GGL) and Vanillin (VNL) and Its Corrosion Study. *International Journal of Chemical Sciences*, 10(1), 524–538.

[43] Zhau-XIA, N., Fa-He, C., Wei, W., Zhau, Z., Jian-Qing, Z. & CHU-NAN, C. 2007. Electrodeposition of Ni-SiC Nanocomposite Film. *Transaction of Nonferrous Metals Society of China*, 7, 9–15.

[44] Helle, K. & Walsh, T.F. 1997. Electrodeposition of Composite Layers Consisting of Inert Inclusions in a Metal Matrix. *Transactions of the Institute of Metal Finishing*, 75, 53–58.

[45] WANG, S. 2004. Studies of Electroless Plating of Ni–Fe–P Alloys and the Influences of Some Deposition Parameters on the Properties of the Deposits. *Surface Coatings and Technology*, 186, 372–376.

[46] Haciismailoglu, M. & Alper, M. 2011. Effect of Electrolyte pH and Cu Concentration on Microstructure of Electrodeposited Ni-Cu Alloy Films. *Surface & Coatings Technology*, 206, 1430–1438.

[47] Naik, Y.A., Venkatesha, T.V. & Nayak, P.V. 2002. Electrodeposition of Zinc from Chloride Solution. *Turkish Journal of Chemistry*, 26, 725 – 733.

[48] Fayomi, O.S, Tau, V.R., Popoola, A.P.I, Durodola, B.M., Ajayi, O.O., Loto, C.A. & Inegbenebor O.A. 2011. Influence of Plating Parameter and Surface Morphology on Mild Steel. *Journal of Materials Environment Science*, 2(3), 271–280.

[49] Xia, X., Zhitomirsky, I., Mcdermid, J.R. 2009. Electrodeposition of Zinc and Composite Zinc-yttria Stabilized Zirconia Coatings. *Journal of Materials Processing Technology* 209, 2632–2640.

[50] FAYOMI, O.S.I, ABDULWAHAB, M. & POPOOLA, A.P.I. 2013. Properties Evaluation of Ternary Surfactant-induced Zn-Ni-Al$_2$O$_3$ Films on Mild Steel by Electrolytic Chemical Deposition. *Journal of Ovonic Research*, 9(5), 123–132.

[51] Fayomi, O.S.I. And Popoola, A.P.I. 2015. Anti-Corrosion and Tribo-Mechanical Properties of Co-deposited Zn–SnO$_2$ Composite Coating. *Acta Metallurgica Sinica (English Letters)*, 28, 521-530.

[52] Fayomi, O.S.I., Aigbodion, V.S. & Popoola, A.P.I. 2015. Properties of TiC/TiB Modified Zn-TiC/TiB Ceramic Composite Coating on Mild Steel. *Journal of Failure Analysis and Prevention*, 15, 54–64.

[53] Malatji, N. & Popoola, A.P.I. 2015. Electrodeposition of Ternary Zn-Cr$_2$O$_3$-SiO$_2$ Nanocomposite Coating on Mild Steel for Extended Applications. *International Journal of Electrochemical Science*, 10, 3988 – 4003.

[54] Praveen, B.M. & Venkatesha, T.V. 2008. Electrodeposition and Properties of Zn-nanosized TiO$_2$ Composite Coatings. *Applied Surface Science*, 254, 2418–2424.

[55] Sajjadnejad, M., Mozafari, A., Omidvar, H. & Javanbakht, M. 2014. Preparation and Corrosion Resistance of Pulse Electrodeposited and Zn-SiC Nanocomposite Coatings. *Applied Surface Science*, 300, 1–7.

[56] Praveen, B.M. & Venkatesha, T.V. 2009. Generation and Corrosion Behavior of Zn-Nano Sized Carbon Black Composite Coating. *International Journal of Electrochemical Science*, 4, 258–266.

[57] Malatji, N., Popoola, A.P.I., Fayomi, O.S.I. & Loto, C.A. 2015. Multifaceted Incorporation of Zn-Al$_2$O$_3$/Cr$_2$O$_3$/SiO$_2$ Nanocomposite Coatings: Anti-corrosion, Tribological, and Thermal Stability. *The International Journal of Advanced Manufacturing Technology*, 1 – 7.

[58] Borkar, T. 2010. Electrodeposition of Nickel Composite Coatings. M.Sc. dissertation, Oklahoma, Oklahoma State University.

[59] Yu, S.R., Liu, Y., Li, W., Liu, J.A. & YUAN, D.S. 2012. The Running-in Tribological Behavior of Nano-SiO$_2$/Ni Composite Coatings. *Composites: Part B*, 43, 1070–1076.

[60] Zheng, H. & AN, M. 2008. Electrodeposition of Zn–Ni–Al$_2$O$_3$ Nanocomposite Coatings under Ultrasound Conditions. *Journal of Alloy and Compounds*, 459, 548–552.

[61] Ghaziof, S. & Gao, W. 2015. The Effect of Pulse Electroplating on Zn-Ni Alloy and Zn-Ni-Al$_2$O$_3$ Composite Coatings. *Journal of Alloys and Compounds*, 622, 918–924.

Chapter 8

Dynamic Characterization of Open-ended Pipe Piles in Marine Environment

Dezi Francesca, Gara Fabrizio and Roia Davide

Additional information is available at the end of the chapter

http://dx.doi.org/10.5772/62055

Abstract

This chapter is focused on the experimental investigations that can be carried out to dynamically characterize open-ended pipe piles in marine environment. Different test typologies, such as impact load test, free vibration test, forced vibration test, and ambient vibration test, are presented and described with the purpose to provide the right tools to analyze the dynamic behavior, at both small and large strains, of single piles or a system of piles. The appropriate instrumentation, with the suitable protection from marine environment and pile driving installation procedure, is also illustrated. Furthermore, the most common signal processing techniques useful for handling the experimental raw data are addressed together with the analysis techniques for the evaluation of the modal parameters: natural frequencies, damping ratios, and mode shapes. Finally, a part of the experimental campaign carried out by the authors on near-shore open-ended pipe piles is reported as a case study.

Keywords: Dynamic characterization, near-shore pipe piles, free vibration test, impact load test, soil-pile interaction

1. Introduction

The dynamic behavior of piles under lateral loading has received much attention in recent years due to its application in machinery foundations and structures exposed to dynamic loads such as wind and earthquakes. However, the dynamic behavior of piles and pile groups is far from being completely understood, as the soil-pile and the pile-soil-pile interactions are very complex phenomena especially in the case of near-shore and offshore structures. Structures that are usually founded on single pile or pile groups (such as offshore and near-shore platforms, wind turbines, docks, jetties, wharfs, mooring structures, and bridge piers) might become subjected to significant dynamic effects due to wave loads, wind actions, mooring and

impact forces from ships, as well as earthquake motions. When subjected to dynamic load, the behavior of the soil-foundation system depends on the characteristics of the surrounding soil, the configuration of the foundation, the inertia of the superstructure, and the nature of the dynamic excitation. In particular, the stiffness and damping characteristics of a soil-water-pile system (impedance function) exhibited during an earthquake motion or any dynamic action on superstructures founded on piles depend on the mechanical and geometric properties of the soil, pile, and their mutual interaction and must be included in the structural model if accurate predictions are desired. In this sense, the estimation of soil-water-pile system stiffness and damping is a key step not only in the project phases, when a variety of different foundation configurations should be compared and a choice on the basis of the structure's functional requirements should be done, but also in the structural modeling and in the verification procedures. For both research and design purposes, the soil-foundation interaction may be approached with a direct method, modeling the whole dynamic soil-pile system with numerical methods or using theoretical approaches, i.e. distinct methods to evaluate kinematical and inertial effects, with different degrees of refinement. It is worth pointing out that the results of these methods are very sensitive to many parameters that define the dynamic characteristics of the soil-pile system. Experimental results carried out on full- or small-scale in situ and laboratory tests are essential for the accurate calibration and validation of these models. Over the past decade, the understanding of dynamic interaction between pile foundations and the surrounding soil has advanced significantly and a large number of theoretical studies have been published following the numerical and analytical approaches. In the field of experimental tests, a variety of small-scale model tests have been performed to improve the understanding of the effects of soil and pile parameters on the system response. On the contrary, very few dynamic tests with lateral excitation on full-scale piles in field conditions (and even less on near-shore/offshore piles [1, 2]) have been reported in the literature. Full-scale dynamic pile response tests are essential as a point of reference with which to compare results of scale-model tests and numerical analyses. In situ tests have the advantage of providing "real" soil and pile stress conditions, whereas laboratory tests offer the opportunity of performing parametric studies under known soil conditions; obviously, taken together, field and laboratory tests assure an accurate and adequate knowledge completing each other.

It is worth noting that, since a seismic soil motion cannot be reproduced in field tests, theoretical methods for seismic analysis cannot be validated by directly comparing numerical and experimental results. However, a large part of these theoretical methods, including those dedicated to the kinematical effects, can be easily extended to simulate the dynamic behavior of soil-pile systems not only under the incoming earthquake motion but also under dynamic lateral loads at the pile head. Therefore, a calibration and validation effort of these methods can be carried out by means of a variety of dynamic field tests.

Results of combined forced and free vibration (snap-back) tests have been reported by Blaney and O'Neill [3] for an instrumented steel closed-ended pipe pile capped with a rigid mass in a deposit of stiff to very stiff, moderately fissured, overconsolidated clay. The frequency down-sweep loads were applied by a large inertial vibrator rigidly connected to the pile cap and the loading frequencies were in the range of seismic or low-frequency machine loadings. Crouse et al. [4] presented the results of combined forced and free vibration tests on a pipe pile in soft saturated peat. In these highly compressible soils, the measurement of the strength properties (provided by standard laboratory and in situ testing methods) for use in pile design has proven not to be meaningful [5], so large- or full-scale static and dynamic load tests of a pile or pile group could be important for the characterization of the behavior of piles. Sa'don et al. [6] performed a series of dynamic tests ranging from low excitation (using an instrumented impact hammer and a low-mass loading of an eccentric mass shaker) to high dynamically induced force from the eccentric mass shaker on hollow steel pipe piles in stiff clay. This study has investigated the nonlinear response of the soil-pile system due to the strain softening of the soil and the formation of a gap between the pile shaft and the surrounding soil. Dynamic and slow cyclic tests have also been conducted in saturated silty sand by Jennings et al. [7]. Cyclic loads were applied using a push-pull jack mounted between the piles, whereas dynamic loads were generated using a variable frequency shaking machine on top of one of the piles. This research focused on evaluating the coefficient of subgrade reaction from testing. Ting [8] reported the results of a full-scale experimental study on a single free-headed pile embedded in saturated beach sand subjected to lateral harmonically varying dynamic loading. At large deflection levels, partial liquefaction and formation of a gap occurred around the pile head, considerably reducing the natural frequencies of the soil-pile system.

This chapter deals with experimental investigations for the dynamic characterization of open-ended pipe piles in marine environment. The main objective is to analyze the dynamic behavior, at both small and large strains, of single piles or a system of piles. Different test typologies, such as impact load test, snap-back test (free vibration test), forced vibration test, and ambient vibration test, are described. The appropriate instrumentations (strain gauges, accelerometers, displacement transducers, and pore pressure transducers) with the suitable protection from marine environment and pile driving installation procedure are also illustrated, giving suggestions about the minimum number and the possible placement configurations of sensors. Furthermore, the most common signal processing techniques useful for handling the experimental raw data are addressed together with the analysis techniques for the evaluation of the modal parameters: natural frequencies, damping ratios, and mode shapes. Finally, a part of the experimental campaign carried out by the authors on near-shore open-ended pipe piles at the "Mirabello" harbor in La Spezia (Italy) is reported as a case study. Piles are vibrodriven into soft marine clay and subjected to horizontal loading, i.e. impact load, snap-back. The modal properties of the soil-water-pile system are identified by means of experimental modal analyses; the values of natural frequencies are estimated from the peaks of frequency response functions, whereas the mean values of damping ratios of the system are evaluated by studying the free vibrations with the logarithmic decrement method. The results of the experimental modal analyses obtained from free vibration tests are presented with regard to the single pile (soil-water-pile system) by analyzing the measurements of the strain gauges installed along the pile both in the portion immersed in water and in the portion embedded in soil. The effects of nonlinearities developing in the soil for increasing force levels of free vibration test are commented; moreover, the linear response of the system is analyzed by comparing the results of free vibration tests with those obtained from the impact load tests. Finally, a comparison among the impedance functions obtained experimentally and numerically (with a 3D finite-element model) is reported.

2. Test typologies

Several test typologies can be useful to identify the modal parameters of a soil-water-pile system; most of them come from the fields of aerospace and mechanical engineering and have been recently developed for applications in civil engineering fields. Those tests can be classified based on the input that can be artificially and expressly created for the test or can be naturally present in the ambient; in most cases, artificial input is applied at the pile head and can be measured, whereas ambient input is usually due to waves coming from soil, water, or air surrounding the pile and cannot be easily measured. In the sequel, the most useful and effective test typologies are described: stepped-sine and sweep-sine tests, impact load and snap-back tests among the typologies requiring an artificial input, and ambient vibration test as an example of tests based on natural input.

The most popular typologies of testing are the stepped-sine and sweep-sine tests [9], also known as forced vibration tests, where the excitation force contains one single frequency at a time (harmonic signal) and either sweeps with a fixed frequency step or varies slowly but continuously within an established frequency range [10], allowing the structure to engage in one harmonic vibration at a time. The input is applied with mechanical or hydraulic shaker located at the pile head (Figure 1a), whereas the response can be recorded with strain gauges or accelerometers installed along the pile shaft. This testing is effective for studying the dynamic behavior of a soil-pile system under both low and high vibration levels characterizing even the nonlinearities of the system [9].

Another method, very popular because of its execution simplicity and low cost of the instrumentation required, is the impact load testing. In this case, the input is impulsive and is applied by means of a hammer (Figure 1b) instrumented with a load cell detecting the magnitude of the impact. The response can be recorded with accelerometers or strain gauges located along the pile length. The investigated frequency range depends on the stiffness of the hammer tip and of the pile surface; generally, a soft tip concentrates the impact energy at a low-frequency range, whereas a hard tip permits the investigation of a wider frequency range. The main limit of impact load testing is that only the linear elastic behavior of the soil-water-pile system can be investigated unless a heavy loading system is used as in the case of Statnamic test [11].

The snap-back test (or free vibration test) can be viewed as a variant of the impact load test [10]. In this testing procedure, the load is gradually applied at the pile head by means of a steel cable pulled with a hydraulic jack (Figure 1c). Then, a quick release of the system can be obtained in two different ways: by cutting the cable with a blowtorch or by inducing the failure for traction of a steel pin placed along the cable between the jack and the pile; after the failure of the pin, which can be opportunely calibrated to obtain the desired load level, the pile undergoes a number of steadily decreasing oscillations around its equilibrium position. The load applied before the release is usually recorded with a load cell located in the loading apparatus, between the jack and the pile. The snap-back test may be performed at increasing levels of force allowing the investigation of the nonlinear behavior of the soil-water-pile system; on the other hand, a snap-back test can excite only a mode shape or a combination of mode shapes, which are generally relevant to the first natural frequencies of the system while the higher modes cannot be investigated.

In the ambient vibration testing, the system is excited only by natural vibrations (e.g. micro-tremors, wind, wave load, and anthropic activities) and artificial excitation is not needed (Figure 1d). This is the main advantage of this technique, which consequently requires small, light, and very portable instrumentation entailing a great reduction of costs and times for the test preparation. On the other hand, in order to record the very low amplitude vibrations induced by natural excitation, specific low noise accelerometers are required. Similar to the impact load test, the natural excitation is assumed to have a flat spectrum such as that of a white noise; therefore, the dynamic characteristics of the system can be investigated in a wide range of frequency even if, because of the very low level of vibrations, only the linear behavior of the system can be studied. In order to investigate the dynamic behavior of the system under a higher level of excitations coming from the ambient, particularly, propagating through the soil, two different strategies can be considered: large-scale blasting inducing artificial ground motion or natural soil motion induced by earthquakes. In the latter case, a permanent moni-toring system must be installed on the system and should be equipped with a trigger system to start recording at the beginning of the seismic event.

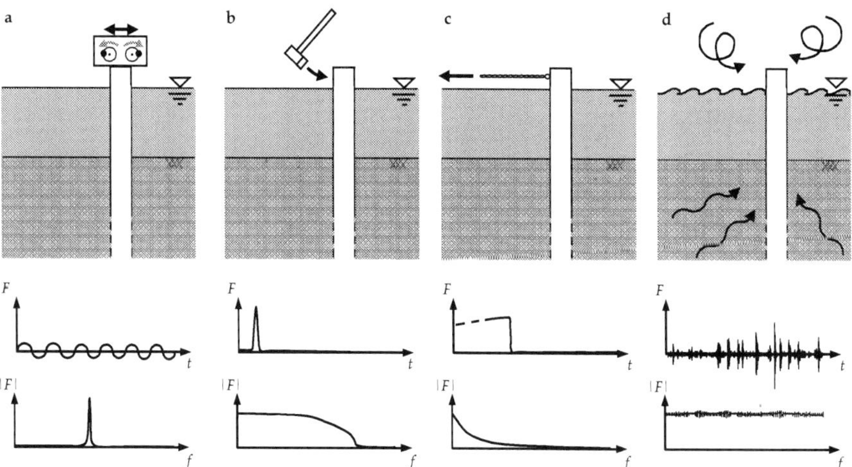

Figure 1. Different typologies of tests on an offshore pipe pile and relevant time histories and frequency content of the input: (a) stepped-sine test, (b) impact load test, (c) snap-back test, and (d) ambient vibration test.

3. Pile instrumentation

This chapter deals with some issues related to the instrumentation necessary for the dynamic characterization of open-ended pipe piles in marine environment. First, the quantities that can be measured and the instrumentation layout are discussed. Then, various typologies of transducers available in commerce and their characteristics are illustrated. Finally, some protection systems available to prevent malfunctioning caused by the marine environment are reported.

3.1. Measured quantities and instrumentation layout

Different measurement strategies can be adopted to characterize the dynamic behavior of open-ended pipe piles in marine environment: typology of sensors and instrumentation layout mainly depend on the objectives of the experimental campaign and also on the dynamic characteristics investigated, whereas the number of sensors depends essentially on the desired refinement of the expected results, e.g. the definition of the mode shape along the pile or the accuracy and reliability of the values of frequencies and damping ratios.

In order to identify the dynamic characteristics of a system, dynamic quantities such as accelerations and/or displacements should be measured; to do this, accelerometers and/or displacement transducers may be installed on the pile surface. In general, acceleration measurements are commonly preferred because acceleration is a absolute quantity and the accelerometer does not require a fixed external reference (necessary in the case of displacement transducer); furthermore, the numerical integration procedures required to obtain displacements from acceleration data are more stable and less sensitive to noise than the derivative procedures required in the opposite case. If possible, both accelerations and displacements should be measured; for example, in the case of snap-back tests, this permits to better catch both the quasi-static part of the test and the oscillation part of the signal. Instead of displacements or accelerations, even strains at the pile surface could be recorded by means of strain gauges; in this case, assuming the pile behavior to be linear, relative displacements can be derived according to the Euler-Bernoulli beam theory. Furthermore, it is often very important to record the pore pressure at the soil-pile interface during the pile installation and the dynamic test; this can be gained by inserting a pore pressure transducer in the pile shaft in such a way that the sensing element is in contact with marine soil.

The location where the sensors will be placed along the pile shaft plays a crucial role in the choice of the most adequate type of sensor. Generally, while sensors can be easily installed in the pile portion over the sea level (if any), greater difficulties must be overcome for the installation and protection of sensors along the submerged part of pile and, especially, along the part embedded in soil. In the latter case, it is not possible to directly measure the pile displacements by means of displacement transducers, whereas accelerometers and strain gauges should be used, paying particular care in their protection not only against aggressive environmental conditions but also against mechanical stresses, especially in the case of vibrodriving installation of the pile.

The layout of sensors along the pile depends on many aspects that involve the distribution of sensors longitudinally along the pile shaft and along the perimeter of the transversal cross-section. In order to evaluate the pile motion, arrays of transducers can be located along the whole pile or just along the section of interest. For a better definition of the mode shapes, the measuring points should be located narrower in the proximity of discontinuities, such as at the water-soil interface, at the interface of soil layers having different stiffnesses, or in correspondence to changes in the pipe pile cross-section (Figure 2a). The length of the array depends on the input level as well as on the sensitivity of the sensor; in the case of input applied at the pile head, higher input energy levels and higher sensor sensitivity allow the monitoring of the pile to a greater depth.

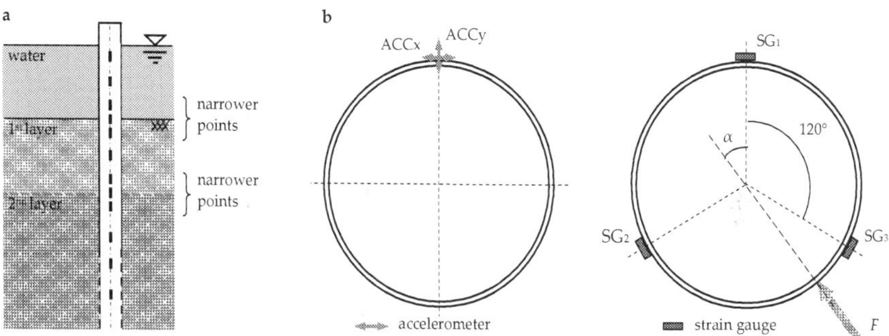

Figure 2. Instrumentation layout: (a) along the pile and (b) on a pile cross-section.

In order to capture the cross-section deformation of the pipe pile, several sensors should be installed around the circular perimeter; the number and layout of sensors depend on how many cross-sectional modes are of interest. However, excluding the sections close to the point where the input load is applied (stiffeners of the transversal cross-section may be added), the pile cross-section can be usually assumed to be rigid in its plane. In this case, to correctly evaluate the direction of the pile lateral motion and therefore of the applied horizontal input (e.g. the hammer impact direction), more than one sensor should be installed around a cross-section. In particular, neglecting the vertical component, if accelerometers are used, the input direction is known when two horizontal orthogonal acceleration components in one point are measured. On the other hand, if strain gauges are used, the pile motion direction can be calculated from the strain values measured by three sensors installed at the same cross-section and spaced 120° apart; assuming that the pile remains in the elastic range, the angle α between the direction of the load F (Figure 2b) and the line connecting SG1 to the center of the cross-section can be calculated according to the following formula:

$$\alpha = \arctan\left[\frac{\left(\varepsilon_{SG_3} - \varepsilon_{SG_2}\right)}{\sqrt{3}\,\varepsilon_{SG_1}}\right] \tag{1}$$

Finally, it is worth noting that a good estimation of the applied loading (e.g. the load exerted by the jack in a snap-back test) can be calculated from the value of strains recorded by two strain gauges SGi and SGj installed along the same vertical in the portion of the pipe pile above the ground where the bending moment (and therefore the longitudinal strain) varies linearly with depth; the actual applied load is thus obtained from the following formula:

$$F = \frac{EW}{\Delta h \cos \alpha} \left(\varepsilon_{SGj} - \varepsilon_{SGi} \right) \tag{2}$$

where E is the elastic modulus, W is the section modulus, and Δh is the distance between the two strain gauges.

3.2. Sensor typologies

3.2.1. Strain transducers

Strains can be measured with different transducer typologies; the most common are electrical strain gauges, vibrating wire strain gauges, and fiber optic strain gauges (Figure 3).

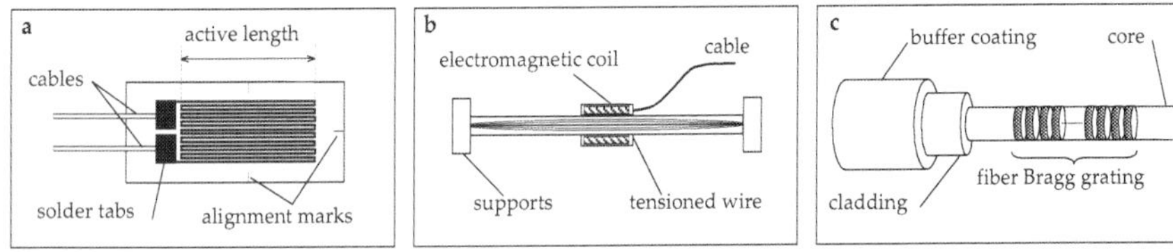

Figure 3. Strain transducers: (a) electrical, (b) vibrating wire, and (c) fiber Bragg grating.

The electrical strain gauges, also known as foil strain gauges, are the most common sensors for strain measurements. A grid made of fine electric resistance wire printed on a polyamide foil constitutes the sensor element (Figure 3a). This sensor typology uses the relationship between the applied strain ε ($\varepsilon = \Delta L/L$) and the relative change of strain gauge resistance ΔR [12], derived by the second Ohm's law and described by the equation:

$$\frac{\Delta R}{R_0} = k\varepsilon \tag{3}$$

where R_0 is the initial strain gauge resistance (usually 120 or 350 Ω) and k is the gauge factor checked experimentally (usually for metal strain gauge is about 2). To measure the resistance changes, a Wheatstone bridge circuit, whose basic configuration is reported in Figure 4a, is usually adopted. The four arms of the bridge are formed by the resistors R_1 to R_4 having the same nominal resistance value; the input voltage V_i is applied between point 1 and 4, whereas the output voltage V_o is measured between points 2 and 3. If all resistances are equal, then V_o is zero, whereas, if at least one resistance changes, then the bridge is unbalanced and the following value of output voltage can be measured:

$$\frac{V_o}{V_i} = \frac{1}{4} \left(\frac{\Delta R_1}{R_1} - \frac{\Delta R_2}{R_2} + \frac{\Delta R_3}{R_3} - \frac{\Delta R_4}{R_4} \right) \tag{4}$$

Substituting Equation (3) in Equation (4), the well-known expression is obtained:

$$\frac{V_o}{V_i} = \frac{k}{4}\left(\varepsilon_1 - \varepsilon_2 + \varepsilon_3 - \varepsilon_4\right)$$

(5)

where ε_i is the strain measured by the strain gauge applied at the ith arm of the bridge. Three configurations are available for the measurements with strain gauges (Figure 4b–d) commonly called quarter bridge, half-bridge, and full bridge if one, two, or four resistances, respectively, are substituted with strain gauges. The latter two configurations permit the compensation of the temperature effects. In the case of large distances among measuring points and data acquisition system, the influence of lead resistances is not negligible and proper correction factor or shunt calibration [13] should be adopted. To obtain a reliable and correct measurement, a particular care is required in the installation of a strain gauge; the correct procedure is standardized and can be summarized in different stages involving the surface preparation, strain gauge bonding, soldering of the cables, and protection of the measuring point (Figure 5a–d, respectively).

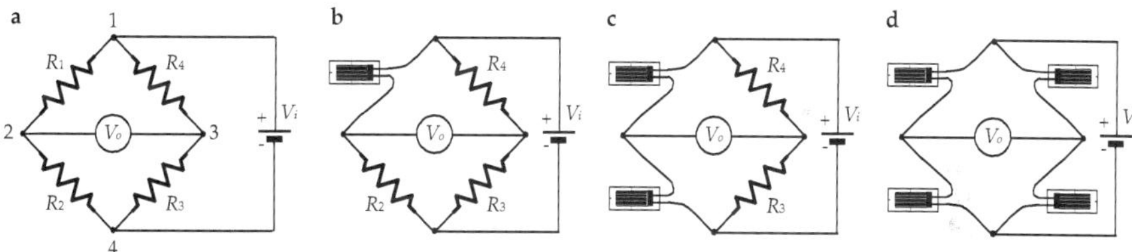

Figure 4. Wheatstone bridge: (a) basic configuration, (b) quarter bridge, (c) half-bridge, and (d) full-bridge configurations.

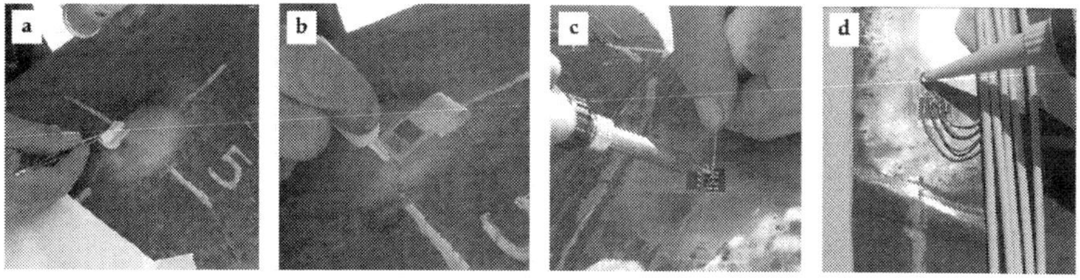

Figure 5. Installation of electrical strain gauge: (a) surface preparation, (b) strain gauge bonding, (c) cable soldering, and (d) protection.

In general, electrical strain gauges are considered the best cost-effective measurement solution and a well-supported technology for applications in typical short-range and controlled operational environments. However, for the monitoring of pipe piles in marine environment, the drawbacks are the not long life span and the increase of electromagnetic interference due to the voltage drop on long lead wires, which make this technology not optimal for long-term monitoring or for applications in which the measuring system cannot be installed close to the sensors.

The vibrating wire strain gauges permit the determination of strains by measuring the consequent change in the natural frequency of transverse vibration of a tensioned wire. The frequency is measured by plucking the wire with an electromagnetic coil connected through a cable to data acquisition system (Figure 3b). The installation on a pipe pile can be carried out welding the sensor ends at the pile shaft. The temperature effects can be compensated, thanks to a termistor (for reading temperature) located on the sensor case. The main advantages are the long-term stability and the high resistance to water intrusion and lighting damage, whereas the drawbacks are the sensor size and gauge active length, usually ranging 50–250 mm, which make the sensor not suitable when the strain at a precise point must be measured, and then, more importantly, the limitation to static measurements, which limits their use to the quasi-static part of tests, such as the pile lateral pulling phase before the release in a snap-back test.

The strain gauges based on the optical fiber technology are the most innovative strain sensors and may be the most powerful ones. The optical fiber is usually composed of three layers: fiber core (constituting the sensing element), cladding and jacket. The perturbation in strain causes corresponding changes in terms of amplitude, phase, frequency, wavelength, and polarization in the optical properties of the transmitted light [14]. Among different technologies, the sensors most used in civil engineering are the Fabry-Perot strain gauge sensor, based on white-light cross-correlation principle, which allows the local measurement of the strain between two fixed points along the fiber with high resolution, and the fiber Bragg grating (FBG) sensor (Figure 3c) in which a series of localized changes in the refractive index of the glass core of the fiber constitute the sensing part. FBG sensors are especially fit for strain measurements of pipe piles in marine environment due to high resolution, quasi-distributed sensing, temperature compensation, good anti-corrosion ability, and long-term stability; these last two characteristics permit to extend their use even to long-term monitoring [15]. The main drawbacks are the costs related to the optical fiber and especially to the demodulation system, which, however, are expected to decrease in the near future.

3.2.2. Acceleration transducers

An accelerometer is a device that converts a mechanical acceleration into a proportional electrical signal, usually voltage output. Commonly, an accelerometer is composed of a case and a seismic mass connected with it by means of an elastic sensing element (Figure 6a). Several accelerometer typologies are available in commerce: piezoelectric, piezoresistive, or capacitive accelerometers depending on the physical phenomenon characterizing the sensing element. Recently, the so-called MEMS (micro electro-mechanical systems) accelerometers have been developed; these sensors are based on the same physical phenomena of the above-mentioned sensors but are realized exploiting microfabrication techniques (Figure 6b). Finally, the servo (force balance) accelerometers are available; these sensors are composed of a case with seismically suspended mass balanced by a magnetic coil, which generates a restoring force to maintain it in the null position; when the null detector (e.g. a capacitance-type transducer) reveals a displacement of the mass, a servo amplifier increases the coil current proportionally to the external applied acceleration [16].

To suitably select an accelerometer, many parameters have to be considered: the sensitivity, i.e. the relationship between sensor voltage output and acceleration; the frequency range, i.e. the range over which the sensitivity is nearly constant; the maximum value of acceleration measurable by the sensor; and the broadband resolution, i.e. the smallest acceleration that the sensor can detect. The maximum value of measurable acceleration and the broadband resolution may often depend even on the signal conditioners and data acquisition systems utilized.

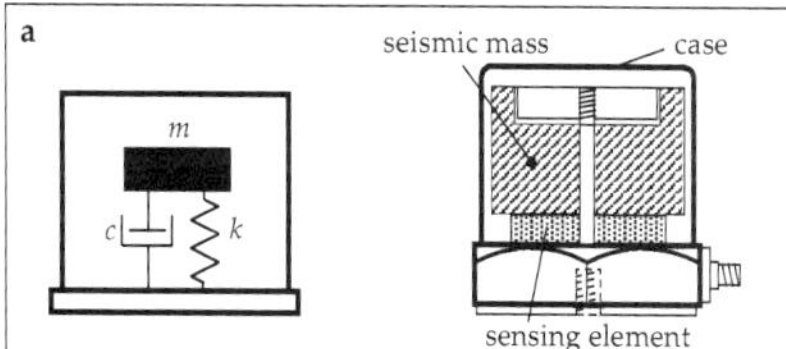

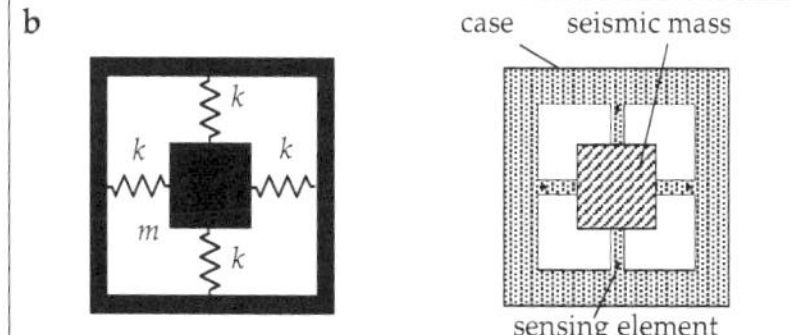

Figure 6. Mechanical and physical sketch of (a) piezoelectric and (b) MEMS piezoresistive accelerometers.

The required characteristics should be determined according to expected frequency values obtained by means of preliminary simulations with analytical or finite-element models and checked, if possible, with preliminary tests. Furthermore, environmental conditions (such as temperature) and sensor size have to be taken into account to choose the better sensor type, especially in the case of sensors that will be located in the submerged or embedded part of the pile.

In Table 1, the typical characteristics regarding the different technologies of sensors available in commerce are listed [17]. When accelerations are characterized by low-frequency content, piezoresistive, capacitive, or servo accelerometer is the better solution. In the case of tests with low-intensity input, piezoelectric or servo accelerometers have to be preferred, thanks to their high sensitivity and good broadband resolution. Finally, when sensor costs must be minimized and/or the smaller size is required, the MEMS technology furnishes the best solution.

Finally, it is worth remembering that, even if the mounting method affects the useful frequency range of an accelerometer [17] reducing the upper limit (kHz as magnitude order), all common available methods (cyanoacrylate, dental or epoxy cement, stud mounting, wax, permanent magnet, and adhesive) are suitable for investigating low-frequency range such as that typical for first natural frequencies of an open-ended pipe pile in marine soil under lateral loads. Generally, the cement mountings are used for accelerometers applied permanently in the submerged and embedded part of the pile, whereas permanent magnets or wax are preferred for accelerometers located over the sea level due to easy application or removal of the sensor.

Accelerometer type	Sensitivity [mV/g]	Frequency Range [Hz]	Broadband resolution [µg]	Size
Piezoelectric accelerometer	0.05-10.000	0.5-50.000	1	small or medium
Piezoresistive accelerometer	0.0001-10	0-10.000	1000	medium
Capacitive accelerometer	10-1.000	0-1.000	50	small
Servo accelerometer	1.000-10.000	0-100	< 1	big
MEMS accelerometer	0.01-10.000	0-10.000	1000	micro

Table 1. Typical accelerometer characteristics.

3.2.3. Displacement transducers

The measurement of displacements at the pile head can be essential, especially during the quasi-static part of a snap-back test. Among a great number of different displacement transducers available in commerce, those most common and able to capture both dynamic and static values are potentiometers, inductive transducers, and linear variable differential transformers (LVDTs). The latter, one of the most known and utilized sensor, is characterized by a very fine resolution (in good environmental conditions), has a frequency range up to some kilohertz, and guarantees high robustness. An LVDT consists of a primary solenoidal coil and two secondary coils; a movable ferromagnetic core is placed parallel to the axis of the cylindrical case and it is attached to the object to be measured, whereas the case is fixed externally to the vibrating system. The primary coil is excited by an alternating current inducing a voltage in secondary coils, which is proportional to the displacement of core.

3.2.4. Pore pressure transducers

When the value of the pore water pressures acting at the soil-pile interface has to be investigated, two different sensor typologies could be used: standard size pressure transducers or miniature pressure transducers. The latter is usually preferred because the small size facilitates the mounting of the transducer in the pile thickness reducing compliance errors and the chance of damage and increasing the frequency response of the measuring system [18]. A pore pressure transducer consists of a thick diaphragm mounted in a case and externally protected by a porous disk. The diaphragm can be equipped with sensing elements or it can be itself the sensing element. In the case of piezoresistive pore pressure transducer, widely used in the geotechnical field, strain gauges are built-in or applied on the diaphragm and connected to a Wheatstone bridge to obtain an amplification of the signal and the temperature compensation.

3.3. Protection of sensor and cabling

The correct selection of the more appropriate systems and products to protect instrumentation and connecting cables is a key point in every experimental campaign, especially when such operation is performed in situ and, particularly, in marine environment. In order to make the correct choice among several products and covering agents available in commerce, various considerations should be done regarding the ambient conditions, the duration of the measurement, and the possible interference of the protection with the specimen [19]. The ambient conditions of both sites where the pile is instrumented and the site where it will be installed should be carefully taken into account. Sensors applied on pile embedded in marine environment are subjected to several types of aggressive agents [20]: mechanical stresses and erosion due to the effects of waves and tides, climatic agents due to variations of temperature, and chemical attacks carried out by sulfate and chloride attacks. Furthermore, the handling and installation phases of the pile could cause important stresses on the pile shaft due to impacts, loads, and frictions. The required lifetime of the instrumentation must be determined based on the duration of the measurements. Tests performed to characterize the soil-water-pile system can be made only on time; they can be periodically repeated or a continuous monitoring may be carried out to investigate variations in time of the dynamic parameters. Especially, in this last case, the protection must be designed considering a proper lifetime of the system. Finally, an accurate isolation from electromagnetic fields allows the reduction of noise, whereas a protection by the moisture avoids signal drifts permitting a greater accuracy and precision of the measurements.

Several precautions should be taken in the application of protective coatings. The installed transducer must be in a perfect condition before being covered. Trapped humidity, perspiration, and flux residue from soldering can lead, sooner or later, to inaccurate measurements or, in some cases, even to complete failure. Efficient protective covers are not only a seal against outside humidity, but they also seal in trapped humidity. For that reason, laboratories or protected ambient should be preferred during sensor installation; furthermore, the first coating layer should be applied as soon as possible after sensor bonding to reduce the contamination of the measuring point with impurities and humidity. When the application of the transducer has to be carried out in humid ambient condition due to bad weather (especially when the pile instrumentation is carried out in situ), the measuring point should be dried with hot air gun, taking care not to exceed the safe operating temperature of the transducer. The area surrounding the installed transducer should maintain the same quality over the whole useful lifetime of the instrumentation. Scratches and impurities allow the penetration of aggressive agents that could damage the sensor; this area should be carefully cleaned and the covering agent should adhere perfectly the surface and cover an area some centimeters larger than the transducer to protect it adequately. In particular, this area should include the cable ends (even underneath) and the cable connections (intrabridge wiring and jumper leads), which, in some cases, can constitute preferential pathways for moisture. Sometimes, one type of covering agent is not sufficient, and multiple coating layers can be combined by applying the first coating directly on the surface of the transducer and then applying the other layers, taking care that each coating covers a portion larger than the previous one (Figure 7).

Transducers applied on a pile embedded in marine environment are exposed to different conditions depending on their location along the pile shaft: over the sea level, submerged in the sea, and embedded in the soil. Different protections are needed in each case. For the transducers installed over the sea level, a protection from humidity and rain is necessary and it can be obtained with a sealed box, for normal size accelerometers, or with protective coating for MEMS accelerometers and strain gauges. In order to seal a measuring point, several products can be used: polyurethane varnish provides a protection against dust, ambient humidity, and oil (usually, it is used as the first insulating layer under the other covers); nitrile rubber varnish is useful in the substitution of polyurethane varnish as a protection in ambient with oil, petrol, and liquefied gases, excluding oxygen; the silicone rubber increases the mechanical protection but is not resistant to oil.

For the strain gauges or MEMS accelerometers installed along the submerged pile section, the protection against marine environment can be obtained with one of the products listed above coupled with butyl rubber (or plastic putty) to guarantee a better waterproofing; an aluminum layer is placed above all as a diffusion additional barrier enhancing the long-term protection.

Finally, sensors installed in the embedded part of pile can be protected with the same covering agents used in the submerged part. When pile handling or installation phases represent the potential causes of stresses on the pile shaft, the measuring points and cables can be protected with additional mechanical protections, e.g. metallic profiles or foam materials.

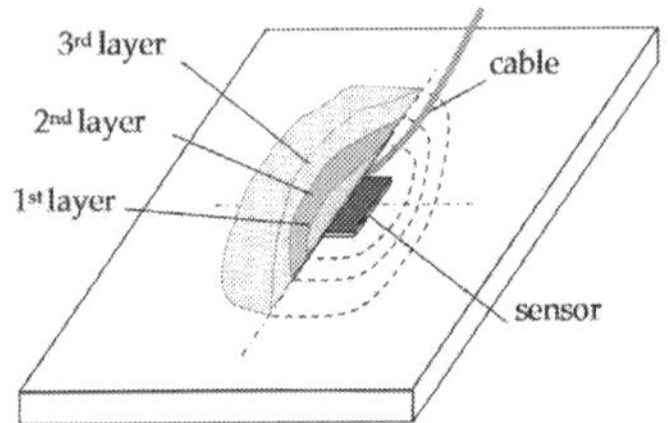

Figure 7. Multilayer coating.

In the case of optical fibers and Bragg grating sensors, they could be located along ridges or grooves already present or realized on purpose along the pile to provide a good mechanical protection. Furthermore, in order to ensure a higher chemical and mechanical protection, the optical fibers could be inserted into a tubing covered with a thick layer of resin; some special precautions can be taken for the measuring point over the Bragg grating sensors by means of different coating protective layers.

4. Signal processing techniques

Before performing the modal analysis, all the recorded data must be suitably processed by means of signal processing techniques. Several operations can be carried out on the raw signals to reduce noise, to limit bothering due to leakage and aliasing, and to eliminate spurious frequency contents. In this part, the operation useful for test typologies discussed above will be summarized.

The windowing consists of the application of weighting function to the time histories in order to reduce the amplitude of initial part, final part, or both (Figure 8a). The use of windowing is a practical solution to the leakage problem associated with the use of discrete Fourier transform of data. Regarding sinusoidal signals or random signals, such as those derived by stepped- or sweep-sine testing and ambient vibration test, the use of Hanning, Hamming, or Kaiser-Bessel window forces the signal to zero at the ends allowing to overcome leakage problem in the computation of the Fourier response function. For transient signals, obtained by impact load or snap-back tests, the exponential window can be used to attenuate the signal in the final part, especially for low damped systems. A boxcar limited and centered on the impact can be applied in order to eliminate the noise from the initial and final parts of the signal. The window functions must be applied with care, as they produce signal modifications that sometimes lead to erroneous results (i.e. the exponential window adds a damping contribution to the signal decay).

Filtering operation can be seen as a windowing carried out in the frequency domain rather than in time domain [10]. Filters are applied when some frequency contribution of the signal must be attenuated (Figure 8b). The main characteristics of a filter are the cutoff frequencies, that define what frequencies are attenuated and what frequencies can pass through the filter, and the order of the filter that controls the roll-off features near these critical frequency regions. Different filters are available: low-pass filters, high-pass filters, band-pass filters, and band-stop filters are the most suitable for our purposes. Low-pass filter is applied during the decimation of the signal in order to eliminate the frequency content over the cutoff frequency and thus avoid aliasing phenomena (so it can be considered an anti-aliasing filter). Furthermore, the low-pass filter is used to cutoff the high frequencies due to the cross-sectional deformation of the pipe pile. Band-pass filter is useful when only a contribution in a finite frequency range have to be analyzed. Finally, band stop filter can be used to eliminate spurious frequencies. As the windowing, filtering operations cause signal modifications that have to be carefully evaluated.

The decimation is used to reduce the sampling frequency of an acquisition (Figure 8c). When the original data are sampled with a too high rate, decimation can be performed in order to decrease the number of data and to consequently speed up the successive analyses. The decimation is a two-step procedure requiring previously a low-pass filtering with a cutoff frequency below the Nyquist frequency (half the final sampling frequency) and then a downsampling of the data at the final sampling frequency. The decimation factor is the ratio between initial and final sampling frequencies.

Offsets or spurious trends can affect the signals due to many reasons, first of all, the temperature (Figure 8d). The offset can be removed simply by subtracting the mean value of time series, whereas spurious trends need of a more complex procedure such as the fitting of the data with a low-order polynomial using a least-squares procedure and then subtracting the obtained values to the original time history.

The averaging is mainly used when the input applied is characterized by low intensity (i.e. for impact load test and ambient vibration test) and the signals are characterized by a low signal-to-noise ratio. By averaging different time series, it is possible to remove the spurious random noise from the signals [10].

The discrete Fourier transform (DFT) is may be the most powerful tools in signal processing, allowing an early raw evaluation of the signal properties and even the modal parameters of a pile or other structures (i.e. with peak picking method). The DFT permits to convert a time series into the frequency domain obtaining the frequency spectrum of the signal. The frequency resolution Δf of the computed spectrum depends on the time length T of the original signal from the well-known relationship $\Delta f = 1/T$. The inverse discrete Fourier transform IDFT performs the inverse operation furnishing the time history relevant to a frequency spectrum. In the signal processing software, the DFT and IDFT are efficiently calculated, thanks to the fast Fourier transform (FFT) algorithm based on the Cooley and Tukey's study [21].

Finally, the frequency response function (FRF) represents the dynamic behavior of the system and can be defined simply as the ratio between the Fourier transform of the system response $X(\omega)$ and of the applied input $F(\omega)$. Different FRF estimators, available in the literature [9], are very useful when highly noisy data have to be processed. These estimators provide the reduction of the noise in the output

$$H_1 = \frac{S_{fx}}{S_{ff}} \tag{6}$$

or in the input

$$H_2 = \frac{S_{xx}}{S_{xf}} \tag{7}$$

with $S_{ff} = F(\omega)F^*(\omega)$, $S_{xx} = X(\omega)X^*(\omega)$, $S_{fx} = F(\omega)X^*(\omega)$, and $S_{xf} = X(\omega)F^*(\omega)$, the auto- and cross-spectra, respectively (* denotes complex conjugate). Several other estimators to reduce noise in both input and output signals are available in the literature (see, for example, [22–24]).

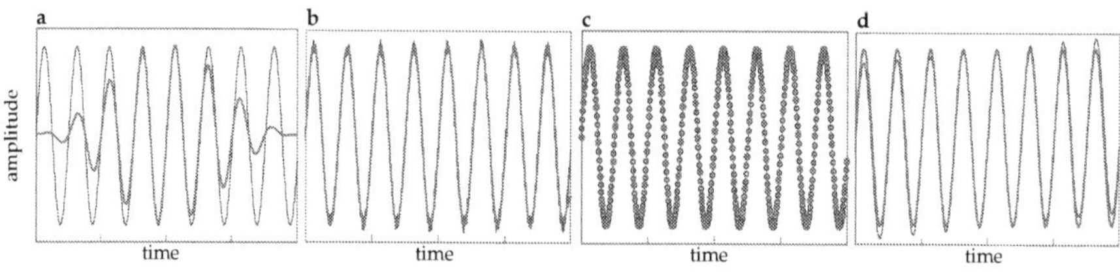

Figure 8. Signal processing operations: (a) windowing (Hanning window), (b) low-pass filtering, (c) decimation, and (d) removal of spurious trend (original and processed signals in blue and red, respectively).

5. Experimental modal analysis

Experimental modal analysis provides the extraction of the modal parameters (natural frequencies, damping ratios, and mode shapes) of a linear time-invariant dynamic system starting from measured vibration data [9]. The theoretical basis of the technique is based on the recognition of the relationship between the vibration response at one location and excitation at the same or another location as a function of excitation frequency. This relationship can be represented, in the frequency domain, by the FRF or, in the time domain, by the impulse response function (IRF). The FRF is the Fourier transform of a response measurement normalized by the Fourier transform of the input and the IRF is the free vibration response of the structure. The passage from FRF to IRF can be performed by means of the IDFT.

In the last decades, several methods have been developed in the field of the experimental modal analysis. The classification of these methods can be based on the working domain, time or frequency, on the number of the degrees of freedom used by the model, single degree of freedom (SDoF) or multidegree of freedom (MDoF), and on the number of inputs and outputs, single input and single output (SISO), single input and multiple output (SIMO), and multiple input and multiple output (MIMO).

Many of the basic methods working in the frequency domain use different plots of FRFs in order to identify the modal parameters in a simple way. Peak-picking method, the simplest SDoF method working in the frequency domain, estimates the natural frequency from the peak amplitude of FRF. The half-power bandwidth permits the damping ratio estimation from the width of the peak in the graph of FRF amplitude. Furthermore, circle fit method, [25, 26], estimates the modal parameters working in the Nyquist plot (Re-Im). Finally, the most advanced methods working in the frequency domain are the rational fraction polynomial (RFP) method [27], Dobson's method [28], inverse method [29], least-squares complex frequency (LSCF) [30], and poly-reference least-squares complex frequency (pLSCF) [31] methods. These methods are based on a curve fitting of the measured FRF (or FRFs for MDoF systems) using an analytical model of the measured structure. A possible analytical expression of the FRF for a MDoF with input applied in j and response measured in i is

$$FRF_{ij}(\omega) = \sum_{r=1}^{m} \frac{{}_{r}A_{ij}}{\omega_r^2 - \omega^2 + 2\xi_r \omega \omega_r i} + R_{ij}(\omega) \tag{8}$$

where ${}_{r}A_{ij}$ is the residue, ω_r is the natural angular frequency, and ξ_r is the damping ratio (viscous) of the rth mode, respectively, and R_{ij} is the residual term to take into account the other modes.

The time domain methods use time response data that can be represented by the IRF for impact load or snap-back tests or by the recorded data during the operational condition for ambient vibration tests. Crossing time and logarithmic decrement are the simplest methods for the estimation of natural frequencies and damping ratios, respectively. These two methods can be performed by consulting the IRF graphs after having isolated each single frequency contribution by means of band-pass filters. The crossing time method uses the zero passing of the signal in order to evaluate the period (the time interval between two crossing is equal to half of the period), whereas the logarithmic decrement estimates the damping using the free decay of the signal (the damping is equal to the natural logarithm of the ratio between the ith and $i+n$th maximum of the signal divided by $2\pi n$). The most powerful time domain methods are based on the recent developments in the control theory. Many of these are implemented in commercial software products for experimental modal analysis: Ibrahim time domain (ITD) method [32–34]; least-squares complex exponential (LSCE) method [35], working with IRF; and random decrement (RD) method [36–38], autoregressive moving average (ARMA) method [39, 40], and stochastic subspace identification (SSI) method [41], working with measured random response. In Table 2, the most useful methods are listed with their main characteristics.

Method	Domain	Degrees of freedom	Type of estimates	Number of input and output
Peak Picking	Frequency	SDoF	Local	SISO
Circle Fit	Frequency	SDoF	Local	SISO
Inverse	Frequency	SDoF	Local	SISO
Dobson's method	Frequency	SDoF/MDoF	Local	SISO/SIMO
Rational Fraction Polynomial	Frequency	MDoF	Global	SISO/SIMO
Least-Square Complex Frequency	Frequency	MDoF	Global	MIMO
poly-reference Least-Squares Complex Frequency	Frequency	MDoF	Global	MIMO
Crossing time	Time	SDoF	Local	SISO
Logarithmic Decrement	Time	SDoF	Local	SISO
Ibrahim Time Domain	Time	MDoF	Global	SIMO
Least-Square Complex Exponential	Time	MDoF	Global	SIMO/MIMO
Random Decrement	Time	SDoF	Local	SISO
Autoregressive Moving Average	Time	MDoF	Global	SIMO
Stochastic Subspace Identification	Time	MDoF	Global	MIMO

Table 2. Classification of different experimental modal analysis methods.

6. Case study: Lateral loading tests on near-shore pipe piles in marine environment

This section is devoted to a concrete case study: an experimental campaign carried out on near-shore open-ended pipe piles at the "Mirabello" harbor in La Spezia, Italy. The general aim of this campaign was to detect the complex dynamic soil-pile interaction, taking into account the influence of the specific construction technique that can affect significantly soil properties close to the piles. In particular, the experimentation was performed in order to (i) develop new field test procedures to gain information on the soil dynamic properties and on pile-soil-pile dynamic response that can be used directly in seismic design of pile foundations and (ii) calibrate and validate analytical and numerical procedures and develop new empirical methods for soil-pile interaction analysis. Some details about the site together with the pile instrumentation system and the relevant chemical and mechanical protections are here reported. More details can be found in the work of Dezi et al. [42, 43].

6.1. Site, test field, instrumentation, and test set-up

Site: The site selected for the tests was next to "Porto Lotti" at the tourist port "Mirabello" in La Spezia, Italy. This area comprises several piers and about 1000 new berths for boats and ships from 14 to 60 m long. The superstructure consists of a grid of precast prestressed beams, which supports a cast in situ concrete slab. The foundation is composed of about 500 vibro-driven steel pipe piles ranging from 609 to 711 mm in diameter, from 11 to 13 mm in thickness, and from 20 to 45 m in embedded length. Several site explorations (1973, 1981, 1991, 1992, and 2008) were carried out before the construction of the tourist port. A large volume of soil was investigated by means of laboratory tests such as the odometer test, in situ tests conducted up to a maximum depth of about 50 m, including the dilatometer Marchetti test (DMT), cone penetrometer test (CPT), piezocone test (CPTU), standard penetration test (SPT), vane tests (FV), and multichannel analysis of surface waves (MASW).

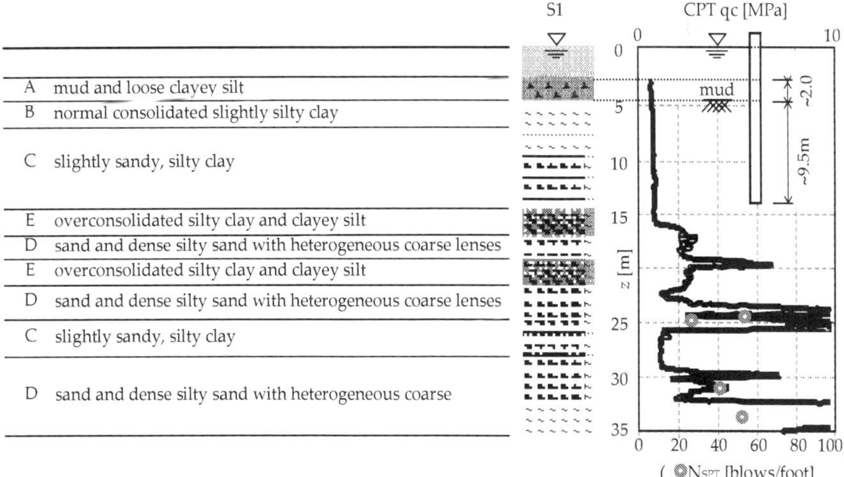

A	mud and loose clayey silt
B	normal consolidated slightly silty clay
C	slightly sandy, silty clay
E	overconsolidated silty clay and clayey silt
D	sand and dense silty sand with heterogeneous coarse lenses
E	overconsolidated silty clay and clayey silt
D	sand and dense silty sand with heterogeneous coarse lenses
C	slightly sandy, silty clay
D	sand and dense silty sand with heterogeneous coarse

Figure 9. Subsoil profile.

The soil stratigraphy can be subdivided into six main layers. The upper layer, Layer A, consists of mud and loose clayey silt, with water content greater than the liquid limit. Normal consolidated slightly silty clay was found in the second main layer, Layer B, whereas the succeeding layer, Layer C, is composed of slightly sandy silty clay. Layer D consists of sand and dense silty sand, including scattered shell fragments and heterogeneous coarse lenses. Layer E includes overconsolidated silty clay and clayey silt. Under level D to the end of the borehole (about a depth of 50 m), there is a compact layer of clayey sandy gravel. The results of these investigations are summarized in Figure 9, where the soil stratigraphy and the CPT performed nearby the test site are reported, and in Table 3, where the main soil properties of each soil layer are presented.

Soil	Borehole		CPT		DMT		FVT	
type	P.P. [kPa]	V.T. [kPa]	qc [MPa]	fs[kPa]	Cu [kPa]	OCR	cu [kPa]	cur [kPa]
A			0.37	13	14		20	5
B		22	0.45	14	19		25	8
C	68	44	0.82	18	40	3.00	47	12
D			7.90	100				
E	120	50	2.25	120	82	5.00	108	22

P.P. := resistance to penetration measured by pocket penetrometer, V.T. = shear strength measured by pocket scissometer, cur = undrained residual shear strength, qc = cone resistance, fs = sleeve friction, Cu = undrained shear strength, OCR = over consolidation ratio, cur = undrained residual shear strength.

Table 3. Geotechnical parameters.

Test field: As shown in Figure 10a, the test field consisted of three steel pipe piles vibrodriven for a depth of 9.5 m into soft marine clay. Piles were 15.5 m long and the pile head elevation was 1.0 m above mean sea level (m.s.l.). Pile P1 was characterized by a diameter of 711 mm and thickness of 11 mm, whereas piles P2 and P3 were characterized by a diameter of 609 mm and thickness of 10 mm; the head of the piles was kept free and the head of pile P1 was stiffened with two steel profiles, welded in a crux shape. The piles were arranged in an "L"-shaped horizontal layout characterized by different distances between pile P1, located at the "L" corner, and piles P2 and P3. It is important to point out that characteristics and location of piles P2 and P3 were practically irrelevant to comprehend the single pile behavior, although it was useful to investigate the pile-soil-pile dynamic interaction. It is worth noting that, based on the results of the geotechnical investigations, the soil strength profile can be reasonably considered as uniform over the pile embedment length.

Pile instrumentation: As regards the sensors, their characteristics were chosen on the basis of preliminarily finite-element analyses performed to define the amplitude-frequency response of the soil-water-pile system. The sensors used to instrument the piles were electrical strain gauges, installed along the pile P1 to measure the longitudinal strains, accelerometers located close to the top of the piles to measure the vibrations of the pile head, and a pore pressure transducer on the pile shaft. In detail, as shown in Figure 10b, a total of 19 strain gauges were placed along three generatrices of the pile P1, spaced 120° apart to capture the cross-section average strains (elongation and curvature of the pipe); 11 strain gauges were located along the main generatrix and 4 along each of the two secondary generatrices. Narrower measure points were considered for the pile section where maximum curvatures were expected. A uniaxial accelerometer was located at 0.3 m from the top of the pile in such a way that the measured acceleration is in the same direction of the lateral input. The measurement chain was also composed of amplifiers, signal conditioners, one spectrum analyzer, two data acquisition systems, and a computer with dedicated software.

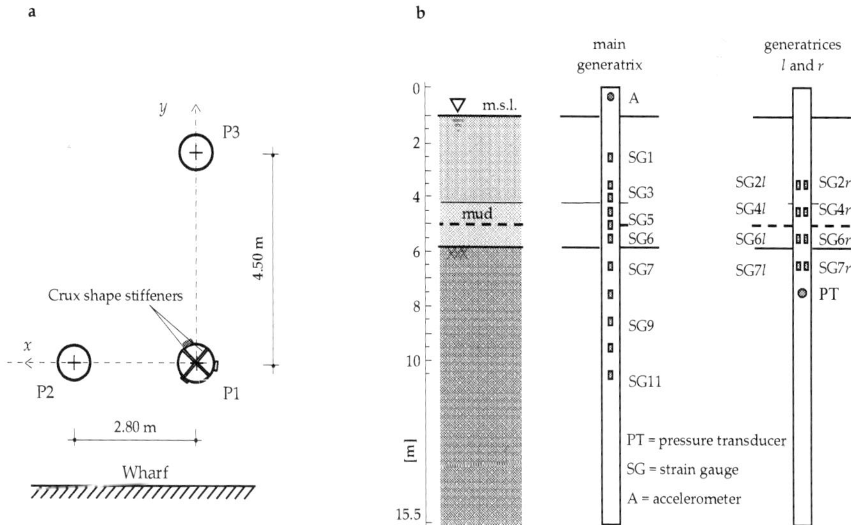

Figure 10. (a) Plan view of the test field and (b) strain gauges placement along the pile.

During installation, the measuring instruments were carefully protected both chemically and mechanically from the marine environment and pile vibrodriving. Cables and strain gauges were first waterproofed and protected with three products, a polyurethane paint and silicone rubber covered by aluminum foil coupled with kneading compound (Figure 11a), and then protected by means of UPN steel profiles welded along the three generatrices, sealing the space between pile and UPN with polyurethane foam (Figures 11a and 11b).

Several hours after pile vibrodriving and during the first test series, all the strain gauges, except one, were working properly. The pressure transducer worked only for several hours due to waterproofing malfunction. In the second series (after 10 weeks from pile installation), another strain gauge appeared not be working correctly. Despite this, the procedure adopted in this experimental campaign for the strain gauge installation was observed to be very effective to protect the gauges in the aggressive marine environment over the whole period of the tests (about 3 months).

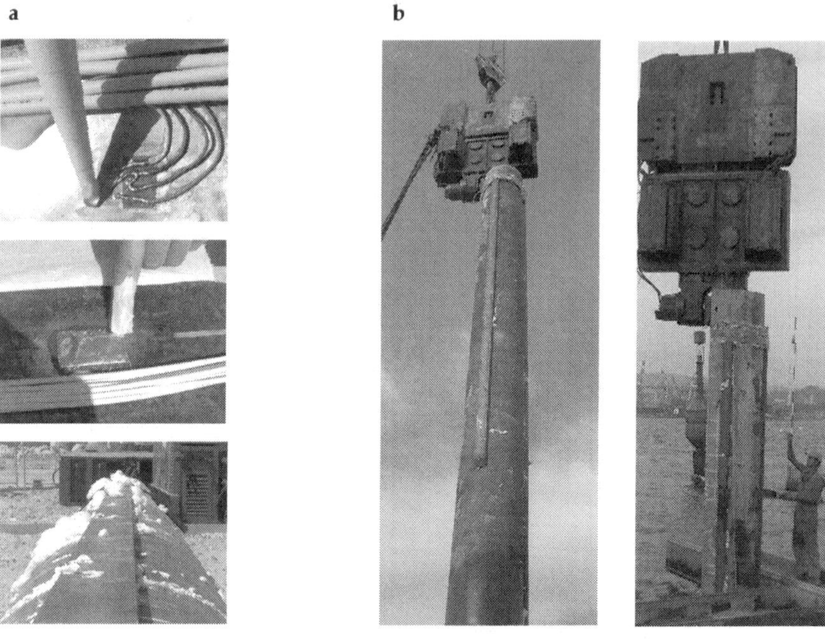

Figure 11. (a) Chemical and mechanical protection of strain gauges and (b) instrumented pile during vibrodriving installation.

6.2. Test typologies

Two test series were conducted at two different time instants: the first 1 week after the pile installation and the second after 10 weeks. The tests were repeated to evaluate differences with time in the dynamic behavior of the soil-water-pile system.

Two different typologies of test were performed: impact load and free vibration tests (Figures 12a and 12b). Impact load tests were conducted at two different time instants: the first series 1 week after the pile installation and the second series after 10 weeks. The snap-back tests were performed after the second series considering different force levels.

Finite-element simulations and several preliminary in situ tests demonstrated that the impact of the hammer with medium-low hard tip induced values of accelerations at the pile heads and longitudinal strains along the pile, measured by the transducers, greater than the ambient-noise. With this hammer, a maximum impact force of about 50 kN was reached. A typical time history of a medium-intensity impact measured by the load cell of the hammer is shown in Figure 13a in which a zoom of the peak is also reported. The corresponding frequency spectrum, shown in Figure 13b, is characterized by a constant force level up to about 100 Hz. Therefore, this hammer permitted to investigate the soil-water-pile behavior in a wide range of frequencies, ranging between approximately 0.5 and 100 Hz. A sampling rate of 10 kHz was chosen to achieve high resolution in time domain, and an acquisition time duration of 2 s was considered to catch the entire duration of the pile oscillation.

The free vibration tests were performed to investigate the dynamic behavior of the system at different strain levels by varying the level of force that is usually imposed with a standard hydraulic actuator. In this experimental study, the load was applied using a double-acting jack with a capacity of about 400 kN, placed on the wharf and connected to the pile P1 by means of steel cables. The wharf provided the reaction for the device. The quick release was achieved, thanks to the failure for traction of a steel pin placed along the steel cable between the jack and the pile P1. The cross-section of the pin was opportunely calibrated to obtain the desired load level. After the pin failure, the pile underwent a number of steadily decreasing oscillations around its equilibrium position.

An analogical pressure transducer was used to sense the hydraulic oil pressure in the pump actuating the jack. Unfortunately, due to its low sensitivity and to the friction between the ram and cylinder wall (amplified by environmental conditions), the adopted measuring system was affected by raw approximation. A more precise estimation of the actual loading exerted by the jack was obtained from the values recorded by two strain gauges located in the portion of the pile above the ground where the bending moment (therefore the longitudinal strain) varies linearly with depth. The actual applied load was thus obtained from the analytical procedure presented above. In Table 4, the value of the force producing the traction failure of the pin (maximum quasi-static force) and the quick release of the pile obtained for each test is reported.

Before the tests, the sea water and soil levels inside and outside the pipe were measured, the ambient noise mainly due to marine waves was registered, and preliminary tests were conducted to set up the data acquisition system, entity of the impact, and proper gains for the accelerometer signals.

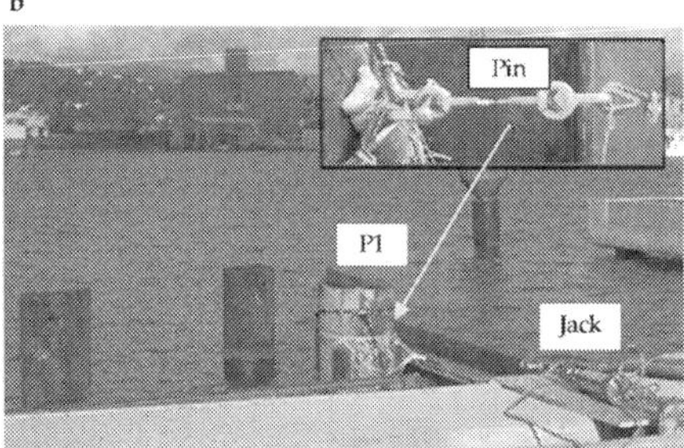

Figure 12. Test field: (a) impact load test and (b) snap-back test.

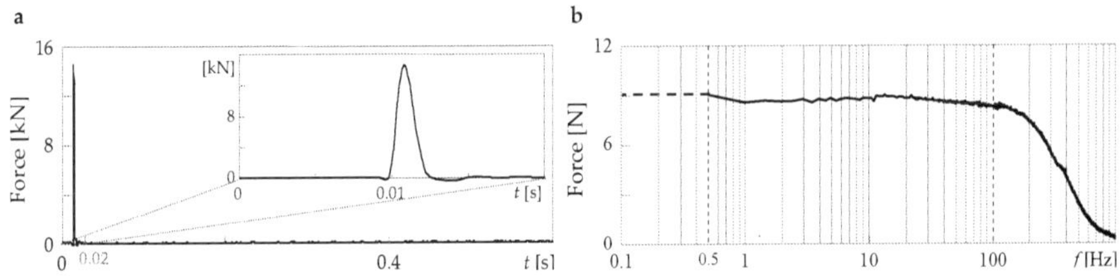

Figure 13. Impact load: (a) time history and (b) frequency spectrum (semilog plot).

Free vibration test	Fy-1	Fy-2	Fy-3	Fy-4	Fy-5	Fxy-1	Fxy-2	Fxy-3	Fxy-4
Force calculated from SGs [kN]	2.8	6.9	18.0	24.0	58.1	2.7	6.7	16.5	26.6

Table 4. Free vibration tests: forces applied before the release.

6.3. Test configurations

The impact load tests of both series were carried out considering three different test configurations, varying the direction (x, y, or diagonal, identified with xy index) of the horizontal impact at the head of pile P1 and the measuring direction of the accelerometer at the pile head. In particular, Hi denotes a test in which the hammer impacts the pile along the *i*th direction, while the acceleration is measured in the same direction. In Hx, Hy, and Hxy tests, the hammer impacted the pile along the x-, y-, and xy-directions, respectively, and the accelerometers were positioned in several configurations. For each test configuration, sets of 10 horizontal hammer impacts were applied and the time histories of the impact load, acceleration at the pile head, strains, and pore pressure along the pile were recorded.

The free vibration tests were carried out with two different configurations, the pile P1 released (i) along the y-direction, aligned with the line connecting piles P1 and P3, and (ii) along the xy-direction, angled at 45° with respect to the y-direction, namely, Fy and Fxy tests, respectively. Tests along the x-direction were not performed because of difficulties in adequately arranging the traction system. A time acquisition of 4 s, including a pretrigger of 2 s, and a sampling rate of 5 kHz were used. Tests were repeated for different load amplitudes: 5 tests in the y-direction and 4 tests in the xy-direction for which different calibrated pins were used. Tests were conducted over 1 day (after the impact load test), firstly Fy tests and then the Fxy ones.

6.4. Experimental results at very small-small strain

Results of impact load tests and free vibration tests at low force level are here reported in terms of strain gauge measurements with the aim to discuss the dynamic behavior of the soil-water-pile system both at very small and small strains, respectively. First, the qualitative features of all the signals and the differences among signals recorded at various measuring points along the pile are discussed and then the modal properties of the soil-pile system in terms of frequencies and damping ratios will be reported and commented.

The time histories of the longitudinal strains recorded on pile P1 with SG5 and SG9 for an impact load of the test Hy during the first series are shown in Figure 14 with a green line and with a black line after signal filtering. The initial part of the signals, especially for strain gauges nearer the pile top, is characterized by high-frequency content due to cross-sectional deformation of the pipe. Therefore, the experimental signals were filtered by a Butterworth low-pass filter with a cutoff frequency of 100 Hz to nearly eliminate the effects due to cross-sectional deformation and noise, which are mainly characterized by high-frequency content.

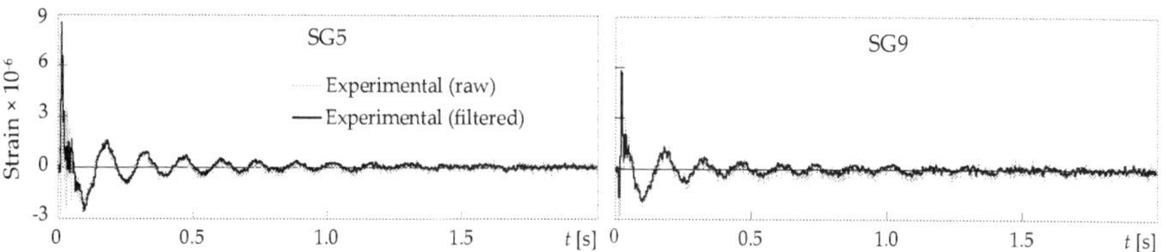

Figure 14. Test Hy: time histories of strain gauge signals.

The measurements recorded by some strain gauges along the main generatrix for one of the 10 impacts of the test Hy during the first and second test series are reported in the left column of Figure 15 in blue and green, respectively. Each strain gauge signal exhibits a first peak with high amplitude due to the impact. The peak is characterized by a time delay due to the traveling time of the impulse, which increases with the distance of the measuring point from the hammer impact point. After the peak, the measurements show a damped harmonic oscillation at the frequency of the first bending mode of the system. With reference to the harmonic oscillation, the strains, which are proportional to the longitudinal bending moment, reach the maximum values for sensors located just below the soil surface (SG7) while decreasing in depth and near the pile head, where the bending moments are smaller; oscillations are practically null at SG1 and SG11, where strains are too small to be captured by sensors, considering their sensitivity and noise. The trends of the strain signals at the pile head are less smooth than those recorded at other locations due to residual ambient noise and radial-circumferential modes. Lower values of strains are registered in the second test series probably due to a lower energy content of the impacts. Disregarding the negligible differences just after the first peak, the response obtained from the two different series is qualitatively very similar. However, it can be observed that the period of the damped harmonic oscillation is shorter for the test of the second series.

The time histories of the first snap-back test (right column of Figure 15) show nearly constant values before the release due to the fact that the load is imposed in a quasi-static manner. After the quick release, free damped oscillations of the pile are manifested at nearly the first natural frequency of the soil-water-pile system (obviously, this result is also evident when plotted in the frequency domain). It worth noting that the values of strains obtained from the two test typologies differ of about one order of magnitude. However, the results are comparable and very similar from a qualitative point of view. The damped harmonic oscillations are practically proportional, and maximum values of strains, proportional to pile bending moment, are attained in the pile section located just below the soil surface (SG7) for both tests.

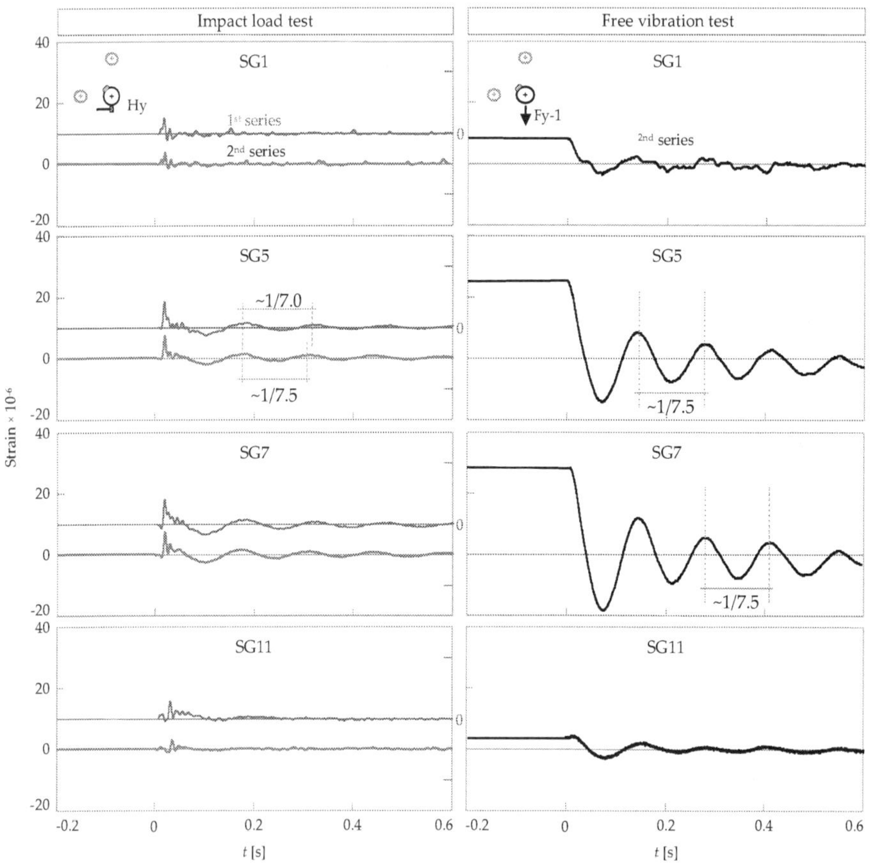

Figure 15. Time histories of SG signals for impact load tests (filtered) and free vibration test.

The modal properties of the soil-water-pile system, such as natural frequencies and damping ratios, were obtained by means of experimental modal analyses using the measured excitation applied to the pile head and the system response measured by strain gauges at various locations. Both the excitation and response time histories were transformed into the frequency domain to define FRFs. Natural frequencies were obtained by means of the peak picking method, selecting the frequencies corresponding to the peak values of the FRF amplitude. As an example, the FRFs relevant to SG5 obtained for the two tests are reported in Figure 16. The left graph shows the averaged FRF relevant to the 10 signals recorded by SG5 during Hy tests; the two clear peaks identify the first two natural frequencies of the soil-water-pile system. It is worth noting that the second series are characterized by higher values of the first and second natural frequencies. Analogously, the right graph reports the FRF of the signal recorded by SG5 for the snap-back test Fy-1; the FRF is not smooth because it was obtained from one signal and not averaged. In this case, the peak defining the first natural frequency is clearly identified, whereas the peak relevant to the second natural frequency is not evident due to the fact that the pile was released from a deformed shape similar to the first mode shape and, consequently, superior modes were only slightly excited. It is also worth noting that the first natural frequency obtained from test Fy-1 is coincident with that obtained from the second series of Hy test.

As regards damping ratios of the soil-water-pile system at small strain, they were estimated from the strain gauge signals by means of the logarithmic decrement, working in time domain. The procedure was applied considering the first eight peaks to obtain a mean value representing the damping of the system during almost the entire oscillation of the pile. Figure 17 shows the natural frequencies and the damping ratios relevant to the first mode. It is important to observe that the second campaign is characterized by higher values of the first natural frequency and by slightly lower values of the damping ratio relevant to the first mode. The increase of the natural frequency values and the decrease of damping ratio values, obtained 9 weeks after the first test series (10 weeks after the vibrodriving), demonstrate an increase of the system stiffness (therefore, minor strains and, then, minor damping), which may be attributed to the soil reconsolidation subsequent to the pile vibrodriving installation. In fact, the vibrodriving technique induces excess pore pressure during installation and subsequent consolidation in the soil surrounding the pile.

In the second test series, very close values of frequency along the pile were obtained for both tests along the y-direction; this confirmed that the system behaves in elastic manner in both cases even if the strain level induced by snap-back tests is about one order greater than that of impact load tests. With reference to mean damping ratios, the average value among all the SGs obtained from the free vibration test is very close to the one obtained from impact load test, 7.2% for both Fy-1 and Hy. It can be noted that the values of damping ratio relevant to each SG signal tend to increase with depth.

This study suggests that both test typologies, impact load and free vibration tests, are effective to evaluate the linear behavior, identifying the dynamic properties, of a near-shore pile driven in soft clay. Other test typologies could have been functional for the same purpose, such as the ambient vibration and/or forced vibration test. The first typology permits to evaluate natural frequencies and relevant damping ratios at very small strain, whereas the second one furnishes a higher excitation level that guarantees a more accurate measurement of the dynamic characteristics at low strain level. Furthermore, forced vibration test allows the excitation of higher-order modes that cannot be identified with other test methods.

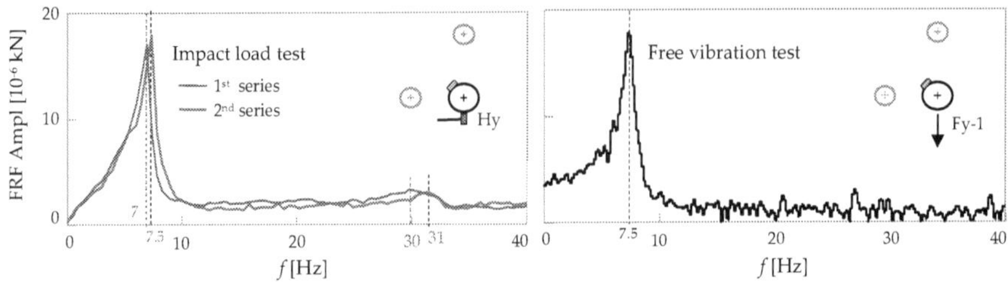

Figure 16. FRFs of strain gauge signals for impact load tests and free vibration test.

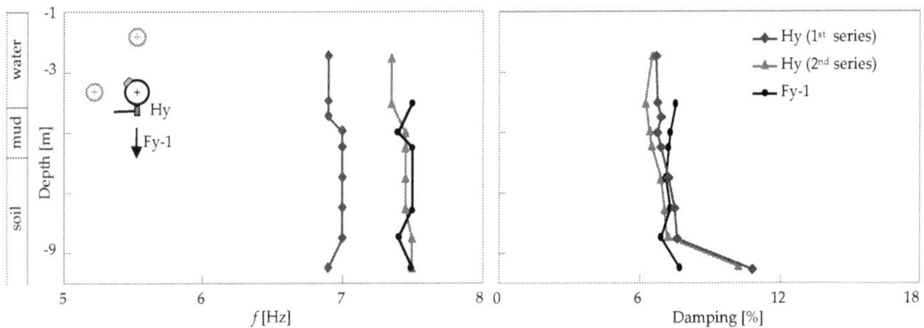

Figure 17. Natural frequency and damping ratios relevant to the first mode for impact load and snap-back test.

6.5. Results at higher strain level: free vibration test at increasing load level

To study the dynamic behavior of the soil-water-pile system at higher strain and to investigate the effects of soil nonlinearities, free vibration tests at four higher load levels were performed; the maximum force (58.1 kN) is about 20 times the minimum one (2.8 kN) as reported in Table 5. In this section, some results in terms of strains (measured values and FRFs) and dynamic properties of the system (frequencies and damping ratios) are reported.

Figure 18 shows the strains (and bending moments that are proportional to strains) measured along the pile by SGs for different loads just before the quick release. The continuous line is obtained, for each level of force, by interpolating the experimental data and allows obtaining an estimation of the maximum value of strain attained along the pile and its location (Table 5). As the load level increases, a progressively moving down of the location of the maximum strains is observed. This may be due to the formation of gap at the soil-pile interface.

With regards to the instrumentation used in this study, the findings obtained in the linear and nonlinear fields showed that, thanks to a meticulous and accurate protection, the electrical strain gauges can provide reliable measurements even if the experimental campaign is quite extensive (few months) and in marine environment, permitting to obtain interesting results both in time and frequency domain.

Furthermore, during this experimental campaign, an abundant number of sensors were used and preliminary FEM simulations were conducted to define the correct location of sensors. For this reason, the definition of the dynamic properties of the system in the linear and nonlinear fields was successful. Thus, it is very important to define correctly the number and the location of sensors because, some time, limited discrete measuring points can be insufficient to capture completely the nature of the strain variation. To overcome this problem, optical fiber cables can be used instead of traditional strain gauges, providing practically continuous strain information along the pipe [44].

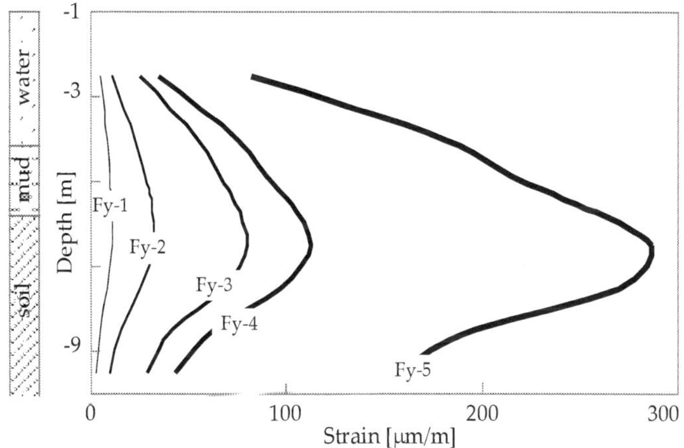

Figure 18. Strains and bending moments along the pile for different load levels of free vibration tests in the y-direction.

Free vibration test	Fy-1	Fy-2	Fy-3	Fy-4	Fy-5
Force calculated from SGs [kN]	2.8	6.9	18.0	24.0	58.1
Maximum pile strain [μm/m]	11.34	32.16	79.36	110.34	278.08
Location of maximum strain [m]	-5.8	-6	-6.1	-6.2	-6.5

Table 5. Free vibration tests: applied forces, maximum strains attained along the pile, and their locations.

Figure 19 shows the mean values of the first natural frequencies and damping ratios obtained from free vibration tests at different load levels considering all the SGs of the main generatrix. The decrement of the natural frequency and the increment of the damping ratios with the load level are clearly evident in all the cases. The results could have been obtained even with forced vibration tests that permit the evaluation of the dynamic behavior at different strain levels depending on the characteristics of loading system. In particular, forced vibration test, even if more expensive and time-consuming, generally allows a more accurate evaluation of the dynamic parameters.

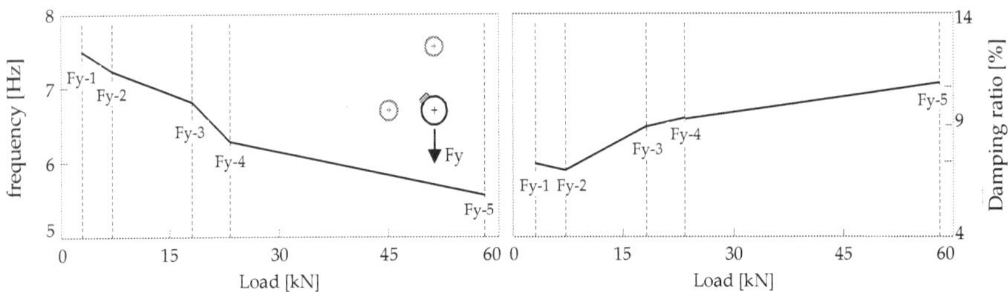

Figure 19. Mean value and standard deviation of natural frequencies and damping ratios vs. load for tests in the y-direction.

6.6. Impedance function of the soil-water-pile system

Impedance function represents the complex stiffness of the soil-water-foundation system under dynamic loads. In this research, the horizontal impedance of the soil-water-pile system was identified as the ratio, in the frequency domain, between the impact load applied at the pile head along a direction orthogonal to the pile axis and the resulting displacement at the pile head along the same direction. Displacements were derived from the recorded accelerometer signals by means of a double discrete integration in the time domain. Thanks mostly to the used high-fidelity modern sensors and data acquisition system equipment, this numerical procedure provided stable results in a wide frequency range.

A typical time history of a raw signal and the averaged FRF of the accelerometer signals, relevant to the Hy test (first test series), are shown in Figures 20a and 20b, respectively. Peaks of the FRF graph identify the natural frequencies of the system: the first two (at 7.0 and 31.1 Hz) are relevant to the first and second flexural modes; the third peak (at 52 Hz), characterized by high amplitude, is in correspondence to the first radial-circumferential mode that is predominant at the pile head, close to the hammer impact point.

Figure 21 shows the real and imaginary parts of the impedance function obtained from the Hy test of the two different series. The curves behave in a similar way and slight differences can be observed. It is worth noting that the frequency at which the real part goes to zero is higher for the curve relative to the tests of the second series. This is in accordance with the increase of the first flexural frequency of the system, in the y-direction, from 7 to 7.5 Hz due to the increase of the system stiffness.

Finally, the obtained results at small and very small strains are compared with those derived from the dynamic soil-pile interaction 3D finite-element model developed in ABAQUS [45].

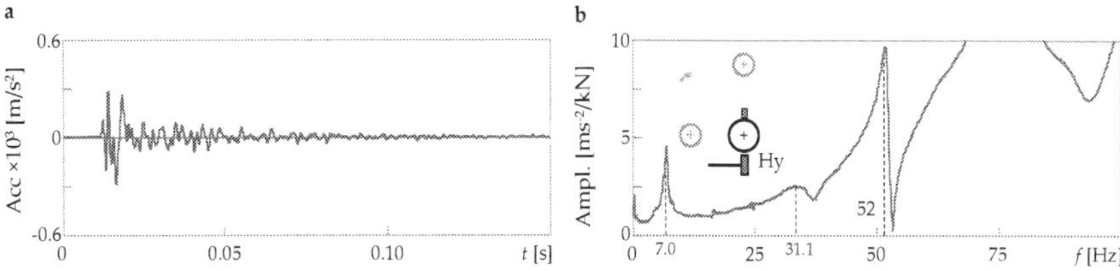

Figure 20. (a) Time histories of an accelerometer signal and (b) averaged FRF of accelerometer signals.

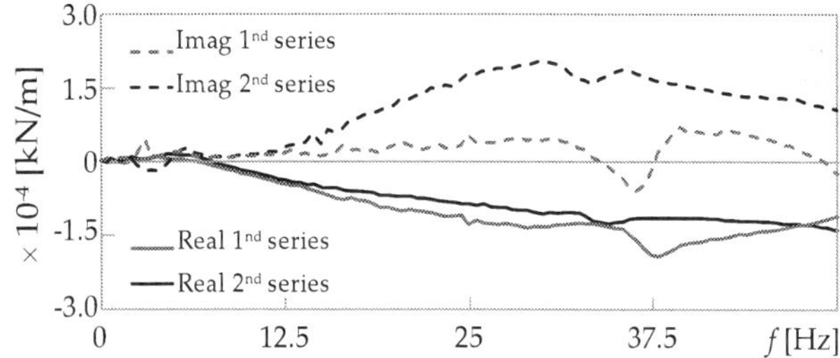

Figure 21. Horizontal impedance function (Hy tests): comparison between the two test series.

Soil domain was modeled as a parallelepiped (dimensions are reported in Figure 22) suitably partitioned into horizontal sublayers. 8-nodes solid elements were used for the soil domain, 4-nodes shell elements were used for the pipe piles, and 8-nodes infinite elements were used for the far field in order to avoid waves reflection/refraction along the boundaries. The mesh was suitably refined near the hammer impact point and in the soil surrounding the piles. With the energy of the impact being very low, the formation of gap along the soil-pile interfaces was excluded and the behavior of the system was considered linear elastic. Water surrounding the submerged portion of the piles was considered as an added mass and the impact of the hammer was reproduced as a uniform pressure applied on a surface equal to the hammer impact zone, varying with time. Dynamic analyses were performed in the time domain, with a direct implicit integration procedure. The mechanical characteristics of the pile P1 and the soil are reported by Table 6.

Figure 23 shows the comparison between experimental and theoretical impedance functions. A good agreement between the experimental results and the theoretical predictions is observed within a significant frequency range, which largely covers the range of interest in engineering practice. Many other theoretical models could be used for the comparison with the experimental results: single pile model based on boundary elements or single pile model on Winkler foundation. Furthermore, in order to evaluate the behavior of the soil-water-pile system at higher strain levels, more advanced finite-element models should be developed with nonlinear constitutive relations in order to take into account the soil nonlinearities revealed by the snap-back test at increase load levels.

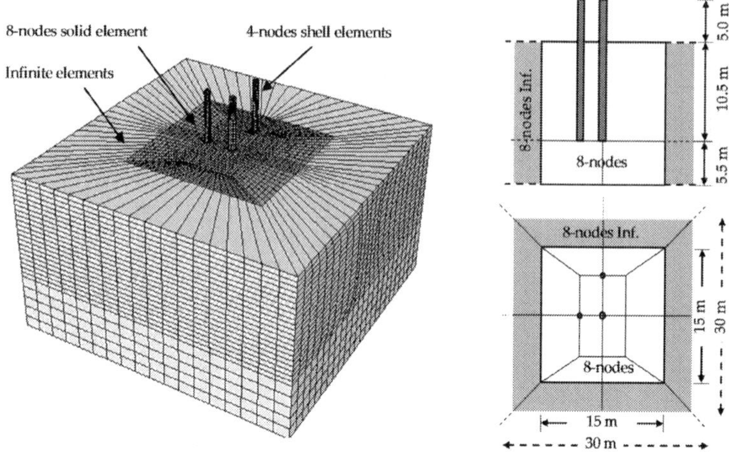

Figure 22. 3D finite-element model.

	ϱ [t/m^3]	E [kN/m^2]	ν
P1 emerged/submerged/embedded	7.8/38.6/35.5	$2.0 \cdot 10^7$	0.3
Soil	1.68	$1,7 \cdot 10^4$	0.49

Table 6. Pile and soil properties.

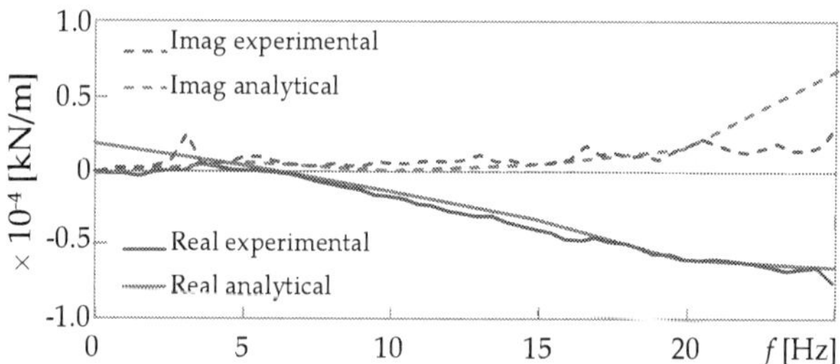

Figure 23. Horizontal impedance function: comparison between experimental and analytical curves.

7. Conclusions

In this chapter, several in situ dynamic testing methods for the dynamic characterization of open-ended pipe piles in marine environment have been described. Different test typologies, the ones without any control on the input (such as ambient vibration test) and the ones where the excitation is artificially induced (such as impact load test and free vibration test), have been illustrated. Some attention has been given to the appropriate sensors to be used and to the suitable protection from marine environment and pile driving installation procedure. Furthermore, the most common signal processing techniques useful for handling the experimental raw signals have been addressed together with the analysis techniques for the evaluation of the modal parameters: natural frequencies, damping ratios, and mode shapes. In the last section, a concrete case study has been reported: an experimental campaign was carried out on near-shore open-ended pipe piles at the "Mirabello" harbor in La Spezia, Italy. During this campaign, impact load and snap-back tests have been performed and the modal properties of the soil-water-pile system have been identified by means of experimental modal analyses at both low and high strain levels. Some considerations can be formulated:

- Impact load test permits to evaluate the behavior of piles in marine environment at very low level of strain, whereas snap-back tests permit at both low and high levels; forced vibration test allows a more accurate evaluation of the modal parameters, whereas ambient vibration tests are especially suited for long-term monitoring.

- Accelerometers at the pile head allow the identification of natural frequencies and damping ratios, whereas strain gauges (electrical or fiber optic for long-term monitoring) along the pile length can be used to identify also the mode shapes.

- The accurate choice of the more appropriate instrumentation together with the most powerful protective systems is crucial in order to obtain successful results.

- Finally, for the estimation of the modal parameters of piles embedded in marine environment, conventional signal processing techniques and modal analysis procedures can be applied.

Author details

Dezi Francesca[1*], Gara Fabrizio[2] and Roia Davide[3]

*Address all correspondence to: francesca.dezi@unirsm.sm

1 Department of Economy Science and Law, University of San Marino, Republic of San Marino

2 Department of Civil and Building Engineering and Architecture, Università Politecnica delle Marche, Ancona, Italy

3 Department of Civil and Building Engineering and Architecture, Università Politecnica delle Marche, Ancona, Italy

References

[1] Byrne B, Mcadam R, Burd H, Houlsby G, Martin C, Gavin K, et al. Field testing of large diameter piles under lateral loading for offshore wind applications. In: 15th European Conference on Soil Mechanics and Geotechnical Engineering. Edinburgh, UK; 2015.

[2] Prendergast LJ, Gavin K, Doherty P. An investigation into the effect of scour on the natural frequency of an offshore wind turbine. Ocean Eng. 2015;101:1–11.

[3] Blaney GW, O'Neill MW. Measured lateral response of mass on single pile in clay. J Geotech Eng. 1986 Apr;112(4):443–57.

[4] Crouse CB, Kramer SL, Mitchell R, Hushmand B. Dynamic tests of pipe pile in saturated peat. J Geotech Eng. 1993 Oct;119(10):1550–67.

[5] Landva A. In situ testing of peat. In: ASCE Conference on Use of in Situ Tests in Geotechnical Engineering n 6. Blacksburg: ASCE; 1986. p. 191–205.

[6] Sa'don NM, Pender MJ, Orense RP, Abdul Karim AR, Wotherspoon L. Dynamic field tests of single piles. In: NZSEE Annual Technical Conference. 2010. Paper 18.

[7] Jennings DN, Thurston SJ, Edmonds FD. Static and dynamic lateral loading of two piles. In: 8th World Conference on Earthquake Engineering. San Francisco, CA; 1984. p. 561–8.

[8] Ting JM. Full-scale cyclic dynamic lateral pile responses. J Geotech Eng. 1987 Jan; 113(1):30–45.

[9] He J, Fu Z-F. Modal analysis. Oxford; Boston: Butterworth-Heinemann; 2001. 291 p.

[10] Ewins DJ. Modal Testing: Theory, Practice, and Application. 2nd ed. Baldock, Hertfordshire, England; Philadelphia, PA: Research Studies Press; 2000. 562 p.

[11] Halling MW, Womack KC, Muhamad I, Rollins KM. Vibrational testing of a full-scale pile group in soft clay. In: 12th World Conference on Earthquake Engineering. Auckland, New Zealand; 2000. Paper 1745.

[12] Hoffmann K. Applying the Wheatstone Bridge Circuit.

[13] Hoffmann K. An Introduction to Stress Analysis and Transducer Design Using Strain Gauges. 261 p.

[14] Li H-N, Li D-S, Song G-B. Recent applications of fiber optic sensors to health monitoring in civil engineering. Eng Struct. 2004 Sep;26(11):1647–57.

[15] Wang T, Yuan Z, Gong Y, Wu Y, Rao Y, Wei L, et al. Fiber Bragg grating strain sensors for marine engineering. Photonic Sens. 2013 Sep;3(3):267–71.

[16] Harris CM, Piersol AG, editors. Harris' Shock and Vibration Handbook. 5th ed. New York: McGraw-Hill; 2002. 1 p.

[17] Wilson JS, editor. Sensor Technology Handbook. Amsterdam; Boston: Elsevier; 2005. 691 p.

[18] Iskander M. Behavior of Pipe Piles in Sand Plugging & Pore-Water Pressure Generation During Installation and Loading. Berlin: Springer Berlin; 2014.

[19] Hoffmann K. Practical Hints for the Installation of Strain Gages.

[20] Reddy DV, Edouard J-B, Sobhan K, Rajpathak SS. Durability of reinforced fly ash-based geopolymer concrete in the marine environment. In: 36th Conference on Our World in Concrete & Structures. Singapore; 2011.

[21] Cooley JW, Tukey JW. An algorithm for the machine calculation of complex Fourier series. Math Comput. 1965 Apr;19(90):297.

[22] Rocklin GT, Crowley J, Vold H. A Comparison of H1, H2, and HV frequency response functions. In: 3rd International Modal Analysis Conference. Orlando, FL, USA; 1985.

[23] Wicks AL, Vold H. The hs frequency response function estimator. In: 4th International Modal Analysis Conference. Los Angeles, CA, USA; 1986.

[24] Wicks AL, Mitchell LD. Methods for the estimation of frequency-response functions in the presence of uncorrelated noise, a review. 1987 Jul;2(3):109–12.

[25] Kennedy CC, Pancu CDP. Use of vectors in vibration measurement and analysis. J Aeronaut Sci Inst Aeronaut Sci. 1947 Nov;14(11):603–25.

[26] Bishop RED, Gladwell GML. An investigation into the theory of resonance testing. Philos Trans R Soc Math Phys Eng Sci. 1963 Jan 17;255(1055):241–80.

[27] Richardson MH, Formenti DL. Parameter estimation from frequency response measurements using rational fraction polynomials. In: 1st International Modal Analysis Conference. Orlando, FL, USA; 1982.

[28] Dobson BJ. A straight-line technique for extracting modal properties from frequency response data. Mech Syst Signal Process. 1987 Jan;1(1):29–40.

[29] Dobson BJ. Modal Analysis Using Dynamic Stiffness Data; 1984 Jun. Report No.: TR-84015.

[30] Guillaume P, Verboven P, Vanlandiut S. Frequency-domain maximum likelihood identification of modal parameters with confidence intervals. In: 23th International Conference on Noise and Vibration Engineering; 1998.

[31] Guillaume P, Verboven P, Vanlandiut S, Van der Auwaerer H, Peeters B. A poly-reference implementation of the least-squares complex frequency-domain estimator. In: 21th International Modal Analysis Conference. Kissimmee, FL, USA; 2003.

[32] Ibrahim SR, Mikulcik EC. A time domain modal vibration test technique. Shock Vib Bull. 1973;43(4):21–37.

[33] Ibrahim SR, Mikulcik EC. The experimental determination of vibration parameters from time responses. Shock Vib Bull. 1976;46(5):187–96.

[34] Ibrahim SR, Mikulcik EC. A method for the direct identification of vibration parameters from the free response. Shock Vib Bull. 1977;47(4):183–98.

[35] Brown DL, Allemang RJ, Zimmerman R, Mergeay M. Parameter estimation techniques for modal analysis; 1979. Available from: http://papers.sae.org/790221/.

[36] Cole Jr H. On-the-line analysis of random vibrations. American Institute of Aeronautics and Astronautics; 1968. Available from: http://arc.aiaa.org/doi/abs/10.2514/6.1968-288.

[37] Cole HA. Method and apparatus for measuring the damping characteristics of a structure [Internet]. Google Patents; 1971. Available from: http://www.google.com/patents/US3620069.

[38] Cole HA. On-line Failure Detection and Damping Measurements of Aurospace Structures by Random Decrement Signature. NASA; 1973 Mar. Report No.: NASA-CR-2205.

[39] Gersch W, Nielsen NN, Akaike H. Maximum likelihood estimation of structural parameters from random vibration data. J Sound Vib. 1973 Dec;31(3):295–308.

[40] Gersch W. On the achievable accuracy of structural system parameter estimates. J Sound Vib. 1974 May;34(1):63–79.

[41] Van Overschee P, De Moor B. Subspace Identification for Linear Systems: Theory, Implementation, Applications. Boston: Kluwer Academic Publishers; 1996. 254 p.

[42] Dezi F, Gara F, Roia D. Dynamic response of a near-shore pile to lateral impact load. Soil Dyn Earthq Eng. 2012;40:34–47.

[43] Dezi F, Gara F, Roia D. Experimental study of near-shore pile-to-pile interaction. Soil Dyn Earthq Eng. 2013;48:282–93.

[44] Doherty P, Igoe D, Murphy G, Gavin K, Preston J, McAvoy C, et al. Field validation of fibre Bragg grating sensors for measuring strain on driven steel piles. Geotech Lett. 2015;5:74–9.

[45] ABAQUS. Hibbitt, Karlsson & Soresen, Inc.; 2009.

Chapter 9

Relationship between Bulk Metal Concentration and Bioavailability in Tropic Estuarine Sediments

Sofia E. Koukina, Nikolay V. Lobus, Valeriy I. Peresypkin and Olga M. Dara

Additional information is available at the end of the chapter

http://dx.doi.org/10.5772/62155

Abstract

Major (Si, Al, Fe, Ti, Mg, Ca, Na, K, S, P), minor (Mn) and trace (Li, V, Cr, Co, Ni, Cu, Zn, As, Sr, Zr, Mo, Cd, Ag, Sn, Sb, Cs, Ba, Hg, Pb, Bi and U) elements, their chemical forms and the mineral composition, organic matter (TOC) and carbonates (TIC) in surface sediments from the Cai River estuary and Nha Trang Bay were first determined along the salinity gradient. The abundance and ratio of major and trace elements in surface sediments are discussed in relation to the mineralogy, grain size, depositional conditions, reference background and SQG values. Most trace-element contents are at natural levels and are derived from the composition of rocks and soils in the watershed. A severe enrichment of Ag is most likely derived from metal-rich detrital heavy minerals such as Ag-sulphosalts. Along the salinity gradient, several zones of metal enrichment occur in surface sediments because of the geochemical fractionation of the riverine material. The parts of actually and potentially bioavailable forms (isolated by four single chemical reagent extractions) are most elevated for Mn and Pb (up to 36% and 32% of the total content, respectively). The possible anthropogenic input of Pb in the region requires further study. Overall, the most bioavailable parts of trace elements are associated with easily soluble amorphous Fe and Mn oxyhydroxides. The sediments are primarily enriched with bioavailable metal forms in the riverine part of the estuary. Natural (such as turbidities) and human-generated (such as urban and industrial activities) pressures are shown to influence the abundance and speciation of potential contaminants and therefore change their bioavailability in this estuarine system.

Keywords: Vietnam, Nha Trang Bay, estuary, surface sediments, major elements, trace elements, bioavailable metal forms

1. Introduction

Estuaries and continental shelf areas comprise 5.2% of the earth's surface and 2% of the oceans volume. However, they carry a disproportionate human load [14, 51]. Coastal and estuarine

waters receive contaminants via local anthropogenic activities and riverine inputs [9]. Chemicals that are released into aquatic systems are generally bound to particulate matter, which eventually settles and becomes a part of the water-sediment system. Therefore, aquatic sediments are the ultimate sink for many pollutants including metals. Toxicity and other adverse effects in sediments are functions of the total concentration and bioavailability of the sediment-associated chemicals. Specific metal forms are useful environmental indicators because the fractionation of the total metal contents may indicate their origin [12, 13, 42].

Currently, numerous natural and anthropogenic hazards expose the tropic aquatic environments to multiple pressures [2, 47]. The Cai River and the Nha Trang Bay form one of the major estuarine systems in the South China Sea, which is inhabited by unique biota. This area, which was wilderness not long ago, now experiences a significant anthropogenic load from the local people activities and particularly the quickly growing touristic industry. Moreover, aquaculture, which is currently one of the most rapidly developing production sectors in coastal Vietnam, contributes to the degradation of estuarine and coastal ecosystems. In the recent decades, significant research efforts have been undertaken to evaluate the biodiversity and ecosystem community structure in the Nha Trang Bay [5, 36-38]. However, notably few data are available relative to the contamination levels and trends. The mercury concentrations in the riverine, estuarine and marine sediments of the region were reported to be at or below the natural levels [24, 25]. The aquaculture activity has been reported to contribute to the ecosystem with heavy metals in the farming environment in Nha Trang Bay and Van Phong Bay [35]. Elevated metal concentrations were obtained in macrophytes that were collected in estuarine waters in the city of Nha Trang [8]. The study on the sediments from coastal lagoons of Khanh Hoa Province reported generally low metal levels that were derived from rocks and soils of the watersheds [42].

The first geochemical study of the Nha Trang Bay covered the distribution of TOC, n-alkanes and the bulk concentration of major and trace elements in the surface sediments [1, 17, 25, 38]. Further analysis and speciation of metals in the sediments of the Cai River estuary and Nha Trang Bay will allow one to track the fate of potential contaminants. Therefore, the objective of the present study is to summarize the data on the abundance, distribution and bioavailability of major and trace elements in the surface sediments of the Cai River estuary under multiple stresses. Because the chemical form or speciation is the most important factor that influences bioavailability, the specific goal of this study is a comparative analysis of the ecologically significant chemical forms of metals.

2. Material and methods

2.1. Environmental setting

The Cai River, its estuary and the adjacent part of the Nha Trang Bay belong to the Central Southern Coastal Region of Vietnam (Khanh Hoa Province) (Figure 1). The regional climate is monsoonal subequatorial with a yearly average temperature of 23–24°C. The rainy season lasts from September to December or January. Rainfall reaches 1,300 to 1,400 mm/year. In the last decades, climate change events such as anomalous precipitations and droughts were observed in the region [37].

The relief is mountain-hilly. The main soils of the region are yellow- and red-coloured ferralitic soils on acid and alkaline magmatic (mainly basalts and andesite-basalts) rocks, whereas the river deltas are covered with alluvial soils [34, 36]. The natural terrain is mainly covered with broad-leaved, mixed-leaved and mangrove forests. The soils of the agricultural terrain are used for crops of rice, maize, sweet potato and cassava.

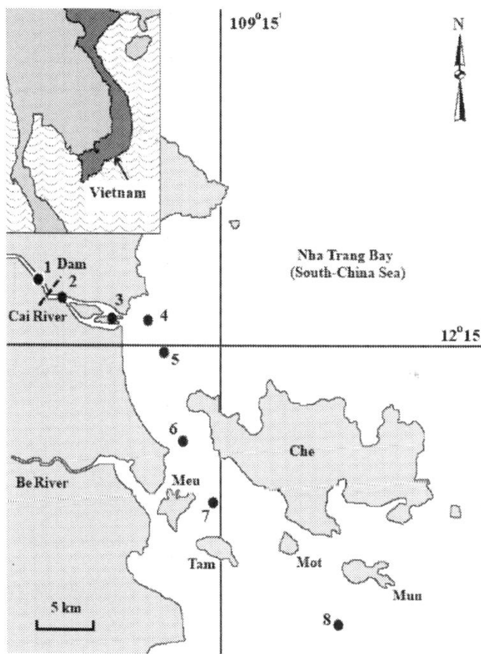

Figure 1. Location of the study sites.

The length of the Cai River is 80 km, its catchment area is 1,450 km², and the annual river discharge is approximately 2 km³. The fresh-water discharge of the Cai River is highly seasonal and varies from 10–15 m³/s in the dry season to 600–700 m³/s in the rainy season. The main river discharge is deflected to the South coast of the Nha Trang Bay along the ancient riverbed (channel) [36]. The fresh river water ($S < 0.1‰$) and saline South China Sea water ($S \approx 36‰$) form a major water-mixing zone. The fill dam that was constructed in 1998–1999 eight kilometres upstream from the river mouth restricts the water exchange and marks the riverine boundary of the water-mixing zone (Figure 1). The Cai River estuary and Nha Trang Bay can be divided into three sub-zones: I – river ($S < 0.1‰$), II – transitional waters (estuary) ($0.1‰ > S > 32‰$) and III – sea (bay) ($S > 32‰$). In the transitional waters, the salinity (S) increases from the river to the sea and from the surface to the bottom. The water column is highly stratified with the pronounced horizontal and vertical salinity gradients [25, 36]. The hydrological regime of the estuarine zone of the Cai River is complicated by daily tides (the tidal range reaches 2 m), seasonal fluctuations in the fresh-water discharge and monsoonal seawater upwelling. The daily tides lead to regular upstream invasions of the near-bottom saline water lenses, which are limited by the dam. The seasonal variability in the Cai River discharge significantly influences the estuarine zonation. Thus, in the rainy season, the fresh water flows seawards beyond the Che Island (Figure 1), whereas most river sediment is carried far to the Nha Trang Bay with the river water in the upper layer [36].

The major part of the Cai River estuary is located inside the Nha Trang city (300,000 inhabitants, 100,000 tourists in the season). The estuarine water flows through the residential and industrial zones with their ports. The outflow of waste and sewage sludge and the shipping, fuelling and vessel production and repair are sources of the local anthropogenic contamination of the estuary. The estuarine and coastal zones of the Nha Trang Bay are a farming environment for aquaculture, where green mussels, babylonia snails, abalone, sea cucumber, red snapper, sea bass and lobster are cultivated. The agrochemicals and veterinary drugs that are widely used in aquaculture production may be an additional source of anthropogenic heavy metals in the Nha Trang Bay area [35].

The climate variability (seasonality and precipitation), extreme weather events (droughts, storms and upwelling) and human activities (deforestation, urbanisation, land use, damming, tourism, coastal construction, transportation and fisheries) expose the Nha Trang Bay to multiple pressures that contribute to further degradation of the estuarine and coastal ecosystems in the region [14].

2.2. Field work

The water and sediment samples were collected in the Cai River estuary and Nha Trang Bay on 6–10 July 2010 at eight locations along the salinity gradient. The sampling stations were located in the riverine (st. 1), transitional (sts. 2–4) and marine (sts. 5–8) sub-zones of the dry season (Figure 1).

The surface and bottom water samples were obtained using a plastic Niskin bottle. The temperature, alkalinity and salinity of the water samples were measured on board immediately after collection using portable conductivity apparatuses HI 98129 Combo and HI 98302 DIST 2 (Hanna Instruments, Germany).

The surface sediment samples were obtained by scuba-divers with a manual plastic piston corer, which was designed at the P.P. Shirshov Institute of Oceanology (Russia). Three samples were collected at each location. On board, the visual estimates of grain size and the colour and relative proportion of the components were made and recorded [30]. The upper layer (0–6 cm) of the samples was retrieved using stainless steel spatulas. Then three upper layer samples were mixed into a one composite sample. The composite sample was transferred into pre-cleaned polyethylene containers and frozen until arrival in the laboratory. One portion of the composite sample was kept frozen until the grain size and mineralogy analyses. One portion was dried to constant weight at 60°C until the major-, trace-element, TOC and TIC analyses. One portion was dried to constant weight at 40°C until the Hg analysis. The sampling, sampling transportation and preparation procedures were performed using standard clean techniques that were described elsewhere [25, 26, 30].

2.3. Analytical methods

Three samples representative of the surface sediment at the riverine (st. 1), transitional (st. 4) and marine (st. 8) locations (Figure 1) were subjected to extensive grain size and mineral composition analyses. The grain size analysis was performed by wet sieving. The mineral composition of the total samples was determined on the X-ray diffractometer D8 ADVANCE (Bruker AXC, Canada) [31].

The total Si content in the samples was determined using the X-ray-fluorescence method on the SPECTROSCAN MAKC-GV WDXRF spectrometer (Spectron, Russia). The total carbon (TC) contents in the samples were determined by dry burning at 900ºC in oxygen flow and the total inorganic carbon (TIC) contents were determined by dry burning at 200ºC with H_3PO_4 with the analyser TOC 5000-V-CPH (Shimudzu Co., Japan). The total organic carbon (TOC) contents were determined as a difference between TC and TIC contents in the samples [38].

For the total Al, Fe, Ti, Mg, Ca, Na, K, S, P, Mn, Li, V, Cr, Co, Ni, Cu, Zn, As, Sr, Zr, Mo, Cd, Ag, Sn, Sb, Cs, Ba, Pb, Bi and U content analysis, the samples were subjected to the total acidic dissolution in $HNO_3+HF+HClO_4$ in an open system with further determination of element contents using the ICP-AES and ICP-MS methods (Thermo Scientific, USA). The detailed sample decomposition and analytical procedures are described elsewhere [15]. The Hg content was determined in the dry samples using a pyrolyse method on the RA-915$^+$ spectrometer with background correction and a two-chamber atomiser PYRO-915$^+$ (Lumex, Russia) [25].

To assess the chemical form of selected metals (Fe, Mn, Li, Cr, Zn, Cu, Pb, Ni and Ag) in the sediments, the samples (excluding sediments with the highest sand (st. 5) and carbonate (st. 7) content) were subjected to single chemical reagent (single-step) extraction procedures. The strong-acid-soluble metals were extracted using 2 M nitric acid, weak-acid-soluble (labile) metals were extracted using 25% acetic acid, dithionite-soluble metals were extracted using sodium dithionite at pH 6.5–7 (Mehra-Jackson extraction) and oxalate-soluble metals were extracted using ammonium oxalate–oxalic acid buffered solution at pH 3.2 (Tamm extraction). To isolate the weak- and strong-acid-soluble metals, 15 ml of 25% acetic acid and 15 ml 2 M nitric acid, respectively, were added to 1.1 g of dry sample in polypropylene vials and shaken in a mechanical shaker for 6 h with acetic acid and 1 h with nitric acid. Then, each extract with the sediment was filtrated into a 25 ml glass volumetric flask. The sediment on the filter was washed with 10 ml of distilled water, and the wash water was added to the flask [30]. To isolate the non-silicate iron compounds and the microelements that are associated with this phase, 20 ml of 0.3 M sodium citrate and 2.5 ml of 1 M sodium hydrocarbonate were added to 1.1 g of dry sample in a centrifuge vial and heated on a water bath up to 80°C; 0.5 g of dry powder sodium dithionite was added, shaken and heated for 15 min at 80–85°C; then, 5 ml of fat NaCl solution was added and centrifuged for 10 min at 3,000 t/min. Then, the extract with the sediment was filtrated into a 250 ml glass volumetric flask. The sediment was subsequently subjected to two more repeated extractions, and the extracts were added to the 250 ml volumetric flask [50]. To isolate the amorphous iron oxides and their associated microelements, 50 ml of ammonium oxalate–oxalic acid buffered (Tamm) solution was added to 1.1 g of dry sample in the 250 ml flat bottom flask, shaken for 1 h and filtrated to a 250 ml glass volumetric flask. The sediment on the filter was washed with 10 ml of distilled water that was mixed with a small amount of oxalic acid, and the wash water was added to the flask. Then, the filter with sediment was added to the sediment in the flat-bottom flask and subjected to one more repeated extraction, and the extract was added to the 250 ml volumetric flask [49]. The metal contents in the extracts were further determined using an atomic absorption spectrometer (AAS) Hitachi 180-8 (Hitachi Co., Japan) in the Analytical Centre of Moscow State Lomonosov University.

The relative accuracy of the analytical determinations was within the standard deviations that were established by the certified reference materials (CRM) SDO-1 (Russia) (for SiO_2, TOC and TIC), SRM 521-84P (Russia) (for Na_2O, Al_2O_3, S_{total}, K_2O, CaO, MnO, Fe_2O_3, Li, V, Cr, Zn, As, Sr, Zr, Mo, Ag, Sn, Ba, Pb), AGV-2 (USA) (for MgO, P_2O_5, TiO_2, Co, Ni, Cu, Sr, Sb, Cs, U) and Mess-3 (Canada) (for Cd and Hg).

3. Results

3.1. Water column characteristics

In July 2010, the salinity of the surface fresh-water layer along the river–sea (sts. 1–8) transect varied from 0 to 35‰, and the salinity in the near-bottom water layer varied from 0 to 36‰ (Table 1). The frontal zone of the contact of fresh and saline waters occurred downstream from the fill dam (sts. 2–3, Figure 1), where the vertical salinity gradient was 2.3–3.7‰ per 1 m depth, and the horizontal salinity gradient was 1.5–9‰ per 1 km distance. The temperature (T) of the water column varied in narrow ranges and decreased from the river to the sea from 30°C to 29°C in the surface water layer and from 29°C to 26°C in the near-bottom water layer. The pH of the water column increased from neutral in the riverine waters (pH 7, st. 1) to low-alkaline in the transitional and sea waters (pH 8–9, sts. 2–8).

3.2. Textural and mineral composition of surface sediments

The on-board textural estimates indicate that the surface sediments of the studied part of the Cai River–Nha Trang Bay estuarine system mainly consist of brownish sandy mud (sts. 1–5) and fine-grained greyish mud (sts. 6–8). The oxidised region (one to three or four centimetres of sediment) is typically orange–brown because of Fe and Mn oxide accumulation. Resuspension of the surface sediment layer is observed in most of the transitional and marine locations (sts. 2–6).

The percent of the content of sand- (63 μm–2 mm), silt- (2–63 μm) and clay- (<2 μm) sized material and the mineral composition were determined in three sediment samples, which were collected at the riverine (st. 1), transitional (st. 4) and marine (st. 8) locations (Table 1). The coarsest sediment is from riverine station 1. Downstream in the transitional sub-zone, at station 4, the sediment contains less sand and more silt. The most fine-grained sediment, with the highest clay content, is from marine station 8.

The surface sediments of the Cai River–Nha Trang Bay estuarine system mainly contain quartz (33–67% of the total dry weight), kaolinite (4–11%), illite (8–12%), albite (5–11%), gibbsite (3–7%) and clinochlore (2–5%), which are present in each of the three studied samples. Microcline (5–8%) and aragonite (1–3%) are only present in the riverine and bay sediments (sts. 1 and 8). Hornblende (3–4%) and gypsum (1–2%) are only present in the estuarine and bay sediments (sts. 4 and 8). A significant content of orthoclase (11%) is only found in the estuarine sediment (st. 4). Calcite (5%) is only found in the bay (st. 8) (Table 1). By analysing the admixtures and individual mineral grains, traces of zircon, tremolite, ilmenite, sphalerite, pyrite, epidote, biotite, muscovite and the Ag- and Ag-Pb sulphosalts of the freieslelebenite and pyrargyrite groups were identified.

Station	Depth (m)	Salinity (‰) surface/bottom	Grain size fractions (% of total content)			Mineral fractions (% of total content)			
			sand (0.064-2 mm)	silt (0.021-0.063 mm)	clay (0.001-0.020 mm)	detrital minerals	clay minerals	carbonates	others
1	2.2	0 / 0	31	29	40	77	14	1	8
4	12.5	27 / 34	11	47	42	55	27	0	18
8	42.3	35 / 36	1	37	62	57	21	8	14

Table 1. Surface sediment texture and minerology.

3.3. Abundance and distribution of major and trace elements in surface sediments

The total contents of Si, Al, Fe, Ca, Mg, Na, K, Mn, Ti, P, S, TOC and TIC are reported in Table 2. The mean contents of the major elements are within the range of the Clark contents in shale, pelagic clays and average world riverbed sediments. However, the studied sediments are enriched with S and depleted in P in comparison to the reference values [22, 44]. Sedimentary Si varies in a narrow range and slightly decreases seaward along the river–sea transect. Sedimentary Al generally increases seaward and is elevated in the sediments from stations 2, 4 and 6. The distribution of Fe, Ti and Mg in the river–sea transect is similar to that of Al with a strong positive correlation (the Spearman rank order correlation coefficient (r_s) ranges from 0.7 to 0.9 at $p \leq 0.05$ and $n = 8$). Na, K, P and S show independent distributions along the salinity gradient in relation to the other studied elements but tend to increase seaward. The distribution of the inorganic carbon content in the sediments along the river–sea transect is characterised by a general increase seaward and a maximum in the sediment from station 5, which is rich in shell detritus. Sedimentary Ca is distributed similarly to inorganic carbon with the strongest positive correlation $(r_s = 0.98, p \leq 0.05, n = 8)$. Sedimentary organic carbon (TOC) is uniformly low and exhibits no significant correlation with other elements that were studied.

The mean content of the major part of the studied trace elements (Li, V, Cr, Co, Ni, Cu, Zn, As, Sr, Zr, Mo, Cd, Sn, Sb, Cs, Ba, Hg and Pb) in the sediments from the Cai River estuary and Nha Trang Bay is lower or corresponds to the reference values for shale, pelagic clays and the average world riverbed sediments (Table 2). The exceptions are U, Bi and Ag. The mean content of U exceeds the reference values by 1.5–2 times; the mean content of Bi exceeds the reference values by 3–7 times; and the mean content of Ag exceeds the reference values by 40–80 times.

Element	Mean	SD	Range	Shale [a]	Pelagic clay [a]	River sed [b]
Si	30.1	1.52	27.6 – 31.8	27.5	25	35
Al	8.7	1.61	6.2 – 11.8	8.8	8.4	4.3
Fe	3.1	0.88	1.7 – 4.3	4.72	6.5	2.5
Ti	0.3	0.09	0.16 – 0.4	0.46	0.46	0.31
Mn	0.04	0.01	0.02 – 0.1	0.085	0.67	0.05
Na	1.6	0.63	0.4 – 2.4	0.59	2.8	0.79
K	1.8	0.25	1.4 – 2.1	2.66	2.5	1.10
Mg	0.7	0.39	0.16 – 1.3	1.5	2.1	0.57
Ca	1.8	2.1	0.26 – 5.8	1.6	1.0	1.7
P	0.04	0.02	0.01 – 0.1	0.07	0.15	0.08
S	0.33	0.17	0.11 – 0.6	0.24	0.20	0.02
Li	47.1	15.0	25.3 – 65.8	66	57	20
V	73.5	21.5	43.3 – 101	130	120	50
Cr	61.5	40.5	24.7 – 150	90	90	50
Co	7.7	2.8	4.4 – 12.5	19	74	15
Ni	33.1	27.6	10.6 – 98.2	50	230	25
Cu	36.8	22.1	18.1 – 73.8	45	250	20
Zn	85.6	19.8	63.8 – 118.2	95	170	60
As	15.2	4.9	7.1 – 23.6	13	20	6
Sr	172	182.7	36.9 – 544	170	180	150
Zr	63.5	17.7	41.8 – 96.7	160	150	250
Mo	2.6	1.4	0.82 – 4.3	2.6	27	1.5
Cd	0.1	0.04	<0.03 – 0.2	0.3	0.42	0.4
Ag	4.1	4.84	0.2 – 12	0.07	0.11	0.1
Sn	4.5	0.94	3.1 – 6.2	3.0	4.0	4.0
Sb	1.4	0.9	0.5 – 2.9	1.5	1.0	2.0
Cs	7.5	1.2	5.6 – 8.5	5.0	6.0	4.0
Ba	259	56.8	190 – 358	580	2300	290
Hg	0.045	0.01	0.03 – 0.06	0.18	0.10	0.05
Pb	55.2	18.9	33 – 86.2	20	80	15
Bi	1.4	0.4	0.8 – 2.1	0.43	0.53	0.2
U	4.0	0.8	2.8 – 5.1	2.7	2.6	3
TOC	0.95	0.4	0.5 – 2	-	-	1.4
TIC	0.58	0.9	<0.01 – 2.5	-	-	0.4

a - cited from Li, 1991;
b - cited from Savenko, 2006.

a-cited from Li. 1991;

b-cited from Savenko, 2006.

Table 2. Major and trace elements contents in sediments (in µg g⁻¹, except for Si, Al, Fe, Ti, Mn, Na, K, Mg, Ca, P, S, TOC and TIC in % dry weight).

Figure 2 illustrates the trace element levels compared to the SQG (Effects Range-Low (ERL) and Effects Range-Median (ERM)) values along the Cai River–Nha Trang Bay transect (Long et al., 1995). The Ag content exceeds the ERL level in the sediments at most locations and is four times higher than the ERM value at stations 3 and 4. The Ni content corresponds to or exceeds the respective ERL value in the sediments at most transitional and marine locations (sts. 3–8) and is much higher than the ERM level in a single sediment sample at the transitional station 4. The Pb and Cu contents are much lower than the respective ERM values at all

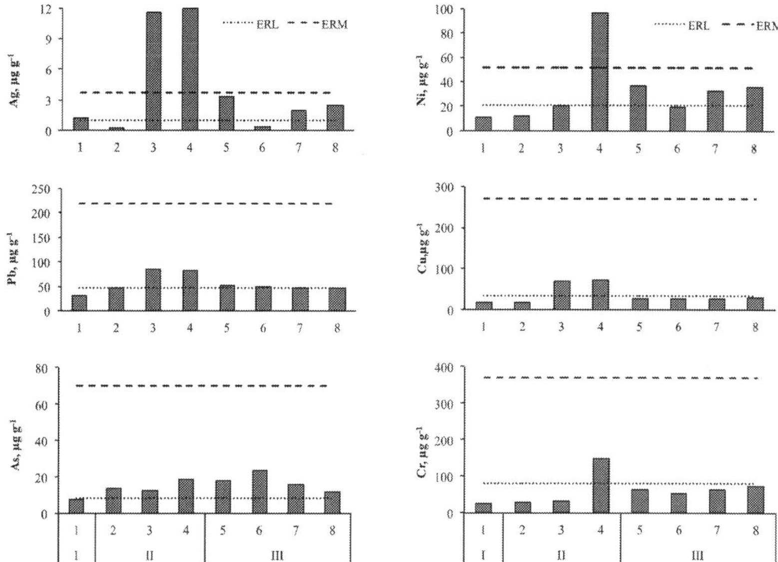

Figure 2. Trace element levels in surface sediments and sediment quality guidelines values (Effects Range-Low (ERL) and Effects Range-Median (ERM); 1–8 – stations; I – river, II – estuary, III – sea).

locations, and correspond to the respective ERL values at most locations while exceeding the ERL levels at stations 3 and 4. The As content is below the ERM level but exceeds the ERL value in the sediments from most transitional and marine locations. The Cr content is below the ERM level at all locations but exceeds the ERL level in a single sediment sample from station 4 in the transitional sub-zone. The Hg and Cd contents in the surface sediments at all locations are below the ERL and ERM levels.

The distribution of metal/Al ratios along the Cai River–Nha Trang Bay transect is provided in Figure 3. The aluminium normalisation was used with the Spearman correlation coefficients (r_s), which were calculated for the inter-relationships among the studied elements. The results revealed associations of elements that are characterised by a similar geochemical behaviour in the sediments along the salinity gradient. Sedimentary Fe, Mg, Ti, Li, Co, Cs, Zn and V vary in relatively narrow ranges and tend to increase seaward. The Spearman correlation coefficients (r_s) for the inter-relationships among these elements range from 0.8 to 0.95 ($p \leq 0.05$ and $n = 8$). Sedimentary Mn also exhibits a significant positive correlation with Li, Co, V and Cs ($r_s = 0.8$). Sedimentary As, Sn, Bi, U and Mo are generally uniformly low (r_s range from 0.7 to 0.8) but tend to increase at stations 2 to 6 where resuspension processes at sediment disturbance events occur. The distribution of Ba, Sr and Ca is largely controlled by the total inorganic carbon (TIC) content in the sediments (r_s range from 0.97 to 0.98). The strong positive correlation of Mn with Ca and TIC ($r_s = 0.85$) indicates that Mn is largely associated with carbonates in the studied sediments. The distribution of Ni, Cr, Zr, Cu, Pb, Sb and particularly Ag was charac-

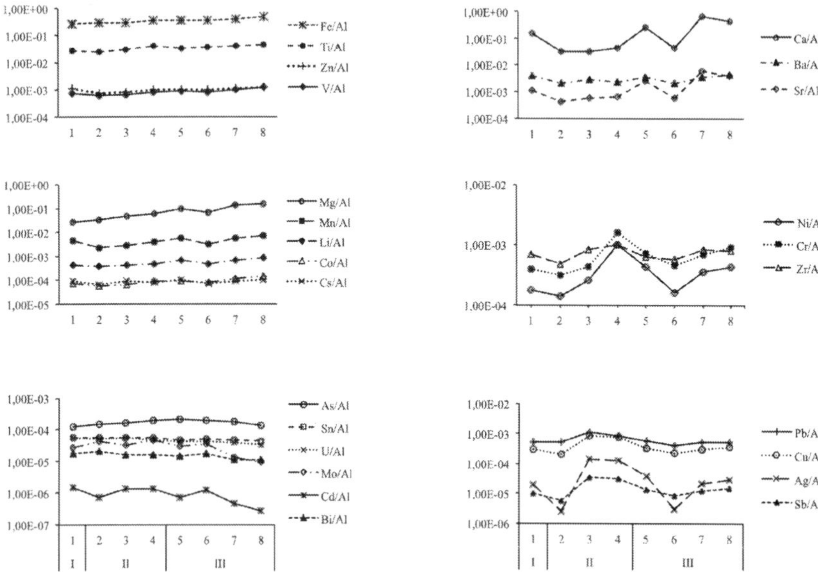

Figure 3. Me/Al ratio in surface sediments (1–8 – stations; I – river, II – estuary, III – sea).

terised by anomalously high concentrations in the estuarine part of the river–sea transect (sts. 3–4). The Spearman correlation coefficients (r_s) for the inter-relationships among these elements range from 0.6 to 0.9.

3.4. Speciation of major and trace elements in surface sediments

The total contents of the dithionite-soluble, oxalate-soluble, weak-acid-soluble and strong-acid-soluble forms of Fe, Mn, Li, Cr, Zn, Cu, Pb, Ni and Ag are provided in Table 3 and Figure 4. In this work, sodium dithionite served to mobilise both crystalline and amorphous Fe oxyhydroxides from the sediments [19]. Ammonium oxalate served to mobilise the easily soluble amorphous Fe-oxyhydroxides and acid-soluble fulvates [49, 50]. Acetic acid removed the labile metals in ion exchange positions, the easily soluble amorphous compounds of iron and manganese, the carbonates and the metals that are weakly held in organic matter [30]. Nitric acid removed the potentially mobilisable metals that are specifically sorbed onto clay particles, bound to easily soluble and resistant iron and manganese minerals or organic compounds, and integrated with discrete hydroxides, carbonates and sulphides [19, 21]. Additionally, the silicate-bound Fe was calculated as the difference between the total and dithionite-soluble forms and the crystalline Fe was determined as the difference between the dithionite-soluble and oxalate-soluble forms [50].

Fraction	Fe	Mn	Zn	Pb	Cr	Cu	Li	Ni	Ag
Dithionite-soluble	1.4±0.3*	101±41.8	2.9±1.3	0.7±0.3	2.5±0.8	≤0.10	≤0.10	≤0.10	≤0.10
	1.0 – 1.9**	44 – 160	0.52 – 4.2	0.26 – 1.3	1.4 – 3.6				
Oxalate-soluble	0.5±0.1	105±47.9	3.8±0.8	4.3±2.3	1.6±0.7	2.1±0.7	1.2±0.7	≤0.10	≤0.10
	0.3 – 0.6	39 – 174	2.3 – 4.5	0.6 – 7.7	0.7 – 2.4	0.8 – 3.1	0.5 – 1.8		
Strong acid-soluble	0.3±0.1	121±77.5	8.1±2.9	18.4±3.9	1.7±0.7	2.6±0.6	0.9±0.9	1.1±0.7	0.2±0.01
	0.1 – 0.4	39 – 256	4.7 – 12.5	12.7 – 23	0.9 – 2.8	1.7 – 3.2	0.1 – 2.6	0.3 – 2.5	0.01 – 0.02
Weak acid-soluble	0.2±0.06	93.8±50.4	6.5±3.1	11.4±1.9	1.5±0.7	0.6±0.4	0.8±0.4	≤0.10	≤0.10
	0.06 – 0.2	38 – 193	3.1 – 12.5	7.6 – 13.5	0.7 – 2.7	0.2 – 1.3	0.1 – 1.2		

* - over the line - mean±SD, ** - under the line – range, n=6).

*-over the line – mean ±SD, **- under the line – range, n=6

Table 3. Metal form contents in surface sediments (in µg g^{-1} except for Fe in % of dry weight)

The total Fe content and the total and percent contents of its silicate-bound form (17–73% of the total content) increase, whereas the total and percent contents of the non-silicate-bound (27–83% of the total content) and crystalline forms (11–66% of the total content) decrease in the sediments along the salinity gradient. The total content of the amorphous form increases from the river to the sea, whereas its percent content varies insignificantly (14–17% of the total content) and is the highest in sediments in the riverine and frontal zone (sts. 1–2) and at the most seaward location (st. 8). The percent content of the weak-acid-soluble Fe, which is mostly comprised of easily soluble amorphous oxides, was constantly low along the salinity gradient (4–6% of the total content). The percent content of the strong-acid-soluble Fe increases from 7% to 13% of the total Fe and silicate-bound Fe contents. The total Mn content tends to increase in sediments along the salinity gradient. The total and percent contents of dithionite-soluble Mn (19–56%), oxalate-soluble Mn (19–61%), weak-acid-soluble Mn (18–42%) and strong-acid-soluble Mn (19–54%) are the highest in the coarsest sediments of the frontal zone (sts. 1–2). Seaward, the contents of the studied forms of Mn decrease in the sediments in the transitional sub-zone (sts. 3–4) and increase again in the marine end-member of the transect (st. 8). All extracts that were used appear to release comparably high amounts of Mn from the sediments (28–36% on average).

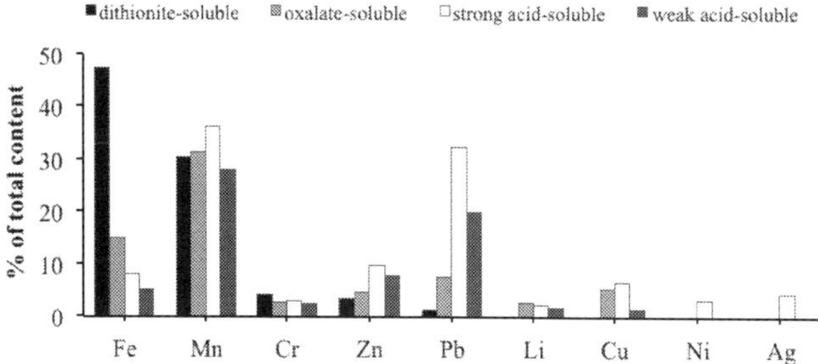

Figure 4. Metal forms in surface sediments (mean percentages).

The contents of the dithionite-soluble form were negligible or below the detection limit for Li, Cu, Ni and Ag. This form comprises 2–12% (4.2% on average) of the total content for Cr, 1–4%

(1.2%) for Pb and 0.5–6.5% (3.4%) for Zn (Figure 4). The contents of the oxalate-soluble form were negligible or below the detection limit for Ni and Ag. This form comprises 1–5% (2.6% on average) of the total content for Cr, 1–16% (7.5%) for Pb, 4–7% (4.5%) for Zn and 0.5–2% (1%) for Li. The contents of the weak-acid-soluble form were negligible or below the detection limit for Ni and Ag. This form comprises 1–4% (2.4 % on average) of the total content for Cr, 13–28% (19.8%) for Pb, 4–11% (7.7%) for Zn, and 0.4–2% (1.8%) for Li (Figure 4). The contents of the strong-acid-soluble form were negligible or below the detection limit for Ag. This form comprises 1–5% (2.6% on average) of the total content for Cr, 19–49% (32%) for Pb, 9–12% (9.5%) for Zn, 0.4–4% (2.1%) for Li and 1–7% (3.2%) for Ni. The percent contents of dithionite-soluble Pb, dithionite-soluble Zn, oxalate-soluble Pb, oxalate-soluble Zn, oxalate-soluble Cr, oxalate-soluble Cu and strong-acid-soluble Cu are the highest in the riverine and frontal zone (sts. 1–2) and generally decrease seaward with non-silicate Fe. The percent content of the strong-acid-soluble Pb is the highest in the riverine and marine sediments (sts. 1–2 and 8) and depleted in the transitional sub-zone at stations 3–4. The total and percent contents of the strong-acid-soluble Zn generally increase along the salinity gradient. The percent content of strong-acid-soluble Cr is uniformly low with a minimum in the sediment that is characterised by a maximum of total Cr (st. 4).

Figure 5 illustrates the distribution of the ecologically most significant weak-acid-soluble (labile) fraction along the river–sea transect. Mn and Pb have the highest percent contents of the labile form and exhibit similar spatial distributions. The sediments are enriched with labile Mn and Pb in the frontal zone and the marine end-part of the river–sea transect. This result contributes to the special affinity of Pb to Mn oxides in sediments. Percent contents of the labile forms of Fe and Zn are elevated in sediments that are enriched with crystallised Fe-oxides and detrital aluminosilicates (sts. 3–6). Cu and Cr exhibit the lowest contents of the labile form. The distribution of the labile Cu and Cr is complicated by a pronounced minimum in the sediment at station 4, which coincides with the maximum in the total Cu and Cr contents. Therefore, in the studied sediments, Cu and Cr are most likely bound to the residual mineral phase that is comprised of detrital heavy minerals (such as hornblende).

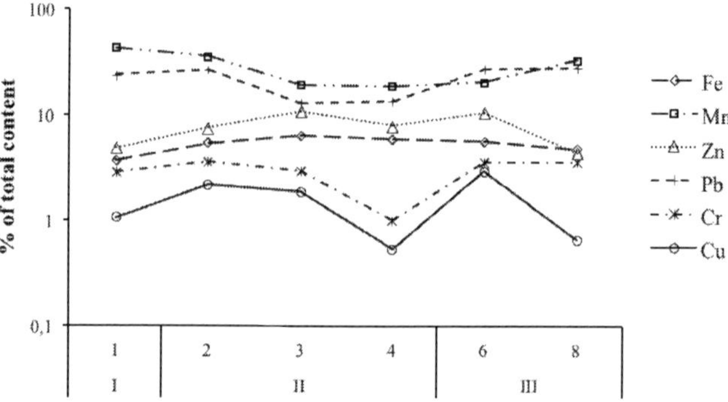

Figure 5. Weak acid-soluble metal form in surface sediments (1–8 – stations; I – river, II – estuary, III – sea).

4. Discussion

4.1. Abundance and distribution of major and trace elements in surface sediments

The hydrology of the riverine sub-zone of the studied part of the Cai River–Nha Trang Bay estuarine system is strongly influenced by the fill dam, which was constructed more than 15 years ago to prevent a saline water invasion into the Nha Trang city water supply system. The surface fresh-water layer flows seaward over the dam, while the upstream penetration of the near-bottom saline water lenses is blocked. The dam serves as a mechanical barrier that sharply decreases the river flow velocity and marks the frontal zone of the contact of fresh and saline waters. Seaward from the dam, the transitional waters are highly stratified with pronounced horizontal and vertical salinity gradients. In the marine sub-zone, the water-mass column becomes homogenized with a 'calming down' of river-flow currents.

The mean grain size of the sediments that were studied generally decreases from the river to the sea. The total content of detrital minerals (quartz, microcline, orthoclase and albite) generally decreases, whereas the total content of clay minerals (kaolinite, illite and clinochlore) and carbonates (calcite and aragonite) generally increases from the riverine to the transitional and bay sediments. This is due to the natural fractionation and deposition of the material of different grain sizes at sites determined by the hydrodynamic conditions of the water-mixing zone. The coarsest river material enriched in detrital minerals is deposited in the riverine part of the estuary at the sharp decrease of the river flow velocity enhanced by the dam. In the transitional waters at the near-bottom resuspension processes, the mixed-sized sediments enriched in silt and clay minerals are formed. In the marine part of the estuary, with a homogenization of the water column, most of the fine-grained material enriched in clay minerals and carbonates is deposited.

The observed distribution of major elements in the sediments illustrates the grain size and mineral fractionation processes. Si content is the highest in the riverine sediment and generally decreases seaward with the quartz content. Al, Fe, Ti, Mg and Mn (to a lesser extent) increase seaward in the sediments with the clay-sized materials. The strong positive correlation of Fe, Ti, Mg with Al and Li implies that the Fe-rich aluminosilicates (such as clinochlore and hornblende) are the dominant host minerals for Fe and the associated elements, although discrete oxide and detrital sulphide minerals may contribute to the accumulation of Fe and other elements at some sites. The abundance and distribution of Na and K are largely controlled by the Na- and K-aluminosilicate (such as albite, microcline and orthoclase) content in the sediments. The distribution of P is largely controlled by the abundance of minerals of the feldspar and amphibole groups (such as hornblende). The distribution of S is controlled to a greater extent by the abundance of sulphates (such as gypsum) and detrital and/or authigenic sulphides. The strongest positive correlation of TIC with Ca implies that carbonates (such as aragonite and calcite) are the dominant host minerals for these elements. TOC content is relatively low and shows no significant correlation with the studied major and trace elements in the sediments along the river–sea transect. This may be due to the intensive microbial decomposition of particulate organic matter, which occurs in the water column during estuarine sedimentation processes [40, 41]. The post-depositional diagenetic reactions, which are enhanced by resuspension processes at sediment disturbance events (such as tides, storms and upwelling), may also contribute to a destruction of sedimentary organic matter and the formation of organic-poor sediments [9, 23, 39]. The organic geochemistry and sources of organic matter in the sediments of the Nha Trang Bay and Cai River were studied by V.I. Peresypkin and N.V. Lobus. The observed TOC and *n*-alkanes distribution showed that sediments from the studied basin tend to be planktonogenic-terrigenous with major input of organic matter from terrestrial plant remains. The presence of petroleum hydrocarbons in some samples in the Nha Trang Bay indicated the input of petroleum products from industrial wastes [26, 38].

The distribution of trace elements in sediments is strongly influenced by the water-column stratification because of the natural fractionation and deposition of materials of different grain sizes at sites which are determined by hydrodynamic conditions [16, 46]. In estuaries, the flocculation and coagulation of riverine microcolloids are initiated when the salinity increases. These processes are accompanied by a rapid scavenging of dissolved trace elements from the water column. Further deposition of newly formed aggregates contributes to the enrichment of sediments with trace elements [33, 43]. To normalise the obtained geochemical data for the grain-size effects and identify the enrichment zones along the salinity gradient, the metal/Al ratios were calculated (Figure 3) [11, 30]. The observed distribution of the normalized trace element contents reflects the association with and/or inclusion of Fe, Mg, Ti, Li, Co, Cs, Zn, V and Mn (to a lesser extent) in the lattices of ferromagnesium silicates (such as hornblende) and clay minerals (such as kaolinite, illite and clinochlore), which constitute the bulk of the fine-grained sedimentary material. These elements are most likely controlled by the accumulation of their most fine-grained aluminosilicate host minerals and materials in the sea floor depression of the marine sub-zone [29]. As, Sn, Bi, U and Mo tend to increase in the transitional and marine sub-zones at stations 2 to 6 where resuspension processes at sediment disturbance events occur. An earlier study showed that Fe/Mn-oxides control the immobilisation of Mo, As and U in the solid phase in modern turbidities [7]. The kinetics of the particle concentration effect (PCE) is likely higher near the estuarine turbidity maximum (ETM), where colloidal-bound trace metals coagulate over shorter periods [3]. Ba, Sr and Ca distribution is largely controlled by the total inorganic carbon (TIC) content in the sediments. These elements form low-soluble carbonates in aquatic environments [44]. The strong positive correlation of Mn with Ca and TIC ($r_s = 0.85$) indicates that Mn is largely associated with carbonates in the studied sediments. The distribution of Ni, Cr, Zr, Cu, Pb, Sb and particularly Ag was characterised by anomalously high concentrations in the transitional part of the river–sea transect (sts. 3–4). The local sediment enrichment in Ni, Cr, Zr, Cu, Pb, Sb and Ag might occur because of a point of anthropogenic contamination from the harbour, domestic and industrial wastes. Thus, metal-based antifouling systems may play a major role in the trace heavy-metal contamination at the harbour [4]. The enhanced content of detrital heavy minerals may also result in anomalous natural metal concentrations [18]. Thus, the Zr concentration in the sediments is most likely heightened because of the enhanced content of Zr-rich minerals such as zircon [45]. To estimate the extent of heavy-metal contamination in the sediments, it was necessary to establish the natural background metal levels, separate the background levels from the anthropogenic inputs and account for the natural variability in the composition of the sediments [29, 32].

There are few published literature data on the total metal levels in coastal sediments from Central Vietnam and Khanh Hoa province in particular. Our estimates of Li, Mn, Ni, V, Pb, Cr, Zn As, Cd, U and Hg are higher or correspond to the measured metal levels in sediments from the coastal lagoons Thuy Trieu and Dam Nai [42]. The reported Hg levels in the present study correspond to the data of N.V. Lobus, which revealed the negative geochemical anomaly of Hg in Central and Southern Vietnam [25]. Due to the lack of available data on metal concentrations of regional sediments, the Clark values for texturally equivalent reference sediment materials (shale, pelagic clays and world river bed sediments) were used as a measure of natural background levels [30]. Comparing to the reference Clark concentrations, the measured levels of U, Bi and especially Ag in the sediments are heightened. U and Bi contents are relatively constant along the river–sea transect and exhibit a negligible elevation in the turbidity zone. Presumably, these disparities were related to regional differences in sediment chemistries rather than to any enhanced contamination. On the contrary, Ag content significantly exceeds the reference values in most of the studied samples and exhibits a sharp maximum in the transitional sub-zone (sts. 3–4). The results of the comparative study of trace element contents and SQGs also indicate that among the studied elements, Ag may contribute to the toxicity of surface sediments [27]. Because element speciation contributes to toxicity and bioavailability in natural systems, the comparative analysis of ecologically significant forms of the studied elements was essential [9, 10].

4.2. Speciation of major and trace elements in surface sediments

Major and trace elements are bound to a variety of sediment fractions that range from easily extractable (and bioavailable) to resistant residual mineral phases [13, 48]. According to the comparative extractability from sediments, Ag, Ni, Li, Cr and Cu are low-labile and mainly occur in the residual phase. These metals were mainly extracted in the detrital fraction, which emphasises the importance of natural weathering and erosion in drainage basins. Therefore, the previously obtained anomalous concentrations of Ag in the sediments of the Cai River estuary and Nha Trang Bay are most likely related to the enhanced content of metal-rich detrital heavy minerals such as Ag-sulphosalts of the freieslelebenite and pyrargyrite groups that may originate from the detrital sulphide minerals of regional bedrocks. The pronounced enrichment of sediments in detrital Ag, Ni, Cu and Cr in the transitional sub-zone (sts. 3–4) may reflect the small-scale fractionation processes such as local deposition of material of particular grain size and mineralogical composition. Fe and Zn are moderately labile and occur in the less resistant phases such as crystallised Fe/Mn oxides and organic compounds that may be a threat in the long term. Mn and Pb are labile, held in ion exchange positions, bound to easily soluble amorphous Fe/Mn compounds and weakly held in organic matter. This result supports the special affinity of Pb to Mn oxides in soils and sediments [9]. The high levels of acid-soluble Pb (28–32% of the total content) compared to previously studied estuarine and coastal sediments suggest a contamination problem in the Nha Trang Bay, which arises from the Cai River discharges, while the elevated level of easily reducible Pb fraction (7.5% of the total content) also contributes to the anthropogenic input of Pb [18, 20]. The sources of the elevated contents of detrital Ag and labile Pb in the studied sediments need further study.

The speciation of Fe and Mn illustrates the natural fractionation processes in the estuarine system that was studied. Fe in the soils of the Cai River catchment area is largely present as non-silicates, which account for 85–97% of the total Fe content, and the crystalline forms are dominant (90% of the total Fe content) [34]. Sedimentation of the rich in non-silicates riverine material and flocculation of the newly formed amorphous Fe oxyhydroxides lead to the accumulation of Fe in the sediments in non-silicate form in the riverine and frontal zone (sts. 1–2) of the estuary. In the bay, most of the fine-grained clay material that largely comprises Fe-rich aluminosilicates is deposited (st. 8). The most bioavailable parts of Mn are concentrated in coarse riverine sediments (in the form of individual Mn oxides) (st. 1) and fine-grained marine sediments (in the form of easily soluble amorphous Mn-containing Fe-oxyhydroxides and carbonates) (st. 8). The speciation of trace elements revealed that riverine and frontal zone sediments (sts. 1–2) are rich in actual and potential bioavailable forms of Mn, Pb, Cr and Cu because of scavenging and co-precipitation by Fe/Mn oxyhydroxides and organic microcolloids. Marine sediments (sts. 6–8) are rich in bioavailable forms of Mn and Pb because of the association with amorphous Fe-oxyhydroxides and carbonates. The transitional sediments are depleted in bioavailable forms in most of the studied metals.

The contents of oxalate-soluble (amorphous) forms were higher than the contents of dithionite-soluble (non-silicate) forms at all sites for Pb and at most sites for Cu and Zn. Therefore, the most bioavailable parts of Pb, Cu and Zn are bound to amorphous Fe and Mn oxyhydroxides and acid-soluble organic compounds. Higher percentages of Cr are obtained in the more resistant reducible (dithionite-soluble non-silicate) fraction, which indicates the inclusion of Cr in crystalline iron compounds. The contents of the strong-acid-soluble forms significantly exceed those of the weak-acid-soluble forms at most sites for Pb and, to a lesser extent, for Cu and Zn. Therefore, the potentially mobilisable Cu, Zn and Pb may be bound to clay minerals and to resistant organic compounds. Assuming that the mean determined amounts of the strong-acid-soluble and weak-acid-soluble forms are a measure of the potential metal bioavailability in sediments, the studied elements can be arranged in the following increasing order of average potential bioavailability: $Ag<Ni<Co<Li<Cu<Cr\ll Fe<Zn\ll Pb<Mn$ (Figures 4 and 5). This sequence is true for sediments in different sub-zones of the water-mixing zone: river, estuary (transitional waters) and sea (bay).

The most bioavailable parts of the studied trace metals are associated with easily soluble amorphous Fe and Mn oxyhydroxides. This result supports the fact that Fe and Mn oxyhydroxides are readily complex trace metals and control the bioavailability in sediments [3, 6, 43]. With the widespread occurrence of Fe and Mn oxyhydroxides and because macrofauna primarily reside in aerobic environments, the role of oxyhydroxides in controlling metal bioavailability is tremendously important and requires further investigation.

5. Conclusions

The major-element composition of the sediments from the Cai River estuary and Nha Trang Bay generally corresponds to the reference background values. The Si content tends to decrease, whereas the contents of Al, Fe, Ti, Mn, Mg, Na, K, S and P increase from the riverine to marine sediments because of a general grain size decrease.

According to the sediment quality guidelines (ERL/ERM) and the reference background values, most of the studied trace elements were below the threshold levels, whereas the total content of Ag significantly exceeded the natural levels. Along the salinity gradient, several zones of metal enrichment occur in the surface sediments because of the geochemical fractionation of the riverine material. Fe, Mn, Mg, Ti, Na, K, Li, Co, Cs, Zn and V are largely controlled by the accumulation of their most fine-grained aluminosilicate host minerals at the sites which are determined by hydrodynamic conditions. As, Sn, Bi, U, Cd and Mo are most likely controlled by the co-precipitation with the dissolved and particulate materials of the river discharge in the transitional waters under turbidity processes. Ca, Ba and Sr are largely controlled by the carbonate abundance in the sediments. Ni, Cr, Zr, Cu, Sb and particularly Ag show a pronounced enrichment at the sites, which is most likely determined by the local accumulation of metal-rich detrital heavy minerals.

The combined method of single-chemical-reagent extractions revealed labile to relatively stable associations of metals in the surface sediments. The studied elements can be arranged in the following increasing order of bioavailability: Ag<Ni<Co<Li<Cu<Cr<<Fe<Zn<<Pb<Mn. The low-labile Ag, Ni, Co, Li, Cr and Cu mainly occur in the residual phase. The severe enrichment of Ag most likely originates from metal-rich detrital heavy minerals such as Ag-sulphosalts. The moderately labile Fe and Zn are largely associated with crystallised Fe/Mn oxides and resistant organic compounds. The labile Mn and Pb are mainly concentrated with easily soluble amorphous Fe/Mn compounds and carbonates. The elevated levels of acid-soluble and easily reducible forms of Pb (up to 36% of the total content) may indicate an anthropogenic origin of Pb in the estuarine system, which requires special study. In the river–sea transect, the riverine- and frontal zone sediments are mostly rich in bioavailable forms of Mn, Pb, Cr and Cu, whereas the marine sediments are rich in bioavailable forms of Mn and Pb.

Mineralogy, grain size and depositional conditions generally determine the abundance and distribution of elements in the studied sediments. Natural (such as turbidities) and human-generated (such as urban and industrial activities) pressures affect the abundance and speciation of potential contaminants (such as trace metals) and may change their bioavailability in the studied estuarine system. Overall, iron and manganese oxyhydroxides largely control the trace-metal bioavailability in the sediments.

Acknowledgements

The authors thank their colleagues from the Russian-Vietnamese Tropical Research and Technology Center for the technical and scientific assistance throughout the study.

This study was performed with the support of the Russian Foundation of Fundamental Research (RFFI Grant 14-05-31059 mol_a) and the state contract item №0149-2014-0036.

Author details

Sofia E. Koukina*, Nikolay V. Lobus, Valeriy I. Peresypkin and Olga M. Dara

*Address all correspondence to: skoukina@gmail.com

P.P. Shirshov Institute of Oceanology of the Russian Academy of Sciences, Moscow, Russia

References

[1] Baturin, G.N., Lobus, N.V., Peresypkin, V.I., Komov, V.T. 2014. Geochemistry of channel drifts of the Kai river (Vietnam) and sediments of its mouth zone. Oceanology 54 (No. 6), 788–797.

[2] Bayen, S. 2012. Occurrence, bioavailabiliy and toxic effects of trace metals and organic contaminants in mangrove ecosystems: A Review. Environment International 48, 84–101.

[3] Bianchi, T.S. 2007. Biogeochemistry of Estuaries. Oxford Press, NY, USA, p. 706.

[4] Briant, N., Bancon-Montignya, C., Elbaz-Poulichet, F., Freydier, R., Delpouxa, S., Cossa, D. 2013. Trace elements in the sediments of a large Mediterranean marina (Port Camargue, France): Levels and contamination history. Marine Pollution Bulletin 73, 78–85.

[5] Britaev, T.A. (ed.). 2007. Benthic fauna of the Bay of Nhatrang, Southern Vietnam. KMK, Moscow, Russia, p. 248.

[6] Burton, G.A. Jr. 2010. Metal Bioavailability and Toxicity in Sediments. Critical Reviews in Environmental Science and Technology 40 (No. 9–10), 852–907.

[7] Chaillou, G., Schäfe,r J., Blanc, G., Anschutz, P. 2008. Mobility of Mo, U, As, and Sb within modern turbidites. Marine Geology 254, 171–179.

[8] Chernova, E.N., Sergeeva, O.S. 2008. Metal concentrations in Sargassum algae from coastal waters of Nha Trang Bay (South China Sea). Russian Journal of Marine Biology 34 (No. 1), 57–63.

[9] Eggleton, J., Thomas, K.V. 2004. A review of factors affecting the release and bioavailability of contaminants during sediment disturbance events. Environment International 30, 973–980.

[10] Fedotov, P.S., Kördel, W., Miró M., Peijnenburg, W.J.G.M., Wennrich, R., Huang, P.-M. 2012. Extraction and Fractionation Methods for Exposure Assessment of Trace Metals, Metalloids, and Hazardous Organic Compounds in Terrestrial Environ-

ments. Critical Review. Environmental Science and Technology 42 (No. 11), 1117–1171.

[11] Gensemer, R.W., Playle, R.C. 1999. The bioavailability and toxicity of aluminum in Aquatic environments. Critical Reviews in Environmental Science and Technology 29 (No. 4), 315–450.

[12] Guor-Cheng, F., Hung-Chieh, Y. 2010. Heavy metals in the river sediments of Asian countries of Taiwan, China, Japan, India and Vietnam during 1999-2009. Environmental Forensics 11, 201–206.

[13] Hass, A., Fine, P. 2010. Sequential selective extraction procedures for the study of heavy metals in soils, sediments, and waste materials – a critical review. Critical Reviews in Environmental Science and Technology 40 (No. 5), 365–399.

[14] Jennerjahn, T.C., Mitchell, S.B. 2013. Pressures, stresses shocks and trends in estuarine ecosystems – An introduction and synthesis. Estuarine, Coastal and Shelf Science 130, 1–8.

[15] Karandashev, V.K., Turanov, A.N., Orlova, T.A., Lezhnev, A.E., Nosenko, S.V., Zolotareva, N.I., Moskvina, I.R. 2008. Use of the inductively coupled plasma mass spectrometry for element analysis of environmental objects. Inorganic Materials 44, 1491–1500.

[16] Koukina, S.E., Calafat-Frau, A., Hummel, H., Palerud, R. 2001. Trace metals in suspended particulate matter and sediments from the Severnaya Dvina estuary, Russian Arctic. Polar Record 37, 249–256.

[17] Koukina, S., Lobus, N., Peresypkin, V., Baturin, G., Smurov, A. Multi-element study of sediments from the river Khai river – Nha Trang Bay estuarine system, South China Sea. 2013. Geophysical Research Abstracts 15, EGU2013-1629.

[18] Koukina, S.E., Vetrov, A.A. 2013. Metal forms in sediments from Arctic coastal environments in Kandalaksha Bay, White Sea, under separation processes. Estuarine, Coastal and Shelf Science 130, 21–29.

[19] Krishnamurti, G.S.R. 2008. Chemical methods for assessing contaminant bioavailability in soils. In: Naidu R. (Ed.). Chemical Bioavailability in Terrestrial Environments. Elsevier, Oxford, UK, pp. 495–520.

[20] Kukina, S.E., Sadovnikova, L.K., Calafat-Frau, A., Palerud, R., Hummel, H. 1999. Forms of metals in bottom sediments from some estuaries of the basins of the White and Barents seas. Geochemistry International 37 (No. 12), 1197–1202.

[21] Ladonin, D.V. 2002. Heavy Metal Compounds in Soils: Problems and Methods of Study. Eurasian Soil Science 35 (No. 6), 605–613.

[22] Li, Y.H. 1991. Distribution patterns of the elements in the ocean. Geochimica et Cosmochimica Acta 55, 3223–3240.

[23] Lisitzin, A.P. 1994. The marginal filter of the ocean. Oceanology 34 (No. 5), 735–747.

[24] Lobus, N.V. Soderzhanie rtuti v donnih otlozheniyah rek i vodokhranilish Yuzhnogo Vietnama. 2012. Toxicologichesky Vestnik (Toxicology Review) 2, 47–52 (In Russian, with English Abstract).

[25] Lobus, N.V., Komov, V.T., Nguen, T.H.T. 2011. Mercury concentration in ecosystem components in water bodies and streams in Khanh Hoa province (Central Vietnam). Water Resources 38 (No. 6), 799–805.

[26] Lobus, N.V., Peresypkin, V.I., Shulga, N.A., Drozdova, A.N., Gesev, E.S. 2015. Dissolved, particulate and sedimentary organic matter in the Cai River – Nha Trang Bay system (South-China Sea). Oceanology 55 (No. 3), 339–346.

[27] Long, E.R., MacDonald, D.D. 1998. Recommended Uses of Empirically Derived, Sediment Quality Guidelines for Marine and Estuarine Ecosystems. Human and Ecological Risk Assessment 4 (No. 5), 1019–1039.

[28] Long, E.R., MacDonald, D.D., Smith, S.L., Calder, F.D. 1995. Incidence of adverse biological effects within ranges of chemical concentrations in marine and estuarine sediments. Environmental Management 19, 81–97.

[29] Loring, D.H., Dahle, S., Naes, K., Dos Santos, J., Skei, J.M., Matishov G.G.1998. Arsenic and other trace metals in sediments from the Pechora Sea, Russia. Aquatic Geochemistry 4, 233–252.

[30] Loring, D.H., Rantala, R.T.T. 1992. Manual for geochemical analyses of marine sediments and suspended particulate matter. Earth-Science Reviews 32, 235–283.

[31] Moore, D.M., Reynolds, R.C. 1997. X-ray Diffraction and the Identification and Analysis of Clay Minerals. Oxford University Press, New York, USA, 400 p.

[32] Millward, G.E., Turner, A. 1995 Trace metals in estuaries. In: Trace Elements in Natural Waters (Salbu, B., Steinnes, E., eds.), CRC Press, Boca Raton, FL. pp. 223–245.

[33] Morris, A.W., Howland, R.J.M., Bale, A.J. 1986. Dissolved aluminium in the Tamar Estuary, southwest England. Geochim et cosmochimacta 50, 189–197.

[34] Motuzova, G.V., Hong Van, N.T. 1999. The geochemistry of major and trace elements in the agricultural terrain of South Vietnam. Journal of Geochemical Exploration 66, 407–411.

[35] Nghia, N.D., Lunestad, B.T., Trung, T.S., Son, N.T., Maage, A. 2009. Heavy Metals in the Farming Environment and in some Selected Aquaculture Species in the Van Phong Bay and Nha Trang Bay of the Khanh Hoa Province in Vietnam Ngo. Bulletin of Environmental Contamination and Toxicology 82, 75–79.

[36] Pavlov, D.S., Novikov, G.G., Levenko, B.A. 2006. Osobennosti struktury I funktzionirovaniya pribrezhnyh ecosystem Yuzhno-Kitaiskogo moria. GEOS, Moscow, Russia, p. 280 (in Russian, with English abstract).

[37] Pavlov, D.S., Zvorikin, D.D. (Ed.). 2014. Ecologia vnytrennich vod Vietnama. KMK, Moscow, Russia, p. 435. (in Russian).

[38] Peresypkin, V.I., Smurov, A.V., Shulga, N.A., Safonova, E.S., Smurova, T.G., Bang, C.V. 2011. Composition of the organic matter of the water, suspended matter, and bottom sediments in Nha Trang Bay in Vietnam in the South China Sea. Oceanology 51 (No. 6), 959–968.

[39] Romankevich, E.A. 1984. Geochemistry of organic matter in the ocean. Springer, Berlin, Germany, 334 p.

[40] Romankevich, E.A., Vetrov, A.A. 2013. Masses of Carbon in the Earth's Hydrosphere. Geochemistry International 51 (No. 6), 431–455.

[41] Romankevich, E.A., Vetrov, A.A., Peresypkin, V.I. 2009. Organic matter of the world ocean. Russian Geology and Geophysics 50 (No. 4), 401–412.

[42] Romano, S., Mugnai, C., Giuliani, S., Turetta, C., Huu C.N., Belucci, L.G., Nhon, D.H., Caporaglio, G., Frignani, M. 2012. Metals in sediment cores from nine coastal lagoons in Central Vietnam. American Journal of Environmental Sciences 8, 130–142.

[43] Saulnier, I., Mucci, A. 2000. Trace metal remobilization following the resuspension of estuarine sediments: Saguenay Fjord, Canada. Applied Geochemistry 15, 203–222.

[44] Savenko V.S. 2006. Chemical composition of world river's suspended matter. Khimichesky sostav vzveshennikh nanosov rek mira. GEOS, Moscow, Russia, 175 p. (in Russian, with English abstract).

[45] Shahid, M., Ferrand, E., Schreck, E., Dumat, C. 2013. Behavior and Impact of Zirconium in the Soil-Plant System: Plant Uptake and Phytotoxicity. Reviews of Environmental Contamination and Toxicology 221, 107–127.

[46] Shevchenko, V.P., Dolotov, Y.S., Filatov, N.N., Alexeeva, T.N., Fillipov, A.S., Noethig, E.-M., Novigatsky, A.N., Pautova, L.A., Platonov, A.V., Politova, N.V., Ratkova, T.N., Stein, R. 2005. Biogeochemistry of the Kem' River estuary, White Sea (Russia). Hydrology and Earth System Sciences 9 (No. 1-2), 57–66.

[47] Szczucinski, W., Statteger, K., Scholten, J. 2009. Modern sediments and sediment accumulation rates on the narrow shelf off central Vietnam, South China Sea. Geo-Marine Letters 29, 47–59.

[48] Tack, F.M.G., Verloo, M.G. 1995. Chemical Speciation and Fractionation in Soil and Sediment Heavy Metal Analysis: A Review. International Journal of Environmental Analytical Chemistry 59 (No. 2-4), 225–238.

[49] Vodyanitskii, Yu.N. 2001. On the Dissolution of Iron Minerals in Tamm's Reagent. Eurasian Soil Science 34 (No. 10), 1086–1096.

[50] Vorobyova, L.F. (Ed.) 2006. Teoriya I praktika khimicheskogo analiza pochv. GEOS, Moscow, Russia, p. 400. (in Russian, with English Abstract).

[51] Wolanski, E. 2007. Estuarine Ecohydrology. Elsevier, Amsterdam, Netherlands. p. 157.

Chapter 10

Stable Isotope Techniques to Address Coastal Marine Pollution

Azhar Mashiatullah, Nasir Ahmad and Riffat Mahmood

Additional information is available at the end of the chapter

http://dx.doi.org/10.5772/62897

Abstract

Stable isotopes of carbon ($\delta^{13}C$), sulfur ($\delta^{34}S$), oxygen ($\delta^{18}O$), hydrogen ($\delta^{2}H$), nitrogen ($\delta^{15}N$), and radioactive isotope of hydrogen (tritium) have been applied in combination with conventional techniques (chemical) to investigate Karachi coastal water pollution due to Layari and Malir rivers, which mainly carry the domestic and industrial wastewater of Karachi Metropolitan. Heavy metal contents of the Manora Channel and southeast coastal waters were higher than the Swedish guidelines for the quality of seawater. By contrast, heavy metal concentrations in coastal sediments were found to be significantly higher than that of seawater. Mn and Ni contents in sediments of entire coast (Manora Channel, southeast and northwest coast) were above USEPA guidelines except at Buleji site, whereas Cr, Zn, and Cu levels only in Manora Channel sediments were higher than USEPA guidelines. The higher heavy metal contents of Manora Channel water and sediments can be attributed to an influx of a major portion of untreated industrial and/or domestic wastewater. Layari and Malir river water was observed to be depleted in $\delta^{13}C_{(TDIC)}$ and $\delta^{34}S$, which showed heavy influx of sewage into these rivers. Manora Channel water was also depleted in $\delta^{13}C_{TDIC}$ and $\delta^{34}S$ during low tide environment, showing a large-scale domestic wastewater mixing with seawater. Southeast coastal water was found to be slightly enriched in $\delta^{13}C_{(TDIC)}$ and $\delta^{34}S$ and exhibited mixing of relatively small quantity of sewage with the seawater as compared to the Manora Channel. $\delta^{13}C_{(TDIC)}$ and $\delta^{34}S$ contents of northwest coastal water were close to the values meant for normal seawater. $\delta^{13}C$ and $\delta^{15}N$ contents of Karachi coastal seaweed ranged from -31.1‰ to -4.9‰ PDB and from 6.1‰ to 17.8‰ air, respectively. The average $\delta^{15}N$ values (10.2‰ air) of *Ulva* spp. collected from nonpolluted northwest coast was higher as compared to the average $\delta^{15}N$ contents (8.0‰ air) of *Ulva* from the Manora Channel, suggesting that nitrogen isotopic ratios of *Ulva* spp. could be a good indicator of sewage pollution.

The results of a two-component isotope mass balance equation using $\delta^{13}C$ and $\delta^{34}S$ values for Layari and Malir rivers and coastal water indicated that tide conditions and distance of sampling site from the pollution source were the main factors to control the transport and dissemination of Layari river pollution into the Manora Channel. High tide environment slowed down the Layari river water mixing with seawater coupled with a gradual decrease in pollution levels from the Layari River outfall zone to the Manora Lighthouse.

Keywords: isotope, carbon, sulfur, oxygen, mixing, sediments, seawater, coastal, algae, metals, Karachi, marine, pollution

1. Introduction

Marine pollution due to anthropogenic activities has now become a worldwide environmental concern [33]. Several researchers [9, 53, 60, W.Q.C., 1972;] have reported the influence of the indiscriminate discharge of untreated industrial effluent and municipal wastewater on the marine environment in terms of danger to habitats, serious risk to marine life, deterioration of aesthetic values, and limited access to coastal areas. Hence, monitoring of marine coastal environment is essential to understand the origin, distribution, fate, and behavior of marine pollutants to formulate a viable management strategy [17]. A variety of techniques ranging from conventional methods to most sophisticated isotopic techniques are available for monitoring the marine coastal pollution [15]. Isotope analysis is an engineering tool that can be used to characterize the pollutants to trace the contributions of pollutants from different sources within the mixing zones of estuaries, coastal water, and shelf water [30].

Like other coastal regions of the world, the Karachi coast, especially the Manora Channel, is heavily polluted due to untreated industrial wastewater and Metropolitan municipal sewage, which are indiscriminately discharged into coastal waters through Layari and Malir rivers [68]. According to a report [32], only 20% of total annual wastewater produced in Metropolitan Karachi is treated and the rest is discharged directly into coastal waters. Sea pollution is further enhanced due to oil spills from cargo ships and oil tankers. The dredging of channel round the year also adds pollution to coastal water in terms of suspended sediments load. This situation demands to characterize the coastal water in order to determine pollution load, its extent, and type. Past attempts to investigate the pollution load of Karachi coastal water [31, 5, 52, 54, 57, 67] were only confined to the use of physicochemical and hydrological techniques. Very recently, [2], however, attempts have been made to apply nuclear techniques to study radionuclide pollution along Pakistan coast. The present study was aimed to apply stable isotope techniques in conjunction with classical pollution monitoring tools to characterize the pollution type and its transport, the mixing of pollutants with sea water, and the origin of salinity in coastal aquifer.

1.1. Coastal areas of Pakistan

A 990-km-long strip of coastal area of Pakistan is stretched from Southeast (Ron of Katch) to Northwest (Gwader) along the Arabian sea [28]. It is geographically divided into two main zones, Baluchistan coastal belt (745 Km) and Sindh coastal belt (245 Km), as shown in Figure 1. Sindh coastal belt includes Indus Delta and Karachi Coast. The Baluchistan coastal belt is scarcely populated and is relatively pollution free. Sindh coastal belt, however, suffers very serious environmental problems because of greater population and industrial activities in Metropolitan Karachi, which is the largest city of Pakistan and is located at latitude 24° 48′ N and longitude 66° 59′ E on the coast of the Arabian Sea.

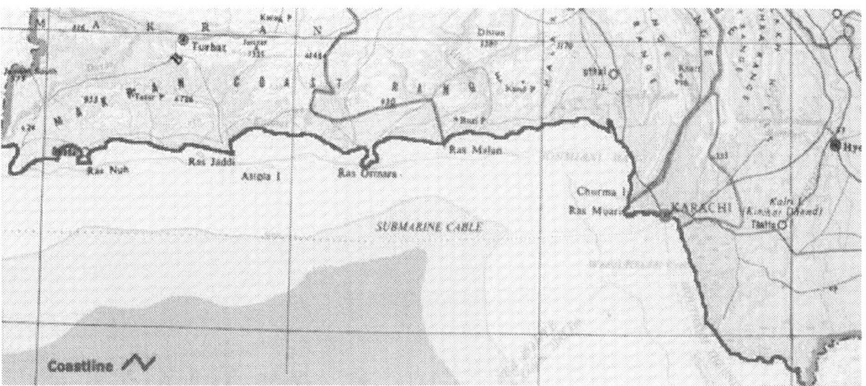

Figure 1. Coastal map of Pakistan.

1.2. Description of pollution sources of Karachi coast

Metropolitan municipal sewage and industrial effluent are two major sources of coastal water pollution. The untreated effluent of more than six thousand industrial units scattered in six big industrial estates along with 300 Mgd municipal wastewater is discharged into Karachi coastal waters through Malir and Layari rivers [67]. Layari River passes right through the center of the Karachi Metropolitan while Malir river flows mainly through eastern part of the city (Figure 2). Both rivers act as an open sewage drain, receiving highly polluted wastewater of industrial and domestic origin. In accordance with a report [32], the Layari River discharges 130,000 tons of solid nitrogen, 160,000 tons of organic matter, 800 tons of nitrogen compounds, 90 tons of phosphate compounds, and 12,000 tons of suspended solid every year in the Manora Channel.

1.3. Stable isotope techniques

Naturally occurring isotopes are found in both stable and radioactive forms in the environment and are known as environmental isotopes [13]. Isotopes of an element exhibit similar chemical

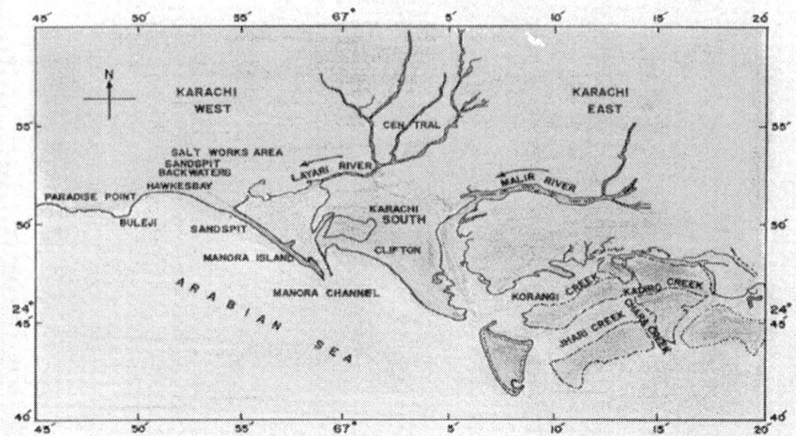

Figure 2. Map showing fall of Layari and Malir rivers into Karachi coast.

properties but differ in physical properties. The difference in the physical properties (bond strength and velocity) gives rise to fractionation among isotopes of an element in an environmental matrix. Fractionation takes place by natural processes, such as evaporation, condensation, and diffusion (kinetic isotope effects), or by ordinary mixing processes. Isotope fractionation leads to isotope variation, which makes it possible to use isotopes as tracers for the study of diverse nature of pollution problems [16, 22, 25, 39]. Figure 3 illustrates the natural variations of ^{18}O, ^{13}C, ^{15}N, and ^2H in environmental matrices.

Stable isotopes such as isotopes of light elements (hydrogen, carbon, nitrogen, oxygen, sulfur, and chlorine) in combination with conventional techniques are most commonly used in the environmental studies [30, 40, 43, 58].

2. Objectives of the study

The main objectives of the study were as follows:

1. To establish baseline inventories of stable isotopes (carbon, sulfur, hydrogen and oxygen), chemical parameters (physicochemical, heavy metal) in marine coastal waters, and/or sediments along Karachi coast for pollution monitoring

2. To determine the extent of mixing of polluted waters of Layari and Malir rivers into the Karachi coastal waters using stable isotopes of carbon (^{13}C), oxygen (^{18}O), and sulfur (^{34}S)

3. To study the potential of stable carbon ($\delta^{13}C$) and nitrogen ($\delta^{15}N$) compositions of seaweeds and mangroves as pollution tracers in coastal area of Karachi

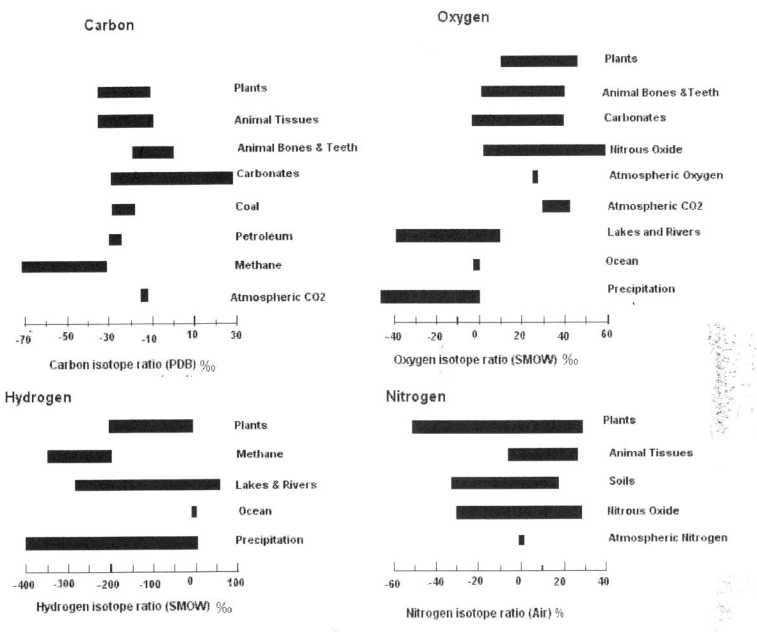

Figure 3. Variations in the carbon, oxygen, hydrogen, and nitrogen isotope ratios of different materials [42].

4. To determine potable water quality and to identify sources and dynamics of groundwater salinization of Karachi coastal aquifer using stable isotope of oxygen ($\delta^{18}O$) and hydrogen (δD) along with hydrochemical data

3. Materials and methods

3.1. Location of sampling sites

For sampling, Karachi coast was broadly divided into three zones: (i) Manora Channel, (ii) southeast coast, and (iii) northwest coast (Figure 4). Seawater from the Manora Channel was sampled from several sites, including Layari River outfall zone, Fish Harbor, KPT Shipyard Butti, KPT Shipyard, Kaemari Boat Basin, Bhaba Island, Bhit Island, Boat Club, Pakistan Naval Academy, and Manora Lighthouse (Figure 5). Sampling sites along the southeast coast include Marina Plaza, Casino, Naval Jetty, Marina Club, Ghizri area, and Ibrahim Haideri Fish Harbor, whereas along the northwest coast, sampling sites are Manora island sea side, PNS Himalaya, Kakka Pir, Buleji, Power house, and Sunari point. Figure 6 shows the locations of these sites. The location of sampling points was determined with Garmin GPS-100 Personal Navigator™ (M/S Garmin, 11206 Thompson Avenue, Lenexa, KS 66219).

Figure 4. Location map of Karachi coast.

Figure 5. Location of sampling sites: the Manora Channel.

(a)

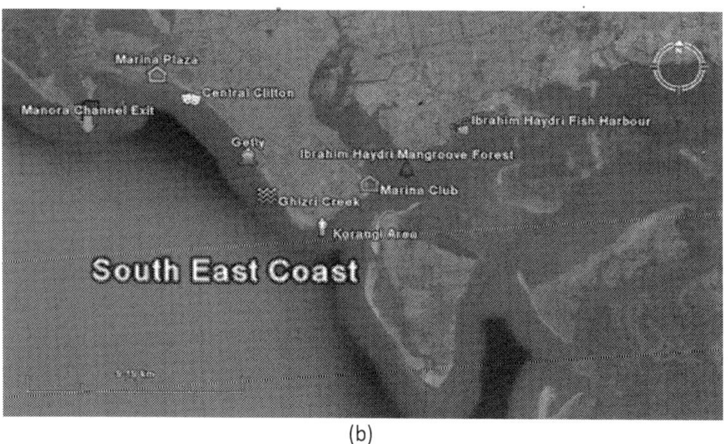

(b)

Figure 6. Location of sampling sites: (a) southeast and (b) northwest coast.

3.2. Sample collection and preservation

The period of field sampling was spanned over 2 years (April 2002–September 2004). This period was selected in order to cover overall variation in pollution transport pattern due to monsoon system. To investigate invasion of seawater, water samples were collected from Layari River, Malir River, Indus river, Hab dam, Karachi sea, shallow aquifer (depth <50 m), and deep aquifer (depth >50 m). All the water samples were preserved in accordance with standard procedure [51, 45, 64].

3.3. Chemical analysis

Samples of seawater, river water, and groundwater were characterized in terms of electrical conductivity (EC), turbidity, and pH.Turbidity was measured with a portable turbidity meter. Each instrument was duly calibrated before use. Major ions were Cl^-, SO_4^{-2}, and HCO_3^{-1}. Chloride contents were determined by ion selective electrodes with Orion Microprocessor Ion Analyzer/901. Carbonate and bicarbonate contents were determined by titration. Sulfate concentrations were determined by a spectrophotometric method using Hitachi 220-A Double Beam Spectrophotometer. Standard procedures [977.15 and 986.15 methods] after [34] and [7] were used for metal analysis of seawater and sediment by atomic absorption spectrophotometer and inductive couple plasma optical emission spectrometer. Metal analyses were performed using inductive couple plasma optical emission spectrometer (ICP-OES, Model 3580).

3.4. Stable isotope analysis

The stable isotope analyses were performed using a modified Varian Mat GD-150 Mass Spectrometer. Stable isotope data are reported as standard mean ocean water (SMOW) for ^{18}O and ^2H analyses, Pee-Dee Belemnite (PDB) for ^{13}C analysis of total dissolve inorganic carbon, and Canyon Diablo Troilite (CDT)for ^{34}S. The overall analytical errors are $\pm 0.01‰$ for $\delta^{13}C$, $\pm 0.1‰$ for $\delta^{18}O$ and $\delta^{34}S$, and $\pm1‰$ forδ^2H measurements. To ensure precision, the standard deviation of the mass spectrometer was also computed, and the standard deviation of each sample was ensured to be within permissible limit. For isotope analysis on mass spectrometer, water/sediments/plant samples are converted into gas phase. Since different sample preparation systems are used for analysis of C, S, N, O, and H, these systems were accordingly modified/redesigned in laboratory.

3.5. Determination of polluted water mixing with coastal water

A two-component isotope mixing equation was used to compute polluted river water mixing with coastal water. The mixing fraction (*f*) of polluted Layari river water (source A) and/or Malir river water (source A) with nonpolluted seawater (source B) of Karachi coast is computed as follows:

$$f_A = (\delta^{13}C_M - \delta^{13}C_B) \ / \ (\delta^{13}C_A - \delta^{13}C_B) \tag{1}$$

where $\delta^{13}C_A$ is the $\delta^{13}C$ (TDIC) of polluted river water, δ^3C_B is the $\delta^{13}C$ (TDIC) of nonpolluted Karachi sea water, and $\delta^{13}C_M$ is the $\delta^{13}C$ (TDIC) of contaminated sea water. To calculate the percent fraction of polluted river water into seawater, Equation 2 was obtained:

$$\%f_A = [(\delta^{13}C_M - ^{13}C_B) \ / \ (\delta^{13}C_A - \delta^{13}C_B)] \times 100. \tag{2}$$

$\delta^{13}C$ was replaced with $\delta^{34}S$ and $\delta^{18}O$ for computing mixing through $\delta^{34}S$ and $\delta^{18}O$ isotopes values.

4. Results and discussion

4.1. Chemical characteristics of coastal water and sediments

Heavy metals are considered as one of the hazardous pollutants in natural environment due to their toxicity, persistence, risk (direct/indirect) to human beings and aquatic life [1, 10, 11, 62], and long-term damage to the environment [21, 62]. Metal elements are added to water bodies of our environment either through natural processes and/or anthropogenic activities. Heavy metals like Cu, Cr, Ni, Mn, and Zn are listed among metals known to be essential for aquatic life [59]. However, these metal elements have well-known toxic effects if they are present above permissible limits [62]. Heavy metal load in seawater/sediments is especially measured in this study because of their continuous addition to seawater due to the indiscriminate discharge of municipal sewage and industrial effluent through Layari and Malir rivers [8, 37, 38, 46, 68].

4.1.1. Heavy metal concentration of polluted rivers and coastal waters

Heavy metal levels in Layari and Malir river water and Karachi coastal waters are summarized in Table 1. Cr contents of Malir river water are relatively higher as compared to Layari River, which could be due to a continuous discharge of untreated effluents of tannery industrial units into Malir River. Higher Cu and Zn concentrations in Layari River can be explained best due to the influx of wastewater of industrial units of cable, electrical appliances, electroplating, textile, and glass [36]. Elevated levels of Ni and Mn in water of Layari and Malir rivers owe to the inflow of untreated effluents such as automobile batteries, electroplating, car painting dying, and glass industries [67].

Sampling sites	Metal element concentration (ppb)				
	Cu	Cr	Mn	Ni	Zn
Layari River	124 ± 8	300 ± 13	190 ± 6	180 ± 10	360 ± 7
Malir River	110 ± 7	400 ± 11	200 ± 8	190 ± 11	180 ± 5
Manora Channel					
Layari River outfall	75 ± 4	110 ± 5	130 ± 6	67± 4	220 ± 7
Karachi Fish Harbor	64 ± 7	87 ± 4	120 ± 3	66 ± 3	210 ± 8
KPT Shipyard	56 ± 5	98 ± 5	88 ± 4	65 ± 3	220 ± 7
KPT Shipyard (Butti)	54 ± 4	82 ± 5	86 ± 5	64 ± 3	200 ± 8
Bhaba Island	54 ± 3	78 ± 4	88 ± 5	45 ± 2	125 ± 4
Bhit Island	45 ± 2	76 ± 3	56 ± 3	42 ± 2	127 ± 4
Boat Club	22 ± 1	55 ± 5	43 ± 4	21 ± 1	82 ± 3
Pakistan Naval Academy	21 ± 1	50 ± 4	39 ± 4	18 ± 1	78 ± 4
Manora Lighthouse	19 ± 1	50 ± 4	40 ± 3	18 ± 0.5	67 ± 6
Southeast coast					

Sampling sites	Metal element concentration (ppb)				
	Cu	Cr	Mn	Ni	Zn
Marina Plaza	20 ± 2	54 ± 4	40 ± 2	11 ± 0.9	42 ± 3
Casino	21± 2	50 ± 4	47 ± 7	11 ± 0.7	44 ± 4
Naval Jetty	25 ± 3	49 ± 3	56 ± 5	11 ± 0.6	40 ± 6
Marina Club	29 ± 2	59 ± 2	75 ± 7	12 ± 0.6	56 ± 3
Ghizri area	76 ± 4	77 ± 5	120 ± 9	69 ± 4	143 ± 8
Ibrahim Haideri	49 ± 3	72 ± 5	110 ± 7	34 ± 3	90 ± 6
Northwest coast					
PNS Himalaya	10 ± 0.8	59 ± 4	10 ± 0.6	11 ± 1	35 ± 3
Sandspit	10 ± 0.8	53 ± 5	23 ± 2	9 ± 0.9	42 ± 4
Kakka Pir	11± 0.9	56 ± 7	29 ± 2	7 ± 0.5	30 ± 2
Buleji	8 ± 0.6	45 ± 4	28 ± 1	8 ± 0.8	16 ± 2
Power House	10 ± 0.7	30 ± 5	27 ± 3	9 ± 0.5	10 ± 1
Sunari point	8 ± 0.4	30 ± 3	25 ± 2	10 ± 0.4	10 ± 1
LOD* (ppb)	1.5	1.5	2	1.5	1.0

Table 1. Heavy metal contents of polluted river water and Karachi coastal water

Swedish guidelines for safe seawater for aquatic life are divided into three broad classes depending upon the degree of environmental disturbance (Table 2). Class 3 shows safe limit of heavy metal content in seawater. Class 4 indicates an increased risk of environmental disturbance, and class 5 represents a high risk of environmental disturbance even at shorter exposure.

Classes	Metal element concentration (ppb)			
	Cu	Cr	Ni	Zn
SNV* Class 3	5–15	3–9	15–45	20–60
SNV Class 4	15–75	9–45	45–225	60–300
SNV Class 5	>75	>45	>225	>300

SNV*—values derived from Swedish guidelines.

Table 2. Swedish environmental guidelines for toxicity of metals [Naturvårdsverketa, 2001]

4.1.2. Heavy metal contentsof sea sediments

Surficial sediments are a feeding source for biological life, a transporting agent for pollutants, and an ultimate sink for organic and inorganic matter settling. In heavily polluted sediments, the anthropogenically introduced components by far exceed the natural components and pose a risk to the marine ecosystem [4, 66]. Many researchers [5, 31, 36, 55, 56] attempted to evaluate heavy metal pollution of Karachi coastal sediments. Heavy metals in sediments are generally much higher as compared to seawater because heavy metals entering from other sources into seawater ultimately deposit onto sediments [19, 41]. As shown in Table 3, Mn and Ni contents in coastal sediments of Karachi are much higher than EPA guidelines at all sites except at one site (Kakka Pir).

Sampling sites	Metal element contents (ppm)				
	Mn	Cr	Ni	Zn	Cu
Manora Channel					
Layari River outfall	300 ± 29	293 ± 23	48 ± 6.1	537± 27	96± 8.4
KarachiFish Harbor	300 ± 30	102 ± 12	25 ± 2.1	581± 19	54± 4.4
KPT Shipyard	600 ± 21	319 ± 21	56± 6.2	666± 17	65± 57
KPT Shipyard (Butti)	300 ± 22	92 ± 7.8	41 ± 5.1	83± 3.4	27± 3.8
Bhaba Island	300 ± 32	80 ± 6.6	27 ± 7.2	95± 5.9	52± 4.9
Bhit Island	500 ± 41	70 ±8.2	30 ± 3.2	96± 5.8	45± 3.2
Keamari Boat Basin	300 ± 34	82 ±6.9	39 ± 7.3	524± 32	27± 2.2
Pakistan Naval Academy	300 ± 21	29 ± 2.7	28 ± 4.9	29± 7.9	25± 3.9
Manora Lighthouse	300 ± 23	24 ± 1.7	27 ± 4.1	25± 3.2	21± 2.5
Southeast coast					
Marina Plaza	400± 41	13± 1.9	18± 1.7	21± 2.1	21 ± 1.9
Casino	600± 28	33± 2.3	26± 2.1	33± 2.7	19 ± 1.4
Naval Jetty	500± 21	25 ± 2.3	23± 2.2	31± 3.1	24 ± 1.7
Marina Club	400± 24	26 ± 3.2	46± 4.2	59± 2.9	34 ± 2.6
Ghizri area	900± 51	85 ± 6.2	56± 4.5	161± 21	65 ± 3.2
Ibrahim Haideri	900± 18	74 ± 2.9	43± 2.7	121± 12	87 ± 3.6
Northwest coast					
PNS Himalaya	500 ± 23	23 ± 4.4	17 ± 1.9	49 ± 3	13 ± 1.4
Sandspit	400 ± 34	33 ± 4	21 ±2.3	49 ± 4	14 ± 1.7
Kakka Pir	600 ± 45	50 ± 4	18 ± 2	80 ± 5	13 ± 1.9
Buleji	400 ± 21	18 ± 1.6	21± 3	29 ± 4	15 ± 2.1
Power house	1300 ± 54	20 ± 1.8	26 ± 4	92 ± 6	16 ± 3.4
LOD	10	0.5	0.5	0.5	1.0
EPA Guidelines	200	60	20	84	19

Values are an average of three samples.

Table 3. Heavy metal contents of coastal sediments

In sediments of the Manora Channel, levels of Cr, Zn, and Cu are also higher except at Manora Lighthouse and Pakistan Naval Academy. Heavy metal load also appeared to be decreased as the distance of sampling site is increased from the entry point of Layari River into the Manora Channel. However, exceptionally higher levels are recorded at KPT Shipyard sediments, which may be due to submarine disposal of effluent of KPT Shipyard in addition to untreated domestic and industrial effluents entering the sea. Layari River outfall zone and Karachi Fish Harbor sediments also show considerably high levels of heavy metals, which owes to waste-water of industrial activity such as boat building and fish processing at the harbor. The Manora Channel sediment is clayey, which has high absorption or trapping capacity for heavy metals [62]. Heavy metal load in southeast coastal sediments is considerably less as compared to sediments of the Manora Channel. Metal contents are appeared to be substantially higher than EPA guidelines at the Ghizri area and Ibrahim Haideri due to continuous influx of untreated domestic and industrial waste into the sea at these sites. A comparison of heavy metal concentrations of three coastal zones is given in Figure 7, which shows that sediments of the Manora Channel are heavily contaminated as compared to southeast and northwest coastal sediments. This may be due to the heavy influx of untreated municipal wastewater and industrial effluent into the Manora Channel. A comparison of heavy metal contents of Karachi coastal sediments with other coastal sediments of the world (Table 4) reveals that the findings of this study are in agreement with the results of a previous study [36] and other studies of the world with an exception of South California harbor sediments where Mn concentration is many times higher than other sea sediments and other metal element contents are significantly low as reported in the other studies

Heavy metal	Heavy metal (ppm)				
	Karachi Coast[1]	Karachi Coast[2]	Istanbul Harbor[3]	Golden Horn Estuary[4]	South California[5]
Mn	300–1300	300	100–500	300–500	800–4000
Cr	24–319	196	11–509	243–485	11–71
Ni	27–56	45	1–41	98–167	2–29
Zn	25–666	120	48–237	98–167	14–104
Cu	21–96	78	5–80	333–390	3–30

[1]Present study.

[2][36].

[3][4].

[4][29].

[5][65].

Table 4. A comparison of heavy metal contents of present study with other coastal sediments of the world

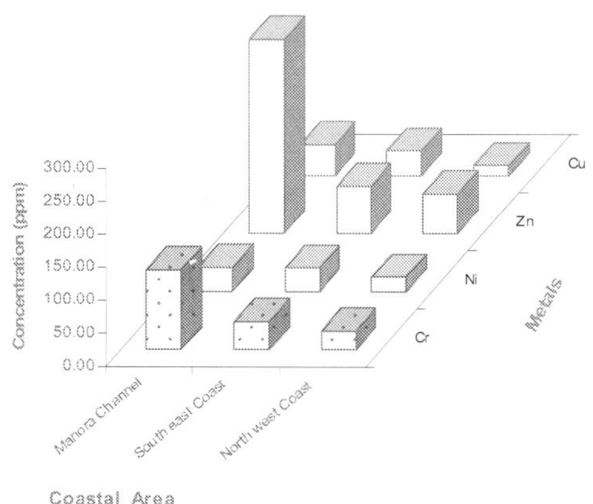

Figure 7. Comparison of average concentrations of metals in coastal zones of Karachi.

4.2. Stable isotope characteristics of Karachi coastal water, sediments, and biota

Stable isotope analysis of coastal water, sediments, and biota was undertaken in order to understand the mechanism of the land base terrestrial pollution transport to coastal ecosystem.

4.2.1. $\delta^{13}C_{(TDIC)}$ analysis of polluted rivers and coastal water

The $\delta^{13}C_{(TDIC)}$ value for normal seawater is generally in the range of ±1 per mill PDB [13], and $\delta^{13}C_{(TDIC)}$ values of waste drains are in the range of -16 to -10 per mill PDB [23, 24]. As shown in Table 5, Layari and Malir river water is depleted in $\delta^{13}C_{(TDIC)}$. The depleted values of $\delta^{13}C_{(TDIC)}$ in water of Layari River (- 8.2‰ PDB) and Malir River (-8.8‰ PDB) owe to the heavy influx of sewage.

River	$\delta^{13}C_{(TDIC)}$ (‰ PDB)			
	April 2002	December 2003	September 2004	Average
Layari River	-8.1	-8.2	-8.0	-8.0
Malir River	-8.8	-8.9	-8.7	-8.9
LOD (‰)	0.1	0.1	0.1	-

Values are an average of five samples, RSD <2%.

Table 5. $\delta^{13}C_{TDIC}$ characteristics of Layari and Malir riverwater

δ^3C_{TDIC} value is a function of sewage pollution load and can be used in tracing land based pollution flow in marine water. Numerous researchers [12, 23, 26, 27, 61] have used δ^3C as a tool to track sewage pollution in the aquatic environment. This study shows that variations in

$\delta^{13}C_{TDIC}$ contents of the Manora Channel seawater are strongly associated with tide conditions, distance from the entrance of Layari River into the sea, and seasonal variation (Table 6). It is apparent that the Manora Channel water is relatively less depleted in $\delta^{13}C_{TDIC}$ during high tide as compared to low tide conditions, which means that high tide environment retards the mixing of the municipal sewage into seawater. Hence, channel water is relatively more depleted in $\delta^{13}C_{TDIC}$ during low tide as it facilitates the diffusion of sewage pollution with seawater. The diffusion of sewage pollution appears to be gradually decreased as the distance of sampling site is increased from Layari River joining point with seawater and is shown by a progressive reduction in $\delta^{13}C_{TDIC}$ values (Figure 8).The $\delta^{13}C_{TDIC}$ values of seawater at PNA and Manora Channel Lighthouse are close to $\delta^{13}C_{TDIC}$ values of normal unpolluted seawater, indicating less influx of sewage. Seasonal variation in $\delta^{13}C_{TDIC}$ values is attributed to strong currents and high flow of seawater in summer, which hinders the mixing of sewage pollution as indicated by relatively less depleted values of $\delta^{13}C_{TDIC}$ as compared to weak currents and moderate seawater flow in winter. Southeast coastal water is enriched in $\delta^3C_{(TDIC)}$ as compared to the Manora Channel, which indicates less influx of sewage pollution. However, considerable depleted values of $\delta^3C_{(TDIC)}$ are observed at the Ghizri area where Malir River empties its pollution load (Figure 9). However, seawater of northwest coast exhibits fairly constant $\delta^{13}C_{TDIC}$ values (-0.2‰ to 0.46‰ PDB), which are close to normal seawater value and appears to be independent of tidal and seasonal effect (Figure 10). It indicates substantially low sewage pollution load as compared to the Manora Channel and the southeast coast.

| | $\delta^{13}C$ (‰ PDB) | | | | | |
| | High tide | | | Low tide | | |
Sampling sites	April 2002	December 2003	September 2004	April 2002	December 2003	September 2004
			Manora Channel			
Layari Riveroutfall	-3.4	-5.8	-6.9	-8.0	-8.0	-8.2
Fish Harbor	-3.3	-6.1	-6.2	-6.2	-6.4	-7.3
KPT Shipyard	-3.8	-5.5	-4.8	-7.1	-7.9	-6.3
Bhaba Island	-1.6	-3.2	-5.7	-5.3	-6.6	-6.2
Bhit Island	-1.4	-3.8	-4.0	-3.5	-6.1	-6.3
Boat Club	-1.6	-3.2	-2.4	-2.6	-4.3	-2.8
Pakistan Naval Academy	-1.6	-2.7	-1.8	-2.2	-3.1	-2.9
Manora Lighthouse	-1.4	-2.2	-1.9	-2.0	-2.3	-2.2
			Southeast coast			
Marina Plaza	-0.5	-0.2	-0.3	-0.6	-0.3	-1.2
Casino	-0.8	-0.7	-0.4	-0.8	-0.9	-1.8
Naval Jetty	-1.1	-1.4	-0.7	-0.9	-1.7	-2.5

Sampling sites	$\delta^{13}C$ (‰ PDB)					
	High tide			Low tide		
	April 2002	December 2003	September 2004	April 2002	December 2003	September 2004
Marina Club	-1.1	-2.0	-1.8	-1.6	-2.1	-2.6
Ghizri area	-5.1	-5.4	-5.7	-7.4	-6.2	-6.5
Ibrahim Haideri	-4.8	-3.4	-3.6	-5.2	-4.1	-4.3
Northwest coast						
Manora Lighthouse SS	-0.4	ND	ND	-0.5	ND	ND
PNS Himalaya	-0.4	-0.3	-0.2	-0.5	-0.3	-0.2
Sandspit	-0.3	-0.2	-0.2	-0.4	-0.2	-0.2
Kakka Pir	-0.1	-0.1	-0.2	-0.2	-0.1	-0.2
Buleji	-0.2	-0.2	-0.2	-0.2	-0.2	-0.1
Power House	0.2	-0.1	-0.2	0.1	-0.2	-0.2
Sunari point	0.1	-0.1	-0.2	0.1	-0.2	-0.1

All the values are an average of five. RSD >2%. ND, not determined.

Table 6. $\delta^{13}C$ (TDIC) composition of Karachi coastal water

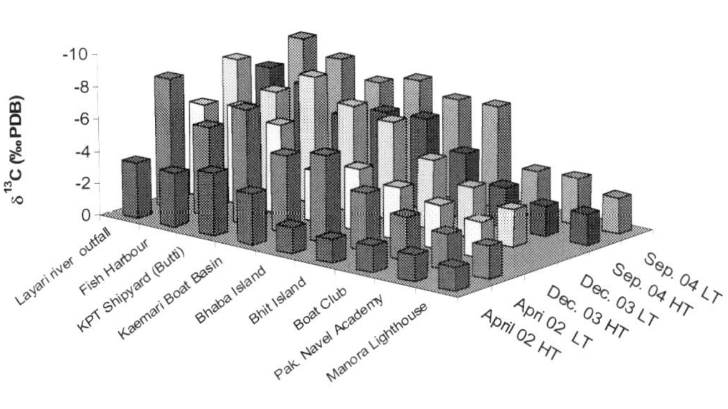

Figure 8. $\delta^{13}C$ (TDIC) in coastal waters of the Manora Channel.

4.2.2. Stable sulfur ($\delta^{34}S$) analysis of polluted rivers and coastal water

Many researchers [3, 35, 48, 49, 50, 63] have applied $\delta^{34}S$ content of seawater for evaluating sewage pollution contribution toward marine ecosystem. The $\delta^{34}S$ value of normal seawater

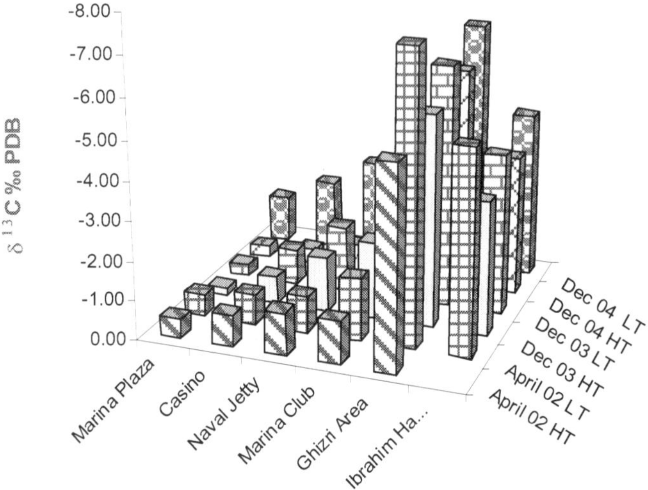

Figure 9. $\delta^{13}C$ (TDIC) in coastal waters of southeast coast.

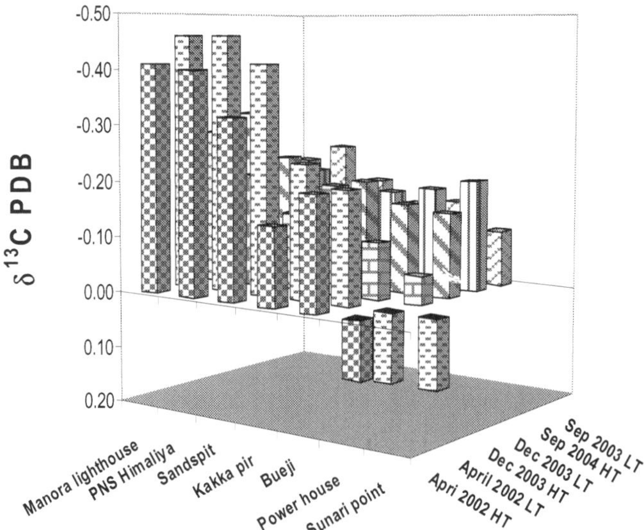

Figure 10. $\delta^{13}C$ (TDIC) in coastal waters of northwest coast.

is around 20‰ CDT, and the seawater is depleted in δ^{34}S with the addition of sewage wastewater. Sampling for δ^{34}S was performed only in April 2002. Layari and Malir river waters (6.6‰ and 10.7‰ CDT respectively) are notably depleted in δ^{34}S (Table 7) because of municipal wastewater (sewage pollution). The δ^{34}S composition of coastal water of Karachi Metropolitan is used to ascertain the seawater contamination sources. It is also evident from Table 7 that the factors controlling variations in δ^{34}S composition of the Manora Channel water are tide conditions and distance from the confluence point of Layari River with sea. The Manora Channel water is enriched in δ^{34}S during high tide as compared to low tide conditions because high tide conditions hamper the mixing of sewage with seawater. During low tide, however, channel water is relatively less enriched in δ^{34}S as it facilitates the mixing of sewage pollution with seawater. Sewage pollution gradually decreases as the distance of sampling site is increased from Layari River entry point with seawater. Southeast coastal waters are depleted in δ^{34}S value at the entrance point of Malir River into seawater at the Ghizri area and Ibrahim Haideri Fish Harbor area. At other sampling sites, δ^{34}S values of seawater are close to the normal seawater value, which demonstrates that the sewage pollution of Malir River diffuses quickly in coastal water due to strong open sea currents. δ^{34}S values of northwest coastal water are apparent to be near to its value for normal seawater in both low and high tide conditions, which shows little domestic wastewater discharge along northwest coastal water.

Sampling locations	δ^{34}S (‰ CDT)	
Layari River	6.6	
Malir River	10.7	
	High tide	**Low tide**
Manora Channel		
Layari river outfall zone	10.8	6.7
Fish Harbor	14	10.2
KPT Shipyard (Butti)	11.4	8.3
Kaemari Boat Basin	17.9	12
Bhaba Island	17.4	11.2
Bhit Island	17.6	14.1
Boat Club	17.5	15.8
Pakistan Naval Academy	17.9	16.0
Manora Lighthouse	18.1	16.2
Southeast cost		
Marina Plaza	19.4	19.3
Casino	19.2	18.8

Sampling locations	$\delta^{34}S$ (‰ CDT)	
Naval Jetty	19	18.2
Marina Club	18.8	18
Ghizri area	15.1	12.3
Ibrahim Haideri Fish Harbor	16.6	14.3
Northwest coast		
Manora Lighthouse (sea side)	19.6	19.7
PNS Himalaya	19.8	19.8
Sandspit	19.9	19.9
Kakka Pir	19.8	19.9
Buleji	20.0	20.0
Power house	20.0	20.0

Table 7. $\delta^{34}S$ composition of Karachi coastal water

4.2.3. Stable oxygen ($\delta^{18}O$) analysis of polluted rivers and coastal water

Oxygen isotope ($\delta^{18}O$) has been extensively used by several researchers [6, 29, Hudson et al. 1995;] to determine mixing of fresh water with seawater as their $\delta^{18}O$ values are quite different. $\delta^{18}O$ value for seawater is around 0‰ [Lear et al., 2000], while freshwater shows a wide range of $\delta^{18}O$ isotopic compositions, generally lower than -5‰. However, seawater is depleted in $\delta^{18}O$ with the mixing of fresh water [13]. Sampling for $\delta^{18}O$ was performed in April 2002. As shown in Table 8, Layari and Malir rivers have $\delta^{18}O$ values of -6.61‰ and -5.94‰, respectively, and are in the range of freshwater values. The $\delta^{18}O$ composition of seawater also varies with sea tides and how far a sampling site is from the points where these rivers empty wastewater into sea. In the Manora Channel, seawater appears to be depleted in $\delta^{18}O$ where fresh water, discharged through Layari River, is mixed with the seawater (Layari River outfall zone). However, seawater is found gradually enriched in $\delta^{18}O$ with increasing the distance from the joining point of Layari River with sea, with an exception of KPT Shipyard where $\delta^{18}O$ values is -2.45‰ during low tide environment. Low values of $\delta^{18}O$ could be due the mixing of KPT Shipyard wastewater. Southeast coastal seawater is less depleted in $\delta^{18}O$ as the Malir river wastewater disperses into seawater quickly. However, at Ghizri area, which is entry point of Malir River into the coast, seawater is depleted in $\delta^{18}O$. Along northwest coast, $\delta^{18}O$ values of seawater remains constant around 1‰ and appears to be independent of tide conditions.

4.2.4. Stable isotope analysis of marine biota

Isotope signatures $\delta^{13}C$ and $\delta^{15}N$ of mangrove and seaweeds from different coastal locations were determined to see possible effect of land based pollution on marine plants in terms of their stable isotope composition and physiological characteristics.

Sampling locations	$\delta^{18}O$ (‰ V SMOW)	
Layari River	-6.61	
Malir River	-5.94	
	High tide	**Low tide**
Manora Channel		
Layari River outfall zone	-3.93	-5.33
Fish Harbor	-0.78	-1.4
KPT Shipyard (Butti)	-0.7	-2.45
Kaemari Boat Basin	-0.75	-1.02
Bhaba Island	-0.65	-0.95
Bhit Island	-0.48	-0.79
Boat Club	0.89	0.72
Pakistan Naval Academy	0.95	0.82
Manora Lighthouse	0.96	0.84
Southeast coast		
Marina Plaza	1.13	1.01
Casino	1.12	1.00
Naval Jetty	1.13	0.89
Marina Club	-0.1	-0.8
Ghizri area	-1.61	-3.4
Ibrahim Haideri	-1.01	-1.77
Northwest coast		
Manora Lighthouse (SS)	1.15	1.13
PNS Himalaya	1.11	1.13
Sandspit	1.15	1.17
Kakka Pir	1.20	1.19
Buleji	1.21	1.22
Power House	1.19	1.17

Table 8. $\delta^{18}O$ composition of Karachi coastal water

4.2.4.1. $\delta^{13}C$ and $\delta^{15}N$ analysis ofmangroves

Stable carbon and nitrogen isotopes composition of mangroves are widely used in environmental studies [Lin and Sternberg, 1992; McKee et al. 2002; Kao et al., 2001]. For example, the leaves of tall and dwarf mangroves often have distinctly different carbon and nitrogen stable isotopic characteristics, which are related to the various environmental conditions such as nutrient status and water use efficiency [Alongi et al., 1992; McKee and Faulkner, 2000; Blasco and Saenger, 1996; Medina and Francisco, 1997]. Dwarf trees are associated with sites that are often in the interior of mangrove islands, while tall trees are, generally, found at the edge of islands. The morphological differences are indicative of phosphorus and nitrogen limitation [McKee et al., 2002].

$\delta^{13}C$ values of mangrove leaves ranged from -28.9‰ to -25.8‰ PDB, which are typical of most C-3 plants [O'Leary, 1981]. The mangroves leaves of the Manora Channel backwaters (receiving domestic waste from Layari River) are depleted in $\delta^{13}C$ as compared to $\delta^{13}C$ of Ghizri Creek mangroves leaves growing in open sea environment (Table 9). McKee et al. (2002) also reported that mangroves leaves growing at domestic waste sites are depleted in $\delta^{13}C$. However, $\delta^{15}N$ values of mangroves in the Manora Channel were quite comparable with $\delta^{15}N$ values of mangroves found in the Ghizri area, which suggests that the $\delta^{15}N$ levels of Mangroves is not a reliable indicator to differentiate between mangroves growing in polluted and nonpolluted sea sites.

Acomparison of $\delta^{13}C$ values of tall and dwarf mangrove plants revealed that $\delta^{13}C$ values of interior tree leaves are higher as compared to leaves of tall trees leaves. The plants with high $\delta^{13}C$ values are reported to have higher water use efficiency [Lin and Sternberg, 1992]. Our results, thus, show that a dwarf mangrove population has higher water use efficiency as compared to tall trees. The findings are also in agreement with earlier study by Lin and Sternberg (1992), where it was shown that dwarf mangrove populations had higher water use efficiency than tall trees. It is generally assumed that water use efficiency is related to the stomatal conductance. Dwarf trees have a lower ratio of the intercellular to atmospheric CO_2 resulting in an enriched $\delta^{13}C$ value and a higher water use efficiency [Farquhar, et al., 1988; Wooller et al., 2003; Lin and Sternberg, 1992].

Location	$\delta^{13}C$ (‰ PDB)		$\delta^{15}N$ (‰ air)	
	Tall tree (n = 5)	Dwarf tree (n = 5)	Tall tree (n = 5)	Dwarf tree (n = 5)
Manora Channel	-28.9 ± 0.3	-27.2 ± 0.2	11.0 ± 0.4	8.3 ± 0.5
Ghizri Creek	-27.5 ± 0.3	-25.8 ± 0.2	10.5 ± 0.3	7.9 ± 0.4

Table 9. Average $\delta^{13}C$ and $\delta^{15}N$ composition of mangrove (*Avecenia marina*) leaves

4.2.4.2. $\delta^{13}C$ and $\delta^{15}N$ analysis of seaweed

Stable carbon and nitrogen isotope compositions were determined in different types of seaweeds as well. $\delta^{13}C$ values of seaweed ranged from -31.1‰ to -4.9‰ PDB. However, $\delta^{13}C$ contents in majority of seaweeds were in the range of -19‰ to -13‰ PDB (Figure 11), which were in good agreement with the reported values for tropical and subtropical algae (Craig, 1953; Black and Bender, 1976; Fry, et al., 1982). $\delta^{13}C$ frequency distribution histograms of green, red, and brown algae are represented in Figure 12. $\delta^{13}C$ values of green algae were in the ranges of -21.14‰ to -4.9‰ PDB with a mean value of -14.1‰ PDB. $\delta^{13}C$ values of brown algae were found to be in the range of -17.82‰ to -8.9‰ PDB with a mean value of -14. 4‰ PDB, whereas $\delta^{13}C$ of red algae ranged from -31.12‰ to -9.49‰ PDB with a mean value of -17.38‰ PDB. The average $\delta^{13}C$ content of red algae (-17.4‰) was low as compared to brown (-14. 4‰) and green algae (-14.1‰). Fry et al. (1982) and Maberly et al. (1992) also reported marine red algae with relatively high $\delta^{13}C$ value. However, the highest $\delta^{13}C$ value was recorded in the green algae *Ulva fasciata* (-4.9‰). Carbon isotope composition of terrestrial plants can be used to depict their photosynthetic pathways. The $\delta^{13}C$ values for terrestrial C_3, C_4, and CAM plants range from -22‰ to -38‰, from -8‰ to -15‰, and from -13‰ to -30‰, respectively [Yeh and Wang, 2001]. Carbon isotope values of marine algae of Karachi coast ranged between -31.1‰ and -4.9‰. However, $\delta^{13}C$ values for majority of seaweed were in the range of -19‰ to -13‰, which apparently corresponds to terrestrial C_4 plants. It is reported in the literature that only one seaweed (*Udotea flabellum*) is known to have C_4 like photosynthesis, while majority of seaweeds are C_3 plants [Reiskind, et al., 1999: Wang and Yeh, 2003]. Thus, seaweed cannot be clearly categorized into C_3 or C_4 plants on basis of $\delta^{13}C$ composition. This can best be explained by the fact that $\delta^{13}C$ values of seaweed may be changed due to extent of CO_2 diffusion or HCO_3^- active transport process. For marine algae, two carbon sources are available for photosynthesis: CO_2 and HCO_3^-. However, it is difficult to guess what proportion of the different carbon sources (HCO_3^{-1} and CO_2) are taken up by the different seaweeds, which obviously complicates the interpretation of the data.

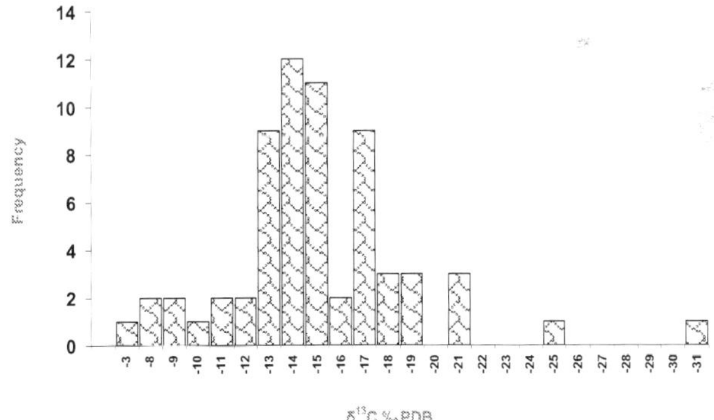

Figure 11. Frequency distribution of $\delta^{13}C$ of seaweed of Karachi coast.

To assess the effect of domestic waste on isotope composition of seaweed, $\delta^{13}C$ values of *Ulva* spp. from contaminated (Manora Channel) and uncontaminated sites (northwest coast) were compared. Average $\delta^{13}C$ values in *Ulva* spp. from polluted site (-14.9‰) and nonpolluted site (-14.5‰) do not display any significant isotopic shift, suggesting that $\delta^{13}C$ composition of *Ulva* spp. was not influenced by domestic wastewater (Table 10). This may be the result of increased primary productivity and/or the lower availability of dissolved inorganic carbon. *Ulva* spp. is green algae that fix atmospheric carbon by photosynthesis and does not utilize dissolved inorganic carbon from dissolve inorganic pool (DIC). Hence, it can be concluded that $\delta^{13}C$ cannot discriminate *Ulva* spp. of polluted and nonpolluted site. These results agree well with the finding of Rogers (1999), where he observed that *Ulva* spp. growing in polluted site was slightly depleted in $\delta^{13}C$ as compared to uncontaminated site but difference was not significant.

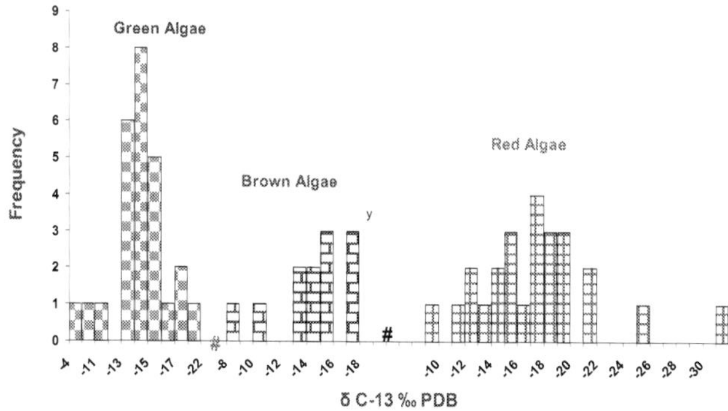

Figure 12. Frequency distribution of $\delta^{13}C$ in green, brown, and red algae of Karachi coast.

Coastal zone	Stable isotope	
	$\delta^{13}C$ (‰ PDB)	$\delta^{15}N$ (‰ air)
Manora Channel ($n = 7$)	-14.9 ± 1.2	7.9 ± 1.0
Northwest coast ($n = 4$)	-14.5 ± 0.34	10.2 ± 0.89

Table 10. Average $\delta^{13}C$ and $\delta^{15}N$ contents of *Ulva* spp. from polluted and unpolluted sites

Unlike $\delta^{13}C$, variation in $\delta^{15}N$ values of seaweeds was small and ranged from 6.1‰ to 17.8‰ air (Figure 13).$\delta^{15}N$ values of green algae varied between 6.1‰ and 17.7‰ air with an average value of 11.0‰ air, brown algae ranged from 8.9‰ to 12.9‰ air with an average 10.8‰ PDB, and $\delta^{15}N$ concentration of red algae was in the range of 7.7‰ to 14.5‰ air with an average value of 11.2‰ air (Figure 14). Red algae were enriched in $\delta^{15}N$ as compared to brown and green algae. $\delta^{15}N$ values of *Ulva* spp. from both polluted, and unpolluted sites are compared in Table 11. The average $\delta^{15}N$ values (10.2‰ air) of *Ulva* spp. from relatively clean water of northwest coast was enriched in $\delta^{15}N$ as compared to average $\delta^{15}N$ content (8.0‰ air) of same species collected from the Manora Channel (polluted water). The $\delta^{15}N$ values of the *Ulva* spp. (10.2‰ air) from the northwest coast (nonpolluted site) are close to $\delta^{15}N$ values of nonpolluted seawater nitrate. This indicates that *Ulva* spp.utilizes nitrate of nonpolluted seawater, similar results are quoted by Monteiro et al. (1997) and Peterson et al. (1985). $\delta^{15}N$ values (8.0‰ air) of *Ulva* from polluted the Manora Channel are close to $\delta^{15}N$ values of sewage, indicating that *Ulva* of polluted site absorb and assimilate sewage-derived nitrogen. Several researchers

[Cabana and Rasmussen, 1996; Monteiro et al., 1997; Hobbie and Fry, 1990; Udy and Dennison, 1997] have also reported that seaweed exposed to sewage discharge are depleted in $\delta^{15}N$. Thus, it can be inferred that nitrogen isotope of *Ulva* spp. could be a good indicator of sewage pollution.

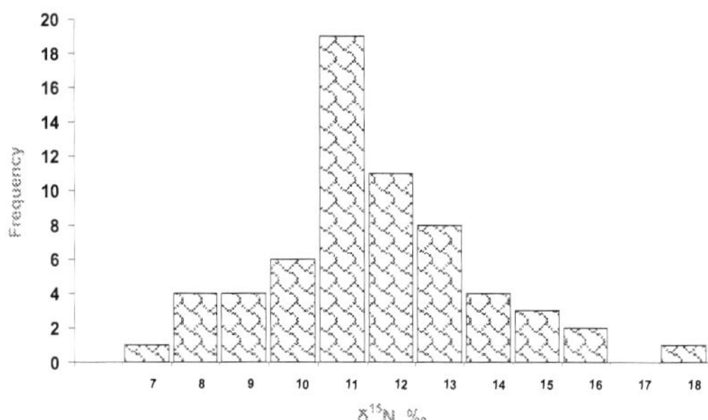

Figure 13. Frequency distribution of $\delta^{15}N$ of seaweed of Karachi coast.

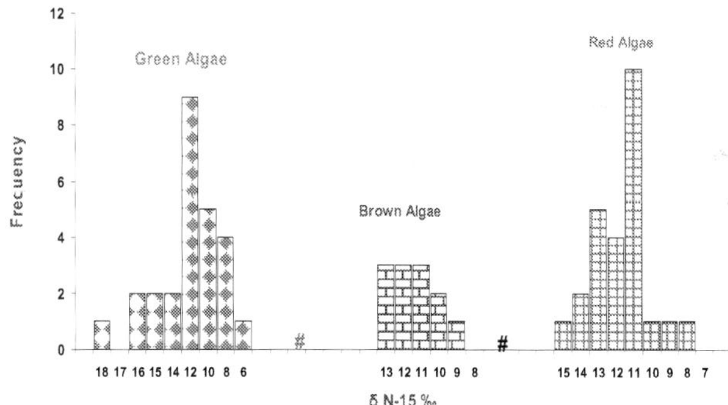

Figure 14. Frequency distribution of $\delta^{15}N$ of green, brown and red Algae of Karachi coast.

4.2.5. Stable carbon isotope composition of sea sediments

$\delta^{13}C$ contents (inorganic and organic) of sediments from Karachi coast are shown in Table 11. $\delta^{13}C_{inorg}$ values of Manora Channel sediments varied from -2.7‰ to -0.6‰ PDB, and $\delta^{13}C_{org}$ contents were in the range of -26.5‰ to -7.0‰ PDB. The sediments from Layari River and Karachi Harbor were appeared to be extremely depleted in $\delta^{13}C_{org}$ (-26.5‰), which indicated domestic sewage as the main source of these sediments. Another reason could be the lithology of the sediments, which was predominantly clay and organic carbon, had affinity toward argillaceous materials [Augley, et al., 2007]. Descolas-Gros and Fontugne (1990) also reported similar $\delta^{13}C_{inorg}$ values in the sediments originating from domestic wastewater. $\delta^{13}C_{org}$ contents of sediments of southeast coast were in the range of -14.9‰ to -8.6‰ PDB. These values do not show any link with sewage material and indicated that sewage material carried through Malir River is diluted and dispersed by sea waves. However, low $\delta^{13}C_{org}$ values recorded in the sediments of Ghizri (-12.86‰ PDB) and Ibrahim Haideri (-14.9‰ PDB) might represent organic matter of dead phytoplankton [Descolas-Gros and Fontugne, 1990]. $\delta^{13}C_{inorg}$ values of sediments pertaining to northwest coast were in the range of -1.0‰ to -0.3‰ PDB, and $\delta^{13}C_{org}$ ranged from -9.4‰ to -7.5‰ PDB, indicating no domestic waste input in sediments.

Sampling zones	$\delta^{13}C$ (inorganic)	$\delta^{13}C$ (organic)
Manora Channel		
Layari River outfall zone	-2.2	-26.5
Karachi Fish Harbor	-2.7	-26.5
KTP Shipyard (Butti)	-1.2	-15.4
Bhaba Island	-1.1	-18.6
Bhit Island	-0.6	-18.5
Boat Club	-0.7	-10.5
Pakistan Naval Academy	-0.6	-11.7
Manora Lighthouse	-0.6	-7.0
Southeast coast		
Marina Plaza	-0.7	-8.6
Casino	-0.7	-9.8
Naval Jetty	-0.7	-9.6
Marina Club	-1.0	-9.5
Ghizri area	-1.9	-12.8
Ibrahim Haideri	-1.6	-14.9
Northwest coast		
Manora Lighthouse (seaside)	-1.0	-9.1
PNS Himalaya	-0.7	-9.4
Sandspit	-0.3	-7.8
Kakka Pir	-0.3	-8.1
Buleji	-0.3	-7.9
Power house	-0.3	-7.5

Values are an average of five.

Table 11. Stable carbon isotope composition (‰ PDB) of Karachi coastal sediments

4.3. Tritium

Tritium (^{3}H) is isotope of hydrogen with relatively short half-life (12.3y) and decays with an emission of a beta particle[Martin, 2000]. Tritium was introduced into the atmosphere primarily as a result of nuclear bomb testing from the period 1951 to 1957. The natural concentration of tritium in seawater is around 0.02 Bq kg^{-1}. However, its concentration up to 100 Bq kg^{-1} has also been reported in some seawaters [McCubbin and Leonard, 2001].

The most important use of tritium is in distinguishing water that entered in water bodies prior to 1952 from water that entered into water bodies after 1952 [Drever, 1997]. Tritium is also used for studying mixing of seawater with fresh water [Mazor, 1991]. As shown in Table 12, tritium (^{3}H) content was higher in water of Layari (7.3 TU) and Malir (7.2 TU) river water as compared to seawater where tritium was in the range of 0.8 – 6.0 TU. Tritium content of seawater in mixing zone of Layari River outfall area was 6.0 TU, which decreased as the increased distance from the mixing zone toward Manora Channel exit (1.6 TU at Manora Lighthouse). Tritium contents in Manora Channel water seawater were higher than southeast and northwest coastal water, which show mixing of Layari River in seawater of Manora Channel.

It has also been reported that tritium content >10 TU in a water sample indicates a post-1952 water, whereas a tritium content lower than 0.5 indicate a pre-1952 water. Tritium content in between 0.5 and 10 represents a mixture of pre-1952 and post-1952 water [Mazor, 1999; Clark and Fritz, 19971]. Tritium contents of Karachi coastal water are between 0.8 and 6.2, which suggested that Karachi seawater is a mixture of pre-1952 and post-1952 water.

Locations	Tritium content (TU)*
Layari River	7.3 ± 0.6
Malir River	7.2 ± 0.6
Layari River outfall zone	6.0 ± 0.7
Boat Club	2.6 ± 0.6
Manora Lighthouse	1.6 ± 0.7
Southeast coast	0.9 ± 0.7
Northwest coast	0.8 ± 0.7
*I TU = 0.118 Bq/kg.	

Table 12. Tritium contents in river water and coastal water

5. Computation of polluted river water mixing with seawater using stable isotopes

A two-component mixing equation was used to assess the pollution contribution to seawater by the Metropolitan wastewater by using stable carbon and sulfur isotopes in addition to the application of oxygen isotope ratio to determine fresh water mixing with seawater. When $\delta^{13}C$ and $\delta^{14}S$ depleted Malir and Layari river water enters into $\delta^{13}C$ and $\delta^{34}S$ enriched seawater, the resulting isotopic signature of mixed water is different from both mixing water bodies. Thus, isotopic values of the two end points (such as polluted water body and seawater) can be used to estimate the contribution of polluted river to marine water through a two-component isotope mixing equation [Spiker, 1980]. Numerous researchers [18; Balesdent, et al., 1988; Amundson and Baisden, 2000; Vitousck, et al., 2003; Currie, 2007;[44]; Iqbal, 1998; Wels, et al., 1990; Raymond and Bauer, 2001; Fry, 2002] round the globe have documented the use of stable isotopes in a two-component mixing equation for various applications, including sewage flow in aquatic environment, food web studies, contribution of C-3 and C-4 plant sources to soil organic carbon, and contribution of different water sources in lake/stream. In this study, an attempt has been made to estimate mixing percentage of polluted water in marine environment through stable isotope signatures.

5.1. Computation of river water mixing into Karachi coastal water using $\delta^{13}C$ (TDIC)

The equation described in Section 3.9 was used to quantify percentage fraction of Layari river water in Manora Channel water and Malir river water in southeast coastal water during the low and high tide conditions of the sea. The dominant factors that appear to control the extent of terrestrial pollution mixing with Manora Channel seawater are tide conditions, distance from the joining point of Layari River with sea and sampling season. High tide environment hinders the Layari river water mixing with seawater coupled with a gradual decrease in pollution levels from entrance point of Layari River into sea. The mixing of river water with seawater is substantially high near Layari River outfall zone and gradually lowers toward Manora Channel exit with the exception at KPT and Bhaba Island where the addition of sewage input from local population may be considered responsible for increased pollution load. However, in high tide environment, considerable mixing (%) of river water in Manora Channel seawater indicated incomplete flushing, which owes to the fact that Manora Channel is a semienclosed bay. Mixing pattern is also appeared to be effected by seasonal environmental variations. In April, strong currents and high flow of seawater in Manora Channel diluted the Layari River added pollution as compared to weak current season and moderate seawater flow in the channel during September and December when volume of seawater in the channel is decreased; thereby, the ratio of Layari river water to seawater in channel is increased. However, at Layari River outfall zone, mixing seems to be independent of seasonal variation, which is clear from the fact that fraction of Layari River in seawater remains more than 97% in all sampling seasons (Figure 15).

Mixing of Malir River with seawater is considerably high at the Ghizri area, which is the entrance point of Malir River in the sea. The pollution load diffuses quickly as the distance is increased from entrance point to other sampling sites. Under low tidal conditions, the Ghizri area and Ibrahim Haideri Fish Harbor contained 70–74% and 46–59% water of Malir River. However, at sampling sites like Plaza, Casino, Naval Jetty, and Marina Club, mixing Malir river water into seawater reduced to 5–29%, which shows quick dispersion of Malir water in the sea by strong open sea currents (Figure 16). It appeared that mixing pattern is not much effected by seasonal changes during low and high tide environments except at Ghizri area.

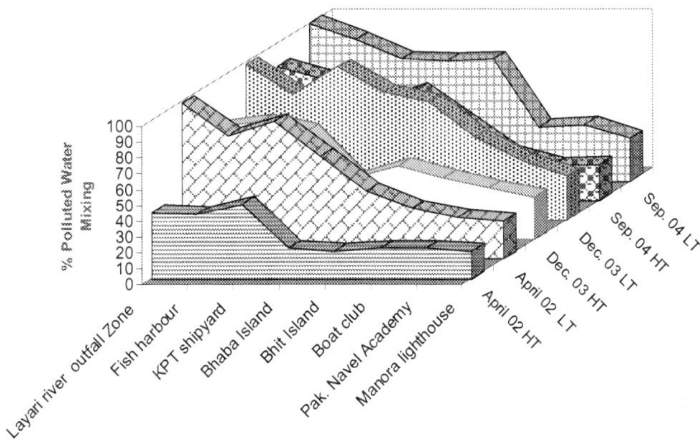

Figure 15. Percent mixing fraction of Layari River into Manora Channel in three sampling phases (HT = high tide, LT = low tide).

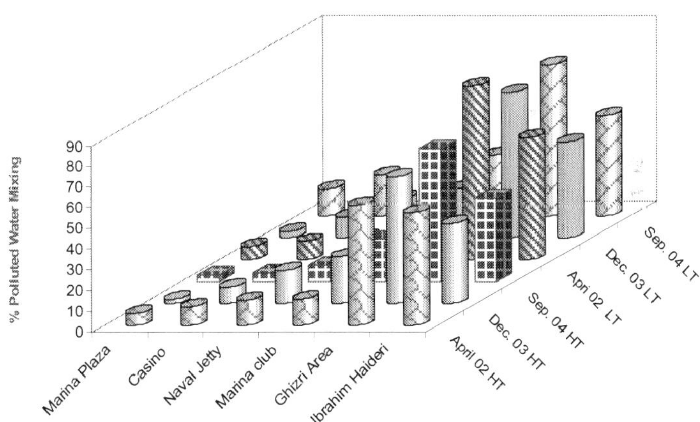

Figure 16. Mixing (%) of Malir River into southeast coastal water (HT = high tide, LT = low tide).

5.2. Computation of polluted river water mixing into Karachi coastal water using $\delta^{34}S$ values

To estimate the percent fraction of polluted river water into coastal water using $\delta^{34}S$ values, the equation in Section 3.9 was modified by replacing $\delta^{13}C$ with $\delta^{34}S$ to get the following two-component isotope balance equation:

$$f_A\left(\%\right) = [(\delta^{34}S_{MIX} - \delta^{34}S_B) / (\delta^{34}S_A - \delta^{34}S_B)] \times 100,$$

where $\delta^{34}S_A$ and $\delta^{34}S_B$ denotes stable sulfur isotope compositions of polluted river water and the nonpolluted seawaters, respectively. The mean $\delta^{34}S$ values of 6.6‰ CDT and 10.7‰ CDT was taken for Layari and Malir river water, respectively, while assuming northwest coastal water as pollution free (20‰ CDT). Tide height and distance of sampling sites from the Layari River outfall are the main factors that play part in transport and spread of Layari River pollution in Manora Channel. Mixing of Layari water with sea water is low in high tide (14–69%) as compared to low tide (28–99%) conditions. There was an optimal mixing of river water with sea at Layari River outfall zone in high and low tide conditions. However, the fraction (percent) of polluted water in seawater appears to be gradually decreased from Layari River outfall zone to Manora Lighthouse (Figure 17).

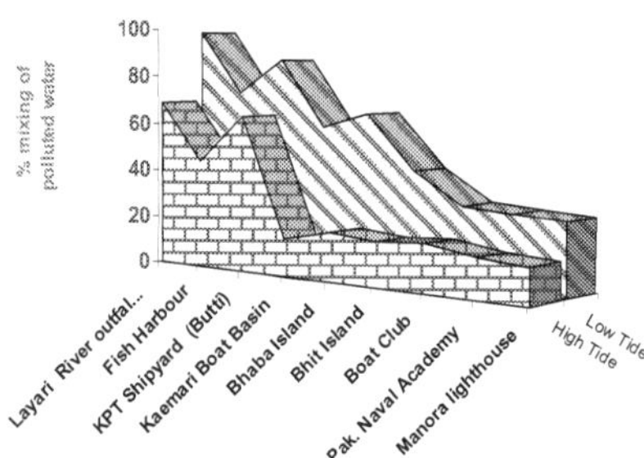

Figure 17. Mixing (%) of Layari river water with Manora Channel water based on $\delta^{34}S$ values.

The computation of Malir river water mixing with seawater computed through $\delta^{34}S$ values show the increase in fraction (%) of river water with seawater was observed in low tide as compared to high tide environment particularly at the Ghizri area and Ibrahim Haideri Fish Harbor. However, at sites like Marina Plaza, Casino, Naval Jetty, and Marina Club, the increase in mixing (percent) was not very significant (Figure 17), which is attributed to dilution of Malir River pollution by open sea currents.

5.3. A comparison of mixing computed through $\delta^{13}C$ and $\delta^{34}S$ values

The results obtained through $\delta^{13}C$ calculation were plotted against the $\delta^{34}S$ calculation (Figure 18). Significant correlation coefficient in Manora Channel ($r^2 = 0.93$) and southeast coast ($r^2 = 0.99$) during low tide indicated that mixing calculation through $\delta^{34}S$ and $\delta^{13}C$ values supports each other and a strong relationship exists between the two isotopes for determining polluted water mixing with seawater.

5.4. Application of $\delta^{18}O$ to determine freshwater–marine water mixing

$\delta^{18}O$ is a useful tool to define mixing between fresh water and marine water mixing [6, 29]. Some researchers [14, 18, 44] used $\delta^{18}O$ contents in a two-component isotope mass balance equation for the calculation of mixing fraction of two sources in a water body.

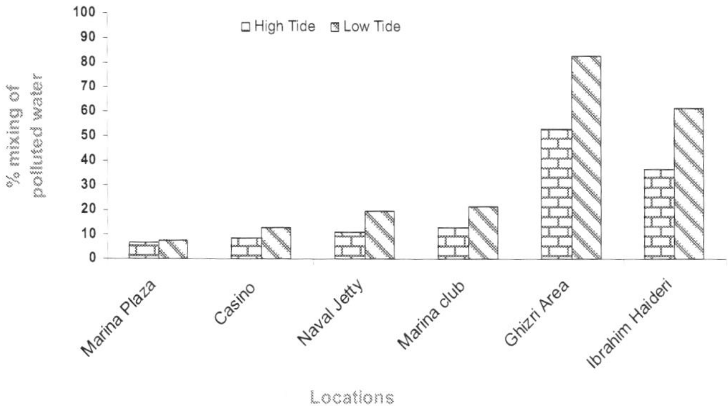

Figure 18. Mixing (%) of Malir river water with southeast coastal water based on $\delta^{34}S$ values.

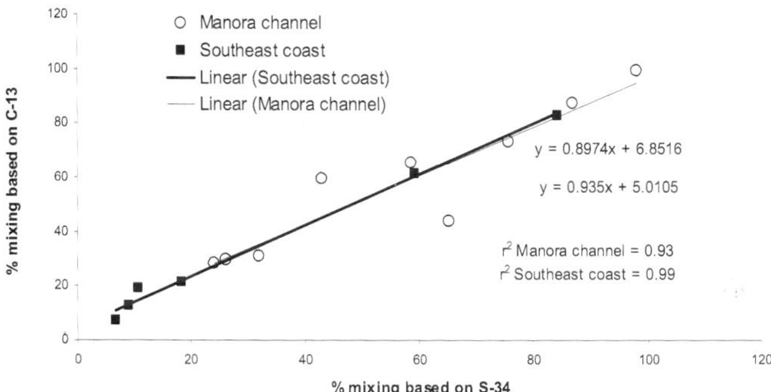

Figure 19. A correlation between percent mixing computed through $\delta^{13}C$ and $\delta^{34}S$ during low tide environment.

In the present investigation, $\delta^{18}O$ values of Layari and Malir river water, unpolluted sea water, and polluted seawater were used in the following two-component isotope equation to evaluate fresh water–seawater mixing:

$$\% f_{A} = [(\delta^{18}O_{MIX} - \delta^{18}O_{B}) \ / \ (\delta^{18}O_{A} - \delta^{18}O_{B})] \times 100,$$

where f_{A} is the fraction of river water in seawater, and $\delta^{18}O_{A,} \delta^{18}O_{B,}$ and $\delta^{18}O_{MIX}$ stand for $\delta^{18}O$ contents of river water, nonpolluted seawater, and polluted seawater, respectively.

It is evident from Table 13 that freshwater–seawater mixing is influenced by tidal height and distance from the fresh water source. Low tide environment is favored, and high tide environment slows the mixing process. A progressive reduction in proportion of freshwater into seawater is observed from Layari River outfall zone to Manora Lighthouse. At Manora Lighthouse, the ratio of freshwater in seawater is substantially low (3–5%). Along southeast coast, mixing of freshwater with seawater is considerably high at the Ghizri area and Ibrahim Haideri Fish Harbor during low tide environment. At sampling stations (Marina Plaza, Casino, and Naval Jetty), the ratio of freshwater into seawater was not significant showing that Malir river water is diluted quickly by strong open sea currents.

Sampling locations	% Fresh water mixing	
	High tide	Low tide
Manora Channel		
Layari river outfall zone	69.32	88.24
Fish Harbor	26.76	35.14
KPT Shipyard (Butti)	25.68	49.32
Kaemari Boat Basin	26.35	30.00
Bhaba Island	25.00	29.05
Bhit Island	22.70	26.89
Boat Club	4.19	6.49
Pakistan Naval Academy	3.38	5.14
Manora Lighthouse	3.24	4.86
Southeast coast		
Marina Plaza	0.99	2.68
Casino	1.13	2.82
Naval Jetty	0.99	4.37
Marina Club	18.31	28.17
Ghizri area	39.58	64.79
Ibrahim Haideri	31.13	41.83

Table 13. Percent mixing of freshwater–seawater

6. Conclusion

The present study has demonstrated that stable isotopes ($\delta^{13}C$, $\delta^{15}N$, $\delta^{34}S$, $\delta^{2}H$, and $\delta^{18}O$) can be effectively used to monitor the marine pollution and to investigate origin of salinity in the coastal aquifer. This study may also provide a precise and accurate isotopic database for researchers interested in seawater pollution and its effect on benthic life.

Author details

Azhar Mashiatullah[1*], Nasir Ahmad[2] and Riffat Mahmood[2]

*Address all correspondence to: mashiatullah@gmail.com

1 Isotope Application Division, Pakistan Institute of Nuclear Science and Technology, Islamabad, Pakistan

2 Institute of Geology, University of the Punjab, Lahore, Pakistan

References

[1] Ackerfors, H. (1971). Mercury pollution in Sweden with special reference to conditions in the water habitat. Proceedings of Royal Society of London, 177, 365.

[2] Akram, M., Qureshi, R. M., Ahmad, N., and Solaija, T. J. (2007). Determination of gamma-emitting radionuclides in the inter-tidal sediments off Balochistan (Pakistan) Coast, Arabian Sea. Radiation Protection Dosimetry, 123, 268–273.

[3] Alewell, C., Mitchell, M. J., Likens, G. E., and Krouse, H. R. (1999). Sources of stream sulfate at the Hubbard Brook Experimental Forest: long-term analyses using stable isotopes. Biogeochemistry, 44, 281–299.

[4] Algan, A. O., Gatay, M. N., Sarikaya, H. Z., Balkis, N., and Sari, E. (1999). Pollution monitoring using marine sediments: a case study on the Istanbul Metropolitan Area. Turkish Journal of Engineering and Environmental Science, 23, 39.

[5] Ali, I., and Jilani, S. (1995). Study of contamination in the coastal waters of Karachi. In: The Arabian, Living Marine Resources and the Environment. M. F. Thompson and N. M. Tirmizi (eds.). Vanguard Books (Pvt) Ltd., 45 The Mall, Lahore, Pakistan, pp. 653–658.

[6] Anderson, T. F., and Arthur, M. A. (1983). Stable isotopes of oxygen and carbon and their application to sedimentologic and paleoenvironmental problems. In: Stable Isotopes in Sedimentary Geology. Society of Economic Paleontologists and Mineralogists Short Course, vol. 10, pp. 1-1–1-151

[7] AOAC. (2000). Official methods of analysis of AOAC international. 17th edition. H. William (ed.). Maryland, USA. Vol. 1. Method 977.15.

[8] Beg, M. A. A. (1994). Pollution in the marine environment of Karachi. Wildlife and Environment, 3, 36–41.

[9] Bernhard, M., and Zattera, A. (1975). Major pollutants in the marine environment. In: Proceedings of the 2nd International Congress on 'Marine Pollution and Marine

Waste Disposal', San Remo 17–21, December, 1973. E. A. Pearson and E. De Fraja Frangipane (eds.). Pergamon Press Ltd., 207 Queen's Quay West, Toronto 1, Canada, pp. 195–300.

[10] Bryan, G. W. (1976). Heavy metal contamination in the sea. In: Marine pollution. R. Johnston (ed.). Academic Press, New York, pp. 186–302.

[11] Bryan, G. W. (1971). The effects of heavy metal (other than mercury) on marine and estuarine organisms. Proceedings of the Royal Society London Series Biological, 177, 389.

[12] Burnett, W. C., and Schaeffer, O. A. (1980). Effect of ocean dumping on carbon-13/carbon-12 ratios in marine sediments from the New York Bight. Estuarine and Coastal Marine Science, 11, 6, 605–611.

[13] Clark, I. D., and Fritz, P. (1997). Environmental Isotopes In Environmental Isotopes in Hydrology. Lewis Publishers, New York, pp. 38.

[14] Coplen, T. B. (1994). Reporting of stable hydrogen, carbon, and oxygen isotopic abundances. Pure Applied Chemistry, 66, 273–276.

[15] Costanzo, S. D., O'Donohue, M. J., Dennison, W. C., Loneragan, N. R., and Thomas, M. (2001). A new approach for detecting and mapping sewage impacts. Marine Pollution Bulletin, 42, 2, 149–156.

[16] Dansgaard, W. (1964). Stable isotopes in precipitation. Tellus, 16, 436–438.

[17] DE Wolf, H., Van Den Broeck, H., Qadah, D., Backeljau, T., and Blust, R. (2005). Temporal trends in soft tissue metal levels in the periwinkle *Littorina littorea* along the Scheldt estuary, The Netherlands. Marine Pollution Bulletin, 50, 463–484.

[18] Dinçer, T., Payne, B. R., Florkowski, T., Martinec, J., and Tongiorti, E. (1970). Snowmelt runoff form measurements of tritium and oxygen-18. Water Resources Research, 6, 110–129.

[19] El Mamoney, M. H. (1995). Evaluation of terrestrial contribution to the Red Sea sediments, Egypt. PhD thesis, Facility of Science, Alexandria University, Egypt.

[20] Ergin, M., Bodur, M. N., Ediger, D., Ediger, V., Yemenicio-glu, S., and Yucesoy, F. (1994). Sedimentation rates in the Sea of Marmora: a comparison of results based on organic carbon-primary productivity and 210Pb dating. Continental Shelf Research, 12, 1371–1387.

[21] Fatoki, O. S., and Mathabatha, S. (2001). An assessment of heavy metal pollution in the East London and Port Elizabeth harbors. Water SA, 27, 2, 233–240.

[22] Faure, G., and Mensing, T. M. (2005). Isotopes: Principles and Applications. 3rd ed. USA: John Wiley & Sons. 897 p.

[23] Finaly, J. C., and Kendall, C. (2007). Stable isotope tracing of temporal and spatial variability in organic matter sources to freshwater ecosystem. In: Stable isotope in

Ecology and environmental Sciences. Robert Michener and Kate Lafta (eds.). Blackwell, pp. 283–333.

[24] Fry, B., and Sherr, E. B. (1989). δ13C measurements as indicators of carbon flow in marine and freshwater ecosystems. In: Stable Isotopes in Ecological Research. P. W. Rundel, J. R. Rundel, and K. A. Nagy (eds.). Springer-Verlag, New York, pp. 196–229.

[25] Gat, J. R. (1996). Oxygen and hydrogen isotopes in the hydrologic cycle. Annual Review of Earth Planetary Science, 4, 225–262.

[26] Gearing, P. J., Gearing, J. N., Maughan, J. T., and Oviatt, C. A. (1991). Isotopic distribution of carbon from sewage sludge and eutrophication in the sediments and food web of estuarine ecosystems. Environmental and Science Technology,25, 3, 295–301.

[27] Gichuki, J., Triest, L., and Dehairs, F. (2001). The use of stable carbon isotopes as tracers of ecosystem functioning in contrasting wetland ecosystems of Lake Victoria, Kenya. Hydrobiologia, 458, 1–3, 91–97.

[28] GoP. (2007). Pollution in Karachi harbor and areas around Pakistan air force bases in Karachi, Senate Committee Report, pp. 1–45.

[29] Hendry, J. P., and Kalin, R. M. (1997). Are oxygen and carbon isotopes of mollusc shells reliable palaeosalinity indicators in marginal marine environments? A case study from the middle Jurassic of England. Journal of Geological Society London, 154, 321–333.

[30] Hunkeler, D., Chollet, N., Pittet, X., Aravena, R., Cherry, J. A. (2004). Effect of source variability and transport processes on carbon isotope ratios of TCE and PCE in two sandy aquifers. Journal of Contaminant Hydrology, 74, 265–282.

[31] IAEA-MEL. (1987). Marine pollution baseline. survey in the Korangi. Phitti Creek, Pakistan. Final Report. ILMR, IAEA, Monaco. 27.

[32] JICA. (2007). Study on Water Supply and Sewerage System in Karachi, JICA, February 2007.

[33] Kennish, M. J. (1997). Practical Handbook of Estuarine and Marine Pollution. CRC Press Marine Science Series, USA.

[34] Kremling, K. (1983). Methods of seawater analysis. K. Grasshoff, M. Ehrhardt, and K. Kremling (eds.). VerlagChemie, 196 p.

[35] Krouse, H. R. (1977). Sulphur isotope abundance elucidate uptake of atmospheric sulphur emissions by vegetation. Nature, 265, 45–46.

[36] Majid, M. (2002). Geochemical Studies of Heavy Metals in the Sea Water along Karachi Makran Coast. PhD thesis, University of Karachi, Pakistan.

[37] Majid, M., Azhar. S., Nadia, M., and Tabassum, M. (1999). Status of trace elements level in blood samples of different age population of Karachi, Pakistan. Turkish Journal of Medical Science, 29, 697–699.

[38] Majid, M., Nausheen, M., Azhar, S., and Tabassum. M. (2000). Changes in blood levels of trace elements and electrolytes in hypertensive patients. Medical Journal of The Islamic Republic of Iran, 14, 2, 115–118.

[39] Majoube, M. (1971). Fractionnement en oxygène-18 et en deutérium entre l'eau et sa vapeur. Journal of Chemical Physics, 197, 1423–1436.

[40] Mancini, S. A., Lacrampe-Couloume, G., Jonker, H., Van Breukelen, B. M., Groen, J., Volkering, F., and Lollar, B. S. (2002). Hydrogen isotopic enrichment: an indicator of biodegradation at a petroleum hydrocarbon field site. Environmental Science and Technology, 36, 2464–2470.

[41] Mansour, A. M., Nawar, A. H., and Mohamed, A. W. (2000). Geochemistry of coastal marine sediments and their contaminant metals, Red Sea, Egypt. A legacy for the future and a tracer to modern sediment dynamics. Sedimentology Journal of Egypt, 8, 231–242.

[42] Mook, W. G. (2000). Natural abundance of the stable isotopes of C, O and H. In: Environmental Isotopes in the Hydrological Cycle Principles and Applications, Vol-I. W. G. Mook (ed.), pp. 60–123.

[43] Morrill, P. L., Sleep, B. E., Slater, G. F., Edwards, E. A., and Lollar. B. S. (2006). Evaluation of isotopic enrichment factors for the biodegradation of chlorinated ethenes using a parameter estimation model: toward an improved quantification of biodegradation.Environmental Science and Technology, ASAP Article 10.1021/es051513e S0013-936X(05)01513-0. Web Release Date: May 19, 2006.

[44] Mortatti, J., Moraes, J. M., Rodrigues, J. C., Victoria, R. L., and Martinelli, L. A. (1997). Hydrograph separation of the Amazon river using 18O as an isotopic tracer. Science of Agriculture, 54, 3.

[45] Mudroch, A., and McKnight, S. D. (1994). Bottom sediment sampling. In: Handbook of Techniques for Aquatic Sediments Sampling, Chapter 4, 2nd edition. Lewis Publishers, Boca Raton, 236 p.

[46] Mujahid, S. M., Majid, M., and Tabassum, M. (1994). Determination of trace elements in blood of CRF patients. Proceeding of SBBP Symposium of Biochemist and Biophysics, 1, 185–187.

[47] Naturvårdsverketa. (2001). Metaller i sjöar och vattendrag, http://www.environ.se.

[48] Novák, M., Kirchner, J. W., Groscheova, H., Havel, M., Cerny, J., Krejci, R., and Buzek, R. (2000). Sulfur isotope dynamics in two Central European watersheds affected by high atmospheric deposition of SOx. Geochimica et Cosmochimica Acta, 64, 367–383.

[49] Nriagu, J. O., Coker, R. D., and Barrie, L. A. (1991). Origin of sulfur in Canadian arctic haze from isotope measurements. Nature, 349, 142–145.

[50] Ohizumi, T., Fukuzaki, N., and Kusakabe, M. (1997). Sulfur isotopic view on the sources of sulfur in atmospheric fallout along the coast of the Sea of Japan. Atmosphere and Environment, 31, 1339–1348.

[51] Parsons, T. R., Yoshiaki, M., and Lalli, C. M. (1984). A manual of chemical and biological methods for seawater analysis. Pergamon Press, Oxford.

[52] Qureshi, R. M., Sajjad, M. I., Mashiatullah, A., Waqar, F., Jan, S., Akram, M., Khan, S. H., and Siddiqui, S. A. (1997). Isotopic investigations of pollution transport in shallow marine environments off the Karachi Coast, Pakistan. Proceedings of the International Symposium on Isotope Techniques in the Study of Past and Current Environmental Changes in the Hydrosphere and the Atmosphere, 14–18 April, 1997, Vienna, Austria, pp. 353–365.

[53] Rabalais, N. N., and Nixon, S. W. (2002). Nutrient over-enrichment of the coastal zone. Estuaries,25, 4B, 639–644.

[54] Rizvi, S. H. N., Masood, T., Khan, S., Ali, M., and Baquer, J. (1995). Circulation pattern in the Karachi coastal waters in relation to the outfall. In: The Arabian Sea—Living Marine Resources and the Environment. M. F. Thompson and N. M. Tirmizi (eds.), pp. 641–652.

[55] Saifullah, S. M., Sarwat, T., Khan, S. H., and Saleem, M. (2004). Iron pollution in mangrove habitat of Karachi, Indus Delta. Earth Interactions, 8, 17, 1–9.

[56] Saleem, M., and Kazi G. H. (1995). Distribution of trace metal in seawater and surficial sediment of Karachi Harbor. In: The Arabian Sea-Living Marine Resources and the Environment. M. F. Thompson and N. M. Tirmizi (eds.). pp. 659–666.

[57] Saleem, M., and Kazi, G. H. (1998). Concentration and distribution of heavy metals (Pb, Cd, Cr, Cu, Ni and Zn) in Karachi shore and offshore sediments. Pakistan Journal of Marine Science, 7, 1, 71–79.

[58] Slater, G. F. (2003). Stable isotope forensics-when isotopes work. Environment Forensics, 4, 13–23.

[59] Smith, D. G. (1986). Heavy metals in the New Zealand aquatic environment: a review. Water Quality Centre, Ministry of Works and Development, Wellington, New Zealand.

[60] Swedmark, M., Broanten, B., Emanuellsson, E., and Grammo, A. (1971). Biological effects of surface active agents on marine animals. Marine Biology, 9, 183–201.

[61] Sweeney, R. E., and Kaplan, I. R. (1980a). Tracing flocculent industrial and domestic sewage transport in San Pedro Shelf, Southern California, by nitrogen and sulfur isotope ratios. Marine Environmental Research, 3, 215–224.

[62] Tam, N. F. Y., and Wong, Y. S. (2000). Spatial variation of heavy metals in surface sediments of Hong Kong mangrove swamps. Environmental Pollution,110, 195–205.

[63] Tucker, J., Sheats, A. E., Giblin, C. S., Hopkinson, P. and. Montoya, J. P. (1999). Using stable isotopes to trace sewage-derived material through Boston harbor and Massachusetts bay. Elsevier Marine Environmental Research, 48, 353–375.

[64] US EPA. (1983). Sample preservation. In: Methods for Chemical Analysis of Water and Wastes, EPA-600/4-79-020. USEPA, Ohio, USA.

[65] Voutsinou-Taliadouri, F., Georgakopoulou-Gregoriadou, E., and Psyllidou-Giouranovits, R. (1995). Geochemical aspects of a gulf influenced by anthropogenic activities (Thermoikos Gulf, N. W. Agean Sea). Rapp. Comm. Int. Mer Medit. 34: 150.

[66] Waldichuk, M. (1985). Biological availability of metals to marine organisms. Marine Pollution Bulletin, 16, 7–11.

[67] WWF (2002). Study of heavy metal pollution level and impact on the fauna and flora of the Karachi and Gwadar coast. Final Project Report, No. 50022801.

[68] Zaigham, N. A. (2004). Unauthorized squatter settlements are one of major sources for polluting surface and subsurface waters in Karachi, Proceedings of the WSSD workshop on human settlement and environment (Pakistan's Response to its Obligations under the WSSD Plan of Implementation), Islamabad, December 14–15, 2004, pp. 100–112.

Chapter 11

Investigation of Po-210 and Heavy Metal Concentration in Seafood Due to Coal Burning – Case Study in Malaysia

Lubna Alam

Additional information is available at the end of the chapter

http://dx.doi.org/10.5772/62171

Abstract

A systematic study on the natural radionuclide Po-210 and heavy metals in marine organisms has been undertaken to establish a baseline data on the contamination of a coal burning power plant area. The organism samples have been collected from the local fish market which was very near to Sultan salauddin abdul aziz power plant and analysed for Po-210 and heavy metals using alpha spectrometer and ICP-MS, respectively. The content of Po-210 and heavy metals in analyzed organisms varies according to the feeding habits and ecological niche.

Keywords: Seafood, coal burning, Po-210, heavy metals, contaminants

1. Introduction

Assessment of radioactive elements in the natural environment and their effects on living organisms has been one of the most important issues in radioecology and radiological protection in recent years. The bioaccumulation of Po-210 refers to a process by which Po-210 accumulates in various tissues of a living organism. Po-210 is a high-energy alpha particle emitter in the uranium decay chain [1] and among natural radionuclides occurring in the ocean, alpha emitters are considered to be the most important [2], because of their high mass and charge, are more damaging and so are accorded a "radiation weighting factor" of 20. Moreover, Moroz and Parfenov [3] described that Po-210 can have a toxic effect even in small concentrations due to its high-energy alpha radiation. Despite its toxic properties, the radionuclide Po-210 is readily assimilated by marine primary producers [4] and is known as a major contributor (90%) of the natural radiation dose from alpha-emitting radionuclides, as received by most marine organisms [1, 5].

Po-210 enters the marine environment mainly through the global fallout of aerosols and gases, through release from the earth's crustal material, through substances released due to activities related to the processing of uranium ore and as a by-product of fertilizers and oil industries. After Po-210 enters the marine environment, they are subjected to different biogeochemical reactions (e.g., dissolution, hydrolysis, complexation, sorption/desorption, coprecipitation, speciation) [6], which depends upon their chemical form and the characteristics of the marine environment [6]. According to Stewart et al. [7], Po-210 displays very strong binding to particle surfaces, including organisms. Moreover, Po-210 is primarily associated with proteins in living organisms and can also penetrate into the cytoplasm of cells [8]. Because of that, Po-210 can effectively assimilate in higher marine organisms. In addition, radionuclides present in the marine environment can be transferred through plankton to higher organisms such as fish and mussels, which are well-known filtration organisms. Therefore, their concentration is increased through the trophic chain with the highest values found in predators [6]. The Po-210 accumulated in marine organisms is generally derived from the food chain [9], and significant differences are noted in its accumulation in different species. Very high levels of Po-210 have been observed in certain organs of marine species such as the digestive tract of fishes, the digestive gland of mollusks and the hepatopancreas of crustaceans; these levels are generally 1–2 orders of magnitude higher than those found in the muscle of these species [10–12].

Similarly, large amount of heavy metals can be deposited in the aquatic system near the coal-burning power plant area as the smallest particles of fly ash are enriched with heavy metals [13] and hold ecological importance because of their toxicity, persistence and bio-accumulation. As a result, there has been growing interest in determining the heavy metals levels in marine environment and attention was drawn to the measurement of contamination levels in public food supplies, particularly fish [14–16]. Additionally, heavy metal concentrations in aquatic organisms along with bio-concentration have been extensively studied in various places around the world [17–24]. Generally, heavy metals are defined as metallic chemical elements that have relatively high density and are toxic or poisonous at low concentrations [25]. In marine environment, heavy metals are potentially accumulated in sediments and marine organisms and subsequently transferred to man via the food chain as described by Pourang et al. [26] and thereby it poses the threat to human health. There are three possible routes by which metal can be derived, namely, (1) from solution, (2) from ingestion of food, and (3) from the ingestion of particulate matter containing metals [27]. Moreover, direct uptake from the sediment particles can be an important additional source of sediment-bound contaminations for sediment-feeding organisms [28].

Although there have been many studies on chemical contamination in marine organisms [29–33], sufficient data are not available for different marine species of the coal-burning power plant area. Therefore, the aim of this work was to investigate the level of Po-210 and heavy metal in marine organisms from the coastal area of Kapar and to compare with the other places of the world.

1.1. Study area

Malaysia, which is located between 2° and 7° north of the Equator, has a tropical climate with warm weather all the year round with temperatures ranging from 21°C to 32°C. Malaysia has about 4,800 km of coastline comprising two distinctly different physical formations, namely the mangrove fringed mud flats and sandy beaches. The Malaysian coastal area includes a land mass totaling approximately 4,162 square kilometers (sq.km), and an EEZ totaling approximately 465,700 sq. km. These areas are endowed with a rich diversity of living and non-living resources which have significant effects on the country's economic growth as well as her environmental well being and social order. The study was conducted at Kapar coastal area, which is very adjacent to Sultan Salahuddin Abdul Aziz power station. This power station is located at the western coast line of west Malaysia, at the Malacca strait, normally known as the largest power station in Malaysia with generating capacity of 2,420 MW and contributes about 23% of the country's energy demand. This power station is the first power station with triple fuel firing capability (gas, oil and coal) in Malaysia. The power plant applies seawater as the source for cooling water and the sea is used as the pathway for transporting coal. On the other hand, the surrounding coastal area is the ultimate recipient of the fly ash which is produced by coal burning. As a result, this area is becoming important for radiochemical analysis and has been preferred for this present study.

2. Materials and methods

Samples of seafood (Table1) popular with the Malaysian population were collected from the local fish market which is very near to Sultan Salauddin Abdul Aziz power plant (Figure1). The seafood types were divided into three groups which are: fish (*Arius maculatus*), crustacean (*Penaeus merguiensis*), and molluscs (*Anadara granosa*). The organism samples were transported to the laboratory for further analysis and kept in freezer.

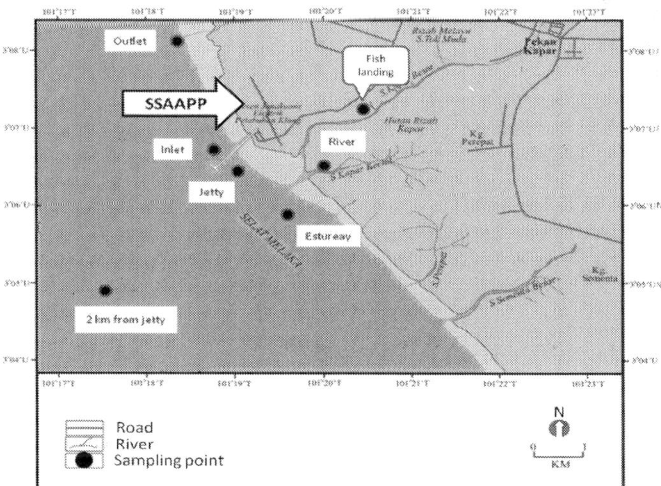

Figure 1. Study area showing the fish market at Kapar coastal area, Malaysia

Category	Species	Classification	
Fish	*Arius maculatus*	Kingdom:	Animalia
		Phylum:	Chordata
		Class:	Actinopterygii
		Order:	Siluriformes
		Family:	Ariidae
		Subfamily:	Ariinae
		Genus:	*Arius*
Crustacean	*Penaeus merguiensis*	Kingdom:	Animalia
		Phylum:	Arthropoda
		Class:	Malacostraca
		Order:	Decapoda
		Suborder:	Dendrobranchiata
		Superfamily:	Penaeoidea
		Family:	Penaeidae
		Genus:	*Penaeus*
Mollusc	*Anadara granosa*	Kingdom:	Animalia
		Phylum:	Mollusca
		Class:	Bivalvia
		Subclass:	Pteriomorphia
		Order:	Arcoida
		Family:	Arcidae
		Genus:	*Anadara*

Table 1. Description of the analyzed species

In the laboratory, the total length (cm) of fish samples was taken from the tip of the snout to the extended tip of the caudal fin. In case of shrimp, the distance from the tip of the rostrum to the end of telson has been considered as the total length. For cockle valve length was measured by using a vernier caliper to the nearest 0.1 mm. The total weight (g) of each sample was measured using a top loading electronic balance.

For the preparation of samples, only edible parts were selected. The soft tissues and muscles from the shells and bones have been separated from molluscs, crustaceans and fishes. The wet weights of the samples have been recorded and then dried in an oven at 60°C overnight to obtain the dry weights. After drying, mass of the dried samples was determined and the fresh weight to dry weight ratio has been calculated. Then, the samples were homogenized with a mortar. Finally, the samples were wrapped with aluminum foil and preserved in leveled plastic bag for radiochemical analysis.

The radiochemical separation method was used to estimate Po-210 in the samples [2, 34]. About 0.5 g of the dried sample was taken and Po-209 of a known activity was added as a yield tracer. Then the samples were digested with nitric acid and perchloric acid. The solution was filtered and gently evaporated to dryness. Then the samples were dissolved in 50 ml of 0.5 M HCl along with a pinch of ascorbic acid to reduce Fe (III) and Po-210 was spontaneously deposited on brightly polished silver discs (2 cm diameter) for a period of 3–4 h at a temperature of 70–90ºC.

Three replicates of organisms sample were analyzed for the measurement of heavy metals such as Cu, Cd, Zn, Pb and Cr, and all the glass wares used for analysis were acid-washed to avoid the possible contamination. About 0.3–0.5 g of dried samples for each replicate were weighted in a beaker using electronic scales. The samples were then digested with a mixture of 30 ml nitric acid (HNO3, GR, 65%, Merck) and 5 ml of concentrated perchloric acid (HClO4;GF; 70%, Merck). After that, 10 ml of concentrated hydrochloric acid (HCl, GR; 37%, Merck) was added in the samples and heated until dryness. After cooling the sample, 2.5 ml of nitric acid was added into the samples. A total of 20 ml of de-ionized distilled water was added into the beaker containing the sample and filtered through filter paper (Whatman, GF/C; diameter 47 mm; pore size 0.45 μm). After that, the filtered solutions were added with de-ionized distilled water until 70 ml to make it to 0.5M HNO3. Determination of heavy metals was carried out using the inductively coupled plasma mass spectrometry (ICP-MS) (Perkin Elmer-Elan 9000).

3. Results and discussions

Arius maculatus collected in this study ranged in size from 10 to 35 cm in total length and 112 to 296 g in total weight. Po-210 concentration in the soft tissue of collected fish samples ranged from 0.17 to 11.86 Bqkg^{-1} (dry weight) with the mean concentration value of 5.42 Bqkg^{-1} (dry weight) (Figure 2). However, the calculated result of the present study has been compared with other reported values of different places. This comparison showed wide variation because where measurements have been made in a wide variety of fish species, consistent interfamily variabilities in Po-210 concentrations are found, and there is strong evidence that the Clupeidae and Scombridae families have significantly higher Po-210 levels than other fish families [35]. Moreover, where bony teleost fish have been compared with cartilaginous elasmobranchs from different oceanic regimes and depths, Po-210 concentrations in the tissues of elasmobranchs are generally lower than in those of teleosts, with the difference between the two taxonomic groups being greater in liver and gonad tissue than in muscle [30, 36]. Similarly, Štrok, and Smodiš [8] described that differences between two fish species can be observed, probably due to their diverse feeding habits. The same results were found in IAEA [37], where the specific activities for pelagic species, such as *S. pilchardus*, were larger than for benthic species (*M. cephalus*).

However, investigations on Po-210 accumulation by fish have thus far shown that the average concentration ranges from 2.6 to 259.0 Bqkg^{-1} [35, 38–42] which is supporting the result of the present study. Po-210 concentration in the present study is within the range of the values reported for Japan, Poland, Sudan, Brazil, England and Baltic Sea. A comparatively higher level of Po-210 concentration was observed in India because of the impact of a nuclear power station which operates at Kalpakkam. In the case of Cuba, the elevated amount of Po-210

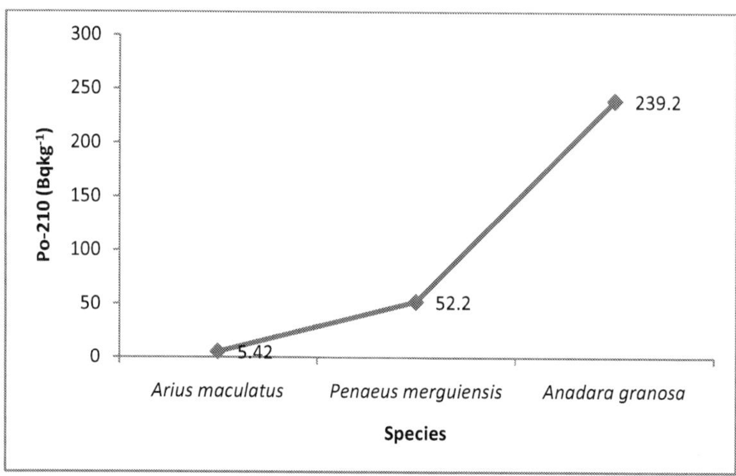

Figure 2. Mean concentration of Po-210 in marine organisms

activities was characterised by global fallout. Similarly, very high values of Po-210 were observed in the USA and Australia. However, the calculated value ranged within the global average value.

The analysed shrimp (*Penaeus merguiensis*) samples varied between 5 to 18 cm in total length and 2 to 15 g in total weight. The Po-210 activity ranged from 16.81±0.75 to 108.02±4.82 Bqkg⁻¹.The total length of collected cockles (*Anadara granosa*) ranged from 2.10 to 4.30 cm whereas the total weight varied between 5 to 19 g. Expressing radiometric data in terms of dry weight, the concentration of Po-210 in the soft part of cockle ranged between 47.20±2.11 and 725.90±32.39 Bqkg⁻¹ with the mean value of 239.20±163.24 Bqkg⁻¹. In the present study, the Po-210 concentration in shrimp is comparable with the values reported for North Atlantic, Baltic Sea and global average value [43–46]. Much higher values were observed for different places of India which might be the impact of nuclear power station [47–50].

The concentration of heavy metals in analysed organisms is presented in Figures 3, 4, 5, 6 and 7. *Anadara granosa* (1.47 ±0.55 µgg⁻¹ dry wt) had the highest concentration of cadmium, followed by *Arius maculatus* (0.22±0.08 µgg⁻¹dry wt) and *Penaeus merguiensis* (0.17±0.10 µgg⁻¹dry wt). In living organisms, copper is considered to be highly toxic, and ranks among the more toxic heavy metals to fish. However, the mean copper concentration was highest in *Penaeus merguiensis* (21.57± 7.78 µg/g dry wt), followed by *Anadara granosa* (6.11±4.20 µgg⁻¹dry wt) and *Arius maculatus* (2.01± 0.57 µgg⁻¹dry wt). In this study, *Anadara granosa* had the zinc concentration of 93.31± 12.56 µgg⁻¹ dry wt, which was higher than that found in *Arius maculatus* (69.80±20.18 µgg⁻¹ dry wt) and *Penaeus merguiensis* (76.40±9.94 µgg⁻¹ dry wt). The higher concentration of lead in fish is probably because of the food source. Comparatively lower lead concentrations were observed in molluscs and crustaceans. Chromium is only moderately toxic to aquatic organisms and in the present study the filter feeding species *Anadara granosa* had the highest concentration of chromium (1.45± 1.14 μgg⁻¹ dry wt), flowed by *Arius maculatus* (0.79±0.26 μgg⁻¹ dry wt) and *Penaeus merguiensis* (0.53±0.04 μgg⁻¹ dry wt).

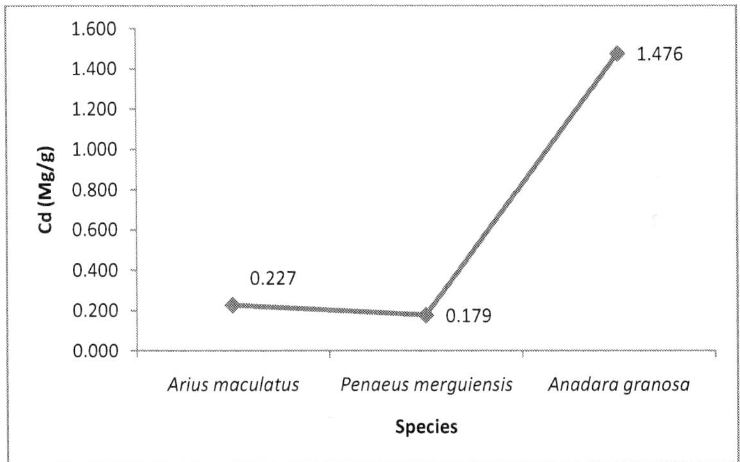

Figure 3. Mean concentration of Cd in marine organisms

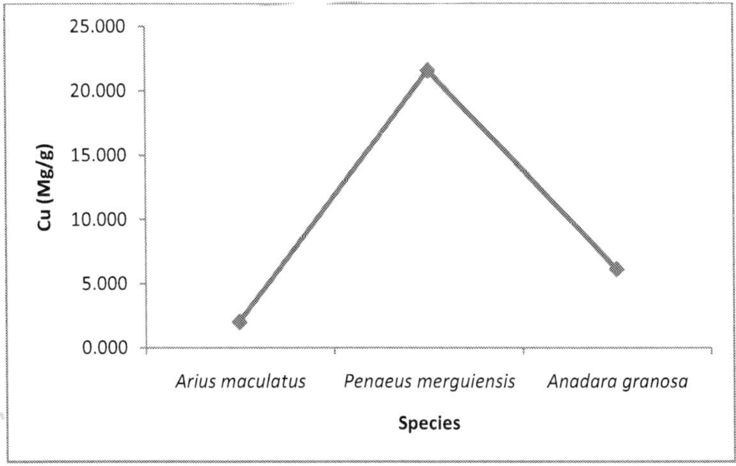

Figure 4. Mean concentration of Cu in marine organisms

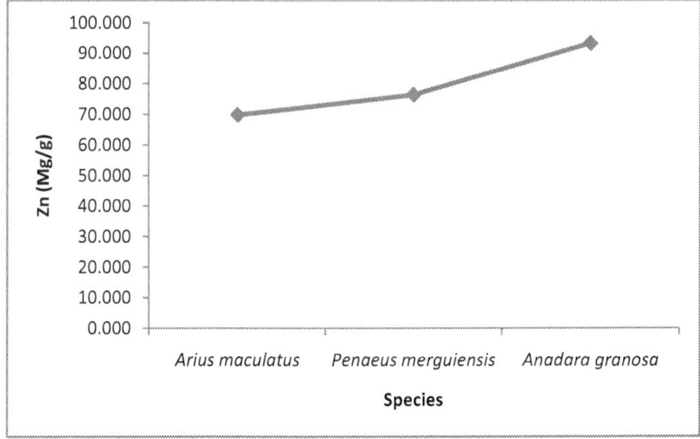

Figure 5. Mean concentration of Zn in marine organisms

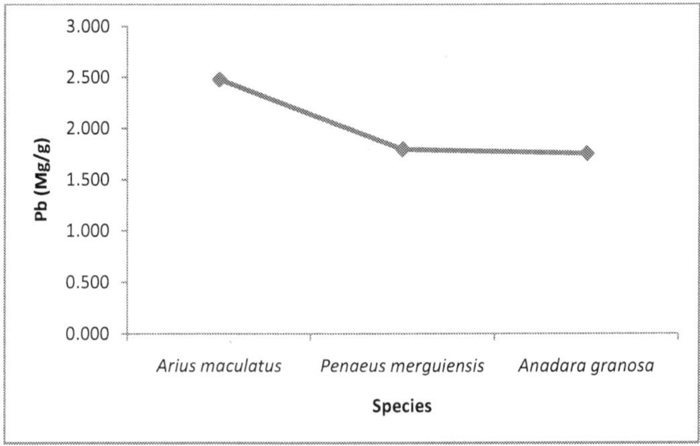

Figure 6. Mean concentration of Pb in marine organisms

Heavy metal concentrations measured in organisms are compared with the reported values of other places and the guideline (Table2). None of the species analysed in this study was found to contain cadmium concentration above the proposed permitted concentration and the values were within the range of other reported values. The calculated values of copper are within the range of previous studies and lower than the guidelines. Zinc concentration in fish is comparatively higher than other reported values but within the range in case of the values reported for Malaysia. However, this value is lower than the safety levels. For crustaceans, comparatively higher value is observed in case of Malaysia, but it is still lower than the safety guideline.

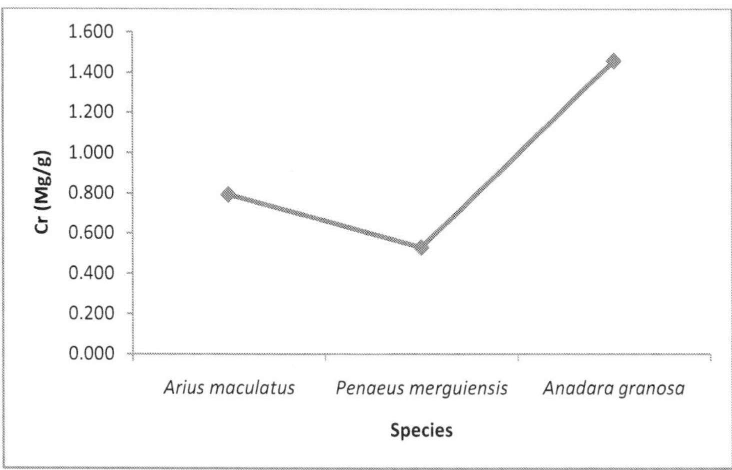

Figure 7. Mean concentration of Cr in marine organisms

On the other hand, in case of molluscs, the measured value was higher than the other reported values but lower than the safety limit which is stated for Malaysian population. Thus, zinc consumption of organisms from Kapar coastal area poses no threat to human. In case of fish and crustaceans, the lead concentrations are higher compared to other reported values, except the value reported for Malaysia. However, this value is lower than the safety limits. On the other hand, the measured value for molluscs in the present study demonstrated higher values than the literature, but the value did not cross the limit of safety which is declared for Malaysian population. However, our value is lower than the reported value of India. On the other hand, the chromium concentration in molluscs is also within the range of other places. Eisler [51] reported that chromium levels in the edible tissues of uncontaminated marine molluscs usually lie between 0.5 and 3.0 μgg^{-1}. Although there is no deleterious health effect of molluscs, the biologically available Cr(VI) is known to be carcinogenic to man and other elements [52].

Organism	Standard/Place	Cd	Cu	Zn	Pb	Cr	Reference
	USEPA limits	2	120	120*	4	8	[53][54]
	WHO	1	30	100	2	50	[55]
	Food standard of Malaysia	1	30	100	2		[56]
Fish	Caspian Sea	0.0032	1.65	20.656	0.0144	0.35	[26]
	China	0.01-0.04	0.06-0.16	2.39-4.49			[57]
	Bahrain	0.03			0.13		[58]

Organism	Standard/Place	Cd	Cu	Zn	Pb	Cr	Reference
	Yugoslovakia	0.01-0.84			0.02-1.7		[59]
	Italian coast	<0.01-0.02			<0.04-0.07		[60]
	Barent Sea	<0.01	0.6	5.6-7.8	<0.1		[61]
	Mumbai, India	0.02	0.31	8.36	0.08	0.78	[28]
	Afyonkarahisar	0.01	0.33-0.6	6.73-10.74	0.02		[62]
	China	0.004-0.021	0.228-1.89	16-130	0.177-0.289		[63]
	Malaysia, Pahang	0.15-0.47	0.13-0.77	1.69-6.76	0.00-1.14		[64]
	Malaysia, Terenganu		0.2	0.1	0.2		[65]
	Peninsular Malaysia	2.4	3.8	58.4			[66]
	Langkawe, Malaysia	0.9	0.01	49.39	1.1		[67]
	India	0.17	3.27	8.74	0.37	0.68	[23]
	Klang, Malaysia	0.14	1.21	41.84	1.50	0.55	Present study
	WHO	2	10	1000	2		[68]
	FAO	0.2	10	1000			[26]
	India	0.095	8.19	17.76	0.50	0.61	[23]
	Senegal	0.1	4.68	13.9	0.5		[69]
Crustacean	Cote	0.25	6.02	17.94			[70
	Cameroon	0.27	4.85	24.5			[71]
	Sabah, Malaysia	1.6-6.1	12.8-159		4.6-32		[72]
	Malaysia	0.2-49.0	32-99	68.19	1.68-54		[73]
	Malaysia	0.1-0.8	0.8-24	5.0-16.0	0.1-5.9		[74]
	Klang, Malaysia	0.10	12.06	42.41	1.00	0.29	Present study
	Food standard of Malaysia	1	30	100	2		[56]
	India	0.258	7.22	42.31	0.41	1.82	[23]
	Thailand	0.28	5.6	16.2	0.18		[75]
Mollusc	Australia	0.2	2	27.7	0.8		[76]
	Australia	2.09	0.73	42.7			[77]
	Malaysia	0.87	0.19	0.2	0.12	0.17	[78]
	Malaysia					0.24-0.41	[79]
	North America					0.1-9.6	[80
	Klang, Malaysia	0.82	3.39	51.63	0.97	0.80	Present study

Table 2. Comparison of heavy metal contents ($\mu g\ g^{-1}$ wet wt) in fish samples the present and guidelines/other studies

4. Conclusion

This study provided a general view of chemical contaminants in the Kapar power plant area where the chemicals were non-uniformly distributed with the groups of organisms. The differences in the level of contaminants in different groups of seafood could be due to the differences in metabolism and feeding pattern. In this study the highest contaminants accumulator species, *Anadara granosa*, is a filter feeder and feeds by straining suspended matter and food particles from the water. Additionally, this species has direct contact with sea sediments and this mode of life may contribute to higher levels of contaminants. On the other hand, the other two species are more mobile and consume food from the water column. Thus, they demonstrated lower contaminants accumulation pattern.

Acknowledgements

Financial assistance provided by the research project "FRGS/2/2014/STWN01/UKM/02/2"is gratefully acknowledged for this publication.

Author details

Lubna Alam*

Address all correspondence to: lubna_762120@yahoo.com

Institute for Environment and Development (LESTARI), Universiti Kebangsaan Malaysia, Selangor, Malaysia

References

[1] Cherry, R.D. and L.V. Shannon, The alpha radioactivity of marine organisms. Atomic Energy Review, 1974. 12: pp. 3–45.

[2] Alonso-Hernandez, C., et al., 137Cs and 210Po dose assessment from marine food in Cienfuegos Bay (Cuba). Journal of Environmental Radioactivity, 2002. 61(2): pp. 203–211.

[3] Moroz, B.B. and Y.D. Parfenov, Metabolism and biological effects of polonium-210. Atomic Energy Review, 1972. 10: pp. 175–232.

[4] Fisher, N.S., et al., Accumulation and cellular distribution of 241Am, 210Po, and 210Pb in two marine algae. Marine Ecology Progress Series, 1983. 11: pp. 233–237.

[5]   McDonald, P., G. Cook, and M.S. Baxter, Natural and artificial radioactivity in coastal regions of UK, in Radionuclides in the Study of Marine Processes, P.J. Kerrsshaw and W.D. S., Editors. 1991, Elsevier Applied Science: London and New York. pp. 329–339.

[6]   Stricht, E.V.D. and R. Kirchmann, Radioecology, radioactivity & ecosystems. Fortemps, Liege, 2001: pp. 219–303.

[7]   Stewart, G.M., S.W. Fowler, and N.S. Fisher, The bioaccumulation of U- and Thseries radionuclides in marine organisms, in Radioactivity in the Environment, M.S. Baxter, Editor 2008, Elsevier.

[8]   Štrok, M. and B. Smodiš, Levels of 210Po and 210Pb in fish and molluscs in Slovenia and the related dose assessment to the population. Chemosphere, 2011. 82(7): pp. 970–976.

[9]   Carvalho, F.P. and S.W. Fowler, A double-tracer technique to determine the relative importance of water and food as sources of polonium-210 to marine prawns and fish. Marine Ecology Progress Series, 1994. 103,: pp. 251–264.

[10] Cherry, R.D., M. Heyraud, and J.J.W. Higgo, Polonium-210: its relative enrichment in the hepatopancreas of marine invertebrates. Marine Ecology Progress Series 1983. 13: pp. 229–236.

[11] Skwarzec, B. and L. Falkowski, Accumulation of 210Po in Baltic invertebrates. Journal of Environmental Radioactivity, 1988. 8(2): pp. 99–109.

[12] Stepnowski, P. and B. Skwarzec, Tissue and subcellular distributions of ^{210}Po in the crustacean Saduria entomon inhabiting the southern Baltic Sea. Journal of Environmental Radioactivity, 2000. 49(2): pp. 195–199.

[13] Fulekar, M.H. and J.M. Dave, Diposal of flyash –an environmental problem. International Journal of Environmental Studies, 1986. 26: p. 191.

[14] Tariq, J., et al., Heavy metal concentrations in fish, shrimp, seaweed, sediment, and water from the Arabian Sea, Pakistan. Marine Pollution Bulletin, 1993. 26(11): pp. 644–647.

[15] Kalay, M., O. Aly, and M. Canil, Heavy metal concentrations in fish tissues from the northeast mediterranean sea. Bulletin of Environmental Contamination and Toxicology, 1999. 63: pp. 673–681.

[16] Rose, J., et al., Fish mercury distribution in Massachusetts, USA Lakes. Environmental Toxicology & Chemistry 1999. 18: pp. 1370–1379.

[17] Amaraneni, S.R., Distribution of pesticides, PAHs and heavy metals in prawn ponds near Kolleru lake wetland, India. Environment International, 2006. 32(3): pp. 294–302.

[18] Dural, M., M.Z.L. Göksu, and A.A. Özak, Investigation of heavy metal levels in economically important fish species captured from the Tuzla lagoon. Food Chemistry, 2007. 102(1): pp. 415–421.

[19] Teodorovic, N., et al., Metal pollution index: proposal for freshwater monitoring based on trace metal accumulation in fish. Tiscia, 2000. 32: pp. 55–60.

[20] Sharif, R., et al., Toxicological evaluation of some Malaysian locally processed raw food products. Food and Chemical Toxicology, 2008. 46(1): pp. 368–374.

[21] Vijayram, K. and P. Geralddine, Are the heavy metals cadmium and zinc regulated in freshwater prawns. Ecotoxicology and Environmental Safety, 1996. 34: pp. 180–183.

[22] Wu, J.P. and H.-C. Chen, Effects of cadmium and zinc on oxygen consumption, ammonium excretion, and osmoregulation of white shrimp (Litopenaeus vannamei). Chemosphere, 2004. 57(11): pp. 1591–1598.

[23] Sivaperumal, P., T.V. Sankar, and P.G. Viswanathan Nair, Heavy metal concentrations in fish, shellfish and fish products from internal markets of India vis-a-vis international standards. Food Chemistry, 2007. 102(3): pp. 612–620.

[24] Hamilton, M.A., et al., Determination and comparison of heavy metals in selected seafood, water, vegetation and sediments by inductively coupled plasma-optical emission spectrometry from an industrialized and pristine waterway in Southwest Louisiana. Microchemical Journal, 2008. 88(1): pp. 52–55.

[25] Connell, D.W. and G.J. Miller, eds. Chemistry and Ecotoxicology of Pollution. 1984, John Wiley and Sons: New York.

[26] Pourang, N. and J.H. Dennis, Distribution of trace elements in tissues of two shrimp species from the Persian Gulf and roles of metallothionein in their redistribution. Environment International, 2005. 31(3): pp. 325–341.

[27] Phillips, D.J.H., The use of biological indicator organisms to monitor trace metal pollution in marine and estuarine environments–a review. Environmental Pollution (1970), 1977. 13(4): pp. 281–317.

[28] Mishra, S., et al., Trace metals and organometals in selected marine species and preliminary risk assessment to human beings in Thane Creek area, Mumbai. Chemosphere, 2007. 69(6): pp. 972–978.

[29] IAEA-TECDOC-838, Sources of Radioactivity in the Marine Environment and their Relative Contributions to Overall Dose Assessment from Marine Radioactivity (MARDOS), in Final Report of a Co-ordinated Research Programme1995, IAEA: Vienna.

[30] Carvalho, F.P., 210Po in marine organisms: a wide range of natural radiation dose domains. Radiat. Prot. Dosim., 1988. 24: pp. 113–117.

[31] ANPA, Environmental Radioactivity Networks in Italy 1994–1997, 1999.: Rome.

[32] Al-Masri, M.S., et al., Transfer of K-40, U-238, Pb-210, and Po-210 from soil to plant in various locations in south of Syria. Journal of Environmental Radioactivity, 2008. 99(2): pp. 322–331.

[33] Connan, O., et al., Solid partitioning and solid-liquid distribution of 210Po and 210Pb in marine anoxic sediments: roads of Cherbourg at the northwestern France. Journal of Environmental Radioactivity, 2009. 100(10): pp. 905–913.

[34] Flynn, W.W., The determination of low levels of Polonium-210 in environmental materials. Analytica Chimica Acta, 1968. 43: pp. 221–227.

[35] Cherry, R.D., et al., Polonium-210 in teleost fish and in marine mammals: interfamily differences and possible association between polonium-210 and red muscle content. Journalof Environmental Radioactivity 1994. 24: pp. 273–291.

[36] Pentreath, R.J., et al. A preliminary assessment of some naturally-occurring radionuclides in marine organisms (including deep sea fish) and the absorbed dose resulting from them. in Proceedings of the third NEA seminar on marine radioecology. 1980. OECD, Paris.

[37] IAEA, Sediment distribution coef f icients and concentration factors for biota in the marine organisms., in Technical Report Series No. 2004, IAEA.

[38] Kauranen, P. and J.K. Miettinen. Polonium and radiolead in some aqueous ecosytems in Finland. in Proceedings of the Symposium on the Biology and Ecology of Polonium and radiolead,. 1970. Sutton, Survey.

[39] Folsom, T.R. and T.M. Beasley. Contributions from the alpha emitter Po-210 to the natural radiation environment of the marine organisms. in Proceedings of International Symposium on Radioactive Contamination of the Marine Environment, Seattle. 1973. IAEA, Vienna.

[40] Shannon, L.V., Marine alpha-radioactivity of Southern Africa. Polonium-210 and Lead-210. Investl. Rep. Div. Sea Fish. S. Africa, 1973. 100: pp. 1–34.

[41] Beasley, T.M., R.J. Eagle, and T.A.Q. Jokela, Polonium-210, Lead-210 and stable lead in marine organisms., in Summ. Rep. Hlth Saf. Lab.,1973, New York Fallout Programme, HASL-273, pp. 2–36.

[42] Cherry, R.D. and M. Heyraud, Evidence of high natural radiation doses in certain mid-water oceanic organisms. Science, 1982. 218: pp. 54–56

[43] Carvalho, F.P., Polonium (210Po) and lead (210Pb) in marine organisms and their transfer in marine food chains. Journal of Environmental Radioactivity, 2011. 102(5): pp. 462–472.

[44] Stepnowski, P. and B. Skwarzec, A comparison of 210Po accumulation in molluscs from the southern Baltic, the coast of Spitsbergen and Saselki Wielki Lake in Poland. Journal of Environmental Radioactivity 2000. 49: pp. 201–208.

[45] Aarkrog, A., et al., A comparison of doses from 137Cs and 210Po in marine food: A major international study. Journal of Environmental Radioactivity, 1997. 34(1): pp. 69–90.

[46] Khan, M.F. and S. Godwin Wesley, Assessment of health safety from ingestion of natural radionuclides in seafoods from a tropical coast, India. Marine Pollution Bulletin, 2011. 62(2): pp. 399–404.

[47] Khan, M. and S. Wesley, Tissue distribution of 210Po and 210Pb in select marine species of the coast of Kudankulam, southern coast of Gulf of Mannar, India. Environmental Monitoring and Assessment, 2011. 175(1): pp. 623–632.

[48] Suriyanarayanan, S., et al., Studies on the distribution of Po-210 and Pb-210 in the ecosystem of Point Calimere Coast (Palk Strait), India. Journal of Environmental Radioactivity, 2008. 99(4): pp. 766–771.

[49] Suriyanarayanan, S., et al., Assessment of 210Po and 210Pb in marine biota of the Mallipattinam ecosystem of Tamil Nadu, India. Journal of Environmental Radioactivity, 2010. 101(11): pp. 1007–1010.

[50] Hameed, S.M.M., et al. Study on the distribution of alpha-emitting radionuclide, polonium-210 in the ecosystem of Kattumavadi Coast (Palk Strait). in The national seminar on atomic energy, ecology and environment 2001. Thiruchirapalli, India.

[51] Eisler, R., Trace Metal Concentrations in Marine Organisms1981, New York: Pergamon Press.

[52] Phillips, D.J.H., et al., Trace metals of toxicological significance to man in Hong Kong seafood. Environmental Pollution Series B, Chemical and Physical, 1982. 3(1): pp. 27–45.

[53] Waquar, A., Levels of selected heavy metals in tuna fish. The Arabian Journal for Science and Engineering, 2006. 31: pp. 89–92.

[54] Broek van den, J.L., K.S. Gledhill, and D.G. Morgan, Heavy metal concentrations in the Mosquito Fish, Gambusia holbrooki, in the manly lagoon catchment, in UTS Freshwater Ecology Report2002, Department of Environmental Sciences, University of Technology: Sydney.

[55] WHO, Heavy metals-environmental aspects. Environment Health Criteria, 1989, World Health Organization: Geneva, Switzerland.

[56] Malaysian Food Act, 1983, MDC: Malaysia.

[57] Onsanit, S., et al., Trace elements in two marine fish cultured in fish cages in Fujian province, China. Environmental Pollution, 2010. 158(5): pp. 1334–1342.

[58] Madany, M.I., A.A.W. Abbas, and Z.A. Alawi, Trace metals concentrations in marine organism from the coastal areas of Baharin, Arabian Gulf. Water Air Soil Pollution, 1995. 91: pp. 233–248.

[59] Ozretic, B., et al., As, Cd, Pb, and Hg in benthic animals from the Kvarner-Rijeka Bay region, Yugoslavia. Marine Pollution Bulletin, 1990. 21(12): pp. 595–598.

[60] Giordano, R., et al., Heavy metals in mussels and fish from Italian coastal waters. Marine Pollution Bulletin, 1991. 22(1): pp. 10–14.

[61] Plotitsyna, N.F. and L.I. Kireeva, Contaminations in marine organisms from the Barents Sea (in Russian), in Material on PINRO research1995, Polar Research Institute of Marine Fisheries and Oceanography (PINRO): Murmansk, Russia, pp. 168–191.

[62] Fidan, A., et al., Determination of some heavy metal levels and oxidative status in <i>Carassius carassius</i> L., 1758 from Eber Lake. Environmental Monitoring and Assessment, 2008. 147(1): pp. 35–41.

[63] Chi, Q.-q., G.-w. Zhu, and A. Langdon, Bioaccumulation of heavy metals in fishes from Taihu Lake, China. Journal of Environmental Sciences, 2007. 19(12): pp. 1500–1504.

[64] Ahmad, A.K. and M.S. Othman, Heavy metal concentration in sediment and fishes from Lake Chini, Pahang, Malaysia. Journal of Biological Sciences, 2010. 10: pp. 93–100.

[65] Kamaruzzaman, Y., M.C. Ong, and K.C.A. Jalal, Levels of Copper, Zinc and Lead in fishes of Mengabang Telipot River, Terengganu, Malaysia. Journal of Biological Sciences, 2008. 8: pp. 1181–1186.

[66] Yap, C.K., A. Ismail, and P.K. Chiu, Concentrations of Cd, Cu and Zn in the fish Tilipia, Oreochromis mossambicus caught from Kelana Jaya pond. Asian Journal of Water Environment and Pollution, 2005. 2: pp. 65–70.

[67] Irwandi, J. and O. Farida, Mineral and heavy metal contents of marine fin fish in Langkawi Island, Malaysia. International Food Research Journal, 2009. 16: pp. 105–112.

[68] Biney, C.A. and E. Ameyibor, Trace metal concentrations in the pink shrimp Penaeus notialis from the coast of Ghana. Water Air Soil Pollution, 1992. 63: pp. 273–279.

[69] Ba, D., Analyse de contaminats chez les organisms Marins d'Importance commercial dans les eaux cotieres du Senegal, 1988, Rapport Atelier WACAF/2 Accra: Ghana.

[70] Metongo, B., Metaux Lourds des organisms marins cote d'Ivoire, 1988, Rapport Atelier WACAF/2, Accra: Ghana.

[71] Mbome, L., Heavy metals in marine organisms form Cameroon, 1988, WACAF/2 Workshop Raport, Accra, Ghana.

[72] Awaluddin, A., M. Mokhtar, and S. Sharif, Accumulation of heavy metals in tiger prawns (Penaeus monodon). Sains Malaysiana, 1992. 21: pp. 103–120.

[73] Patimah, I. and A.T. Dainal, Accumulation of heavy metals in Penaeus monodon in Malaysia, in nternational Conference on Fisheries and the Environment. Beyond 20001993: UPM, Serdang, Malaysia.

[74] Ismail, A., Heavy metal concentrations in sediments off Bintulu, Malaysia. Marine Pollution Bulletin, 1993. 26(12): pp. 706–707.

[75] Huschenbeth, E. and U. Harms, On the accumulation of organochlorine pesticides, PCB and certain heavy metals in fish and shellfish from Thai coastal and inland waters. Arch. Fischereiwiss, 1975. 25: pp. 109–122.

[76] Mackay, N.J., et al., Heavy metals in cultivated oysters (Crassostrea commercialis = Saccostrea cucullata) from the estuaries of New South Wales. Australian Journal of Marine and Freshwater Research, 1975. 26: pp. 31–46.

[77] Phillips, D.J.H., The common mussel <i>Mytilus edulis</i> as an indicator of pollution by zinc, cadmium, lead and copper. II. Relationship of metals in the mussel to those discharged by industry. Marine Biology, 1976. 38(1): pp. 71–80.

[78] Abbas Alkarkhi, F.M., N. Ismail, and A.M. Easa, Assessment of arsenic and heavy metal contents in cockles (Anadara granosa) using multivariate statistical techniques. Journal of Hazardous Materials, 2008. 150(3): pp. 783–789.

[79] Mat, I., Arsenic and trace metals in commercially important bivalves, Anadara gradosa and Paphia unduluta. Bulletin of Environmental Contamination and Toxicology, 1994. 52: pp. 833–839.

[80] Hall, A.R., E.G. Zook, and G.M. Meaburn, National Marine Fisheries Service Survey oftrace elements in the fishery resources, in U S Dep Commerce NOAA Tech. Rept1978, U S Dep Commerce.

Chapter 12

Atmospheric Pollution Causes Deterioration of Sweeteners of Treats and Decreases Competitiveness in the Food Industry of Coastal Baja California, Mexico

César Sánchez Ocampo, Judith Marisela Paz Delgadillo, Gustavo López Badilla and Salvador Baltazar Murrieta

Additional information is available at the end of the chapter

http://dx.doi.org/10.5772/61993

Abstract

The use of pigments and sweeteners in the food industry has been of great importance over the last thirty years, as an aid to obtaining better quality products of certain types of food that require specialized methods of production, such as sweet breads. Since pigments were first used they have increased the competitiveness of the regional, national and international bakeries and treats companies found on the coast of Baja California, in northwest Mexico. In this region of the country, many people consume a considerable quantity of sweet bread, principally in winter. The pigments are made using specialized methods and have qualities relevant to the special conditions required in the storage and manufacturing processes, as well as to the types of food and the packaging associated with them. This is done in order to achieve a favorable appearance, and to create an be aactive product fortoforonsumers. Any bread with defective sweetener or pigmentation levels must be rejected and returned to the manufacturing process or be offered as food to pigs at a much lower market price resulting in economic losses to the company and a decrease in competitiveness. A specialized analysis of the subject was undertaken between 2012 and 2014 with the use of Scanning Electron Microscopy (SEM).

Keywords: Food industry, competitiveness, air pollution, pigments, food industry, SEM analysis

1. Introduction

Most pollution at sea, in coastal areas and around beaches etc. takes the form of oil spills. This has led to several natural disasters resulting in the death of fish, a loss of revenue from tourism

in some areas, and economic losses in the food industries located in coastal zones. The contamination affecting the food industry in these areas causes the deterioration of food products, is injurious effects to health and the damaged food products are consigned to wastage. The visual appearance of the final product represents one of the key factors of food quality and this can be affected and altered by atmospheric pollution generated by contaminantspollutan sulfur in Tijuana city, located on the coast of Baja California. In addition to these chemical agents that modify the nutritional properties of the pigments that are added to food products in this marine zone of the Mexican Republic, variations in climatic factors such as temperature and humidity are also considered in this study. In the major food companies that manufacture sweetened breads, pastries and sweets in this Tijuana cihat fabricate lower quality sweetened breads and foods are occasionally sometimes made as a result of the deterioration of certain types of artificial pigments that are very sensitive to light, heat, and mainly acidity [1]. An investigation was carried out to discover the causes of the loss of quality in aspects of the production process such as candy coating.

1.1. Manufacturing processes and quality factors in the industry

Manufacturing is defined as a set of organized and programmed activities for the transformation of materials, objects or services into articles or services useful to society. The manufacturing engineer utilizes industrial processes as the mechanism for the transformation of materials into useful items for society [2]. It is also seen as the structure and organization of actions that allow a system to accomplish a certain task. In the vast majority of cases, in order to obtain a particular product it will be necessary to perform a lot of individual operations — depending on the scale of observation — including operations from the extraction of the natural resources necessary to sell the product to those involving a job with a particular machine or tool. A manufacturing industry is present under all types of economic systems. In a capitalist economy, manufacturing is usually directed toward the mass production of products for sale to consumers at a profit. In a collectivist economy, manufacturing is directed by a state agency. In modern economies, manufacturing runs under some degree of government regulation [3]. Modern manufacturing includes all the intermediate processes required in the production and integration of the components of a finished product. The industrial sector is closely related to the engineering and industrial design sectors. The process may be manual or may employ the use of machines. For higher volumes of production, the technique of the division of labor is utilized, whereby each worker performs only a small portion of the whole task. This specialization saves time and money and results in a greater speed of production. Although craft production has been part of humanity for a very long time (since the Middle Ages), it is generally considered that modern manufacturing emerged around 1780 with the advent of the British Industrial Revolution, expanding thereafter throughout continental Europe, then to North America and eventually to the whole world [4]. Manufacturing has become a huge part of the economy of the modern world. According to some economists, manufacturing is the sector that creates wealth in an economy while the service sector tends to be a consumer of wealth. The process of quality control involves applying quality manufacturing processes to a product. Techniques used are those such as Statistical Process Control (SPC) applied to samples of the product. By closely controlling the process, the risk of the product being of poor quality is greatly lessened. This technique has the advantage of lower losses through preventing the occurrence of defective product and the associated generation of higher costs. The quality control process operates under the supervision of the quality control department. For some time now, there have been a lot of manufacturing systems based on quality assurance rules [5]. The transition to the serving of a customer in base in organizations, has led to fundamental changes in manufacturing practices. For this reason, the changes have been especially evident in areas such as product design, human resource management and supplier relations. For example, in product design, activities are now integrated into narrow marketing operations, engineering and manufacturing. The practice focuses on empowering employees to collect and analyze information, taking crucial operational decisions and accepting responsibility for continuous improvement, thus spreading the responsibility of the quality control department across in the enre manufacturing processes. The providers have become partners in product design and manufacturing efforts. In this research project, which was developed in the food industry, the manufacturing process is evaluated in the area where the sweetener is applied to the final product of a sweetened bread. The operations in this part of the process are performed using specialized activities, which are regulated by certified companies and government authorities. The proper adherence to the guidelines for the use of bread sweetener results in the presentation of an optimal product to the consumer [6].

1.2. Engineering and competitiveness

Creativity is one of the most valued and demanded qualities by companies in the field of industrial engineering, due largely to the competitiveness of the market. The application of creative thinking to the design process of an engineering project gives added value in each and every one of its phases [7]. To carry out this fusion of creativity and engineering, we need to redefine traditional project methodology. In teaching the concept of creative engineering, the transition of the systems must be taken into account, along with other important factors related to industrial engineering and competitiveness. This approach is a suitable one to promote the genduction of thee thsired quality and quantity of goods in the time required, usi the generation of new ideas, methods and processes available to the industry with a special emphasis on engineering and competitiveness. Product engineering serves as a model to help us better understand about the to use the most creative and innovative engineering methods along with traditional values. Competitiveness is a clear example of the synergy produced between engineering and creativity and is apparent through the project results [8].

1.3. Internal and external factors in the competitiveness of the industry

The factors that define the degree of competitiveness of an industry and in engineering services in manufacturing can be classified as internal and external factors. The internal factors are those that directly depend on the organization and on which the organization can act [9], while external factors are those that do not depend on the organization. Internal factors can be grouped into three areas: quality, efficiency and innovation as mentioned in Table 1.

FACTOR	PRIMARY CHARACTERISTIC	SECONDARY CHARACTERISTIC
Quality	Is a fairly major factor at the time of purchase. Some regional markets demand that products and services have a hallmark of various types of ISO standards, and the rigorous certification of food and health, among other guarantees.	Both the security and speed of delivery, along with after-sales service can be considered as part of the service quality of customer attendance.
Efficiency	It is known that for greater efficiency, the use of appropriate management techniques and technologies are required, as with industrial automation, process automation, and automation of information being the principal operations.	With the invasion of Japanese products into markets of the United States and Europe, and the improvement in the manufacturing processes of many industries, the internal factors of efficiency and quality are presented as the basis of competitiveness.
Innovation	During the 90s, with the appearance on the market of increasingly innovative products, a third factor became important in defining competitiveness. This is innovation and it refers to products or services that meet new needs (many of them introduced by the new product and service).	Product innovation reduces the life cycle of products on the market. An innovative product with an acceptable quality and price will, in general, result in the removal of the predecessor product from the market.

Table 1. Internal factors affecting the competitiveness in the manufacturing engineering.

FACTOR	PRIMARY CHARACTERISTIC	SECONDARY CHARACTERISTIC
EXTERNAL		
Legal framework	The legal framework refers to laws giving security to investment and establishes the rules for industries and service companies.	The legal framework is applied in all areas of industry involved in manufacturing engineering.
Foreign trade policy	Tariffs are the main mechanism of action in a foreign trade policy.	The tariffs vary according to the business relationship, where each country generates activity fees according to their economy.
Tax incentives	The purpose of tax incentives is to promote the development of activities and specific regions, through mechanisms such as the import tax refund to exporters, franchises, subsidies, reduced tax rates, total or partial exemption to a certain tax, and a temporary increase in asset depreciation rates.	The principal tax incentives where you can get benefits for people who file taxes in Mexico are as follows: tax on income (ISR), interim payments, donation of goods, leasing of real estate, companies that hire the disabled or the elderly, public works contracts and payment of educational tuition from elementary to high school level.
Natural phenomena	Natural phenomena that are unpredictable are taken into account in the analysis of	Another negative effect is the economic losses to property and company property damaged

FACTOR	PRIMARY CHARACTERISTIC	SECONDARY CHARACTERISTIC
EXTERNAL		
	competitiveness, because whenever a natural disaster occurs human loss or damage to the workers can occur and this represents a negative effect on the economy of companies.	by natural phenomena such as earthquakes, hurricanes and climate change.
Infrastructure	Infrastructure is essential for export and to meet the domestic market. This is made up of ports, airports, roads, telephone lines, communication networks, water networks, electricity grids, etc. In fact, it is a prerequisite for being competitive in a market economy.	The government sector should always bear in mind the modernization of infrastructure in order to maintain existing national and foreign investment and to attract new investment.
Financial investment	Economic investment also plays an important factor in competitiveness. This can mean modernization of the apparatus of production, and cost reduction services as the principal factors.	Economic investment should be analyzed in great detail to prevent the generation of economic losses due to poor planning.
International Environmental Policy	Environmental policies are a major factor in the competitiveness of enterprises in developing their activities in accordance with the international environmental standards in each region.	Recent international environmental policies establish new standards of consumption and production.
Monetary Policy and Inflation	International economic crises are a relevant factor, since they result in the economic insecurity of investments in emerging markets.	To be more competitive, there needs to be favorable external conditions that allow local products and services to compete with imported ones so far as to impose on international markets.
International Economic Crisis	International crises also affect the economy of domestic and foreign companies where depending of the consumption are presented the low and high levels.	Crises in each region are part of the negative effects on a country area, indicated by the type of consumption of domestic companies.
Labor and Union Conditions	Working conditions and trade unions in each country and region of the world vary depending on the culture of the inhabitants.	Unions can have an impact on the economy of domestic and foreign companies due to their political influence.

Table 2. External factors affecting competitiveness in the manufacturing engineering.

Constant competition and the drive for excellence in new products make the integration of the manufacturing process and automatized systems a necessity. In the process of product innovation, specialized simulation software allowing for repeated testing has generated optimal systems for the development of parts and new products with specialized parameters.

Certainly, the development of many systems mentioned in this section relies on other areas of knowledge. For example, expert systems require the knowledge and experience of specialists in the area of study in question. Engineering support systems require the participation of engineers and experts in their conception and development; in many cases these systems are developed by highly qualified personnel and both undergraduate systems and computer systems [10].

1.4. Relationship between engineering, competitiveness and educational activities

Working on such projects can help to engage the different learning styles of young people and educators have an obligation to stimulate student curiosity in the operation and efficient use of matter and energy. Therefore, physics should be shown as a science that is essential to the scientific and technological progress of mankind and thus contributes to improving the quality of life.

In an analysis of a study into the relationship between engineering and competitiveness, it is stated: "When I read, I'm imagining what is described in the book I'm studying" [11]. This makes me think that the student must remain permanently connected to the physical subjects presented by the teacher, and what better way to stimulate them than with a didactic prototype as shown in Figure 1. Vicarious learning is something that occurs through the observation of others. People and animals can learn just by watching another person or animal, and this behavior challenges the idea that cognitive factors are not necessary to explain learning. If people learn by observing, then we can focus their attention by creating images that to promote remembering, analyzing and decision making to affect learning as is presented in Figure 1. The skills and attitudes of competitiveness are developed when working in an environment in which a suitable nade and orderly effort coexist. The trust between students creates an interaction that motivates them towards project the comof projects, producing in them a proactive change. Engineering activities, especially ones related to topics of physics and chemistry, are in essence a practical science related to the use of many concepts and formulas. There is no single constructivist theory of learning. Most theories in cognitive science include some form of constructivism, as they believe that individuals construct their own cognitive structures as they interpret their experiences in particular situations. The group dynamics used to better understand the relationship between engineering and competitiveness, refers to the set of forces —movement, action, change, interaction, reaction or conversion — that operates in a particular group for the duration of their relationship, giving a style or personality to that group and making them behave in that particular way. The reciprocal interactions of these forces and their result on a given group are its dynamics [11]. Finally, communication skills and student attitudes will be developed when working in an environment of appropriate adequate planning rderly effort.

1.5. Competitiveness in the food industry

The study of the influence of competitiveness in the business world is a matter of ongoing discussion, especially in the food sector where there is a huge number of companies worldwide — a significant percentage of the total of which are manufacturers of sweetened breads and

Figure 1. Educational activities showing the use of engineering tools in industry (2015).

treats. Some companies are internationally recognized, but those with smaller economic resources are in a highly competitive part of the market, and to some experts this is considered to be the limiting factor to growth and competitiveness [8]. The findings of this study include some interesting trends, although they lack certain stan information and more accurate statistical data. The industries of this type, is related with sustainable competitive advantages and managed by companies that are analyzed in the study as vectors of competitiveness. The three vectors considered are: technological innovation, internationalization and funding.

1.6. Food industry

Food processing includes the methods and techniques used to harvest raw materials using clean components and then transform them into commercial food products for human consumption componto produce commercial of different ways in which food can be produced. The manufacture of one foodstuff could take days depending on customer specifications and conditions for the design capacity of the industrial plant in question [4]. The production line is used when the market size is large and where is a range within a selection of products. A large number of the same goods to be produced make it possible to use simple methods that require a complex manufacturing process. All methods include an estimation ng of the number of customers who want to buy the food product in question. Mass production is used when there is a mass market for a large number of identical products, such as chocolate bars, ready meals and canned foods. The product passes from one stage of production to another along a production line. All components of the product are specified by the customer, who chooses them based on their structure, taste, smell and other character-istics of the food. An extensive network of global transportation is required by the food industry to market its numerous products. These include suppliers, manufacturers, ware-houses, retailers and consumers. There are also companies that add vitamins, minerals, and other necessary requirements to the product to replace cover those lost during manufacturing process. The wholesale markets for fresh produce have tended to decline in importance in Latin America and some Asian countries as a result of the expansion of supermarkets that buy directly from farmers through preferred providers, instead of going through markets. The constant and uninterrupted flow of goods from distribution centers to store locations is a critical link in the operation of the food industry. Distribution centers operate more efficiently; performance can be increased, costs reduced and advantage taken of better work if the right steps are taken when a material handling system is set up in a warehouse. With large parts of the population concentrated in urban areas, food shopping is growing and is a basic human requirement according to the nutritional process. This is a relatively recent development that has taken place mainly in the last 50 years [4]. The supermarket is the defining element of sale in the food industry, where tens of thousands of products are gathered in one place and are in continuous year-round supply. Restaurants, cafes, bakeries and mobile trucks are also ways to reach consumers of food. Food prepare another area where the changes in recent decades have been spectacular. The food industry sells fresh products and the bakery industry in particular needs a large amount of raw materials and ingredients to make the products and get them to their consumers. One type of important ingredient is color pigment — used to for better improve thrion and appearance of the product [12].

1.7. Stages in the food industry

The food industry uses a variety of processes in the manufacture of food products with advanced technologies utilized in those production processes. This is in order to create the optimum conditions for customer satisfaction, hygiene and to lower costs in all types of food processing technologies such as those used in sweetened bread and treats. The main stages in the manufacture of foods are [13]:

a. Use of technologies for the monitoring and storage of food and raw material for processing.

b. Evaluation of the types of ingredients for food preservation in the company and outside of it.

c. The use of both simple and complex methods in the processing of food according to the specific conditions requested by the client.

d. Monitoring of the quality of the final product using specialized methods and techniques to suit the needs of consumers.

e. Storage of finished products in specific areas with high levels of security.

f. Analysis of food waste or chemicals used in the process for the manufacture of foodstuffs.

Heavy industry, such as the metals industry has been a factor in the pollution of the coastal zones and was studied to better understand the effects of this industry and the damage it causes to the company being evaluated.

1.8. Sweeteners and Food

Sweeteners are additives such as coloring pigments or substances that impart color when they are added to foods or beverages. They come in many forms, consisting of liquids, powders, gels and pastes. Sweeteners are used in the commercial production of food and domestic cooking. For your safety and general availability, the dyes are also used in a variety of non-food applications including cosmetics, pharmaceuticals, craft projects and medical devices4. People associate certain colors with certain flavors, and the color of food can influence the perceived flavor in many respects. Sometimes the goal is to simulate a flavor that is perceived by consumers as natural, such as adding red coloring to candied cherries (which would otherwise be light brown), but sometimes it is for the purpose of cosmetic appearance, such as in Heinz ketchup green tomato sauce, launched in 1999. Flavor additives are used in foods for many reasons, including [14]:

a. To offset color loss due to the exposure to light, air, extremes of temperature, moisture and storage conditions.

b. For the correction of natural variations in flavor, to enhance colors that occur naturally, and to provide a variety of colors to colorless substances and foods to give them a better appearance.

Flavor and color additives are recognized as an important part of many of the foods that we eat. The flavorings and sweeteners used in the food industry are tested for safety by various agencies around the world, and sometimes the different bodies have different views on food safety with respect to the color additives used [8]. In the United States, the regulatory body is the Food and Drug Administration (FDA) and they approve which additives can be used in foods, drugs and cosmetics — as synthetic food colors do not exist in nature   while in the European Union, all additives, both synthetic and natural, have to be approved for use in food applications. The best known food flavorings and colorings are turmeric and lutein. Most countries have their own regulations and list of food dyes that are approved for use in various applications, including that for the maximum daily intake [15]. The certified flavors and colors are synthetically produced and widely used since they impart a strong uniform color, and are less expensive and more easily mixed to create a variety of hues and flavors. Certified flavors and colors in foods generally do not add undesirable flavors to those foods, while pigments derived from natural sources such as vegetables, minerals or animals are exempt from certification. Additives derived from natural flavors and colors are generally more expensive than certified synthetic colors and may add undesirable flavors to foods. Prime examples of exempt colors are annatto extract, beet, caramel, beta-carotene and grape skin extract. The natural colors of food take on a variety of shades and hues [6]. Around 80% of what we taste is detected as coming from the sense of smell. The trigeminal nerve is responsible for detecting irritants that enter through the mouth or throat and can sometimes determine flavor. The taste of food is not just a concern for cooks, but is also a scientific challenge for the food industry. Flavorings and seasonings, whether natural (spices) or artificial (E numbers), are used to highlight or modify flavors. An increasing number of natural food colors are being produced commercially, in part due to consumer concerns surrounding synthetic dyes. Examples include candies made with caramelized sugar, annatto – a reddish-orange dye from the annatto seed, chlorophyllin – a green dye made from algae, chlorella cochineal –, red dye derived from the insect *Coccus dactylopius*, Betanin extracted from beet, curcuminoids from turmeric, carotenoids from saffron, paprika, lycopene, elderberry juice, a green dye from *Pandanus amaryllifolius*, and a blue food coloring from the butterfly pea *Clitoria ternatea*. These residues often remain in the finished product, but it is not a legal requirement for this to be declared on the product packaging as the residues are part of a group of substances known as carry-over ingredients. Because of their organic nature, natural food colors can sometimes cause anaphylactic shock and allergic reactions in susceptible individuals. Coloring agents known to be potential hazards include annatto and cochineal carmine [17].

1.9. Effects of air pollution on the properties of pigments

The presence of contaminants in the atmosphere is of great concern due to their varied effects according to humidity and temperature. For example, in some seasons contaminants can cause the deterioration of foodstuff and thus change the properties of the pigments used in them [18]. These properties can result in a loss of appearance in the food, even when packaged and packed using a bleaching process. Despite the enforcement of quality standards in the environments in which of foodstuffs are manufactured, and even in systems with robust quality control measures in place, certain types of pollutants were present within the industry where the study was conducted. There are various types of sources that emit these pollutants into the atmosphere, mainly located in the coastal city of Tijuana, which suffers from the emission of sulfides from cars and the thermoelectrical plant that supplies Tijuana with electricity. The highest levels (at 70% Relative Humidity (RH) and 25 °C), that are presented r the majority of the year, cn change the properties of pigments used in the manufacture of sweetened breads and treats so that sometimes the pigment does not properly adhere to the surface of the bread. Sometimes the pigment drops off, or in the process of marketing and transport the external structure of the food product is modified resulting in a poor appearance and the food in question being returned to the company that manufactured it, generating economic losses as shown in Figures 1a and 1b. As shown in figure below, the dark sections are parts of pigment which are not properly attached to food products resulting in economic losses as it becomes necessary to market them at a lower price to the consumer. This decreased the competitiveness of the industry in the period evaluated, and is presented in Figure 2. In the zones with a dark color, the figure shows the deterioration of the sweeteners in bread, where a layer was peeled off the bread, resulting in a damaged appearance and a lower quality product. Specialized persons and managers in the food industry where this phenomenon was evaluated, considered that it was necessary put products made with such defects into the trash resulting in economic losses. This analysis was represented of the study evaluated in the company mentioned.

1.10. Petroleum as a pollutant in seawater

In recent decades, oil has played a major part in the global economy, contributing to socio-economic and technological development in many different cultures and countries. Oil only occurs in certain parts of the planet and excavations and drillings are restricted to these areas.

tured foods. The difference between a food in good condition and a food with unbound pigment sections is shown in the results obtained.

3.1. Analysis of climatic factors

Climatic factors are important in the production and manufacturing processes of the food industry, where high or low levels of relative humidity and temperature can result in a negative effect on products as the sweeteners can be added at some times but not at others. This is a concern to specialized people and managers, who consider that the products manufactured need to be put in low quality sections or may even need to be put in the trash. These low quality products are considered to be defective and may require some extra work to rectify problems with them. This in turn makes it necessary to pay workers overtime and this needs to come from the budget of evaluated company. Also generated space in some areas of the food industry analyzed and this represents an economic spending. On the other hand, we also see a loss of time in the delivery process to the customer, which can generate costs not anticipated at the inception of product operations, in addition to potential legal battles and lost customers. Figures 3 and 4 show information representing the correlation of the principal climatic factors of RH and temperature, with the defective products manufactured in the months of July 2014 (Figure 3) and December (Figure 4). This caused a negative effect on the financial operations of the food industry evaluated. This company is located in Tijuana, Baja California, in northwest Mexico, near of the border with the United States. This area of Mexico is a coastal zone, with changes in the levels of the climatic factors mentioned, principally in the periods of the presence of the phenomenon called the Santa Ana Winds at the start of the spring period (the last days of March and the first days of April) and in the autumn (the last days of October and the first days of November) of each year. In Figure 3, the analysis of July of 4 showed the high quantity of defective products, with 320 as the maximum and values of relative humidity and temperature around 75% and 35 °C respectively. The minimum number of defective products was 12, with values of the climatic factors mentioned above of 15% and 35 °C. Figure 4 represents the analysis of December 2014 where the maximum value for the quantity of defective products was 320 with values of 65% and 35 °C relative humidity and temperature respectively. This was causing the observed effect of the occasional detachment of the sweeteners in the bread. To improve the manufacturing processes of the sweetened bread it was necessary to aveexert control over the climatic factors through the use of an automatic system proposed and designed by the researchers of this study. This automatic system works through the detection of both low and high humidity and temperature together, to trigger the operation of fans and an automated irrigation system within the manufacturing area without affecting the manufactured products.

The automatic system designed to control the climatic factors is discussed below. The system was designed to consist of simple steps and the control device had to have a cost of less than 100 pesos, which is already assessed in companies with improving productivity performance that keep companies from installation cost. The automatic system consists of several simple steps shown below in a block diagram and then explained theoretically (Figure 5):

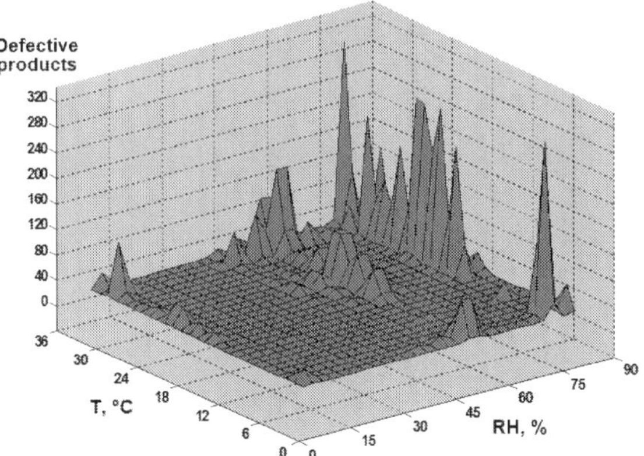

Figure 3. Correlation of climatic factors and defective products in the food industry of Tijuana, Baja California, México (July, 2014).

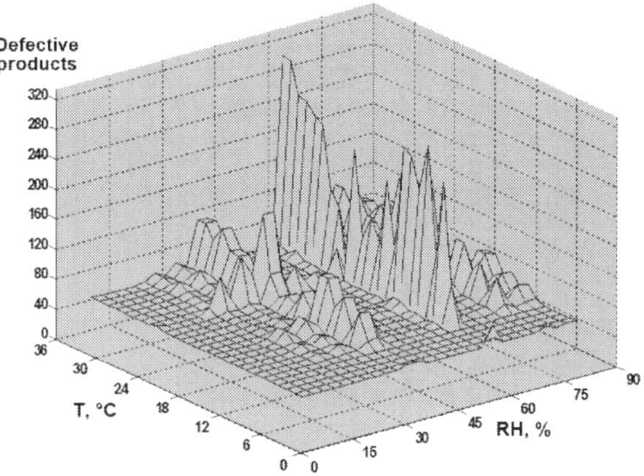

Figure 4. Correlation of climatic factors and defective products in the food industry of Tijuana, Baja California, México (December, 2014).

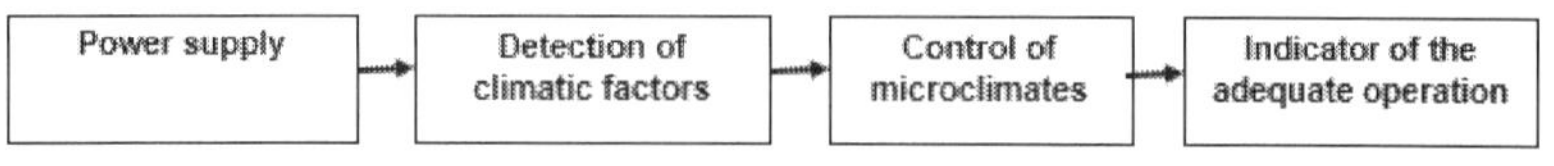

Figure 5. The automatic system designed to control the indoor climatic factors of the food industry under evaluation.

3.2. Correlation of air pollution with the deterioration of pigments

The properties of pigments vary according to the variations in relative humidity and temperature levels as indicated by color in Figures 6 and 7. Red indicates high levels of color loss through fading, followed by yellow, then the light blue and ending with dark blue. All levels are derived from the correlation of the values of relative humidity and temperature. The coloring pigments are also shown along with the ranges in which humidity can have a negative effect and therefore when to take the necessary measures. Various pigments such as caramel, annatto, chlorophyllin, and saffron used by the company where the study was conducted were analyzed. In the winter season, more risks occurred than in the summer but this does not mean that the pigments were fading quickly, but a simulation indicated the possibility that this will occur. In both figures, the colors define the defects in according to the events occurred and the possibility to be considered as a defective products. In Figure 6, blue appeared the most times, indicating that the manufacturing processes was adequate, with some parts of the figure showing green and yellow colors representing products with some defects. Green represents a quality of product that can be considered as adequate while a yellow color indicates a product outside of quality tolerances and considered to be defective. Red represents a low level of quality. Both analyses were made in 2014.

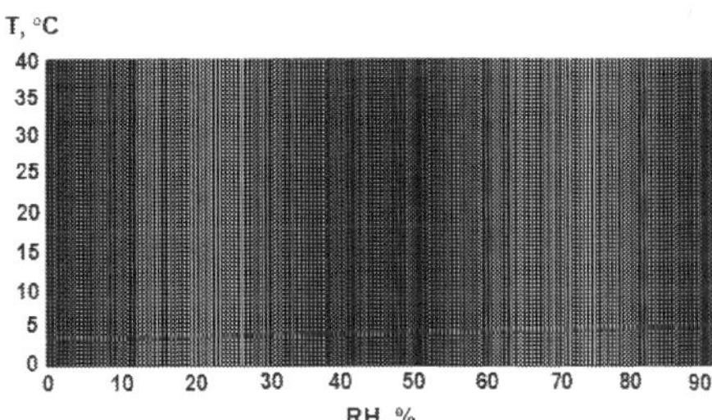

Figure 6. Correlation of RH and temperature with coloration of pigments indices of achiote in the summer season in 2014.

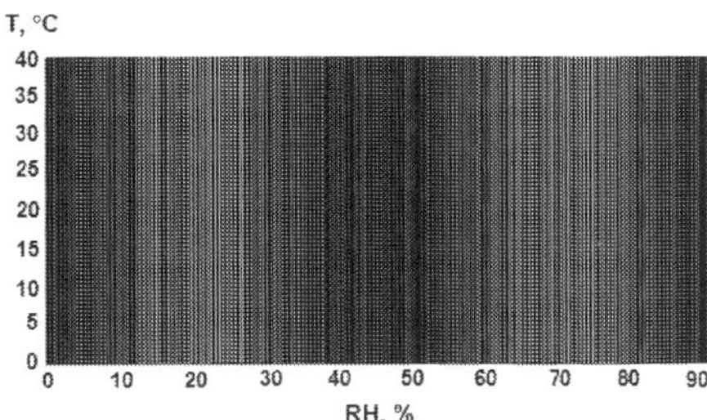

Figure 7. Correlation of RH and temperature with the coloration of pigments indices of achiote in the winter 2014.

Figure 7 shows the same process as Figure 6 but with fewer blue zones and more green, yellow and red zones of the same size.

3.3. Analysis of productivity correlated with atmospheric corrosion

The productivity of a company is very important in the fight to keep it competitive in the world market and to maximize profits. This represents the adequate index of the industrial operations in the manufacturing areas. The productivity yield is affected by with climatic and atmospheric parameters, where air pollution has a negative effect on the adhesion of the sweeteners used in the bread. Figure 8 shows the correlation of climatic factors related to the pollution agents in the indoor area of the food industry, and the quantity of products manufactured by day. The chemical agents were evaluated using colors to represent the level of negative effect on the yield of sweetened bread products manufactured. This analysis was made in the winter when the index of RHrela higher as this was the parameter with the most significant effect on the deterioration of the bread. The air pollutants with the most significant effects were as sulfur dioxide or sulfur derivatives followed by ion chlorides, nitrogen oxides and ozone, that shows the graphic with the colors. The sulfurs and ion chlorides were present in the low productivity zones of the figure. In change where were presented the levels of nitrogen oxides and ozone, productivity was high. Relative humidity Handhith the most significant effect on productivity when they were in the ranges of 50% to 75% and 20 °C to 35 °C respectively.

3.4. SEM evaluation

SEM techniques were used to make a microscopic evaluation of the defective and non-defective breads in order to compare the manufacturing processes using the automatic system with that where the climatic factors were not controlled. Figure 9 represents two daily evaluations of the deterioration of the sweetened bread on a summer day (July) and a winter day (December), where the relative humidity was higher than 80% and the temperature between 20 °C and 30 °C as is common in winter in Mexican cities. Figure 9 shows two examples of pigments in manufactured foods with good color in section (a) and spoiled food color level, causing discoloration and (b) food products deteriorated in different days of the year.

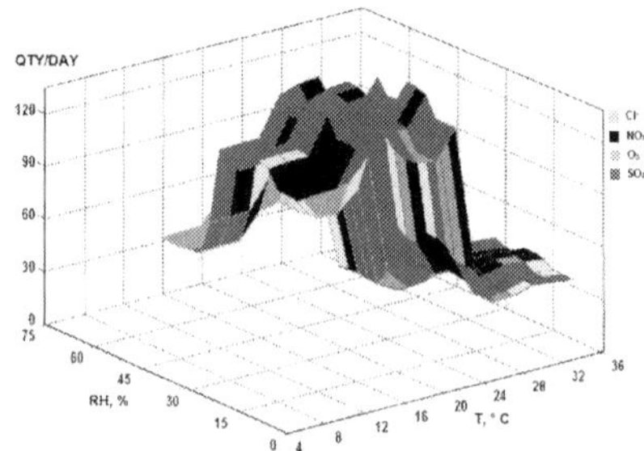

Figure 8. Correlation of RH and temperature with coloration of pigments indices of achiote in the winter 2014.

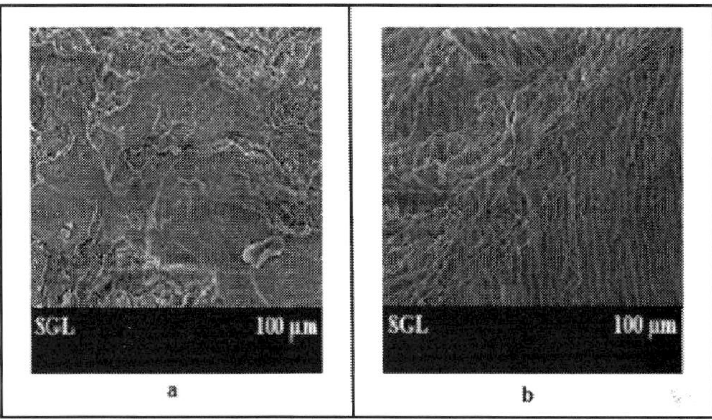

Figure 9. Microanalysis of the surfaces of sweet bread manufactured on: (a) a summer day and (b) a winter day with relative humidity higher than 80% in an indoor area of the industrial plant (2014) at 100 X.

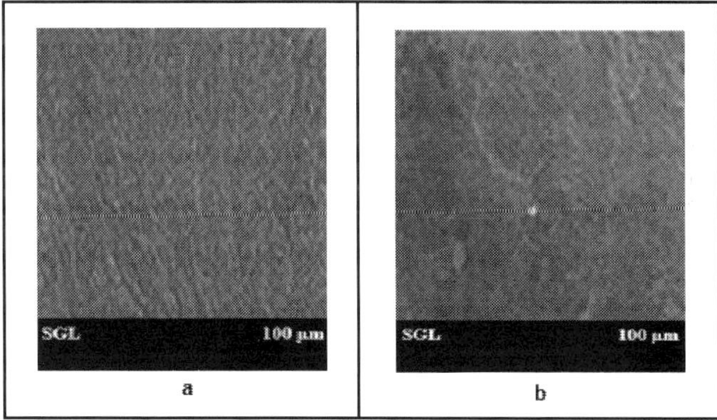

Figure 10. Microanalysis of surfaces of bread with a good appearance in: (a) day of July and (b) day of December in the indoor area of the industrial plant evaluated in 2014 at 100 X.

In change in figure 10, was made an evaluation with the use of the automatic system representing an improvement in the sweetened bread with better adhesion of the sweeteners as shown in a layer which appears smoother. Figure 10 shows a microanalysis of the appearances of sweet bread with ideal adhesion of the pigments in both summer and winter seasons. In both seasons the formation of the sweeteners was uniform and was considered to be of a high level of quality. The automatic system was very helpful in obtaining products with a good appearance that would therefore appeal to customers. This system controlled the relevant climatic factors and sent an electrical signal when the level of air pollutants was higher than that of the air quality standards, with the inefficient operation of the electronic device that make the function of control.

4. Conclusions

General corrosion is the cause of the destruction of most of the materials used in the food industry. The development of corrosion processes in the materials used to make food products leads to the modification of their properties which has a subsequent adverse effect on the manufacturing processes, particularly in the case study of the company that allowed us to investigate their marketed foods. Pigments used as food additives are very important in producing foods with an attractive appearance and that enable quick and easy marketing in the region, before some changes in these coatings start to be observed. Moisture was the factor that had the most significant effect on this process and a humidity control system — which has already been tested in other companies — is recommended. Temperature had an effect to a lesser extent, but it is still important to consider temperature, because if the humidity varies then the characteristics of the temperature change too. Sometimes the pigments have good adhesion and these products have the quality required in this type of food. At other times the sweeteners and pigments deteriorate easily, and some people consider that fresh bread made on the coast of the Baja California area is of low quality. In support of this, a study was done on pigments quickly losing their sweetener properties through variations in climatic factors such as RHrand temperature. Air pollution correlations were developed, showing the effect of levels of rH Hnd temperatures greater than 80% and 10 °C in the winter season, mainly in the months of December and January in Tijuana city.

Author details

César Sánchez Ocampo[1], Judith Marisela Paz Delgadillo[1], Gustavo López Badilla[1*] and Salvador Baltazar Murrieta[2]

*Address all correspondence to: glopezbadilla@yahoo.com

1 Manufacturing Department, Polytechnical University of Baja California, Mexicali, Baja California, México

2 Engineerng SchoolCETYS University, Mexicali, Baja California, México

References

[1] Jackson P., Monroy T. and Robertson J. (2009). "The acidity in food"; Journal of Nutrition; Vol. 5 (2); ISSN: 2116-3267; pp. 32-41.

[2] Derek U., Henderson A., Murray R and Anderson F. (2007). "Manufacturing processes in food transformation"; Journal of Food and applications; Vol. 3 (3); ISNN: 2118-2456; pp. 45-56.

[3] Brady G., Strovesky T., George Y., Wilson R., Carter M. (2010). "Regulations in the food industry. Case of meat food"; Journal of Food and Environment; Vol. 5 (2); ISSN: 2178-3146; pp. 12-24.

[4] López B. G., González R. F., Benavides U. L. and Hernández T. J. (2011); "Evaluación de la transformación industrial desde la Revolución Industrial. Caso de la industria alimentaria"; Revista de Ingeniería y Alimentación; Vol. 3 (5); ISSN: 3167-2156; pp. 56- 71.

[5] Frederick J., Homoto T. and Kerson M. (2010). "The improvement of the regulations in the food industry"; Journal of Food and Health; Vol. 3 (2); ISSN: 2119-3456; pp. 43-58.

[6] Peter H., Pederson K., Durant R. and Jackson H. (2009); "Analysis of consumers of the food industry"; Journal of Food and Economics; Vol. 4 (2); ISSN: 2168-3289; pp. 15-24.

[7] Johnson P., Fisher A., Vorosky T. and Patrick S. (2012). "Evaluation of phases in the food industry"; Journal of Food and Industrial Engineering; Vol. 3 (4); ISSN: 2145-3562; pp. 18-29.

[8] Sánchez, C; Tesis de Doctorado. (2010). "Análisis de un modelo de competitividad aplicado a las comercializadoras de autos en Mexicali, Baja California, México." Septiembre de 2010.

[9] Porter, M. (1991). "La Ventaja Competitiva de las Naciones"; México: Javier Vergara Editors.

[10] Hendrikson H., Kenneth P., Morrison G., Thomas D. and Ellens A. (2013). "The skills to obtain the better competitiveness in the engineering activities"; Journal of Engineering and Competitiveness; Vol. 4 (4); ISSN: 2169-3154; pp. 24-35.

[11] Woolfolk A. (2013). "Educational Psychology 12/E"; Ed. Pearson; ISBN-10:0132613166; pp 672.

[12] . Ray S, Easteal A, Quek SY, Chen XD. (2006). "The potential use of polymer-clay nanocomposites in the food industry"; International Journal of Food and Engineering; Vol. 2(4); ISSN: 2172-4159; pp. 1–11.

[13] Soroka, W. (2002). "Fundamentals of Food Industry"; Institute of Packaging Professionals (IoPP), ISBN-1:930268-25-4; pp. 348.

[14] Avella M, De Vlieger JJ, Errico ME, Fischer S, Vaca P, Volpe MG. (2005). "Biodegradable composite films for the food industry applications". Journal of Food Chemical; Vol. 9(3); ISSN: 2136-2459; pp.67–74.

[15] Brody A, Strupinsky ER, Kline LR. (2001); "Active sources of manufacturing processes for food applications. A scientific report"; Lancaster, Pa.; Technomic Publishing Company, Inc.; pp. 70.

[16] Brody Aaron L., Bugusu Betty, Han Jung H., Sand Koelsh, Mchugh Tara H. (2008). "Innovative Technology for the Food Industry; Journal of Food Science; Vol. 5 (3); ISSN: 2178-2457; pp.56- 69.

[17] Kirwan MJ, Ellenson T., Rondima A. (2009). "Food processes and technology"; Journal of Food and Industry; Vol. 5 (3); ISSN: 2156-2459; pp. 58-72.

[18] Cooksey K. (2010). "Effects of microbial organism in the food industry"; Food and Addition Materials; Vol. 6 (5); ISSN: 2178-3256; pp. 34-47.

[19] Organización Marítima Internacional, ed. (2005). Manual Sobre La Contaminación Ocasionada Por Hidrocarburos – Sección Iv. Imo Publishing. p. 236. ISBN 9789280100822.

[20] López-Badilla, Gustavo; Valdez-Salas, Benjamín; Schorr-Wiener, Michael. (2012)."Atmospheric Corrosion in Indoor of Seafood Industry in the Norwest of Mexico"; Revista Científica, Vol. 16, núm. 2, Abril-Junio; pp. 67-73; ISSN: 1665-0654.

[21] Walsh, Azarm, Balachandran, Magrab, Herold and Duncan Engineers Guide to MATLAB, Prentice Hall, 2010, ISBN-10: 0131991108.

[22] ISO 11844-1:2006. (2006). "Corrosion of metals and alloys - Classification of low corrosivity of indoor atmospheres- Determination and estimation of indoor corrosivity"; ISO, Geneva.

[23] ISO 11844-2:2005. (2005). "Corrosion of metals and alloys - Classification of low corrosivity of indoor atmospheres - Determination and estimation attack in indoor atmospheres"; ISO, Geneva.

[24] ASTM Standards. (2000). ASTM Standards Technical and Measurements Manual.

Chapter 13

Depositional Environment of Phosphorites of the Sonrai Basin, Lalitpur District, Uttar Pradesh, India

Shamim A. Dar and K. F. Khan

Additional information is available at the end of the chapter

http://dx.doi.org/10.5772/62186

Abstract

Phosphates are regarded as one of the most important fertilizer minerals used by man. In Sonrai basin of Lalitpur the phosphorites are found to occur as lenticular and detached bodies throughout the Formation of the Bijawar Group. Individual bodies range from a few meters to about 4 km in length, and width varies from thin bands to about 125 meter with P_2O_5 concentration ranging from 10 to 20%. The Paleoproterozoic Bijawar Group are overlain by the Archaean Bundelkhand Basement Complex and underlain by Vindhyan Supergroup. The occurrence of phosphorites is confined to the Sonrai Formation which consists of massive to brecciated phosphorite within the lower reddish shales, with at least three bands identified. Megascopic study reveals that the brecciated phosphorite is reddish brown in color and fine to medium grained with angular fragments of chert and quartz embedded in a groundmass of iron oxides and secondary silica intercalated with minor veins of chert and iron oxides. The phosphorite horizon in the Lalitpur area is associated with pink to white brecciated massive quartzite, shale, dolomite and limestone of the basal unit. The concentration trends of certain major oxides indicate that the phosphorites are more enriched in CaO, P_2O_5 and SiO_2 than Al_2O_3, Fe_2O_3, TiO_2, Na_2O and K_2O. The concentration trends of trace elements reveal that the phosphorites are moderately enriched in Co, Zn, Zr, Pb, U than in Sc, Ba, V, Cr, Ni, , Rb, Sr, Y and Th. The dispersion patter, correlation coefficient and mutual relationship of significant major oxides represented by plotted diagrams, indicate that SiO_2, CaO, MgO are antipathetically related with P_2O_5. The relationship suggests a gradual replacement among these oxides during diagenesis. High values of P_2O_5 and CaO in the phosphorites indicate more concentration of apatite constituent. The difference in geochemical behavior of CaO and MgO may be due to ionic substitution of Ca^{+2} by MgO^{+2} in the apatite crystal lattice during alkaline environment of the basin. The strong negative relationship between P_2O_5 with Fe_2O_3 in phosphorites may be due to leaching and/mild weathering of iron from the ores and reprecipitation along with P_2O_5 in the pore spaces, cavities/voids, veins, etc in highly oxidizing marine environment of the basin. The minimum evidence of organic matter, absence of sulphide minerals and lower concentration of V, Ni, and

Cu suggest that the phosphorites were deposited in an oxidizing environment with slightly anaerobic to highly aerobic facies.

Keywords: Depositional Environment, Phosphorites, Sonrai basin, Lalitpur, Geochemistry

---

1. Introduction

Phosphorous and potash are the essential elements for plant growth. The importance of phosphate and potash in soil in agriculture has long been recognized. A predominantly agricultural country like India needs a vast amount of fertilizers. Phosphates are regarded as one of the most important fertilizer minerals used by man. Nearly 90% of the total world phosphate output is utilized for the manufacture of a number of fertilizers such as superphosphate, triple-superphosphate, di-ammonium-phosphate and mono-ammonium phosphate, which are essential for agro-industries all over the globe.Rock phosphate also referred to as phosphorites, are sedimentary deposits with high phosphorus (P) concentrations. These rocks are one of the primary sources for P which in turn is a critical and nonrenewable element for fertilizer production upon which global fertility depends. Phosphorite is a non-detrital sedimentary rock, which contains 18–20% phosphate-bearing minerals [1-11].

Typical phosphorite beds contain about 30% P_2O_5 and constitute the primary source of raw materials for most of the world's production of phosphatic fertilizers. Alteration of phosphorites tends to leach carbonates and increase the percentage of P_2O_5. The Phosphoria Formation contains about 34% P_2O_5 near the surface as compared to depth where it is about 28% [12]. Phosphorite deposits generally occur in three principal types: (i) regionally extensive crystalline nodular and bedded deposits; (ii) localized concentrations of phosphate-rich clastic deposits (bone beds); and (iii) guano deposits. Broadly, phosphates can be grouped as: (i) primary phosphates that have crystallized from a liquid; (ii) secondary phosphates formed by the alteration of primary phosphates; and (iii) fine-grained rock phosphates formed at low temperatures from phosphorus-bearing organic material primarily underwater.

In the study area, phosphorites are found to occur as lenticular and detached bodies throughout the Formation of the Bijawar Group. Individual bodies range from a few meters to about 4 km in length and width varies from thin bands to about 125 m with P_2O_5 concentration ranging from 10 to 20%. Estimated reserves of the Precambrian phosphorites of Lalitpur area are 5 million tons with an average grade of 25% based on detailed studies of two prominent lensoid bodies of phosphorites [9-11, 13, 14].

The phosphorites of the Sonrai basin of Lalitpur district in Uttar Pradesh was discovered in (1977) [13]. District of Lalitpur (Latitude 24° 11'N and 25° 13'N and Longitude 78° 11'E and 79° 0'E is surrounded by the adjacent district Jhansi in the north, Sagar (M.P.) in the south, Tikamgarh and Chhatarpur districts in the East and Shivpuri and Guna districts (M.P.) in the West. The western boundary of the district runs along river Betwa for about 96 kms. The southwestern boundary is formed by Narain Nandi, the northeastern by river Jamni for about 65 km and the southeastern boundary for about 40 kms by Dhasan Nadi. The Sonrai basin in Lalitpur district is located at the extreme south-west corner of Uttar Pradesh. The basin is 28 km long and 5 km wide and shows an east–west extension. The study area follows the regional east–west strike and forms a westward plunging syncline on a regional scale with closure on the eastern side. They are faulted against the northern crystalline rocks and the Berwar formations to the south. East and west are covered by the flat-lying Vindhyan conglomerate and sandstone [15, 16]. These deposits have been investigated by many earlier workers from the point of view of their geological and economic importance. However a detailed study of the geochemical characteristics of these rocks has not been done so far. In the present study, new geochemical data of advanced nature on Lalitpur phosphorites are utilized to deal with the major, trace, and rare earth element geochemistry in order to determine their abundance, distribution pattern, and interelement relationship of phosphorites ultimately to interpret their impact on phosphatic mineralization. Therefore, the present investigation may be considered as a first serious attempt to study the geochemistry of the Lalitpur phosphorite deposits in order to present a conceptual genetic model not only for these deposits but also to apply the same in areas of similar geologic and environmental setting.

2. Aim and objective

The present investigation has been, thus, undertaken with a view:

- To evaluate the governing factors involved during the process of phosphorite formation

- To understand the chemical and biological processes in the phosphorite deposition

- To study the mineralogy and chemistry of the various rock units in order to understand the genesis of the various minerals present in the phosphorites and to assess the quality of the deposit

3. Geological background

The Paleoproterozoic meta-sedimentary rocks of Bijawar Group are overlain by the Archean Bundelkhand Basement Complex and underlain by the Vindhyan Supergroup. The Sonrai rift basin preserves volcano-sedimentary units of Late Archaean to early Proterozoic age [17]. The Precambrian phosphorites of Bijawar Group of rocks show characteristics of an epicontinental sea with restricted and very shallow marine environment of formation along some shoals which existed during the iron-rich Precambrian times [18]. The Bijawar basin in Lalitpur district is located at the extreme southwest corner of Uttar Pradesh (Fig. 1) [6, 7, 9-11,15,16]. The basin is 28 km long and 5 km wide and shows an east–west extension.

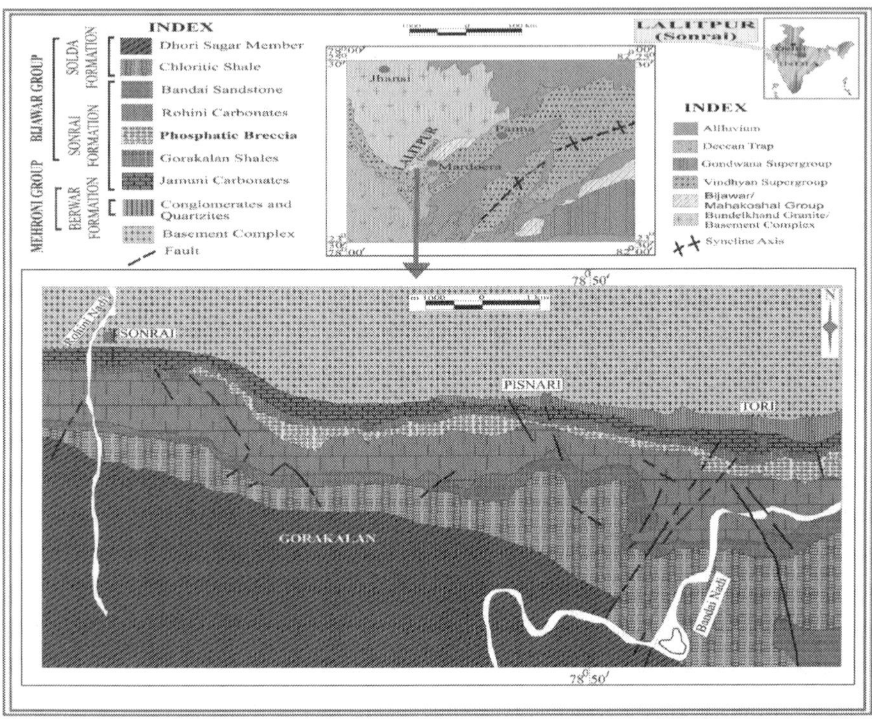

Figure 1. Simplified geological map of phosphorites of the Lalitpur district, Uttar Pradesh, India (modified after [6, 7, 9-11,15,16].

The lithostratigraphy of the area [19] is given in Table 1. The exposed lithostratographic units of the Bijawar Group in the Lalitpur district are divided into three formations: Berwar, Sonrai, and Solda. The Berwar Formation at the base contains banded hematite–quartzite, chloritic shale, quartzite, and pebbly conglomerate. The overlying Sonrai Formation is divided into four members which in ascending order are: (a) Shale Member consisting of dark-grey, olive-grey, reddish and black shales with carbonate and phosphorite bands; (b) Brecciated Quartzite Member comprising brecciated quartzite and lensoid bodies of phosphorites with quartzite and phosphatic matrix cementing angular to subangular fragments of quartzite; (c) Tori Member with ferruginous massive quartzite and massive botryoidal and brecciated phosphorites and (d) Rohini Member with its coeval Kurrat Lava Member. The Rohini Member consists of thinly bedded, light-grey shaly dolomitic and pink sandy carbonate rocks. The uppermost Solda Formation is divided into three members: (a) the Bandai Member at the base consisting of sandstone and grits which are calcareous at places; (b) the overlying Dhauri Sagar Member with tuffaceous shales and (c) the uppermost Hadda Member comprising quartzite and shale intercalations.

VINDHYAN SUPERGROUP		Vindhyan sandstone, etc.	
------------------	---------------------UNCONFORMITY--		
BIJAWAR GROUP	Solda Formation	Hadda Member	Quartzite shale intercalations
		Dhauri Sagar Member	Tuffaceous shale
	Sonrai Formation	Bandai Member	Tuffaceous shale, sandstone and grit
		Rohini Member/ Kurrat lava	Sandy, shaly and dolomitic carbonate rocks Basic volcanics and pillow lava
		Tori Member	Massive quartzite with **phosphorite**
		Brecciated Quartzite Member	**Brecciated quartzite with lensoid bodies of phosphorite**
		Shale Member	Grey, green and red shales with
	Barwar Formation		Banded hematite quartzite, chloritic shale, quartzite, and conglomerate
--UNCONFORMITY-------------------------------------			
BUNDELKHANDGROUP	Bundelkhand Granitoids Complex with pegmatite and quartz veins		

Table 1. Stratigraphy of the Bijawar Group in Lalitpur Basin, India (after 6, 7, 9-11,15,16).

Megascopic study reveals that the brecciated phosphorite is reddish brown in color and fine to medium grained with angular fragments of chert and quartz embedded in a groundmass of iron oxides and secondary silica intercalated with minor veins of chert and iron oxides. The rock is hard, compact, bedded, and ferruginous but nonlaminated. These brecciated phosphorites are gradually merged into the host rocks. Their structural characteristics are very similar to those of the associated rocks. They have undergone folding like other litho units in the area. As such they appear to be intraformational sedimentary breccia only. The rocks consist mainly of clasts of subangular to subrounded cherty materials, quartz (which may be derived from quartz-veins/reefs), quartzites, schistose rock fragments, and other host rocks of the Bijawar Group. The intergranular spaces of these brecciated phosphorites are filled with secondary silica, ironoxides, and calcareous material in the form of matrix, which varies from ferruginous mudstone to a ferruginous cherty mass. The brecciated phosphorites generally contains fragments of apatite and other phosphatic minerals embedded in dark-brown cherty matrix or sometimes ferruginous. Such brecciated phosphorite deposits have also been reported in metasediments of Udaipur area, Rajasthan [20] and Bijawar rocks of Sagar–Chhatarpur districts of Madhya Pradesh. Brecciated rocks rich in apatite and magnetite have also been reported from Singhbhum thrust belt areas of Jharkhand and phosphatic breccia in Proterozoic metasedimentary sequence of Betul district of Madhya Pradesh [21]. The phosphorite horizon in the Lalitpur area is associated with pink to white brecciated massive quartzite, shale, dolomite, and limestone of the basal unit. The phosphorites occur in the form of stratified lensoid bodies and appear to be of sedimentary residual type. These extend in length from 50 to 250 m and range from 5 to 35 m in thickness (Fig. 2a,b,c,d) [9, 11].

Figure 2. (a,b,c) Field photographs showing an exposure of phosphatic breccias; (d)thin section photomicrographs showing pseudo-ooliths and pellets of collophane of phosphorites of the Lalitpur district[9-11].

4. Analytical methods

With the help of geological maps and topographic sheets (54L/15, 54L/11, 54L/16) of the area sampling was carried out and 87 fresh unweathered representative samples (6″x 4″) were collected from the field at regular intervals. The samples were selected for chemical analysis. For preparation of solution, 50 gms of each sample was taken for crushing and powdering. The samples were crushed in a steel mortar to coarser fractions free from secondary association then they were handpicked and dried in the oven at 100ºC to eliminate the hygroscopic moisture. The coarse material was further crushed in porcelain mortar till it became medium to fine grained. Finally the material was powdered in centrifugal ball mill and sieved by standard sieve of -200 mesh. The powder of each sample was packed in bottles and numbered properly for analysis.

The major element geochemical analysis was carried out at National Geophysical Research Institute (NGRI), Hyderabad. The major elements were determined from pressed pellets, which were prepared by using collapsible aluminum cups [22]. These cups are filled with boric acid and about 1 gm of the finely powder rock sample was put on the top of the boric acid and pressed under a hydraulic press at 20 tons pressure to get a pellet. The sample pellets were analyzed using a Philips MagiX PRO model PW–2440 wavelength dispersive X-ray fluorescence spectrometer (XRF) coupled with automatic sample changer PW 2540.

Trace and rare earth elements (REE) were determined by inductively coupled plasma-mass spectrometry (ICP-MS) technique using Perkin-Elmer, SCIEX, Model ELAN DRC-II system at National Geophysical Research Institute (NGRI), Hyderabad. Analytical solutions of the samples for geochemical analysis were prepared by open acid digestion method [25]. 50 mg (0.0500g) powder of rock sample was taken in a clean dried PTFE Teflon beaker. Each sample was moistened with a few drops of ultrapure water. Then, 10 ml of an acid mixture (containing 7:3:1 HF: HNO_3: $HClO_4$) was added to each sample. Samples were swirled until completely moist. The beakers were then covered with lids and kept overnight for digestion after adding 1 ml of 5 µg ml^{-1} Rh solution (to act as an internal standard). The following day, the beakers were heated on a hot plate at ~200°C for about 1 h and the lids were removed and the contents were evaporated to incipient dryness until a crystalline paste was obtained. The remaining residues were then dissolved in 10 ml of 1:1 HNO_3 and kept on a hot plate for 10 min with gentle heat (70°C) to dissolve all suspended particles. Finally, the volume was made up to 250 ml with double distilled water (purified water) (18 MΩ) and stored in polythene bottles. This solution, stored in polythene bottles, was used for the determination of trace and REE elements by inductively coupled plasma–mass spectrometry techniques (ICP-MS). The instrumental and data acquisition parameters are same as given by [23]. The precision of ICP-MS data is better than ±6% RSD for all the trace elements and was obtained in all cases with comparable accuracy. All the available data were standardized against the international reference rock standards NIST-120C for phosphorites. The data of major and trace element analysis are given in Figs. 3 and 4.

5. Results and discussions

The abundance and concentration trend of major, trace and rare earth elements in the phosphorites of the Sonrai basin of Lalitpur district are presented in Figs. 3 and 4. The concentration trends of certain major oxides indicate that the phosphorites are more enriched in CaO, P_2O_5, and SiO_2 than in Al_2O_3, Fe_2O_3, TiO_2, Na_2O, and K_2O. The concentration trends of trace elements reveal that the phosphorites are moderately enriched in Co, Zn, Zr, Pb, and U than in Sc, Ba, V, Cr, Ni, Rb, Sr, Y, and Th.

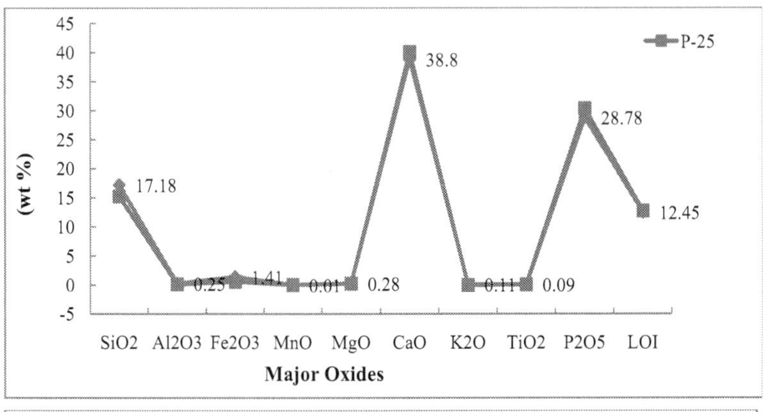
P-25
45
40
35
30
25
20
15
10
5
0
-5
(wt %)
38.8
28.78
17.18
12.45
0.25
1.41
0.01
0.28
0.11
0.09
SiO2 Al2O3 Fe2O3 MnO MgO CaO K2O TiO2 P2O5 LOI
Major Oxides

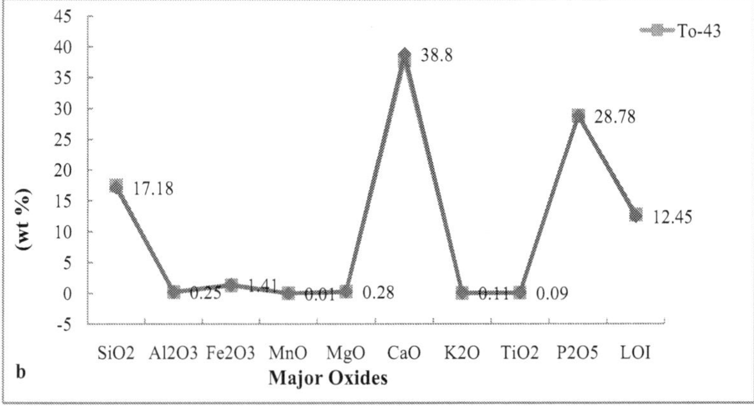
To-43
45
40
35
30
25
20
15
10
5
0
-5
(wt %)
38.8
28.78
17.18
12.45
0.25
1.41
0.01
0.28
0.11
0.09
SiO2 Al2O3 Fe2O3 MnO MgO CaO K2O TiO2 P2O5 LOI
b Major Oxides

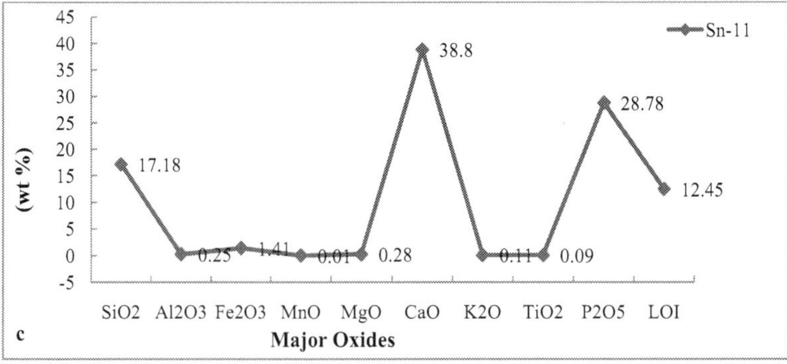
Sn-11
45
40
35
30
25
20
15
10
5
0
-5
(wt %)
38.8
28.78
17.18
12.45
0.25
1.41
0.01
0.28
0.11
0.09
SiO2 Al2O3 Fe2O3 MnO MgO CaO K2O TiO2 P2O5 LOI
c Major Oxides

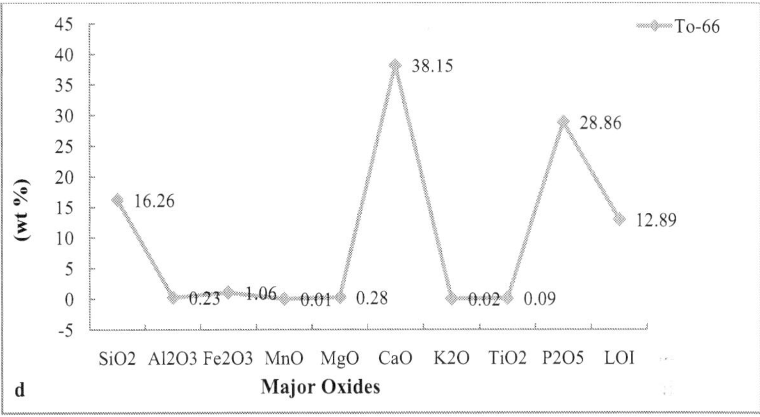

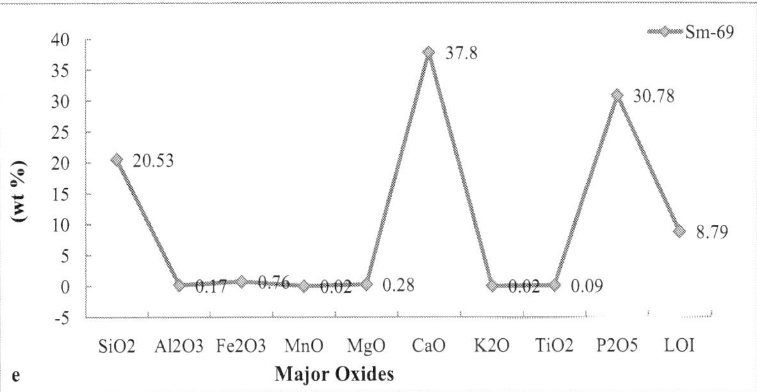

Figure 3. (a, b) Graph showing relative abundance of certain major oxides in phosphorites of the Lalitpur district [6, 7, 9, 10]. (c,d,e) Graph showing relative abundance of certain major oxides in phosphorites of the Lalitpur district [6, 7, 9, 10].

Phosphates are known to be deposited in a wide range of geological environments. (1) Normally the phosphates are deposited in very shallow, near-shore, marine or low-energy environments. The possible environments are supratidal zones, littoral or intertidal zones, and estuarine zones. (2) areas of oceanic upwelling cause the formation of phosphates because the constant stream of phosphate brought from the large, deep-ocean reservoir giving continuous growth of organisms. (3)The genesis or origin of phosphorites is a controversial problem. The complexity of the processes involved in the formation of marine phosphorite deposits has been the subject of study by a large number of investigators for a fairly long time. Several divergent views regarding the deposition and accumulation of phosphorites have earlier been put forward. The origin of phosphorites is not well understood and over the years there have been several hypotheses for their origin that include biological, chemical, and volcano-sedimentary

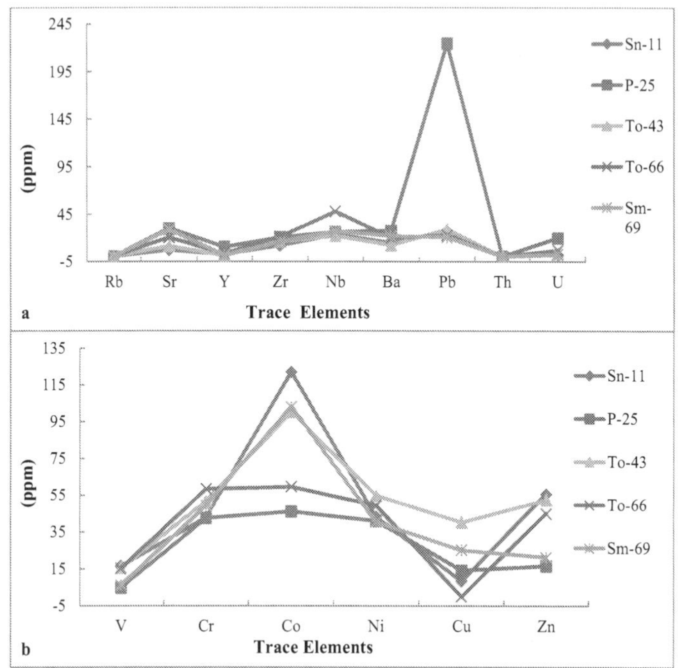

Figure 4. (a,b) Graph showing relative abundance of certain trace elements in phosphorites of the Lalitpur district [6, 7, 9, 10].

[24, 25]. Many phosphorite deposits in the ocean occur in areas of upwelling, where phosphate, regenerated from the recycling of organic matter in the deep ocean is brought to coastal areas and triggers blooms of primary productivity such as along the coasts of Namibia[26] and Peru–Chile [27]. Nonupwelling areas such as estuaries and semi-restricted embayments can also be host to phosphate deposits or authigenic carbonate fluorapatite[26]. During early diagenesis, phosphorus may be released from organic matter and iron oxides under anoxic conditions, may play a vital role in phosphogenesis.

The geochemistry of phosphorites and their constituent mineral apatite have been widely studied owing to their economic importance and the potential utility of their geochemistry to estimate paleomarine chemistry. The variation in major and trace elements' abundance is attributed to a change in environmental condition of deposition. The mobilizations of ferrous ions and subsequent precipitation as ferric oxides and oxyhydroxides have clearly redistributed iron in these phosphorites [28-32]. The same interpretation has been given by [33]for Ogun phosphate rocks.

Higher values of P_2O_5 and CaO in the phosphorites of the area (Figs. 3, 4) indicate more concentration of apatite constituent. This corroborates to a greater extent to the Ogun phosphate rocks of Nigeria (34). Positive correlation of P_2O_5 and CaO (Fig. 5a) might be the result of diagenetic phosphatization in which CO_2 might be removed by PO_4 before the precipitation of phosphorite during high oxidizing conditions of the basinal water which gave rise to carbonate fluorapatite as an end product [6, 8].This relationship also indicates the diagenetic origin of phosphorites during highly oxidizing and alkaline basinal sea water[28, 35]. The positive correlation of P_2O_5 with SiO_2(Fig. 5b) may be due to the presence of detrital minerals like quartz, feldspars, micas and clay bearing minerals. The positive relationship also suggests that much of SiO_2 content is related to apatite. An inverse relationship between MgO and P_2O_5(Fig. 5c) in phosphorites indicates a continuous replacement of MgO by P_2O_5 during diagenesis. Poor correlation between MgO and P_2O_5 in phosphorites of the area indicates that Mg might not have been incorporated in the apatite structure.The strong negative relationship between P_2O_5 with Fe_2O_3(Fig. 5d) in phosphorites may be due to leaching and/mild weathering of iron from the ores and reprecipitation along with P_2O_5 in the pore spaces, cavities/voids, veins, etc., in highly oxidizing marine environment of the Bijawar basin. This relationship between the constituents indirectly indicates the alkaline nature of basinal waters. Like iron, aluminum also decreases with the increase in P_2O_5 concentration (Fig. 5e). It may also be due to ionic substitution of Ca^{+2} by Al^{+3} in the carbonate fluorapatite phase due to weathering of the ores.The presence of low content of Al_2O_3 in the samples of phosphorites of the study area indicated the arid to humid climatic conditions. The low concentration of alumina may be due to minor occurrences of clay minerals and major content of ferruginous minerals like hematite, ferric-oxides, etc that constitute the cementing material of the phosphatic rocks. The presence of low Al_2O_3 in phosphorites indicates the humid to low temperature climatic conditions at the time of precipitation and phosphatization of these sediments in the area.

The term trace element is usually taken to mean those elements present in rocks in concentrations of less than a few thousands parts per million (ppm). Chemical elements like uranium (U), nickel (Ni), cobalt (Co), chromium (Cr), vanadium (V), copper (Cu), lead (Pb), gallium (Ga), etc., the content of which in rocks is rather appreciable and which form many minerals of their own, are commonly known as "trace elements". In recent years, the chemical composition of rocks, more particularly the distribution of trace elements in sediments has been used for distinguishing and identifying precisely their depositional environment. Some of the earlier workers, particularly in the earlier years of this century, have emphasized the distribution and significance of trace elements in geological materials.

Ba, Cu, Ni, and Zn are depleted in Lalitpur phosphorites in comparison to the average shale and phosphorite of Quaternary off southeast coast phosphate stromatolites formed under the influence of upwelling. This indicates that planktonic organic matter may not be the prime source of these elements in the Lalitpur phosphorite deposits. Since the sediments hosting phosphorites of Lalitpur were formed in shallow marine conditions, some of these elements may have recycled back to the overlying water column during sediment diagenesis in oxygenated environments. Pb, Zn, and Mo are below the detection limit in Lalitpur phosphorites suggesting that strongly reducing conditions were not prevailing at the time of phosphorite formation.

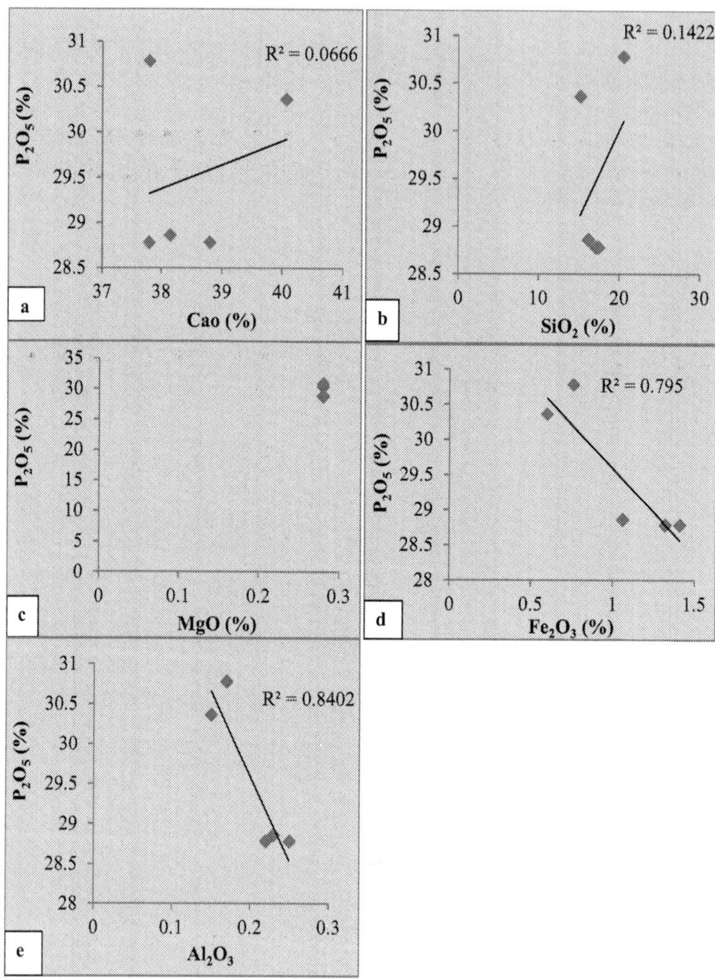

Figure 5. Graphical representation of P_2O_5 in relation to SiO_2, CaO, MgO, Fe_2O_3, Al_2O_3, in phosphorites of the Lalitpur district[6-10].

The lower concentration of copper, which may be meager evidence of pyrite, indicates highly oxidizing conditions of their formation. Due to the absence of sulfate (SO_4) ions in the water the pyrite could not be precipitated at the time of phosphatization. The minor content of copper may be due to mutual ionic substitution of divalent Cu^{+2} by Fe^{+2} and Mg^{+2} during oxidizing conditions of the basin. The meager evidence of organic matter, near absence of sulfide minerals, and lower concentration of V, Ni, and Cu suggest that the phosphorites were deposited in an oxidizing environment slightly anaerobic to highly aerobic facies. In the

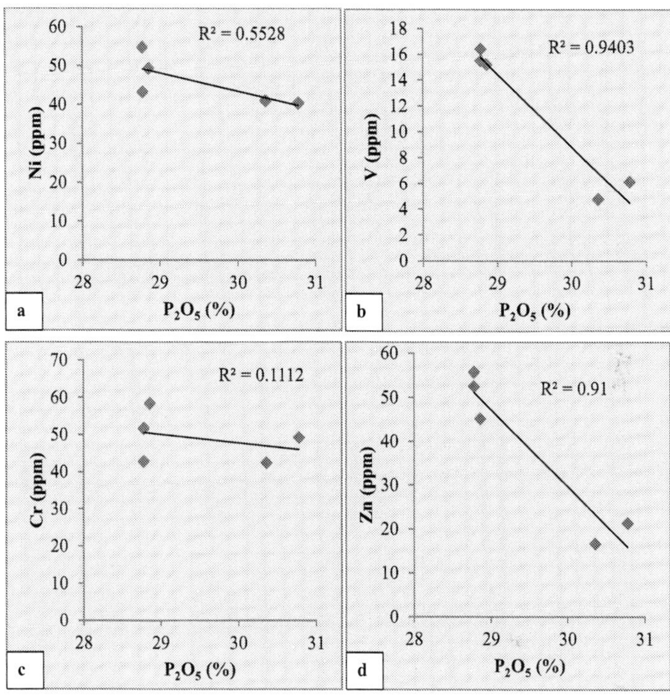

Figure 6. Scatter diagram indicating the geochemical relations between P_2O_5 and Ni, V, Cr, Zn in phosphorites of the Lalitpur district[6-10].

present study, Ni, Cr, V, Zn show negative correlation with P_2O_5 (Fig. 6a,b,c,d) in all the samples of the area which may be due to mutual substitution within the apatite lattice that might have taken place during shallow marine environmental conditions of the basin.

The lower MgO content (0.18%) in the francolite of the phosphorites of the Duwi Formation compared to the Chile and Peru margins authigenic phosphorites proposed by [36]as 1.5% and 2.3%, respectively, can be attributed to the formation of the phosphatic grains in low MgO seawater [32]. The phosphorites of the study area also corroborate the findings of previous investigators showing low MgO concentrations. The same results were observed from Ogun phosphate rocks [33] and phosphorites of the Sonrai basin [6]. The study area shows good concentrations of P_2O_5. The P_2O_5 content of Matoon phosphorite deposits, district Udaipur (Rajasthan) is as high as 36% [37]. [38] reported that sediments of SE Coast of USA contain as high as 40% P_2O_5. The P_2O_5 content may be due to the presence of cryptocrystalline collophane as is commonly present in most Precambrian mudstones [39]. The silica content in Lalitpur phosphorites are in line with those of the Peru margin (8.79 wt%) [40] and Namibian margin phosphorites (19.4 wt%) [26]. The low concentration of TiO_2 has also been reported from the Ogun phosphate rocks, Nigeria [33] which coincides to the TiO_2 concentration of Lalitpur phosphorites indicating the presence of detrital heavy minerals like hematite, ilmenite, and rutile entrained in the sedimentary environment. This corroborates to a great extent the findings of [33]. The Lalitpur phosphorites show low concentration of MnO. The concentration of MnO in phosphorites has been reported by many earlier workers 0.38-0.4 wt% [41]; 0.014-0.021 wt% [42].

The concentration of Pb in phosphorites has been reported by many earlier investigators as10–43 ppm [43] and 3–4 ppm [44]. In the study area, uranium is an important trace element whose content in the phosphorite samples ranges from 1.67 to 129.67. The low concentration of U in few samples of the study area has also been noted in other phosphorite deposits of the world. The reason for low U content in phosphorites is due to its removal from the crystal lattice of carbonate-apatite during catagenesis [3]. The concentration of Zn in phosphorites has been reported by many earlier workers as 11–65 ppm [43]and 52–61 ppm [44]; [3] reported average concentration 614.51 ppm of phosphorites related to sedimentary rocks and 233.4 ppm related to endogenous rocks; others reported 18–33 ppm [45]; 104–130 ppm [35]; 30–290 ppm [46]; and 27.84–50 ppm [12].

6. Conclusion

The Paleoproterozoic metasedimentary rocks of Bijawar Group are overlain by the Archaean Bundelkhand Basement Complex and are underlain by the Vindhyan Supergroup. These were deposited as small intracratonic basins like Sonrai, Hirapur and Gwalior Basins. The lithostratographic units of the Bijawar Group in the Lalitpur district are better exposed and are divided into three formations: Berwar, Sonrai, and Solda Formations. Among these, phosphorites are found only in the Sonrai Formation.

Megascopic study reveals that the brecciated phosphorite is reddish brown in color and fine to medium grained with angular fragments of chert and quartz embedded in a groundmass of iron oxides and secondary silica intercalated with minor veins of chert and iron oxides. The phosphorite horizon in the Lalitpur area is associated with pink to white brecciated massive quartzite, shale, dolomite and limestone of the basal unit. Thin-section photomicrographs showing pseudo-ooliths and pellets of collophane of phosphorites of the Lalitpur district.

The concentration trends of certain major oxides indicate that the phosphorites are more enriched in CaO, P_2O_5, and SiO_2 than in Al_2O_3, Fe_2O_3, TiO_2, Na_2O, and K_2O. The concentration trends of trace elements reveal that the phosphorites are moderately enriched in Co, Zn, Zr, Pb, and U than in Sc, Ba, V, Cr, Ni, Rb, Sr, Y, and Th. The dispersion pattern, correlation coefficient, and mutual relationship of significant major oxides, represented by plotted diagrams, indicate that SiO_2, CaO, MgO are antipathetically related with P_2O_5. The relationship suggests a gradual replacement among these oxides during diagenesis. High values of P_2O_5 and CaO in the phosphorites indicate more concentration of apatite constituent. Poor correlation between MgO and P_2O_5 in phosphorites of the area indicates that Mg might not have been incorporated in the apatite structure. The low magnesium content in the phosphorites indicates that little dolomitization has occurred. The difference in geochemical behavior of CaO and MgO may be due to ionic substitution of Ca^{+2} by MgO^{+2} in the apatite crystal lattice during highly oxidizing, alkaline environment of the basin. The strong negative relationship between P_2O_5 with Fe_2O_3 in phosphorites may be due to leaching and/or mild weathering of iron from the ores and reprecipitation along with P_2O_5 in the pore spaces, cavities/voids, veins, etc. in highly oxidizing marine environment of the Bijawar basin. The low concentration of alumina may be due to minor occurrences of clay minerals and major content of ferruginous minerals like hematite and ferric-oxides, etc. that constitute the cementing material of the phosphatic rocks. The presence of low Al_2O_3 in phosphorites indicates the humid to low temperature climatic conditions at the time of precipitation and phosphatization of these sediments in the area.

The lower concentration of copper which may be meager evidence of pyrite, indicates highly oxidizing conditions of their formation. The minimum evidence of organic matter, absence of sulfide minerals, and lower concentration of V, Ni, and Cu suggest that the phosphorites were deposited in an oxidizing environment slightly anaerobic to highly aerobic facies. In the present study, Ni, Cr, V, Zn show negative correlation with P_2O_5 in all the samples of the area, which may be due to mutual substitution within the apatite lattice that might have taken place during shallow marine environmental conditions of the basin.

Acknowledgements

We are highly thankful to the Chairman, Department of Geology, Aligarh Muslim University, Aligarh, for providing necessary facilities. Thanks are also due to the Director, National Geophysical Research Institute, Hyderabad (A. P.) for providing us with laboratory facilities to carry out the geochemical analysis of the rock samples. UGC is highly thankful for providing financial assistance.

Author details

Shamim A. Dar[1*] and K. F. Khan[2]

*Address all correspondence to: sjshamim@gmail.com

1 Department of Geology, Sri Pratap (S.P) College, M. A. Road Srinagar, Kashmir, India

2 Department of Geology, Aligarh Muslim University, Aligarh, Uttar Pradesh, India

References

[1] Slansky M: *Geology of Sedimentary Phosphates*. North Oxford Academic Publisher Ltd., 1986. pp. 19-35.

[2] Notholt JGA: African phosphate geology and resources: A bibliography, 1979-1988. *J Afr Earth Sci.* 1991; 13:543-552.

[3] Zanin Yu N, Zamirailova AG: Trace elements in supergene phosphorites. *Geochem Int.* 2007; 45(8):758-769.

[4] Daessle LW, Carriquiry JD: Rare earth and metal geochemistry of Land and Submarine phosphorites in the Baja California Peninsula, Mexico. *Marine Geores Geotechnol.* 2008; 26: 240-349.

[5] Filippelli GM: Phosphate rock formation and marine phosphorus geochemistry: The deep time perspective. *Chemosphere.* 2011; 84: 759-766.

[6] Khan KF, Shamim A Dar, Saif A Khan: Geochemistry of phosphate bearing sedimentary rocks in parts of Sonrai block, Lalitpur District, Uttar Pradesh, India. *Chemie der Erde-Geochemistry.* 2012a; 72, 117-125.

[7] Khan KF, Shamim A Dar, Saif A Khan: Rare earth element (REE) geochemistry of phosphorites of the Sonrai area of Paleoproterozoic Bijawar basin, Uttar Pradesh, India. *J Rare Earths.* 2012b; 30(5), 507-514.

[8] Khan KF, Saif A Khan, Shamim A Dar, Z Husain: Geochemistry of phosphorite deposits around Hirapur-Mardeora area in Chhatarpur and Sagar Districts, Madhya Pradesh, India. *J Geol Mining Res.* 2012c; 4(3), 51-64.

[9] Shamim A Dar: Geochemical and Mineralogical Studies of Phosphorites and Associated Rocks in Parts of Lalitpur district, Uttar Pradesh, India. Ph. D. Thesis (Unpublished), Department of Geology, Aligarh Muslim University, Aligarh. 2013.

[10] Shamim A Dar, Khan KF, Khan SA, Mir AR, Wani H, Balaram V: Uranium (U) concentration and its genetic significance in the phosphorites of the Paleoproterozoic Bijawar Group of the Lalitpur district, Uttar Pradesh, India. *Arab J Geosci.* 2014; 7(6), 2237-2248.

[11] Shamim A Dar, KF Khan, Saif A Khan, Samsuddin Khan, M Masroor Alam:Petromineralogical studies of the Paleoproterozoic Phosphorites in the Sonrai basin, Lalitpur District, Uttar Pradesh, India. *Natural Resour Res.* 2015; 24(3), 339-348.

[12] Kharin, GS:Phosphorite Potential of cretaceous and paleogene rocks in the Kaliningrad and Southeastern Baltic regions. *Lithol Min Res.*2009; 44(4), 305-327.

[13] Pant A: Resource status of rock phosphate deposits in India and areas of future potential. In Sheldon RP and Burnett WC (Eds.), *Fertilizer Mineral Potential in Asia and the Pacific.* Honolulu: East-West Resource Systems Institute; 1980. pp. 331-57.

[14] Kothiyal DL, Srivastava VC, Verma RN: Geology and mineral resources of the states of India, Uttar Pradesh and Uttaranchal. Geological Survey of India. Miscellaneous Publication. 2002;30, (13): 26-65.

[15] Prakash R, Swarup P, Srivastava RN: Geology and mineralization in the Southern Parts of Bundelkhand in Lalitpur District, Uttar Pradesh. *J Geol Soc Ind*. 1975; 16: 143-156.

[16] Roy M, Bagchi AK, Babu EVSSK, Mishra B, Krishnamurthy P: Petromineragraphy and mineral chemistry of Bituminous Shale-hosted uranium mineralization at Sonrai, Lalitpur district, Uttar Pradesh. *J Geol Soc Ind*. 2004; 63: 291-298.

[17] Sharma KK: Evolution of Achaean Paleoproterozoic crust of the Bundelkhand craton, northern Indian shield. In: Verma OP and Mahadevan TM (Eds.), Research Highlights in Earth System Science, DST's Spl. 1 Margules data to the garnet-biotite geothermometer. *Am Mineralogist*. 2000; 85: 881-892.

[18] Banerjee DM, Khan MWY, Srivastava N, Saigal GC: Precambrian Phosphorites in the Bijawar rocks of Hirapur-Bassia area, Sagar district, Madhya Pradesh, India. *Min Deposita*. 1982; 17: 349-362.

[19] Pant A, Khan HH, Sonakia A: Phosphorite resources in the Bijawar Group of central India. In: Notholt AJG, Sheldon RP, and Davidson DF (Eds.), *Phosphate Deposits of the World. Phosphate Rock Resources*. International Geological Correlation Programme Project 156: Phosphorites. Cambridge University Press, 1989; 2: pp. 473-477.

[20] Banerjee DM, Saigal GC, Srivastava N, Khan MWY: Element variation pattern in the Precambrian phosphorites and country rocks of Udaipur, Rajasthan, Geological Survey of India, Special Publication; 1984; 17:63-77.

[21] Nagaraju M, Shivakumar K, Verma SC: A note on the uraniferous Phosphatic breccia in Proterozoic metasedimentary sequence, Betul district, Madhya Pradesh. *Ind Mineralogist*. 2011; 45(1): 75-82.

[22] Govil PK: X-ray fluorescence analysis of major, minor and selected trace elements in new IWG reference rock samples. *J Geol Soc Ind*. 1985; 26: 38-42.

[23] Roy P, Balaram V, Kumar A: New REE and Trace element data on two kimberlitic reference materials by ICP-MS. *Geostandards Geoanal Res*. 2007; 31 (3): 261-273.

[24] Cook PJ, Shergold JH: Proterozoic and Cambrian phosphorites-an introduction. In: Cook PJ and Shergold JH (Eds.), *Phosphate Deposits of the World, Proterozoic and Cambrian Phosphorites*, Cambridge University Press, New York, 1990; 1: pp. 1-8.

[25] Papineau D: Global biogeochemical changes at both ends of the Proterozoic: Insights from Phosphorites. *Astrobiology*. 2010; 10(2): 1-17.

[26] Bremner JM, Rogers J. Phosphorite deposits on the Namibian continental shelf. In: Burnett WCand Riggs SR (Eds.), *Phosphate Deposits of the World*, Cambridge University Press, Cambridge, UK, 1990; 3: pp. 143-158.

[27] Veeh HH, Burnett WC, Soutar A: Contemporary phosphorite on the continental margin of Peru. *Science*. 1973; 181: 844-845.

[28] Donnelly TH, Shergold JH, Southgate PN, Barnes CJ: Events leading to global phosphogenesis around the Proterozoic–Cambrian transition. In: Notholt AJG and Jarvis I (Eds.), Phosphorite research and development. *J Geolog Soc Lon*. Special Publication. 1990; 52: 273-287.

[29] Ingall ED, Jahnke R: Evidence for enhanced phosphorus regeneration from marine sediments overlain by oxygen depleted waters. *Geochimica et Cosmochimica Acta*. 1994; 58: 2571-2575.

[30] Jarvis I, Burnet WC, Nathan Y, Almbatydin FSM Attia AKM, Castro LN, Flicoteaux R, Hilmy ME, Hussain V, Qutawnah AA, Serjani A, Zanin Y N., (1994). Phosphorite geochemistry state-of-the-art and environment concerns. *J Eclogae Geologicae Helvetiae*, 87(3): 643-700.

[31] van Cappellen P, Ingall ED: Benthic phosphorus regeneration, net primary production, and ocean anoxia: a model of the coupled marine biogeochemical cycles of carbon and phosphorus. *Paleoceanography*. 1994; 9: 677-692.

[32] Baioumy HM: Iron-phosphorus relationship in the iron and phosphorite ores of Egypt. *Chemie der Erde-Geochemistry*. 2007; 67: 229-239.

[33] Adesanwo OO, Adetunji MT, Dunlevey JN, Diatta S, Osiname OA, Adesanwo JK: X-ray fluorescence studies of Ogun phosphate rocks. *Electronic J Environ Agri Food Chem*. 2009; 8(10): 1052-1061.

[34] Howard PF, Hough MJ: On the geochemistry and origin of the D'Tree, Wonarh and Sherrin Creck phosphorite deposits of the Georgia basin, N. America. *Econ Geol*. 1979; 74(2): 260-284.

[35] Altschuler ZS: The geochemistry of trace elements in marine phosphorites, Part-I. Characteristic abundances and enrichment: Society of Economic Paleontologists and Mineralogists, Special Publication.1980; 29: 19-30.

[36] Baturin GN, Bezrukov PL: Phosphorites on the sea floor and their origin. *Mar Geol*. 1979; 31: 317-332.

[37] Banerjee DM: Lithotectonic phosphate mineralization and regional correlation of Bijawar Group of rocks in the Central India. In: Valdiya, KS (Eds.), *Geology of the Vindhyanchal*. Hindustan Publication Corporation, New Delhi. 1982; pp. 19-54.

[38] Riggs SR: Tectonic models of phosphate genesis. In: *Fertilizer Mineral Potential in Asia and Pacific*.E-W Center. Hawaii. 1980; p. 481.

[39] Bhattacharjee I, Agarwal M, Yadav OP, Roy MK, Saxena VP: A Note on the phosphatic tuffaceous shale hosted uranium mineralization in Semri group of sediments around Mankisar-Kathar Area, Sidhi District, Madhya Pradesh, Geological Society of India. 2008; 72(2): 215-218.

[40] Burnett WC, Veeh HH. Uranium-series disequilibrium studies in Phosphorite Nodules from South America.*Geochimica et Cosmochimica Acta*. 1977; 41(6): 755-764.

[41] Baioumy HM: Forms of iron in the phosphorites of Abu-Tartur area, Egypt. *Chin J Geochem*; 2002; 21(3): 215-226.

[42] Baturin GN, Derkachev AV: Association of phosphates with tuffites in a core of sediments from the Sea of Japan. *Doklady Earth Sci.* 2008; 419A(3): 368-372.

[43] Israili SH, Khan MWY. Trace element studies of phosphorite bearing Gangolihat dolomites, Pittorgarh, U. P. India. *Chem Geol.* 1980, 31: 133-143.

[44] Al-Hwaiti M, Matheis G, Saffarini G: Mobilization, redistribution and bioavailability of potentially toxic elements in Shidiya phosphorites, Southeast Jordan. *Environ Geol.* 2005; 47: 431-444.

[45] Rao VP, Hegner E, Naqvi SWA, Kessarkar PM, Masood SA, Raju DS: Miocene phosphorites from the Murray Ridge, northwestern Arabian Sea. *Palaeogeogr Palaeoclimatol Palaeoecol.* 2008; 260: 347-358.

[46] Imamoglu MS, Nathan Y, Coban H, Soudry D, Glenn C: Geochemical, mineralogical and isotopic signatures of the Semikan, West Kasrik Turkish phosphorites from the Derik–Mazidagi–Mardin area, SE Anatolia. *Int J Earth Sci.* 2009; 98: 1679-1690.

Coastal Management

Chapter 14

Management of Marine Protected Zones – Case Study of Bahrain, Arabian Gulf

Humood A. Naser

Additional information is available at the end of the chapter

http://dx.doi.org/10.5772/62132

Abstract

Coastal and marine environments in Bahrain are characterized by a variety of habitats, including seagrass beds, coral reefs, and mangroves that support some of the most endangered species such as dugongs and turtles. Marine Protected Areas (MPAs) are considered the most advocated approach for marine conservation. Several MPAs have been established in Bahrain. This study explores the ecological and legal contexts of MPAs in Bahrain and evaluates the effectiveness of these MPAs in achieving their conservation goals. Although MPAs are contributing to the protection of critical coastal and marine habitats and their associated flora and fauna, there is yet further need to strengthen efforts on conserving coastal and marine environments in Bahrain. Effectiveness of MPAs in Bahrain could be enhanced by developing management plans, implementing the necessary regulatory measures, and investing in long-term monitoring and research programs. Findings of this study could contribute to wider regional and international experience of the effectiveness of MPAs in protecting important coastal and marine environments.

Keywords: Conservation, biodiversity, marine protected areas, effectiveness, management

1. Introduction

Human-induced pressures including unsustainable fishery activities, coastal development, and pollution are major challenges to the conservation of biodiversity and natural and cultural resources in coastal and marine environments. Marine protected areas (MPAs) are globally recognized as an important regulatory tool for protecting and conserving coastal and marine biodiversity [1]. A marine protected area (MPA) is defined by the International Union for Conservation of Nature (ICUN) as a delineated geographical area, recognized, dedicated, and

managed, through legal or other effective means, to achieve long-term conservation of nature with associated ecosystem services and cultural values [2]. MPAs can improve ecosystem functions and services through maintaining ecological structure and processes that support economic and social uses of marine resources [3]. MPAs can also contribute toward climate change adaptation by protecting ecosystem resilience and essential ecosystem services [4].

According to reference [5], there are multiple objectives for the designation MPAs. Such objectives may include: (1) conservation and protection of natural marine resources to ensure their long-term viability and to maintain their genetic diversity; (2) restoration of damaged or over exploited areas considered as critical to the survival of important species; (3) improvement of the relationship between humans, their environment, and economic activities, by maintaining traditional uses and the sustainable exploitation of resources; (4) improvement of fishing yields, by protecting spawning and recruitment processes, and by enhancing fishing management; (5) resolution of present or anticipated conflicts between coastal area users; (6) improvement of knowledge about the marine environment by dealing with research and educational aspects; and (7) valuation of heritage for local communities through tourism and economic profitability. Therefore, there are growing national and international interests in establishing MPAs to protect biodiversity and to manage marine resources [6].

Various relevant international conventions, including the Convention on Biological Diversity (CBD), the Convention on Wetlands of International Importance (Ramsar Convention), and the World Heritage Convention, serve to advance the number and coverage of MPAs worldwide [7]. Aiming to protect and conserve its natural environment, Bahrain signed the Convention on Biological Diversity in 1992 and ratified it in 1996 [8]. Bahrain recognizes MPAs as a key strategy to protect sensitive ecosystems and marine resources. Therefore, natural coastal and marine protected areas have been established in Bahrain.

A recent study [9] evaluated the management effectiveness of MPAs in the Arabian Gulf and revealed variable levels of performance. This study highlighted a knowledge gap in relation to MPAs' conditions and their management practices in Bahrain. The main objective of the present study is to investigate whether the current environmental management system is enabling MPAs to achieve their intended objectives in protecting sensitive coastal and marine ecosystems in Bahrain. Accordingly, this study (1) explores the ecological and legal contexts of MPAs in Bahrain, (2) evaluates the effectiveness of Bahraini MPAs in achieving their conservation goals, and (3) suggests measures that might enhance the effectiveness of those MPAs in protecting coastal and marine ecosystems in Bahrain.

2. Methodology approach

The methodological approach of the present study (Fig. 1) is composed of four dimensions: (1) providing an ecological description for the regional and national coastal and marine ecosystems; (2) investigating the ecological characteristics of the Bahraini MPAs by reviewing available published and gray literature; (3) collecting in-field information related to the ecological conditions of MPAs in Bahrain; (4) analyzing the legal framework of MPAs in

Bahrain; and (5) soliciting views of conservation practitioners, academics, and other relevant bodies about the importance of MPAs and their effectiveness in Bahrain using a semi-structured interview.

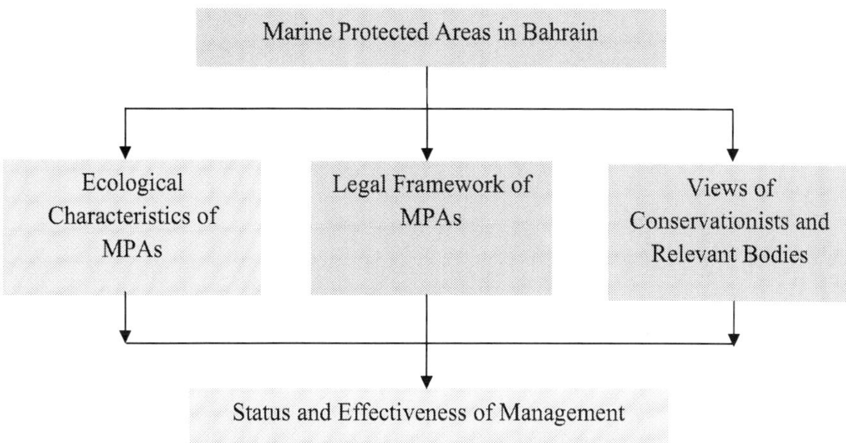

Figure 1. Methodology approach of the study.

3. Ecological setting of the Arabian Gulf

The Arabian Gulf is a semi-enclosed sea situated in the subtropical region of the Middle East (Fig. 2). It constitutes part of the Arabian Sea Ecoregion and represents a realm of the tropical Indo-Pacific Ocean. The Arabian Gulf is a relatively shallow basin with an average depth of 35 m. The marine environment of the Arabian Gulf is naturally stressed due to marked fluctuations in sea temperatures and high salinities. Sea temperatures can exceed a range of 20°C between winter and summer. Summer temperatures can exceed 35°C, making the Arabian Gulf the hottest sea on Earth [10]. The Arabian Gulf is also characterized by extreme salinities that exceed 44 psu in open waters and 70 psu in coastal lagoons. Therefore, marine flora and fauna inhabiting the Arabian Gulf are living close to the limits of their environmental tolerance [11]. Despite these extreme climatic conditions, the Arabian Gulf supports a wide range of coastal and marine ecosystems that contribute significantly to the productivity of marine resources, including seagrass beds, coral reefs, mangroves, and mudflats [12].

Anthropogenically, the Arabian Gulf is considered among the highest impacted regions in the world [13] due to intensive reclamation and dredging activities, industrial and sewage pollutants, hypersaline water discharges from desalination plants, and oil pollution [14]. Therefore, designation of MPAs is arguably critical for the protection of naturally stressed coastal and marine ecosystems in the Arabian Gulf. In relation to this, there are currently about

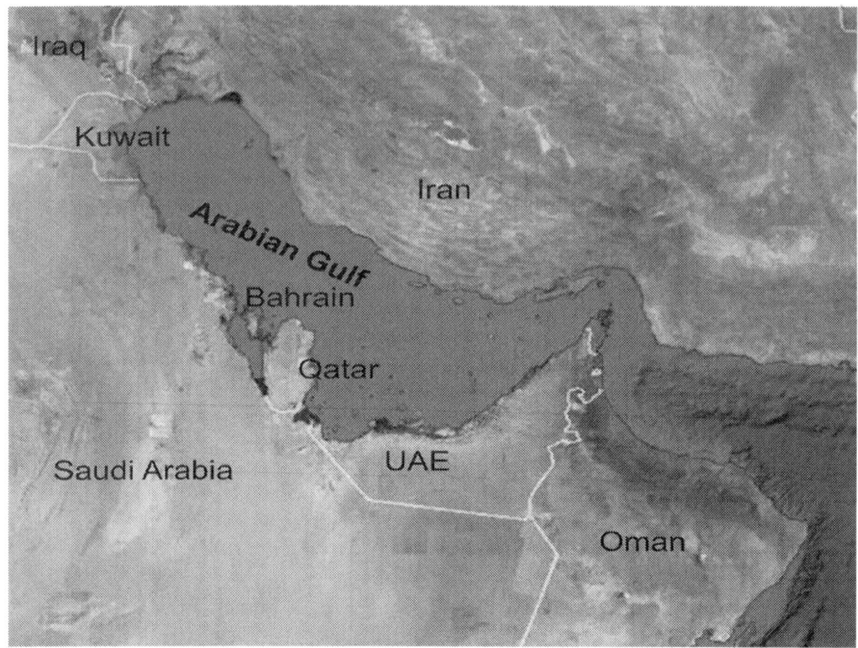

Figure 2. A map showing the Arabian Gulf and bordering countries (Google Earth).

38 officially designated MPAs covering around 18,180 km² in the Arabian Gulf [15]. However, several weaknesses in the MPAs in the Arabian Gulf were identified, including limitation in regulation enforcement, lack of management plans, weak communication with local stakeholders, traditional communities, and local marine resource users [9].

4. Coastal and marine biodiversity in Bahrain

The Kingdom of Bahrain consists of an archipelago of approximately 40 natural islands in addition to several islets, shoals, and patches of reefs, lying centrally on the southern shores of the Arabian Gulf between latitudes 25° 32' and 26° 20' north and longitudes 50° 20' and 50° 50' east. The general aspects of climate in Bahrain are cool winters (14–23°C) with moderate rainfall, and hot summers (26–40°C) with high humidity. The winter season in Bahrain is accompanied by rainfall with an average of approximately 74 mm per annum. The summer season is hot with a substantial increase in humidity due to the high rate of evaporation from the surrounding seawaters [16]. Salinities around Bahrain are generally high due to the effects of high temperatures associated with high evaporation rates. Salinities on the west coast are higher than those on the east coast, with average means of 50–57 psu for the west coast and 43–45 psu for the east coast. This variation in the salinity gradients may be attributed to a complex system of water circulation around Bahrain enforced by reduced water exchange, particularly in south of Bahrain, where salinity could reach 70–80 psu in areas with restricted water flow such as tidal pools and lagoons [17]. The total land area of Bahrain in 2012 is about 777 km² [16]. Despite the limited land area of Bahrain, waters surrounding its islands support several Valued Ecosystem Components (VECs) such as seagrass beds, coral reefs, mangrove swamps, and mudflats that provide important ecological goods and services.

Seagrass beds are highly productive ecosystems that are characterized by important ecological and economic functions. They provide food sources and nursery grounds for turtles, dugongs, shrimps, and a variety of economically important marine organisms. Three species of seagrass occur in Bahrain, namely *Halodule uninervis*, *Halophila stipulacea*, and *Halophila ovalis* [18]. Dense seagrass beds (>75 coverage) mostly occur on sandy substrates. The majority of the seagrass habitats in Bahrain waters are located in the eastern subtidal areas, mainly from south of Fasht Al-Adhm to Hawar Islands. The second largest seagrass beds are found in the western subtidal areas of Bahrain [19].

Coral reefs are characterized by both biological diversity and high levels of productivity. They provide a variety of ecological services such as renewable sources of seafood; maintenance of genetic, biological, and habitat diversity; recreational values; and economic benefits such as utilizing destructive reefs for creating land through the process of reclamation. Coral habitats are mainly restricted to the north and east of Bahrain. These include Fasht Al-Adhm to the east of Bahrain (the largest reef in Bahrain, 200 km²), Fasht Al-Jarim and Khor Fasht 20 km to the north, and Bulthama, 75 km northeast of Bahrain. Additionally, several other smaller reefs are scattered around eastern areas of Bahrain [19, 20]. Live coral cover in Bahrain is generally low, ranging from 5 to 16% [20]. This could be attributed to the massive bleaching events that occurred in the Arabian Gulf in the summers of 1996 and 1998. Bahrain was the worst affected by these events with an estimated overall loss of 97% of live corals [21].

Mangrove habitats are ecologically important coastal ecosystems that provide food, shelter, and nursery areas for a variety of terrestrial and marine fauna. Natural mangrove plants, which are represented by a single species, *Avicennia marina*, can be found only in Tubli Bay. The largest aggregation of mangroves is confined to Ras-Sanad lagoon, which is the most sheltered area on the coast of Bahrain. This lagoon has characteristics that support the growth and development of mangrove plants, including minimal motion of the water, muddy nature of the substrate, and an input of low salinity water from nearby farms and underground springs [22]. Transplanted mangroves can be found in Duwhat Arad area, which is a small intertidal mudflat.

Mudflats are among the most productive marine habitats that contribute significantly to the marine productivity in Bahrain. These habitats typically harbour large numbers of wintering and migratory shorebirds [23]. Mudflats in Bahrain are represented by low-energy areas on the northern and eastern coasts of Bahrain, as well as patches of sheltered lagoons within Hawar Islands.

Recent marine habitat mapping in Bahrain revealed sixteen types of habitats in Bahrain, namely algae, coral, seagrass, sand, mud, rock, mangrove, salt marsh, sabkha, rock and sand, coral-rock-sand, algae-rock-sand, mud and sand, mixed habitats, deep water mud, and deep water mixed habitats. Figure 3 shows the distribution of these marine habitats based on the Marine Atlas of Bahrain [19].

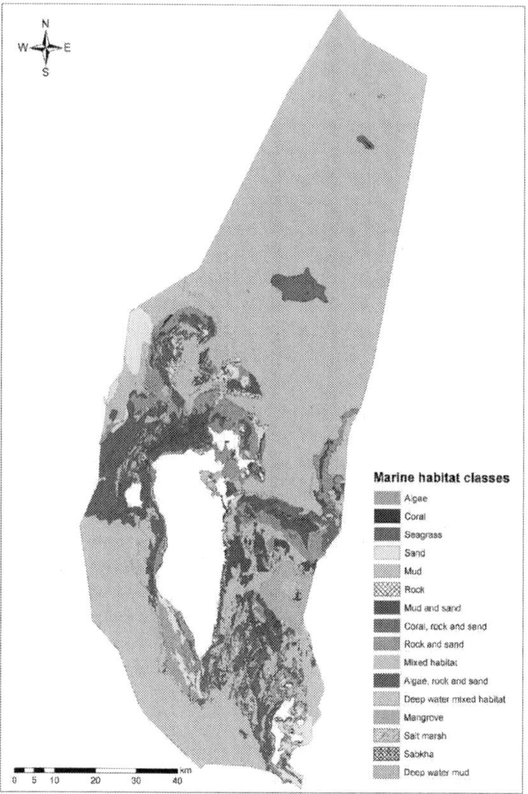

Figure 3. A map showing marine habitat classification in Bahrain [19].

5. Institutional governance of conservation in Bahrain

This section provides a brief overview of the institutional governance of conservation in Bahrain, which will provide a context for analyzing the legal framework of MPAs. The Environmental Protection Committee (EPC), established by Decree No. 7 of 1980 under the Ministry of Health, was the first governmental authority concerned with protection of environment in Bahrain. The EPC was upgraded by the Legislative Decree No.21 of 1996, establishing Environmental Affairs under the Ministry of Housing, Municipalities and

Environment. In 2000, the Environmental Affairs was reorganized under the Ministry of State for Municipalities Affairs and Environmental Affairs. A second governmental body concerned with wildlife protection is the National Committee for Wildlife Protection, which was established by the Legislative Decree No. 2 of 1995. This committee was upgraded to become the National Commission for Wildlife Protection by the Legislative Decree No 12 of 2000.

For the purpose of integration of efforts and resources to achieve an effective protection for the environment, the main governmental bodies concerned with the environment were joined under the umbrella of the Public Commission for the Protection of Marine Resources, Environment and Wildlife (PCPMREW) in 2002. The PCPMREW was established by the Legislative Decree No. 50 of 2002 and reorganized further by Legislative Decrees No. 10 and No. 43 of 2005.

More recently (2012), the Supreme Council for Environment (SCE) was established by the Legislative Decree No. 47 of 2012. The SCE forms the foundation to strategically integrating biodiversity considerations among all governmental and private sectors in Bahrain. The Legislative Decree with respect to Establishing and Organizing the Supreme Council for Environment explicitly indicates that decisions of the Council are binding to all ministries, authorities, and institutions in the Kingdom. Figure 4 shows the chronological development of environmental bodies in Bahrain.

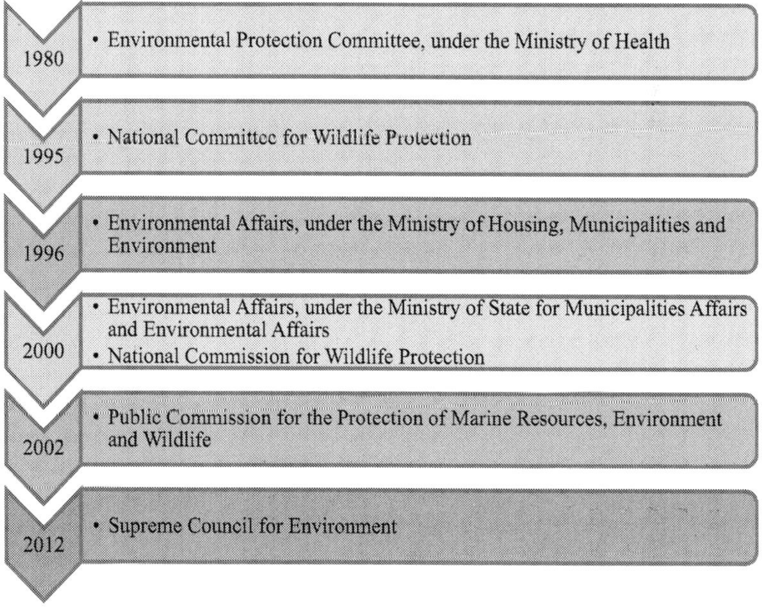

Figure 4. Institutions concerned with conservation in Bahrain.

6. Marine protected areas in Bahrain

Coastal and marine environments in Bahrain are characterized by regionally and globally important habitats, including seagrass beds, coral reefs, and mangroves. Additionally, waters surrounding Bahrain support some of the most endangered species such as dugongs and turtles. These ecosystems maintain genetic and biological diversity and contribute to food security for the region. However, these ecosystems are vulnerable to environmental extremes in the Arabian Gulf and susceptible to the increasing anthropogenic stressors from the rapid economic and industrial developments in Bahrain. Therefore, protecting these fragile ecosystems is a priority in Bahrain. Designating marine protected areas is considered as a key strategy to protect coastal and marine ecosystems in Bahrain. Toward this, there are currently five designated natural marine protected areas in Bahrain, namely Hawar Islands, Tubli Bay, Mashtan Island, Duwhat Arad, and Fasht Bulthama (Fig. 5).

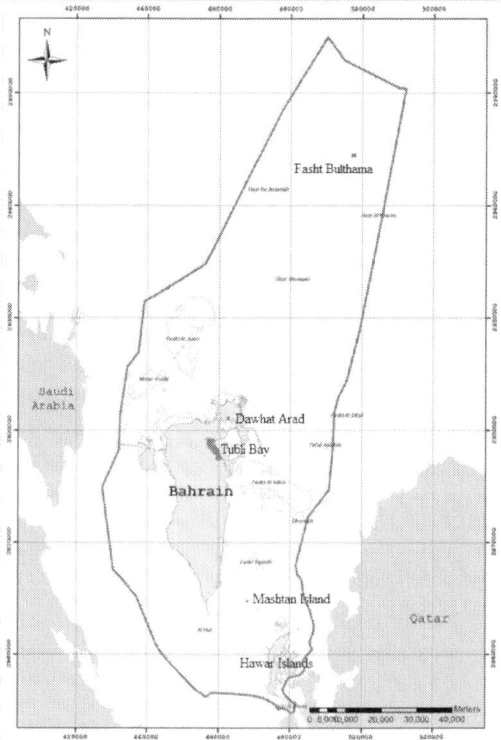

Figure 5. A map showing marine protected areas in Bahrain; Hawar Islands, Tubli Bay, Mashtan Island, Duwhat Arad, and Fasht Bulthama. Modified from reference [24].

6.1. Hawar islands

6.1.1. Ecological characteristics

The archipelago of Hawar Islands (581 km^2), which is located in the Gulf of Bahrain (Fig. 6), is characterized by varied coastal habitats, including muddy, sandy, and rocky shores as well as saline wetlands. These islands are surrounded by shallow waters (average depth 2 m) that promote the growth of extensive seagrass beds and algal mats (Fig. 7). Waters surrounding Hawar Islands support vulnerable species such as dugongs, turtles, and dolphins. For instance, seagrass beds between Bahrain and Hawar Islands support a large population of the globally threatened dugongs (*Dugong dugong*) with an estimated 5,800 individuals. This population of dugongs is considered the largest one known outside Australia [25]. Similarly, significant populations of green turtles and dolphins depend on these seagrass beds, and their associated fish as feeding grounds [26].

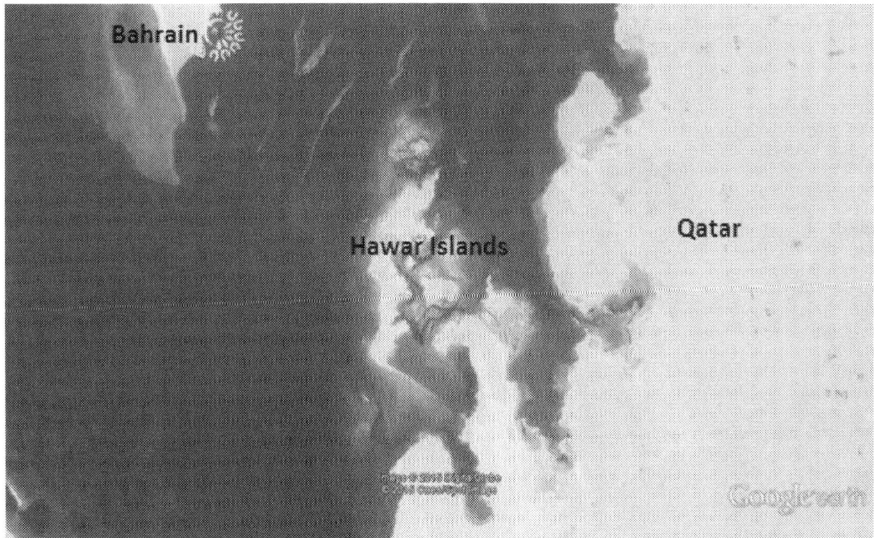

Figure 6. A map showing the geographic location of Hawar Islands (Google Earth).

Hawar Islands are vastly important for seabirds. These islands host the World's largest breeding colony of Socotra Cormorants (*Phalacrocorax nigrogularis*), with a winter population of 200,000 individuals [27]. Additionally, they are a breeding area for many other important seabirds, including western reef herons (*Egretta gularis*) and terns (*Sterna caspia, Sterna repressa, Sterna anaethus, Sterna bengalensis*). These islands are also important as a feeding and roosting area for many migratory seabird species such as greater flamingos (*Phoenicopterus ruber*), seagulls (*Larus genei* and *Larus argentatus*), and several waders (*Calidris minuta, Calidris alba,*

Figure 7. Extensive growth of seagrass beds around Hawar Islands.

Calidris ferruginea, Calidris alpina) [27, 28]. The islands also host endangered and vulnerable birds, such as the sooty falcon (*Falco concolor*) and Ospreys (*Pandion haliaetus*) [27]. Intertidal sand and mudflats of the Hawar Islands are characterized by rich and diverse macrobenthic assemblages. For instance, a baseline ecological survey for the main island revealed a total of 119 species of macrobenthic assemblages [29].

6.1.2. National and international legal protection

Hawar Islands are characterized by a complex of ecological habitats, including seagrass beds, algal mats, sand, and mudflats. Additionally, these islands are home to distinctive populations of flora, mega-fauna (*i.e.*, dugongs, turtles, and dolphins), and avifauna. Therefore, Hawar Islands are under national and international designations. Nationally, Hawar Islands were declared as a wildlife sanctuary by the Prime Minister Order No. 16 of 1996 with respect to Declaration of Hawar Islands and their Territorial Waters as a Protected Area. This designation was in accordance to the Legislative Decree No. 2 of 1995 with respect to the Protection of Wildlife.

The objectives of the Prime Minister Order were the protection of the terrestrial and marine natural habitats and their wild species of Hawar Islands and the protection of rare and endangered species. Several subsequent regulations were issued to support the protection of Hawar Islands, including the Ministerial Order No. 6 of 1996 with respect to the recommendations of the National Commission for Wildlife Protection Related to Hawar Islands and their Territorial Waters. This order prevents all forms of fishing around Hawar Islands and their territorial waters and only allows the use of traditional fishing gear such as hadrah (an intertidal fixed stake trap), cages, and trolls. This order also prevents taking eggs or chicks of breeding birds, including *Falco concolor, Pandion haliaetus, Phalacrocorax nigrogularis, Sterna repressa*, and *Egretta gularis*.

Fishing around Hawar Islands was regulated by the Order of the Public Commission for the Protection of Marine Resources, Environment and Wildlife No. 13 of 2005 with respect to the Regulation of Fishing in Hawar Islands and their Territorial Waters. This order prohibits fishing in commercial quantities and overfishing around Hawar Islands and their territorial waters. It also prevents the use of any tools, machines, or materials that can pose a threat to the marine resources in the region. The order allows the use of the traditional fishing gears (*i.e.*, hadhrah, cages, trolls). However, an amendment of this order based on the Order No. 4 of 2010 restricts the use of hadhrah to a license from the General Directorate for the Protection of Marine Resources.

Internationally, Hawar Islands were designated as a Ramsar site (Convention on Wetlands of International Importance) in 1997 due to the abundance of globally significant, rare, and endangered bird species.

6.2. Tubli Bay

6.2.1. Ecological characteristics

Tubli Bay (Fig. 8), located in the northeast of Bahrain, is a sheltered coastal body that supports a variety of habitats, including mudflats, seagrass beds, algal mats, mangrove swamps, and patches of rocky shores. Tubli Bay is a productive shallow bay that forms an important nursery ground for several commercial fish species and prawns [30]. Additionally, it provides important feeding and roosting grounds for large numbers of wintering and migratory shorebirds. Several wader species were recorded in the bay, including *Ardea cinerea, Arenaria interpres, Calidris minuta, Charadrius alexandrinus, C. hiaticula, C. mongolus, Egretta garzetta, Gallinula chloropus, Himantopus himantopus, Larus ichthyaetus, L. ridihundus, L. genei, Limicola falcinellus*, and *Pluvialis squatarola* [28, 31].

Tubli Bay hosts the last standing natural mangrove ecosystems in Bahrain (Fig. 9). Mangroves, *Avicennia marina*, are confined to few patches of sheltered muddy areas along the coastline of the bay. The main aggregation of mangroves is found in Ras-Sanad area (0.5 km²), southwest of the bay. Water and sediment characteristics at Ras-Sanad area support the growth and development of mangrove plants [22]. Studies investigating the benthic community structure in Tubli Bay revealed a significant role of mangrove plants in maintaining diversity and productivity of marine benthos [32, 33].

Figure 8. A map showing Tubli Bay and Ras-Sanad area, southwest of the bay (Google Earth).

Figure 9. Mangrove plants in Ras-Sanad area, Tubli Bay.

Tubli Bay has been under pressure from several human activities. These include reclamation and dredging activities (Fig. 10) and sewage discharges (Fig.11) from the main sewage treatment plant in Bahrain. The marine area of Tubli Bay has been reduced from 25 to 12 km^2 in 2008 due to intensive reclamation activities. These activities significantly destroyed mangrove stands and reduced their spatial distribution. Historically, mangrove stands were densely dominant on the coastline of Tubli Bay [33]. However, due to coastal development and pollution, mangrove stands have progressively reduced over the years (Fig. 12). Recently, a study investigating the composition of mangrove plant community in Tubli Bay during the years 2005 and 2010 indicated that the current area of mangroves is around 0.31 km^2 [34].

Figure 10. Reclamation activities along the coastline of Tubli Bay.

6.2.2. National and international legal protection

The mangroves in Ras-Sanad areas were declared as a nature reserve by the Environmental Protection Committee in 1986. Tubli Bay was declared as a protected area in 1995 based on the Ministerial Order No. 1 of 1995 with respect to Banning of Reclamation in Tubli Bay. This order explicitly prohibits construction activities in the mangrove area in Ras-Sanad as it is considered as a nature reserve category (*a*) in accordance with the Council of Ministers' decision dated April 16, 1995. The order also prevents construction activities in the remaining area of Tubli Bay as it is considered as a nature reserve category (*b*) in accordance with the same decision of the Council of Ministers.

Figure 11. Sewage discharges to the marine environment of Tubli Bay from the main sewage treatment plant in Bahrain.

Law No. 53 of 2006 with respect to the Declaration of Tubli Bay as a Natural Protected Area was issued to prevent the unsustainable reclamation activities along the coastline of Tubli Bay. This law considers Tubli Bay as natural protected area (Natural Park) and stresses on stopping all types of reclamation and dredging activities in the Bay. The law also directs all the competent governmental authorities to take all necessary measures to protect the environment and wildlife in the bay. This was followed by the Ministerial Order No. 70 of 2011 with respect to Determination of the Reclamation Line in Tubli Bay. This order specifies the final line of the marine area of the bay. Internationally, Tubli Bay was designated as a Ramsar site in 1997 for birds.

6.3. Mashtan Island

6.3.1. Ecological characteristics

Mashtan is a small offshore sandy island (0.2 km^2) located between Bahrain and Hawar Islands (Fig. 13). Waters surrounding this island are characterized by widespread growth of seagrass beds that provide feeding grounds for many species, including dugongs and sea turtles. Additionally, the surrounding environment is characterized by the presence of mixed habitats, including scattered coral reefs and patches of sand and mudflats [19].

Figure 12. Destroyed mangrove plants in Ras-Sanad area due to a large spill of cooking oil from a nearby factory in 1997.

6.3.2. National protection

Mashtan Island is nationally protected based on the Order of the National Commission of Wildlife Protection No.1 of 2002 with respect to the Declaration of Mashtan Island as a Protected Area. The designation of Mashtan Island as a protected area is based on the general policy of the government to protect terrestrial and marine habitats and their associated organisms.

6.4. Duwhat Arad

Duwhat Arad is a sheltered bay (0.5 km^2) located in the northeast of Bahrain (Fig. 14). The bay is a shallow mudflat that forms an important ground for feeding and roosting of important shorebirds, including *Charadrius alexandrinus*, *Charadrius hiaticula*, *Egretta gularis*, *Himantopus himantopus*, *Larus ridibundus*, and *Phoenicopterus ruber* (Fig. 15). Duwhat Arad has been an inshore area of great economic, ecological, and scientific importance that warrants the designation as a protected area [23]. Several mangrove stands of *Avicennia marina* have been transplanted in the bay (Fig. 16). Although the transplanted mangrove plants are found in relatively stunted forms, they contribute to the productivity in the bay. Duwhat Arad is protected based on The Public Commission for the Protection of Marine Resources, Environment and Wildlife Order No. 4 of 2003 with respect to the Declaration of Duwhat Arad as a Marine Protected Area.

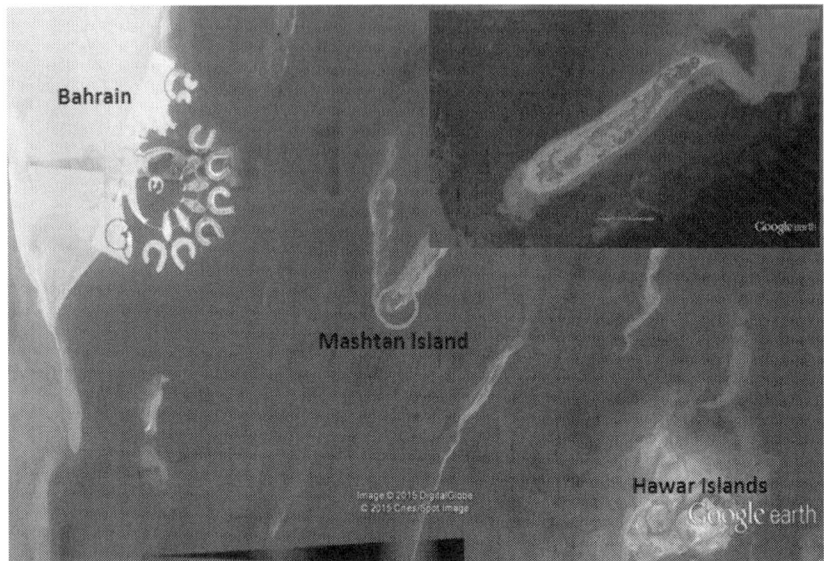

Figure 13. A map showing the location of Mashtan Island (Google Earth).

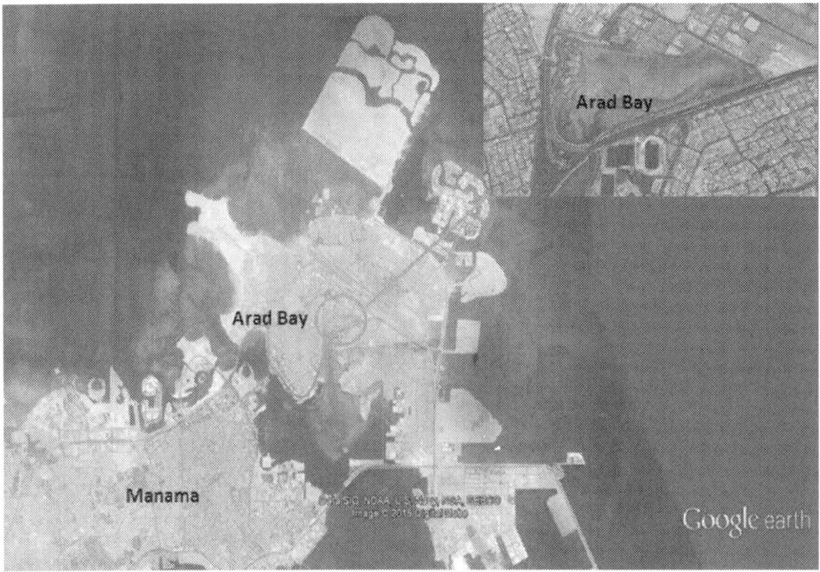

Figure 14. A map showing the location of Arad Bay (Google Earth).

Figure 15. Wader birds in Arad Bay during high tide.

Figure 16. Transplanted mangroves in Arad Bay.

6.5. Hayr and Fasht Bulthama

Fasht Bulthama contains the healthiest corals in Bahraini waters (Fig. 17). This reef is characterized by relatively high levels of coral diversity (22 species) and coral cover (16.3%) [20]. The reef is surrounded by pearl oyster banks characterized by high abundance of the pearl oyster (*Pinctada* sp.). The reef and the oyster banks represent a hub for high fish stocks. Hayr and Fasht Bulthama are protected based on the Order of the Public Commission for the Protection of Marine Resources, Environment and Wildlife No. 8 of 2007 with respect to the Declaration of Hayr Bulthama Area as a Natural Marine Protected Area. This order bans trawling and the use of nets in this protected area.

Figure 17. Fasht Bulthama contains high levels of coral and fish diversity. (© Courtesy of Hani Bader)

7. Views of conservationists on the effectiveness of MPAs in Bahrain

Views of conservationists, environmentalists, and other relevant bodies related to the effectiveness of marine protected areas in Bahrain were solicited using semi-structured interviews. The target population for the interviews consisted of members from the Supreme Council for Environment (1), academic institutions (8), and environmental consulting firms (2). They are experts in several fields related to biology conservation, including marine ecology, marine biology, environmental biology, and environmental management. Additionally, selected graduates (5) who completed a course in conservation biology at the Department of Biology, University of Bahrain were included in the interviews. The interviewees were asked for their views and experience in reference to the effectiveness of the five designated natural marine protected areas in Bahrain.

The contents of the semi-structured interviews were derived from the Marine Protected Area Score Card [36]. This score card is a simple questionnaire that depends on the use of existing knowledge and information from academics, local scientists, or experts. It is composed of six main headings reflecting the stages of marine protected area management proposed by the International Union for Conservation of Nature (IUCN). These stages include context, planning, inputs, processes, outputs, outcomes. The score card comprises 34 main questions under the mentioned stages related to legal status and regulation of MPA, threats to MPA, monitoring of MPA, stakeholder engagement and awareness, objectives of MPA, management plan, research, and communication. However, some of these questions were modified or merged to accommodate the environmental and legislative contexts, and regularity frameworks in Bahrain.

A comprehensive and enforced legislation framework is a key pillar for the success of any MPA. Generally, the five MPAs in Bahrain are legally designated (*i.e.*, gazetted). Two of these MPAs (Hawar Islands and Tubli Bay), are internationally designated as Ramsar sites. There are supplementary regulations for Hawar Islands and Tubli Bay that can further enhance their legal status. Nonetheless, these regulations might be constrained by the limited enforcement and implementation. The designation orders of Mashtan Island, Dawhat Arad, and Fasht Bulthama contain limited information related to the MPAs' objectives, management, and boundaries. Although the MPAs' objectives and boundaries might be known by the environmental authority, it is critically important to define them for relevant stakeholders and the public. For instance, the boundaries surrounding Hawar Islands, Mashtan Island, and Fasht Bulthama are not clearly defined.

Human and financial resources are critical for the success of any conservation initiative. The establishment of the Directorate of Biodiversity at the Supreme Council for Environment in 2012 is a major step forward to strengthen the institutional capacity in protecting biodiversity in Bahrain. However, most of the interviewees indicated that the Directorate of Biodiversity might be understaffed with limited budget, which could affect the implementation of major ongoing conservation programs. For instance, The Supreme Council for Environment has finalized a detailed plan to ecologically rehabilitate the mangroves in Ras-Sanad area. However, limited financial resources are hindering such initiatives.

Globally, MPAs are increasingly exposed to a wide range of anthropogenic threats. Reclamation and dredging, industrial and sewage effluents, hypersaline water discharges from desalination plants, and oil pollution are examples of anthropogenic stressors that may affect the MPAs in Bahrain [37]. The remote locations of Hawar Islands and Fasht Bulthama make them less vulnerable to human disturbance. However, these two MPAs are affected by overfishing. Several interviewees confirmed the intensive fishing activities in Hawar Islands and Fasht Bulthama. Mashtan Island is susceptible to sedimentation due to dredging and reclamation activities south of Bahrain. Avifauna in Arad Bay might be affected by intensive recreational activities. Tubli Bay is mainly affected by reclamation activities and sewage pollution. Reclamation has been more marked in Tubli Bay. This bay has nearly lost 50% of its original marine area. Reclamation activities in the bay have resulted in a loss of many mangrove stands, mudflats, and rocky patches. Excessive nutrient enrichment due to the continuous discharge of domestic sewage to the shallow marine area of the bay has resulted in several eutrophication incidents. Generally, all interviewees agreed that there are limited comprehensive mechanisms to control or limit human activities affecting MPAs in Bahrain.

A successful MPA depends on knowledge-based decision making, where scientific research is integrated into management actions [38]. Studies investigating marine protected areas and their associated ecosystems in Bahrain have included surveys of marine benthic assemblages in Tubli Bay [30, 32, 33, 39, 40], ecological aspects and spatial distribution of mangroves [35, 41, 42], physiological aspects of mangroves [22], microbial communities in Ras-Sanad area [43], ecological surveys for Hawar Islands [29, 44], coral cover in Fasht Bulthama [20], avifauna distribution in Tubli Bay and Arad Bay [23, 28, 31], environmental assessment for Tubli Bay [45], heavy metal contamination in Tubli Bay [46], and the selection of marine protected areas in Bahrain [47]. This latter publication [47] is considered the most comprehensive study detected to prioritize candidate MPAs based on ecological and socioeconomic criteria using a Geographic Information System (GIS). Despite their importance, these studies are rarely referred to in the decision-making process related to MPAs in Bahrain as agreed by most of the interviewees.

Monitoring physical, chemical, and biological aspects of MPAs can provide decision makers with information on the state of biodiversity, and consequently, assist in identifying management goals and assessing priorities for conservation [48]. It was agreed by interviewees that information on the biophysical, sociocultural, and economic conditions associated with MPAs in Bahrain is not sufficient to support planning and decision-making process. Generally, adequate monitoring programs for MPAs in Bahrain are limited. Some of the interviewees argued that the main Universities in Bahrain (University of Bahrain and Arabian Gulf University) have the scientific and technical capabilities to conduct integrated monitoring programs for the MPAs in Bahrain. Toward this end, the University of Bahrain has conducted two major ecological surveys for Hawar Islands in support of sustainable development in these islands. However, such monitoring programs are hindered by limited financial resources. For instance, an interviewee indicated that a proposed study by the University of Bahrain to conduct an integrated environmental assessment for Arad Bay, which could result in formulating a management plan for the bay, is awaiting financial support from concerned environmental authority in Bahrain.

Understanding of MPAs' objectives by local communities is recognized as a key for the success of conservation initiatives [49]. Further, public participation in the designation process is important for effective implantation of regulations and management plans. Knowledge and attitudes of local communities toward marine protected areas in Bahrain were investigated by a recent study [50] using a questionnaire survey. This study reported an agreement (95%) among the surveyed population with respect to the need for MPAs in Bahrain. However, only 55% of the respondents indentified all the MPAs in Bahrain. This highlighted the need for more systematic measures to increase environmental awareness related to MPAs in Bahrain among stakeholders and the public. Most of the interviewees stressed the need to further incorporate conservation aspects in schools' curricula, and to increase awareness programs through different sources of media.

Education plays an important role in capacity building related to conservation biology in Bahrain. Toward this end, a course in conservation biology was introduced at the Department of Biology, University of Bahrain in 2009. This course explores the principles of conservation biology, including biological diversity and its value, threats to biological diversity, conservation at the population and species levels, and management of habitats and ecosystems. The course discusses aspects of biodiversity conservation in Bahrain. Several graduate students skilled in knowledge relating to conservation biology joined relevant institutions. Although such initiatives should reflect positively in capacity building related to conservation biology, most of the interviewees stressed the need for systematically introducing the biodiversity concerns into the higher education institutions in Bahrain.

Generally, there is an assumption among the interviewees that indicated that the effectiveness of MAPs in Bahrain could be improved by taking necessary legislative and management actions. Around 19% viewed MPAs in Bahrain as being effective. This was mainly attributed to the relatively stable ecological conditions in Hawar Islands and Fashit Bulthama. Nearly 37% of the interviewees emphasized that there are increasing signs of ecological and environmental degradation in the MPAs and deemed them relatively effective. Conversely, a large proportion of the interviewees (44%) indicated that MPAs are not effective under the current marine conservation system in Bahrain (Fig. 18). They attributed this to the increasing evidence of further degradation in the mangroves in Ras-Sanad area, the increasing of human activities in Arad Bay that may be limiting wintering waders from using the mudflats and the increasing of overfishing activities in Hawar Islands and Fasht Bulthama that could compromise their ecological integrity.

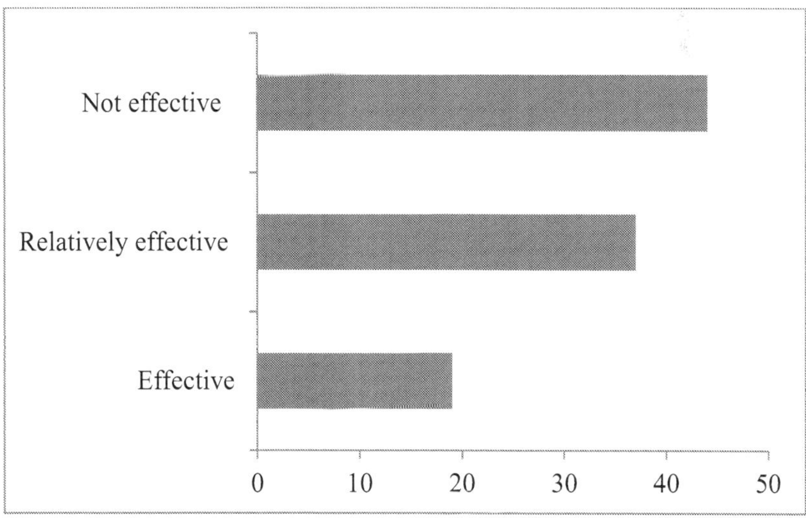

Figure 18. Views of interviewees on the effectiveness of MPAs in Bahrain.

8. Challenges and the way forward

The effectiveness of any MPA lies in its ability to support viable long-term populations; especially for vulnerable and endangered species and to maintain healthy ecosystems. MPAs in Bahrain are contributing to the protection of critical coastal and marine habitats and their associated flora and fauna. However, there are several constraints that may restrict the effectiveness of the MPAs in Bahrain. These include limitation in regulation enforcement, lack of comprehensive management plans, and limitation in participation of stakeholders and local marine resource users in the designation process.

Adaptive management, which is a structured and iterative management process, could be suggested as a measure to enhance the effectiveness of MPAs in protecting coastal and marine ecosystems in Bahrain. Adaptive management is composed of several stages, including collecting information related to MPAs from different sources, formulating and implementing management plans, and adjusting and refining management plans based on the outcomes from monitoring and practices. Figure 19 illustrates a typical adaptive management cycle [51].

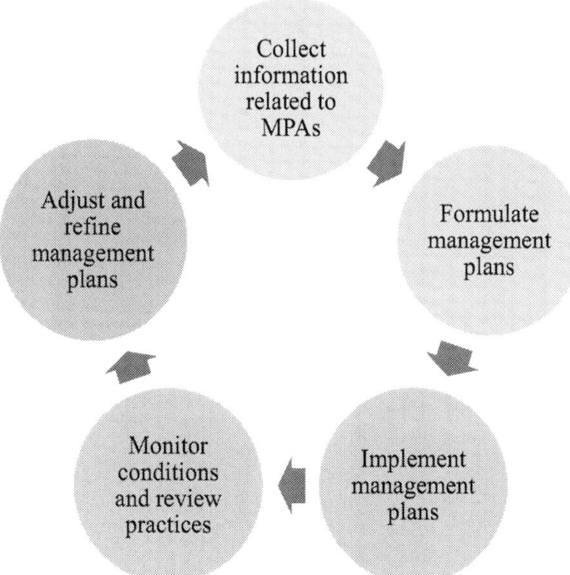

Figure 19. A typical adaptive management cycle [51].

Findings of this study indicate that there are yet further needs to strengthen efforts on conserving coastal and marine environments in Bahrain. Recommendations that may contribute to enhancing the effectiveness of MPAs in Bahrain might include:

- Characterizing the current biological, ecological, legal, social, and economic conditions of MPAs in Bahrain;

- Collecting information related to MPAs in Bahrain from all relevant bodies, including environmental authorities, stakeholders, scientists, and the general public;

- Consulting best currently available scientific knowledge related to MPAs in Bahrain to support decision making and management;

- Conducting studies to assess provisioning, supporting, and regulating cultural goods and services provided by coastal and marine ecosystems in Bahrain;

- Linking ecosystem services and human well-being, which will lead to a better understating of the importance of MPAs by decision makers, stakeholders, and the public;

- Formulating comprehensive management plans for MPAs in Bahrain;

- Incorporating concerns of all relevant stakeholders in the management plans;

- Defining clearly the conservation objectives, individually and collectively, of MPAs in Bahrain;

- Specifying the boundaries of MPAs in Bahrain;

- Formulating strategies to address human threats to MPAs in Bahrain;

- Implanting and enforcing the management plans in cooperation with all relevant stakeholders;

- Designing integrated monitoring programs that allow evidence-based decision making regarding MPAs in Bahrain;

- Encouraging scientific research and providing necessary financial resources;

- Incorporating conservation aspects in the curricula of schools and universities;

- Amending relevant MPAs legislations;

- Allocating human and financial resources to accomplish conservation initiatives;

- Strengthening the institutional and legal capacity of the Biodiversity Directorate at the Supreme Council for Environment;

- Increasing public awareness about ecosystem services and the role of MPAs.

Challenges of marine conservation in Bahrain can be extended to the whole region of the Arabian Gulf [9]. Regional network of MPAs could assist in protecting the fragile ecosystems of the Arabian Gulf. Therefore, further cooperation between local and regional institutions and organizations concerned with MPAs, scientific research, and environmental monitoring is required. In this respect, the Regional Organization for the Protection of the Marine Environment (ROPME) can play an important role in coordinating the efforts of marine conservation in the Arabian Gulf.

9. Conclusion

The marine environment in Bahrain supports a variety of habitats such as seagrass beds, coral reefs, mangroves, and mudflats. In addition to their intrinsic value and role in maintaining biodiversity, these ecosystems provide important ecological, economic, and cultural services to communities in Bahrain. Marine resources including fisheries provide sources of income, employment, and recreation, and contribute to the cultural heritage and food security. Bahrain has recognized the importance of marine biological and environmental resources and the need to protect them. MPAs are the most advocated approach for marine conservation. Several MPAs have been established to protect representative sensitive habitats in Bahrain, including offshore islands surrounded by extensive seagrass beds, mangroves swamps, coral reefs, and mudflats. However, these MPAs are influenced by anthropogenic stressors that may affect their ecological integrity, including reclamation and dredging, industrial and sewage effluents, oil pollution, and overexploitation. Effective management of MPAs in Bahrain is restricted by limitation in enforcement of relevant laws, lack of comprehensive management plans, and limitation in financial resources. Adopting integrated adaptive management approach, and promoting scientific research, education, and environmental awareness can enhance the effectiveness of MPAs in protecting coastal and marine ecosystems in Bahrain.

Acknowledgements

Support provided by the Department of Biology, University of Bahrain is greatly appreciated. The contributions by the interviewees are greatly acknowledged and appreciated. Special thanks are due to Dr. Brendan O'Connor for his constructive comments. Thanks are also extended to Tahani Naser for her technical assistance.

Author details

Humood A. Naser[*]

Address all correspondence to: hnaser@uob.edu.bh; humood.naser@gmail.com

Department of Biology, College of Science, University of Bahrain, Bahrain

References

[1] Fraschetti, S., Claudet, J., Grotud-Colvert, K. (2011). Management transforming from single-sector management to ecosystem-based management: What can marine pro-

tected areas offer? In: J. Claudet (ed.). *Marine Protected Areas: A Multidisciplinary Approach*. Cambridge University Press, Cambridge. pp. 11-34.

[2] Dudley, N. (2008). *Guidelines for Applying Protected Area Management Categories*. World Conservation Union (IUCN), Gland. pp. 86.

[3] Hoagland, P., Sumoila, U., Farrow, S. (2010). Marine protected areas. In: J. Steele, S. Thorpe, K. Turekian (eds.). *Marine Policy and Economics*. Elsevier Ltd. pp. 122-138.

[4] McLeod, E., Salm, R., Green, A., Almany, J. (2009). Designing marine protected area networks to address the impacts of climate change. *Front Ecol Environ*, 7: 362-370.

[5] Claudet. J., Pelletier, D. (2004). Marine protected areas and artificial reefs: A review of the interactions between management and scientific studies. *Aqua Living Res*, 17: 129-138.

[6] Douvere, F. (2008). The importance of marine spatial planning in advancing ecosystem-based sea use management. *Marine Policy*, 32: 762-771.

[7] Green, S., White, A., Christie, P., Kilarski, S., Blesilda, A., Meneses, T., Samonte-Tan, G., Karrer, L., Fox, H., Campbell, S., Claussen, J. (2011). Emerging marine protected area networks in the Coral Triangle: Lessons and way forward. *Conserv Soc*, 9 (3): 173-188.

[8] Fuller, S. (2005). Towards a Bahrain national report to the Convention on Biological Diversity. Prepared for the General Directorate for Environment and Wildlife Protection, Kingdom of Bahrain. pp. 147.

[9] Van Lavieren, H., Klaus, R. (2013). An effective regional marine protected area network for ROPME Sea Area: Unrealistic vision or realistic possibility? *Marine Poll Bull*, 72: 389-405.

[10] Riegl, B., Purkis, S. (2012). *Coral Reefs of the Gulf: Adaptation to Climatic Extremes*. Springer, Netherlands. pp. 379.

[11] Price, A. (2002). Simultaneous hotspots and coldspots of marine biodiversity and implications for global conservation. *Marine Ecol Progr Series*, 241: 23- 27.

[12] Naser, H. (2014). Marine ecosystem diversity in the Arabian Gulf: Threats and conservation. In: G. Oscar (ed.), *Ecosystem Biodiversity -The Dynamic Balance of the Planet*. InTech Publishing, pp. 297-328.

[13] Halpern, B. Walbridge, S. Selkoe, K. Kappel, C. *et al.* (2008). A global map of human impact on marine ecosystems. *Science*, 319: 948-952.

[14] Hamza. W., Munawar, M. (2009). Protecting and managing the Arabian Gulf: Past, present and future. *Aqua Ecosys Health Manag*, 12: 429-439.

[15] Van Lavieren, H., Burt, J., Feary, D., Cavalcante, G., Marquis, E., Benedetti, L., Trick, C., Kjerfve, B., Sale, P. (2011). Managing the growing impacts of development on

fragile coastal and marine ecosystems: lessons from the Gulf. A policy report, UNU-INWEH, Hamilton, ON, Canada. pp. 99.

[16] CIO, 2012. Census Summary Results, 2012. Central Informatics Organization. Kingdom of Bahrain (CIO). www.cio.gov.bh.

[17] Price, A. Vousden, D., Ormond, R. (1985). An ecological study of sites on the coast of Bahrain. UNEP Regional Seas Reports and studies No. 72, Geneva. pp. 10.

[18] Phillips, R. (2003). The seagrasses of the Arabian Gulf and Arabian Region. In: E. Green and Short F. (eds.), *World Atlas of Seagrasses*. UNEP-WCMC. pp. 74-81.

[19] Loughland, R., Zainal, A. (2009). *Marine Atlas of Bahrain*. GEOMATEC Bahrain Centre for Studies and Research, Bahrain.

[20] Burt, J., Al-Khalifa, K., Khalaf, E., Alsuwaikh, B., Abdulwahab, A. (2013). The continuing decline of coral reefs in Bahrain. *Marine Poll Bull*, 72: 357-363.

[21] Rezai, H., Wilson, S., Claereboudt, M., Riegl, B. (2004). Coral reef status in the ROPME sea area: Arabian/Persian Gulf, Gulf of Oman and Arabian Sea. In: Wilkinson, C. (ed.), *Status of Coral Reefs of the World*. Australian Institute of Marine Science, Townsville, pp. 155-169.

[22] Naser, H., Hoad, G. (2011). An investigation of salinity tolerance and salt secretion in protected mangroves, Bahrain. Gulf II: An international conference. The state of the Gulf ecosystem: Functioning and services. Kuwait City, Kuwait. 7-9 February 2011.

[23] Al-Sayed, H. Naser, H., Al-Wedaei, K. (2008). Observations on macrobenthic invertebrates and wader bird assemblages in a tropical marine mudflat in Bahrain. *Aqua Ecosys Health Manag*, 11: 450-456.

[24] Zainal, A., Al-Zayani, A., A.Qader, E., Burshaid, F., Roy, P. (2006). Marine Environmental Geographic Information System (MarGIS II) report. Geomatic, Bahrain. pp. 78.

[25] Preen, A. (2004). Distribution, abundance and conservation status of dugongs and dolphins in the southern and western Arabian Gulf. *Biologic Conserv*, 118: 205-218.

[26] Pilcher, N., Phillips, R., Aspinall, S., Madany, I., King, H., Hellyer, P., Beech, M., Gillespie, C., Wood, S., Schwarze, H., Al Dosaery, M., Farraj, I., Khalifa, A., Boer, B. (2003). Hawar Island Protected Area (Kingdom of Bahrain): Management plan. National Commission for Wildlife Protection, UNESCO, World Heritage. Bahrain. pp. 61.

[27] King, H. (1999). *The Breeding Birds of Hawar*. Ministry of Housing, Bahrain. pp.94.

[28] Mohammed, S. (1993). *Birds of Bahrain and the Arabian Gulf*. Bahrain Centre for Studies and Research Series, BCSR, Bahrain. pp. 228.

[29] Zainal, K. Al-Sayed, H. Ghanem, E. Butti, E., Nasser, H. (2007). Baseline ecological survey of Huwar Islands, The Kingdom of Bahrain. *Aqua Ecosys Health Manag,* 10: 290-300.

[30] Abdulqader E. (1999). The role of shallow waters in the life cycle of the Bahrain penaeid shrimps. *Est Coastal Shelf Sci,* 49:115-121.

[31] Kavanagh, B. (2014). *Birds of Dilmun: An Introduction to the Birds of Bahrain.* Miracle Graphics, Bahrain. pp. 356.

[32] Naser, H. (2010). Testing taxonomic resolution levels for detecting environmental impacts using macrobenthic assemblages in tropical waters. *Environ Monitor Assess,* 170: 435-444.

[33] Naser, H. (2013). Soft-bottom crustacean assemblages influenced by anthropogenic activities in Bahrain, Arabian Gulf. In: G. Sisto (ed.), *Crustaceans: Structure, Ecology and Life Cycle.* NOVA Science Publishers, Inc. New York. pp 95-112.

[34] Abbas, J. (2002). Plant communities bordering the sabkha of Bahrain Island. In: Barth, H. and B. Boer (eds.), *Sabkha Ecosystems: The Arabian Peninsula and Adjacent Countries.* Dordrecht, Kluwer Academic Publishers. pp. 21-62.

[35] Abido, M., Abahussain, A., Abdel Munsif, H. (2011). Status and composition of mangrove plant community in Tubli Bay of Bahrain during the years 2005 and 2010. *Arab Gulf J of Sci Res,* 29:100-111.

[36] Staub, F., Hatziolos, M. (2004). Score Card to Assess Progress in Achieving Management Effectiveness Goals for Marine Protected Areas. World Bank.

[37] Naser, H. (2015). The role of environmental impact assessment in protecting coastal and marine environments in rapidly developing countries: The case of Bahrain, Arabian Gulf. *Ocean Coastal Manag,* 104: 159-169.

[38] Cvitanovic, C., Wilson, S., *et al.* (2013). Critical research needs for managing coral reef marine protected areas: Perspectives of academics and managers. *J Environ Manag,* 114: 84-91.

[39] Al-Sayed, H., Buali, A., Ravenndran, E., Buflasa, A. (1995). Ecological characteristics of the littoral fauna at Tubli Bay, Bahrain. *Qatar Univ Sci J,* 15 (1): 231- 238.

[40] Naser, H. (2011). Human impact on marine biodiversity: macrobenthos in Bahrain, Arabian Gulf. In: Lopez-Pujol, J. (ed.), *The Importance of Biological Interactions in the Study of Biodiversity.* InTech, Croatia. pp. 109-126.

[41] Al-Sayed, H. (1995). Mangrove in Bahrain: A preliminary ecological assessment. Proceeding of the symposium on wildlife protection and development in the Arabian Gulf. National Commission for Wildlife Protection, Bahrain. pp. 12.

[42] Al-Zayani, A. K. (1999). Mangrove of Bahrain. National commission for wildlife protection (NCWP), Bahrain. pp. 74.

[43] Al-Sayed, H., Ghanem E, Saleh, K. (2005). Bacterial community and some physico-chemical characteristics in a subtropical mangrove environment in Bahrain. *Marine Poll Bull*, 50: 147-155.

[44] Buali, A. (2009). Final report of research project on ecological assessment in support of sustainable development in Hawar Islands. University of Bahrain, Bahrain. pp. 187.

[45] Abahussain, A., Al-Sabbagh, M. (2010). Integrated Environmental Assessment of Tubli Bay, Bahrain: Policy Analysis and Future Scenarios. *J Gulf Arab Peninsula Studies*, 36 (136): 245-286.

[46] Naser, H. (2012). Metal concentrations in marine sediments influenced by anthropogenic activities in Bahrain, Arabian Gulf. In: Shao Hong-Bo (ed.), *Metal Contaminations: Sources, Detection and Environmental Impacts*. NOVA Science Publishers, New York. pp. 157-175.

[47] Al-Zayani, A. K. (2003). The Selection of Marine Protected Areas: A Model for the Kingdom of Bahrain. PhD thesis. Centre for Environmental Sciences, University of Southampton, UK.

[48] Collen, B., Pettorelli, N., Baillie, J., Durant S. (2013). *Biodiversity Monitoring and Conservation: Bridging the Gap between Global Commitment and Local Action*. John Wiley & Sons, London. pp. 464.

[49] Kideghesho, J., Roskaft, E., Kaltenborn, B. (2007). Factors influencing conservation attitudes of local people in Western Serengeti, Tanzania. *Biodiversity Conserv*, 16: 2213-2230.

[50] Hazeem, L., Al-Rumaidh, M., Naser, H. (2013). Environmental awareness about marine protected areas in the Kingdom of Bahrain. The International Conference on Women in Science and Technology in the Arab Countries. Kuwait Institute for Scientific Research, Kuwait, April 21-23, 2013.

[51] Primack, R. (2010). *Essential of Conservation Biology*. Sinauer Associates. New York. pp. 585.

Coastal Environment Monitored by Remote Sensing

Chapter 15

Monitoring the Coastal Environment Using Remote Sensing and GIS Techniques

Dong Jiang, Mengmeng Hao and Jingying Fu

Additional information is available at the end of the chapter

http://dx.doi.org/10.5772/62242

Abstract

The coastal zone has been of importance for economic development and ecological restoration due to their rich natural resources and vulnerable ecosystems. Remote sensing techniques have proven to be powerful tools for the monitoring of the Earth's surface and atmosphere on a global, regional, and even local scale, by providing important coverage, mapping and classification of land cover features such as vegetation, soil, water and forests. This chapter introduced the methods for monitoring the coastal environment using remote sensing and GIS techniques. Case studies of port expansion monitoring in typical coastal regions, together with the coastal environment changes analysis were also presented.

Coastal zones are important for economic development and ecological restoration due to their rich natural resources and vulnerable ecosystems. Remote sensing techniques have been shown to be powerful tools in the monitoring of the Earth's surface and atmosphere on global, regional, and local scales. These techniques provide important coverage and mapping and help classify land cover, such as vegetation, soil, water, and forests. The main objectives of this chapter are (1) to introduce the state-of-the-art techniques on monitoring coastal environments using remote sensing and geographic information systems (GIS), including landscape classification, automatic classification and change detection on regional scale, and objective-based classification on local scale, and (2) to present case studies of coastal environment monitoring, including Zhan Jiang Port in the stage of specialization, Gwadar Port and Djibouti Port in the expansion stage, and Ilichevsk Port in the stable stage.

Keywords: Coastal environment, remote sensing, GIS, modelling

1. Introduction

1.1. Main issues of coastal environments

Until now, there has been no strict definition of coastal zones. Generally, coastal zones are the transition zones between the land and the ocean [1]. As defined by the International Geosphere Biosphere Programme (IGBP) via its Land Ocean Interactions in the Coastal Zone (LOICZ) program, the coastal zone is the entire region from the 200 m bathymetric contour in the sea to the 200 m elevation contour on land [2]. The coastal zone represents a comparatively small but highly productive and extremely diverse system with a variety of ecosystems [3]. Hence, the coastal zone is clearly of major economic and social importance. For example, (a) the coastal zone occupies 18% of the surface of the globe, (b) it is the area where approximately one quarter of global primary productivity occurs, (c) it is where approximately 60% of the human population lives, (d) it is where two thirds of the world's cities with populations of more than 1.6 million people are located, and (e) it supplies approximately 90% of the world's fish catch [4]. Therefore, monitoring the coastal zone is meaningful.

Currently, coastal zone monitoring primarily includes the content of the shoreline displacement, landscape change, ecosystem change, and environmental pollution. The total suspended matter (TSM) concentration was estimated from MODIS data using a neural network model in China's eastern coastal zone [5]. The trend of sea surface temperature (SST) was monitored from MODIS data for the Mumbai coast [6]. The spatiotemporal changes in the erosion and deposition of two Mediterranean river deltas over a 25-year period (1984–2009) were analyzed in [7]. Land-use change processes in Port Harcourt, which is a densely populated coastal city located in southeast Nigeria, have been analyzed over 27 years in [8]. The water quality of coastal waters was monitored based on chlorophyll-a in [9]. The mangrove forests in Malaysia were monitored and evaluated their significance to the coastal marine environment in [10]. The coastline changes and erosion-accretion evolution of the Pearl River Estuary were analyzed based on remote sensing images and nautical charts in [11]. The coastal and marine ecological changes and fish cage culture development in Phu Quoc, Vietnam, from 2001 to 2011 were estimated in [12]. The coastal sensitivity of Thiruvananthapuram on the west coast of India was assessed in [13].

1.2. Methods for monitoring coastal environments

As the intersection of land and sea, coastal zones are complex and variable. The traditional means of coastal zone research have certain limitations. Both the monitoring means and the monitoring intensity struggle to meet the demand of real-time monitoring due to coastal zone development, environmental changes, and disasters. Among the modern methods for monitoring terrestrial ecosystems, remote sensing is of primary importance due to its ability to provide synoptic information over wide areas with high acquisition frequencies [3]. Remote sensing has been used in resource development, the planning and management of the coastal zone, the monitoring of shoreline changes, and the understanding of physical processes in the coastal environment with geographic information systems (GIS) [14].

1.2.1. Remote sensing

Remote sensing technology acquires and records information without coming into direct contact with an object. Remote sensing was redefined as the science and technology of Earth observation, including space to Earth observation, aerial observation, and field monitoring[15]. From this perspective, remote sensing data can be divided into data on a global scale, a regional scale, and a local scale.

Global-scale satellites include static meteorological satellites, such as the GOES-8, the GOES-10, and the GMS meteorological satellite. The GOES-8 and the GOES-10 are the stationary satellites of NOAA. Their purposes are daily weather observations. These satellites provide a variety of meteorological and nonmeteorological services and play an important role in the study of global climate change, weather forecasting, disaster prevention, and disaster reduction.

Regional-scale data are generally obtained from moderate-resolution remote sensing images, such as MODIS sensor and Landsat satellites. MODIS is an important sensor equipped with Terra and Aqua satellites. Its multispectral data can reflect the information of land surface condition, ocean color, phytoplankton, biological geography, chemistry, atmospheric water vapor, aerosol and surface temperature, atmospheric temperature, and so on. Launched by the United States, Landsat satellites are primarily used to capture the remote sensing images of land, including soil organisms and plants. These satellites provide accurate and dynamic geographical information sources. Regional-scale data can be used in the macroscopic analysis of coastal zone changes, for example, see [6].

Local-scale satellites are usually used for monitoring in a smaller scope with high resolutions, such as worldview satellite, airborne satellites, and unmanned aerial vehicles (UAVs). The benefits of high spatial and high spectral resolution data are their ability to match the rich spectral and spatial diversities observed in coastal systems. For example, The invasion of *Spartina alterniflora* was monitored using very high resolution UAV imagery in Beihai [16]. The results showed that UAV imagery can provide details on the distribution, progress, and early detection of *S. alterniflora* invasion, and the total accuracy was 94.0%.

Based on the spectral types, satellites can be divided into optical satellites and microwave satellites. Hyperspectral and multispectral data provide more information for identifying targets. Unlike optical satellites, microwave satellites can penetrate through snow, soil, and forest. The benefit of combined optical and SAR-based approaches in improving classifications over some coastal habitats was demonstrated.

Currently, remote sensing data are used in various ways: they serve as input boundary conditions and validation data for numerical simulation models, they are combined with in situ measurements to draw sediment transport maps, and they are assimilated in 3D coastal sediment transport models and in the light forcing of an ecosystem model [17].

1.2.2. Geographic Information System (GIS)

GIS is an important and specific spatial information system. It is the technical system for the collection, storage, management, operation, analysis, display, and description of geographic distribution data for the entirety or a part of the Earth's surface (including the atmosphere) in support of computer hardware and software systems. GIS has spatial analysis capabilities, can store and manage vast amounts of complex spatial data and attribute data, and can use spatial databases for the comprehensive analysis of multiple factors with quantity, quality, and localization. However, it is difficult to achieve attribute data modeling relying solely on GIS software.

Combining GIS and mathematical models can make modeling easier by improving accuracy and solving problems effectively. For example, Anthony et al. used GIS with fuzzy learning vector quantization (FLVQ) for coastal vegetation classification. The classification accuracy of FLVQ was comparable to a conventional supervised multilayer perceptron, trained with backpropagation (KHAT accuracy: 82.82% and 84.66%, respectively; normalized accuracy: 74.60% and 75.85%, respectively), with no significant difference at the 95% confidence level [18]. An eutrophication model for Bohai Bay based on a cellular automata-support vector machine (CA-SVM) was established in [19] by applying the soft computing approach with a large quantity of remote sensing data to the marine environment. Their comparison between the optimized model and the basic model indicated that prediction accuracy was improved by the optimized model. The spatiotemporal patterns of phosphorus concentrations in a coastal bay were explored with machine learning models in [20]. GIS spatial analysis methods, along with mathematical models, can be used to analyze the interaction between coastal zone changes and the effects of human activities. These methods are very effective for coastal zone monitoring.

2. Methodology

2.1. Landscape classification

Remote sensing image classification is an important way to extract information. Different surface features have different spectral characteristics. Landscape classification is the clustering of the same or similar pixels and the assigning of value to each pixel class through the analysis of spectral characteristics based on satellite remote sensing images.

Landscape classification is a complex data processing task. Many factors, such as the spatial resolution of remote sensing images, different sources of data, and different classification methods, may influence the accuracy of landscape classification. The selection of the classification method is critical. Usually, classification methods can be divided into supervised classification approaches and unsupervised classification approaches. Supervised classification approaches identify the class of each pixel by selecting typical and representative training samples that the types are known already and training the classifier to classify the spectral data with the spectral characteristics of training samples. There are many classifiers, such as maximum likelihood, minimum distance, the SVM, the artificial neural network (ANN), and the decision tree. Unsupervised classification approaches are used to partition a spectral image into a number of spectral classes based on the statistical information inherent to the image. Unsupervised classification approaches merge spectral classes into meaningful classes based on the size of the similarity between pixels instead of training samples. The methods of unsupervised classification include, for example, the k-means algorithm, the iterative self-organizing data analysis method (ISODATA), and fuzzy clustering [21].

2.1.1. Regional scale: Automatic classification and change detection

On a regional scale, the data most frequently used in landscape classification and change detection are moderate spatial resolution remote sensing images, such as land-resource satellite data. Some traditional landscape classification methods require considerable labor and prior knowledge, and accuracy cannot be guaranteed. To reduce human intervention to a minimum and to achieve accurate classification rapidly, An efficient automatic landscape classification approach using prior accurate land-cover data as the background experience was proposed in [22-23]. This approach is distinguished from the previous semisupervised findings of landscape classifiers by adopting prior knowledge. It can be simply described in two steps: (1) detecting landscape changed pixels from satellite images compared with a prior landscape map and (2) classifying the landscape of changed pixels based on pattern recognition and changed rules.

- Automatic collection of training samples

The first phase of this method is sampling, with a purpose of obtaining pure pixels of landscape classes. The selected training samples were used to ultimately convey the information to a three-dimensional feature space. The automatic collection of training samples is illustrated in Fig. 1.

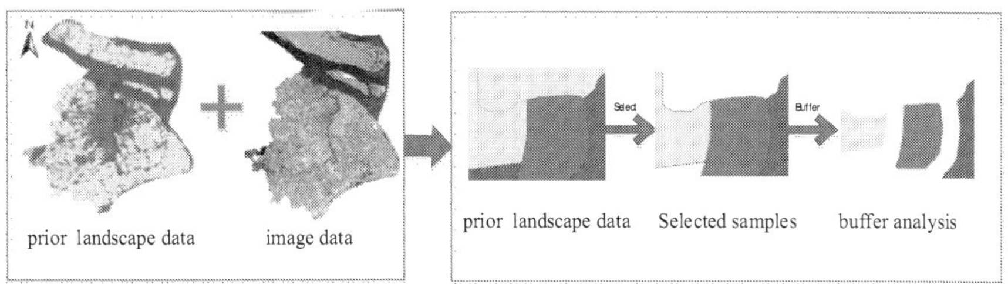

Figure 1. Diagram of automatic collection of training samples.

- Establishment of three-dimensional feature space

Principal component statistical analysis was used to process the data in all spectral bands of each landscape class, extracted by the region of interest. The first three principal components were selected for orthogonal decomposition to construct the three-dimensional feature space of different landscape classes. The three-dimensional spectral feature space of different landscape classes is established in the Fig 2.

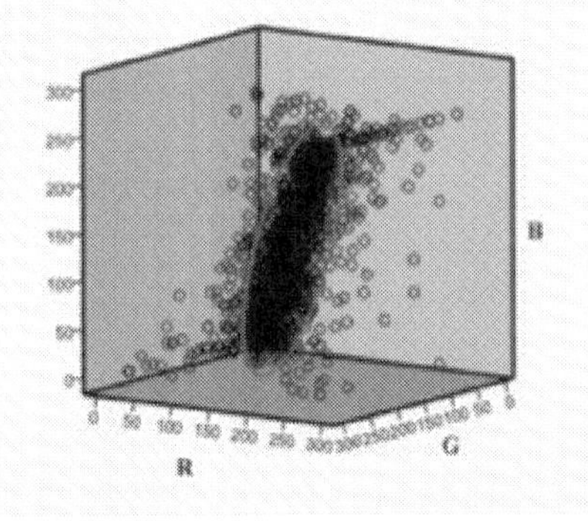

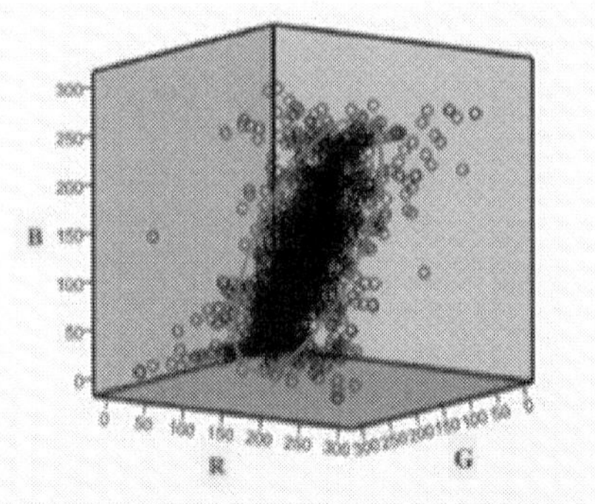

spectral feature space of forest

spectral feature space of grass

Figure 2. Spectral feature spaces of typical landscape types.

- Change detection and classification

For each landscape class, all extracted cell spectral data of the same landscape class were used to calculate the values of the corresponding feature space. The pixels outside the corresponding three-dimensional feature space were considered the changed areas. After the changed pixels of the landscape were obtained, the satellite images and the three-dimensional feature space were employed to classify them based on pattern recognition and changed rules.

- Modification of postclassification results

There is an inevitable problem that a few individual pixels are inconsistent with the surrounding landscape class. The modification of postclassification results solved this problem by dilating operation and eroding operation on the classified results. The "salt-and-pepper" error is also a common problem. A moving window with a size of 3×3 pixels was defined to eliminate noise.

- Accuracy assessment and time-consuming evaluation

A kappa coefficient and a time-consuming evaluation were adopted to evaluate the practicability of classification. A kappa value equal to or greater than 0.61 was considered to be in good agreement. According to the requirements, a less time-consuming combination of P_a [accumulation area threshold (Pa) index] and P_{buffer} (area threshold for buffer analysis) or a combination with a time-lapse rate minimum would form the optimal combination model.

2.1.2. Local scale: Object-oriented classification with high-resolution image data

With the improvements in satellite sensor resolution, high-resolution remote sensing imaging has played an important role in relevant application fields. High-resolution remote sensing images contain rich information, such as spectral information, shape information, and texture information. High-resolution remote sensing images can clearly capture fine features and changes, such as port construction, roads, ecosystem characteristics, and species distribution (e.g., mangroves and invasive plant distribution). Pixel-based classification approaches are primarily based on the spectral information of pixels to extract feature information, and they do not make full use of rich spatial information (shape information and texture information) in the process of classification. According to the characteristics of high-resolution images, an object-oriented classification method was proposed in [24], which makes full use of spectral, shape, and texture information and obtain a high precision of classification results. Object-oriented classification is based on image segmentation to obtain a homogeneous image object and then analyze the spectrum, shape, and texture features to classify and extract the feature information. The main phases of classification are shown in the Fig 3.

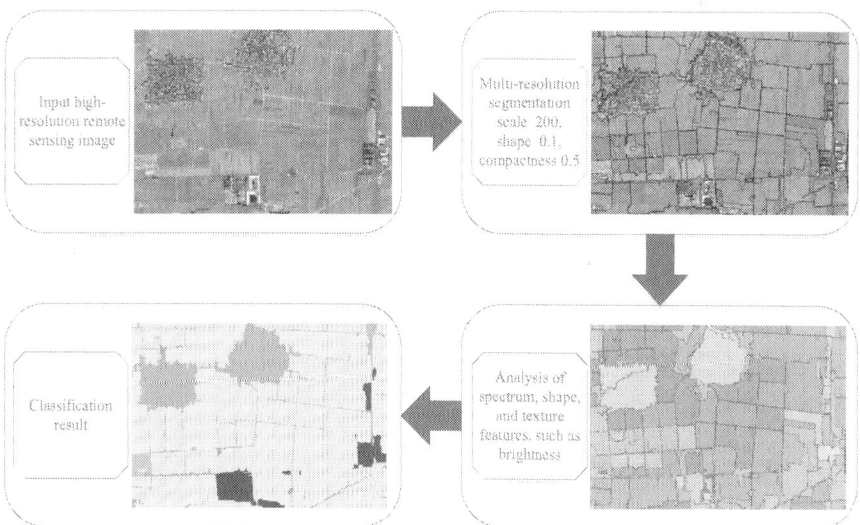

Figure 3. Object-oriented classification.

Object-oriented remote sensing image classification includes two key steps: multiresolution segmentation and classification.

1.   Multiresolution segmentation

The multiresolution segmentation region grows and merges the algorithm and minimizes the average heterogeneity of image objects. The multiresolution segmentation algorithm consecutively merges pixels or image objects and is thus a bottom-up segmentation algorithm based on a pairwise region-merging technique [25]. Multiresolution segmentation algorithms include three factors: the band-weighting factor, the heterogeneity factor, and the segmentation scale. As the segmentation scale increases in size, the objects grow larger. Heterogeneity is composed of spectroscopy heterogeneity, shape heterogeneity, spectral weight, and the shape weight of four variables [26-27]:

$$f = w_l \times h_{color} + \left(1 - w_l\right) \times h_{shape} \tag{1}$$

where h_{color} is spectroscopy heterogeneity, h_{shape} is shape heterogeneity, w_l is spectral weight, and $1\text{-}w_l$ is shape weight.

The multiresolution segmentation is the process of merging image objects in several loops in pairs starting with a single image object of one pixel. In each loop, the single image object acts as a seed to find its best-fitting neighbor in four directions or eight directions, as shown in Fig 4. If the best neighbor's best-fitting neighbor is the seed, then the two pixels are merged into an image object. If otherwise, the best neighbor as a new seed begins to look for its own best-fitting neighbor until a pair of best-fitting image objects are found. The loops continue until the heterogeneity of each pair of image objects is greater than the segmentation scale.

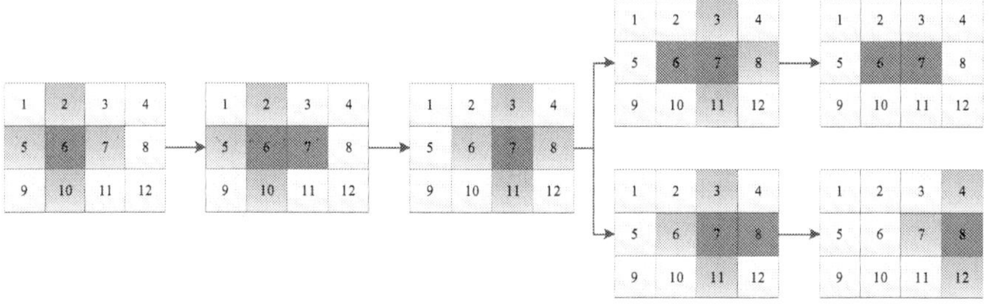

Figure 4. Process of merging image objects.

2.   Classification

After segmentation, the image unit becomes an irregular polygon object composed of homogeneous pixels. The classification based on rules was achieved through comprehensively analyzing object features and establishing proper rule sets. Object features of remote sensing images are the necessary factor in object-oriented classification. There are three characteristics of an object: spectral features, shape features, and texture features.

The spectral features include mean, brightness, and poor standards (see Table 1).

The shape features include area, length/width, and length (see Table 2).

The texture features reflect the characteristics of the spatial distribution of pixels. The gray-level cooccurrence matrix (GLCM) is the most common statistical analysis method. GLCM provides the direction of an image's grayscale, interval, and variation range, but it does not directly provide texture features. Based on GLCM, texture features include, for example, homogeneity, contrast, and dissimilarity (see Table 3).

Spectral features	Formula	Illustration
Mean	$\bar{C}_k(v) = \dfrac{1}{N}\sum_{(x,y,k)\in v} C_k(x,\,y,\,k)$	
Brightness	$\bar{C}(v) = \dfrac{1}{w^b}\sum_{k=1}^{k} w_k^b \bar{C}_k(v)$	
Max. diff.	$d(v) = \dfrac{\max\limits_{i,j\in k} \lvert \bar{C}_i(v) - \bar{C}_j(v) \rvert}{\bar{C}(v)}$	k is a layer, v is an image object, (x,y) are pixel coordinates, C is the value of a pixel, $\bar{C}$ is the mean of a pixel in an object, w^b is the sum of brightness weights of all image layers k, NIR is a near-infrared ray, and R is an infrared ray.
Standard deviation	$\sigma_k = \sqrt{\dfrac{1}{N}\sum_{x,y,k\in v}(\bar{C}_k(x,\,y,\,k) - \bar{C}_k(v))^2}$	
Ratio	$\sigma_k = \dfrac{\bar{C}_k(v)}{\sum_{k=1}^{n} w_k^b \bar{C}_k(v)}$	
Min. pixel value	$\min_v = \min C_k(x,\,y)$	
Max. pixel value	$\max_v = \max\limits_{(x,y)\in v} C_k(x,\,y)$	
NDVI	$\mathrm{NDVI} = (\mathrm{NIR} - R)/(\mathrm{NIR} + R)$	

Table 1. Spectral features

Shape features	Formula	Illustration
Area	$A_v = N \times u^2$	N is the number of image objects; u is the pixel size; λ_1 and λ_2 are eigenvalues; a and b are the length and width of the bounding box, respectively; f is the bounding box fill rate; $Var(X)$ and $Var(Y)$ are the variance of X and Y, respectively; and b_o and b_i are the length of the outer and inner borders, respectively.
Length/width	$\gamma_v^{EV} = \dfrac{\lambda_1(v)}{\lambda_2(v)}$ or $\quad\gamma = \dfrac{1}{w} = \dfrac{a^2 + ((1-f)\times b)^2}{A_v}$	
Length	$l_v = \sqrt{A_v \bullet \gamma}$	
Width	$W_v = \sqrt{\dfrac{A_v}{\gamma}}$	
Border length	$b_v = b_o + b_i$	
Asymmetry	$A_s = \dfrac{\sqrt{\dfrac{1}{4}(Var(X) + Var(Y))^2 + (Var(XY))^2 - Var(X)\bullet Var(X)}}{Var(X) + Var(Y)}$	
Border index	$BI = \dfrac{h_v}{2(l_v + W_v)}$	
Shape index	$SI = \dfrac{b_v}{4\sqrt{A_v}}$	
Compactness	$Com = \dfrac{l_v \bullet W_v}{N}$	

Table 2. Shape features

Texture feature	Formula	Illustration
GLCM_Homogeneity	$GLCM_Hom = \sum_{i,j=0}^{N-1} \dfrac{P_{i,j}}{1+(i-1)^2}$	
GLCM_Contrast	$GLCM_Con = \sum_{i,j=0}^{N-1} P_{i,j}(i-j)^2$	
GLCM_Dissimilarity	$GLCM_Dis = \sum_{i,j=0}^{N-1} P_{i,j}\,|\,i-j\,|$	
GLCM_Mean	$GLCM_Mean = \sum_{i,j=0}^{N-1} P_{i,j} \bullet i$	i is the row number, j is the column number, and $P_{i,j}$ is the normalized value in the cell i,j.
GLCM_Variance	$GLCM_Var = \sum_{i,j=0}^{N-1} P_{i,j} \bullet (i-mean)^2$	
GLCM_Entropy	$GLCM_Ent = \sum_{i,j=0}^{N-1} P_{i,j} \bullet (-\ln P_{i,j})$	
GLCM_Angular Second Moment	$GLCM_ASM = \sum_{i,j=0}^{N-1} P_{i,j}^2$	
GLCM_Correlation	$GLCM_Cor = \sum_{i,j=0}^{N-1} \dfrac{(i-mean) \bullet (j-mean) \bullet P_{i,j}^2}{Var}$	

Table 3. Texture features

3.   Application of object-oriented classification based on remote sensing image

There are two key steps for classification: multiresolution segmentation and feature selection. Multiresolution segmentation is a crucial step transiting from image process to image analysis, and feature selection is the most important factor that influences the accuracy of classification. Taking the Changli Golden Beach National Ocean Natural Preserve as an example, we classified the landscape of this area using the object-oriented classification method.

• **Multiresolution segmentation**

According to the extracted object, the proper parameters are chosen to achieve the optimal segmentation result with scale 75, shape 0.1, and compactness 0.5 as shown in Fig. 5.

Figure 5. Remote sensing image and result of multiresolution segmentation.

- Feature selection

Feature selection is to select the spectral features, shape, and texture features for classification through the analysis of feature information, see the analysis result of feature information in Fig. 6. According to the classification targets (landscapes in Table 4), some features were selected and proper thresholds were set, such as normalized difference vegetation index (NDVI), mean, shape index, and brightness.

Figure 6. Analysis of feature information.

Landscape	Selected features
Aquafarms	NDVI, length/width, shape index
Water area	NDVI, mean, brightness
Forest land	NDVI, mean, ratio
Industrial land	Brightness, standard deviation, compactness
Cultivated land	NDVI, brightness, standard deviation
Residential area	NDVI, standard deviation, GLCM_Contrast
Construction land	NDVI, brightness asymmetry

Table 4. Selected features for different landscapes

- Classification result

As shown in Fig. 7, the landscape mainly consists of aquafarms, water area, forest land, industrial land, cultivated land, residential area, and construction land. The classification result is reasonable through comparative analysis between the results and the original image.

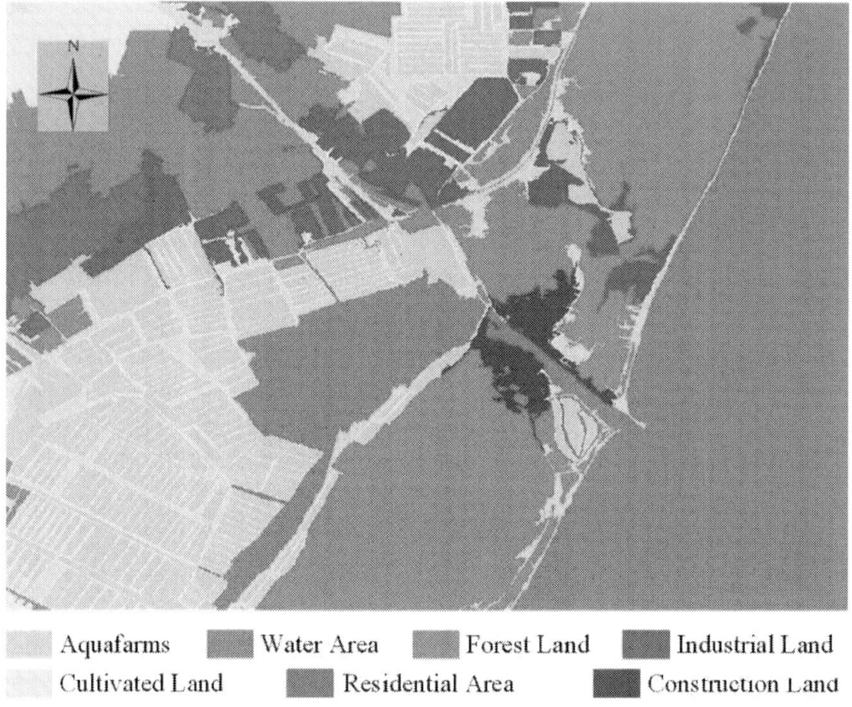

Figure 7. Result of classification.

The object-oriented classification method we used in this study shows a good performance with the total accuracy of 89% and the kappa coefficient of 0.83. This method provides a high precision and a reasonable classification result.

2.2. Retrieval of eco-environmental parameters based on remote sensing technique

The parameters of land surface energy and water balance are important inputs for research on global climate change, crop yield assessment, and ecological environment evaluations. Remote sensing-derived radiation, temperature, and other data sets on a global scale have been used as standardized products in research and applications. Remote sensing information is convenient and easy to access over a large area at a low cost.

2.2.1. Energy- and water-balance parameters: solar radiation and Evapotranspiration (ET)

- Solar radiation

Solar radiation is the Earth surface's most basic and important source of energy as well as the main driving factor of plant photosynthesis, transpiration, and soil evaporation land-surface processes [28]. Solar radiation is responsible for the formation and evolution of important climate driving forces. Changes in solar radiation change the temperature, humidity, precipitation, atmospheric circulation, the hydrological cycle, and other processes. Solar radiation is an important physical and ecological parameter in the land surface and atmospheric energy exchange process. Accurate solar radiation data retrieved from satellite data help improve net radiation, ET, and other precision products [29].

Geostationary meteorological satellite data have often been adopted as the data source for solar radiation and ET retrieval. GMS-5 data are easily acquired with a relatively high temporal resolution (1 h). GMS-5 data consist of three types of bands: (1) the visible (VIS) band with a spatial resolution of 1.25 km and a spectrum range from 0.55 to 1.05 μm, (2) the thermal infrared (TIR) band with a spatial resolution of 5 km and a spectrum range from 10.5 to 12.5 μm, and (3) the water vapor (WV) band with a spatial resolution of 5 km and a spectrum range from 6.2 to 7.6 μm. The VIS and TIR bands were employed for ET retrieval, and the WV band was used for calibration and validation.

Net radiation was calculated as the net result of the short-wave (solar) and long-wave (terrestrial) radiative fluxes, and it is expressed as a daily average:

$$I_n = \left(1 - a\right)I_g + L_n \tag{2}$$

where a is a surface albedo, which could be derived from the VIS data of the GMS-5; I_g is the daily average solar irradiation at the Earth's surface; and L_n is the net long-wave (thermal) radiation loss.

Figures 8 and 9 show the spatial distribution of the global and net solar radiation of Asia in January, April, July, and October 2004, respectively.

Solar radiation is an important factor in maintaining the Earth's climate system and the ecosystem's energy balance and is the main source of energy in the ecosystem, which plays an important role in the process of human development. Many environmental changes are related to solar radiation [30]. The use of satellite remote sensing data, particularly the stationary meteorological satellite data for observing a fixed surface area with the temporal resolution of 1 h, can greatly compensate for the lack of ground observation data. It is important to obtain long and continuous region surface solar radiation to analyze solar radiation's spatial and temporal variation of the surface area. The analysis of surface solar radiation and climate change research will thereby be greatly facilitated.

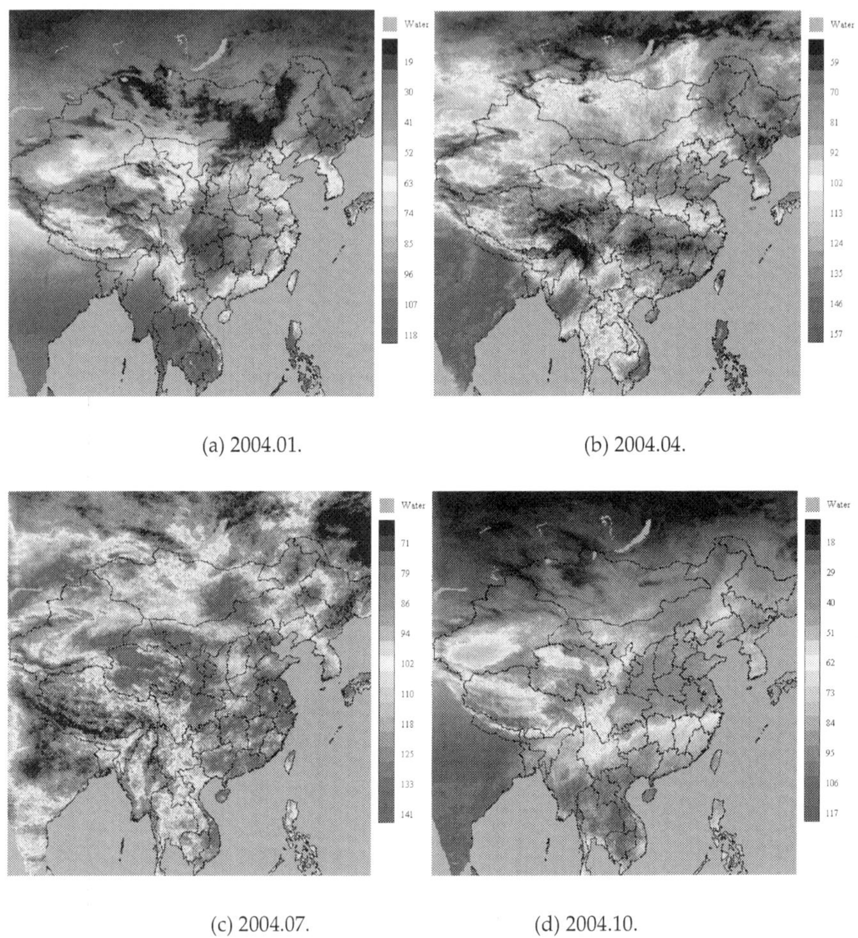

(a) 2004.01.   (b) 2004.04.

(c) 2004.07.   (d) 2004.10.

Figure 8. Spatial distribution of global solar radiation in parts of Asia.

- ET

Evapotranspiration (ET) is usually understood to be the sum of soil evaporation and plant transpiration, which is a soil-plant-atmosphere continuum system and an important process (SPAC) in water movement. ET is essential to crop growth and the development of water and energy sources. Ecological systems are an important link between land surface and hydrological processes, which are closely related to the strength and size of the underlying surface of plants. ET can be said to be both a surface heat-balance component and an important part of water balance. ET is highly variable across space due to changes in precipitation, soil hydro-

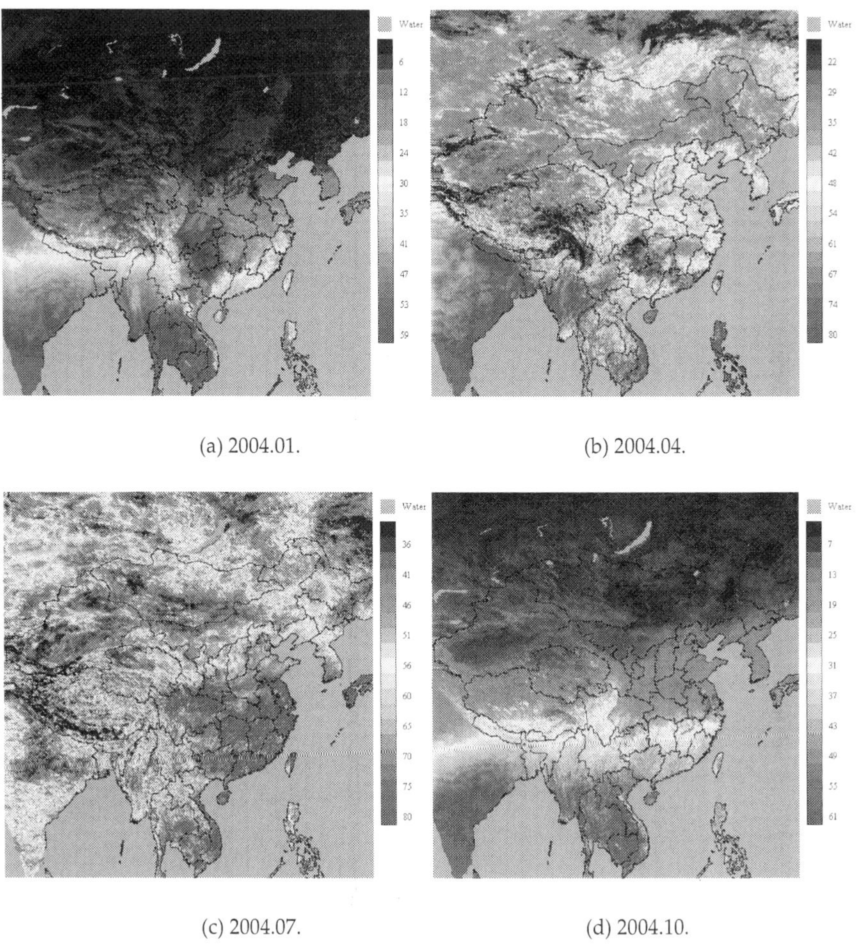

(a) 2004.01.

(b) 2004.04.

(c) 2004.07.

(d) 2004.10.

Figure 9. Spatial distribution of net solar radiation in parts of Asia.

logical parameters, and vegetation type and density. Its strong spatial variability across time is due to climate changes at different times.

Calculation of the sensible heat flux: The sensible heat flux into the atmosphere is proportional to the temperature difference across the atmospheric boundary layer $(T_0\text{-}T_a)$. The simple formulation is:

$$H = \left(a_c + a_r\right)\left(T_0 - T_a\right)$$

(3)

where a_c stands for surface resistance and a_r stands for the resistance of the atmosphere.

Having determined the net radiation (I_n) and the sensible heat flux (H), the latent heat flux (i.e., the actual ET in energy units) can be obtained based on regional energy and water balance:

$$LE = I_n - H - G \qquad (4)$$

Item G is the heat flux into the soil, which is very small on the daily time scale, and may be considered a constant.

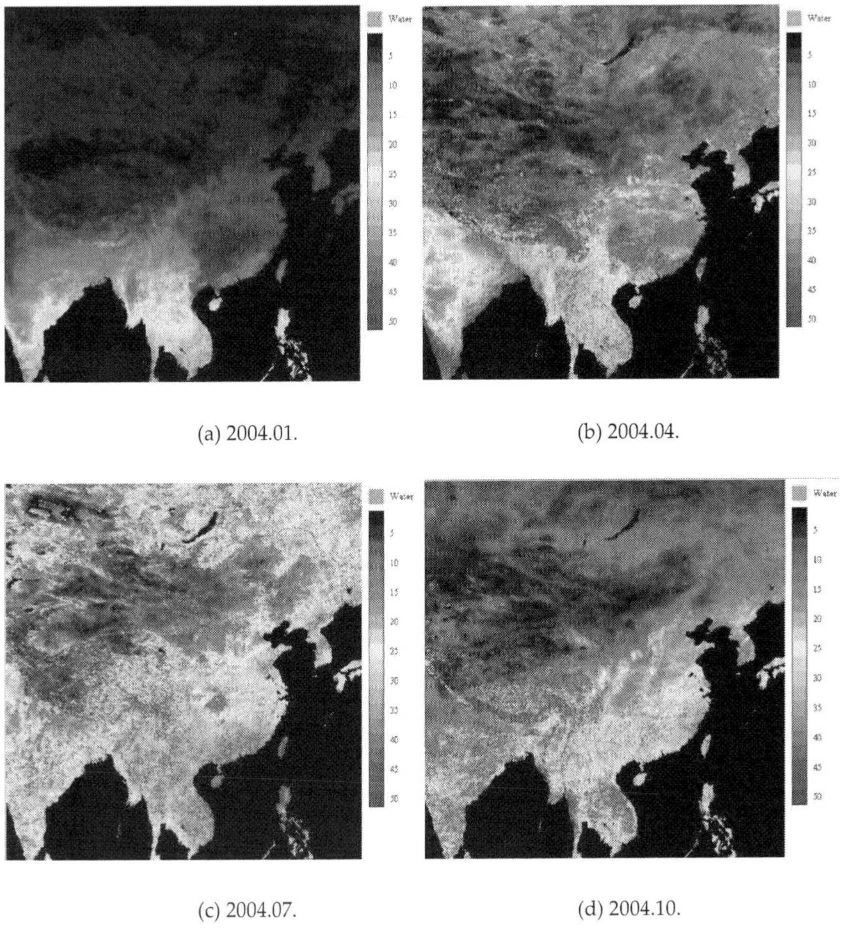

(a) 2004.01.   (b) 2004.04.

(c) 2004.07.   (d) 2004.10.

Figure 10. Spatial distribution of ET in parts of Asia.

Figure 10 shows the spatial distribution of the ET of Asia in January, April, July, and October 2004 and reveals the significant differences between the distributions of different months of relative ET space. Relative ET was higher at low latitudes than at high latitudes and varied across months. Due to the lush plant growth in July, relative ET is generally high in most of Asia. In January, in most of Asia, including the South and Southeast Asian countries, there are low temperatures and generally low relative ET.

Many ecosystem processes, such as soil moisture changes, vegetation growth, and nutrient cycling, are closely linked to the ET process. The ET process is affected by climate, soil characteristics, and vegetation growth status [31-32]. Therefore, the calculation of ET reveals the changes in the time of ET and its impact factors to quantify the contribution of plant transpiration and soil evaporation to ET. ET can not only reveal various land-surface (vegetation, particularly surface) water consumptions [33] but also help us understand the effect of global warming on the actual ET and water balance, land surface ecology, and environment. ET can also improve climate models and accurately simulate climate change, which is important.

2.2.2. Vegetation: VI, photosynthetically active radiation (PAR), and Net Primary Productivity (NPP)

- VIs

VIs are used to evaluate fractional vegetative covers qualitatively and quantitatively, including the leaf area index (LAI), chlorophyll concentration, photosynthetic activity, biomass, and vegetation growth. Most of them are derived from satellite data or spectrometers using the spectral information of vegetation [34-35]. Time-series VIs that use observations of the Earth from a space platform are a valuable way to derive several plant biophysical parameters for ecological, hydrological, and climate models and to study land-use land-cover change dynamics [36]. VI has been one of the primary sources of information for the operational monitoring of the Earth's vegetative cover. So far, more than 150 indices have been defined for different purposes and optimized to assess a process of interest because it is not possible to design an index that is sensitive to only the desired variable and totally insensitive to all other vegetation parameters [37].

[38] presented the most well-known and widely used VI, the NDVI. To compensate for soil background influences and atmospheric effects, the average NDVI of different surface temperatures of diverse surface types was computed in [39]. The improved indices, such as the renormalized difference VI (RDVI), the perpendicular VI (PVI), and the modified simple ratio (MSR), linearized the relationships with vegetation biophysical variables. Then, MSR was suggested as an improvement over RDVI in combination with the simple ratio (SR=NIR/Red) [40]. SR and MSR are considered more linearly related to vegetation parameters.

The soil-line VIs, soil-adjusted VI (SAVI), an improved SAVI (MSAVI), and the soil and atmospherically resistant VI (SARVI) improved the resistance to soil and atmospheric effects.

New VIs based on three discrete bands, the chlorophyll absorption ratio index (CARI), the modified CARI (MCARI), and the triangular VI (TVI), have a strong correlation between leaf chlorophyll concentration and reflectance ratios. New and improved VIs for green LAI predications, MCARI1, MCARI2, MTV1, and MTV2, showed the best linear relationships between NIR reflectance and VIs [35].

• PAR

PAR is solar radiation that can be absorbed through the processes of chlorophyll synthesis and the photosynthesis of plants and then transformed into chemical energy in the waveband of 400 to 700 nm [41-42]. PAR is an important factor for plant growth and development. Many studies have shown that PAR contributes significantly to plant physiology, biomass production, biometeorology, energy budget, and the Earth's climate system [42-43]. The monthly gross primary production (GPP) was estimated and calibrated using the enhanced VI (EVI) and PAR in [44]. The ratio of PAR with global solar radiation has been found generally decreased with various sky conditions in each month at FuKang in northwest China[45]. It was reported in [46] that the PAR and UVR interacted during photoinhibition and could initiate the protein repair of cyanobacterium *Arthrospira platensis*. The PAR losses in the atmosphere were associated with "photochemical factors", and other materials in North China depended on spatial and seasonal changes [47].

Absorbed PAR (*APAR*) is a key parameter for estimating *PAR* absorbed by the green canopy during photosynthesis. The carbon exchange between the crop canopy and the atmosphere is primarily controlled by the amount of *APAR* by the light use efficiency (LUE). *APAR* is determined by the fraction of PAR (*FPAR*) and the total solar surface radiation (SOL; MJ m^{-2}; [48]:

$$APAR(x,t) = SOL(x,t) \times FPAR(x,t) \times 0.5 \tag{5}$$

where the constant 0.5 represents the ratio of the total solar radiation (with a wavelength range of 0.4–0.7 μm) used by the vegetation.

• NPP

NPP represents the organic matter accumulated by plants per unit area and time. From an ecological perspective, NPP measures the rate at which solar energy is stored by plants as organic matter. NPP is influenced by climate, soil, vegetation type, and human activities [49]. For various ecological monitoring activities, NPP is generally regarded as an important factor that provides a comprehensive evaluation of ecosystem status and services, including productivity capability, habitat and wildlife, and ecological footprint [50-51].

NPP is not a directly observable ecosystem characteristic, and it is difficult to measure accurately over large areas [52-53]. A number of NPP models have been developed. Regression-based models such as the Miami model, process-based models such as simple models based on LUE, and mechanistic models based on soil-vegetation-atmospheric transfer (SVAT; [54-55] need more parameter requirements and complexities. Satellite data-driven production efficiency models (PEMs), such as the Carnegie-Ames-Stanford approach (CASA; [48], TURC

[56], and GLO-PEM [57], have been used to analyze the spatiotemporal patterns of NPP over continents and global land surfaces [58-61].

The CASA model is a biogeochemical model that uses a system of first-order linear differential equations to represent the flow of carbon between various pools and to track the long-term changes in terrestrial carbon stocks on a monthly time-step. This model computes the NPP as a function of APAR and LUE [48]:

$$NPP(x,t) = APAR(x,t) \times LUE(x,t) \tag{6}$$

where x represents the grid cell, t represents the period that NPP is accumulated, $LUE(x,t)$ represents the actual LUE [62].

2.3. Integrated analysis of coastal environment supported by GIS

2.3.1. Factors of coastal environment analysis

Coastal zones and shelf seas are influenced by both natural and socioeconomic conditions. An integrated evaluation criteria system was set up, which contained nine factors belonging to three categories: (1) environmental background factors, including elevation, slope, geomorphological types, accumulated temperature, and a wetness index; (2) water/land resources, including precipitation, river density, and land use; and (3) socioeconomic factors, including railway density, road density, and population density. However, human life and development may also directly or indirectly affect costal zones and shelf areas, including mariculture and marine fishing production, sea pollution, and others (e.g., building, sports, and travel).

2.3.2. Model

• Analytical hierarchy process (AHP) method

To evaluate such complex factors, the AHP or Delphi method provides a good solution. The AHP is a decision-making method in qualitative and quantitative analysis, which is associated with the decision scheme of elements decomposed into goals, principles, and levels. The characteristics of the AHP are based on the nature of complex decision problems, influencing factors, and inherent relationship depth analyses using quantitative information with fewer decisions to construct a mathematical thinking process with multiobjective, multicriteria, or nonstructural properties of complex decision-making problems. The AHP is an easy method to use. However, there are many disadvantages, such as the overuse of the data statistic index, making the weight difficult to determine. Quantitative data are fewer in number and not as convincing. Therefore, many new models have emerged to analyze these factors.

• Neural net

Neural net classification was used to analyze land growth due to coastline changes over different periods based on TM imaging. ANNs may be of significant value in extracting vegetation-type information in complex vegetation mapping problems, particularly in coastal wetland environments. Further remote sensing research involving fuzzy ANNs is also needed.

Feature extraction methods are optimized for high-dimensional data analysis by ANNs. One example is the decision boundary feature extraction (DBFE) algorithm [63]. In addition, the growing neural gas (GNG) algorithm [64] could be utilized to enhance outlier detection capability by inserting a new processing unit in the vicinity of the neuron with the highest requanitization error.

In general, all neural algorithms have higher classification accuracies, which can help address complex vegetation mapping problems and improve hyperspectral images and MODIS data classification accuracies.

• Support vector machine (SVM)

SVM is a supervised learning model that is usually used for pattern recognition, classification, and regression analysis. The CA-SVM was established by applying the soft computing approach with a large quantity of remote sensing data to the marine environment.

Two main tasks were undertaken in this study. First, to choose reasonable influence factors as the input parameters of the model, nine series of training and simulation exercises were conducted based on nine different types of input parameter combinations. A reasonable input parameter combination was selected, and the eutrophication model (the basic model) was established by the comparative analysis of the simulation results. Second, according to Shelford's law of tolerance, an optimized model was developed. The model combines nine special models, and each model corresponds to a stage of SST and chlorophyll-a concentration, respectively.

In general, SVM is a more optimized coupling model, and prediction accuracy is improved. This model could provide a scientific basis for the prediction and management of the aquatic environment of Bohai Bay.

• Ecosystem modeling

Ecosystem models are powerful tools for anticipating future conditions and exploring carrying capacity. These models allow for the study of stocks, energy fluxes, and potential interactions in complex coastal ecosystems. However, the application of ecosystem models in the analysis of the coastal environment has been limited partially due to the cost involved in generating and testing the models. Therefore, the use of more generic and flexible models could facilitate the implementation of modeling. A physical-biogeochemical coupled model for studying the long-term processes in shallow coastal ecosystems was presented in [65]. The biogeochemical model was constructed in Simile, a visual simulation environment software that is well suited to accommodate fully hydrodynamic models. Specifically, Simile integrates PEST (a model-independent parameter estimation), an optimization tool that uses the Gauss-Marquardt-Levenberg algorithm and can be used to estimate the value of a parameter or set of parameters to minimize the discrepancies between the model results and a data set [66].

The other critical aspect of modeling exercises is the large amount of data necessary to set up, tune, and ground truth the ecosystem model. Tools such as remote sensing, scenario analysis, and optimization can help fill these data gaps. Recently, satellite remote sensing has been used to feed individual-based models to predict bivalve growth [67-68] even in data-poor environments. GIS tools and optimization have been employed to collect and use data for developing ecosystem models in aquaculture data-poor sites. Similarly, a study has been conducted to evaluate the best site for shellfish aquaculture in Valdivia (Chile), which combines GIS-based models with a farm-scale carrying-capacity model [69-70].

In general, ecosystem modeling and optimization are ideal tools for a marine environment; as described, ecosystem modeling combines existing models, remote sensing data sets, and GIS tools into a fully spatial model to provide an integrated analysis method for optimizing coastal environments from a sustainable ecosystem standpoint.

2.3.3. Integrated analysis supported by GIS

Coastal environmental comprehensive assessment involves many factors, including the spatial pattern of ecological environment quality conditions. However, the above method reflects the quality situation of the coastal environment in longitude, excluding the spatial distribution of the evaluation results. To grasp the overall quality of the coastal environment condition, the in-depth analysis of its spatial pattern is indispensable. GIS technology has a strong spatial analysis function to make up for the deficiencies of these methods. Therefore, it is helpful to improve the level of coastal environmental assessment through integrating GIS and these models.

In the field of suitability assessment for coastal environments, GIS, remote sensing, and numerical modeling techniques have been shown by recent studies to be efficient tools. The land-use suitability analysis was reviewed based on GIS in the United States in [71]. A GIS-based geo-environmental evaluation for urban land-use planning was presented in [72]. An integrated GIS-based analysis system for the land-use management of a lake area in the urban fringe in central China was built up in [73], and the analytic hierarchy process (AHP) method was adopted to derive weights for the evaluating model. Similarly, an integrated evaluation of urban development could be conducted in an operational way using remote sensing data, the GIS spatial analysis technique, and the AHP modeling method [74]. Satellite remote sensing has been used to feed individual-based models to predict bivalve growth [66] even in data-poor environments. A study was conducted in [69] to evaluate the best site for shellfish aquaculture in Valdivia (Chile), combining GIS-based models with a farm-scale carrying-capacity model.

In general, the integration of GIS with digital models provides a new way that makes full use of the powerful spatial analysis function of GIS and the advantage of a multifactor comprehensive evaluation of model for improving the level of coastal environment quality evaluation. Moreover, GIS has great advantages in data processing, management, and analysis, and it can objectively reflect the coastal environment states and problems in horizontal aspect. Many research results indicate that the integrated analysis of the coastal environment supported by GIS has important theoretical and practical significance.

3. Case study

Several typical ports with different stages of development were selected as the study regions: (1) Zhan Jiang Port, China, East Asia; (2) Gwadar Port, Pakistan, South Asia; (3) Djibouti Port, Republic of Djibouti, East Africa; and (4) Ilichevsk Port, Ukraine, Eastern Europe (see Fig. 11).

The ports and their coastal surroundings were selected for a comparative study of the spatiotemporal changes in the coastal environment using time-series satellite imagery and GIS methods.

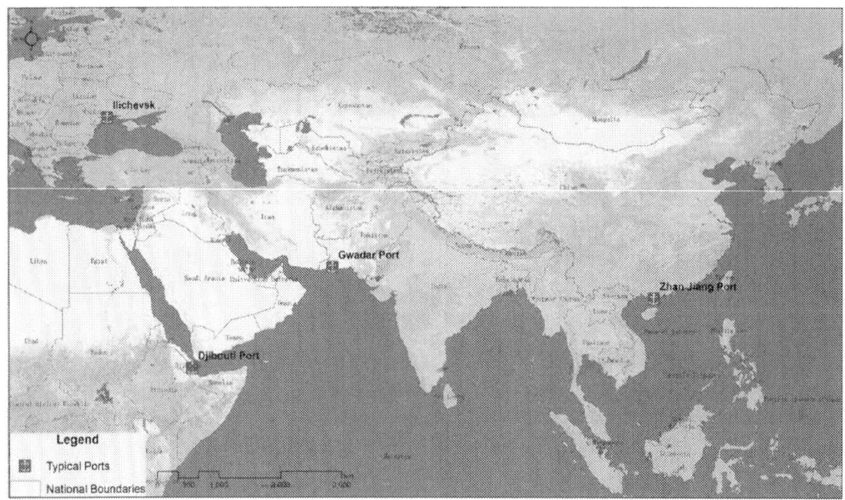

Figure 11. Sketch map of typical port locations.

Figure 12 shows that, in the northern part of Zhan Jiang Port, the cultivated land, water area, and urban land were reduced significantly, whereas unutilized land and construction land increased from 2003 to 2011. The middle part shows a reduction in grassland and an increase in unutilized land. The southern part shows a reduction of construction land and a reduction of unutilized land.

The landscape types of Zhan Jiang Port were based primarily on the water area, cultivated land, and urban land. From 2003 to 2011, the urban land increased the most, with a percentage of 26.05%. The water area was reduced the most.

Overall, the urban land of Zhan Jiang Port increased significantly, whereas the water area was reduced significantly. However, other landscapes did not obviously increase or decrease.

Therefore, the landscape of Zhan Jiang Port from 2003 to 2011 was relatively stable with a small trend of sea reclamation, and Zhan Jiang Port was at the stage of specialization.

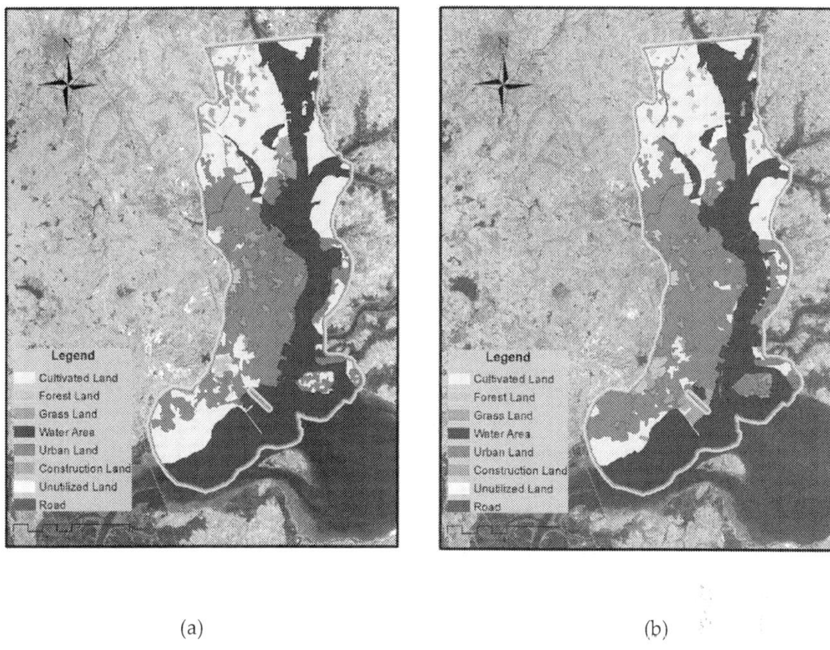

(a)   (b)

Figure 12. Landscape of Zhan Jiang Port in 2003 (a) and 2011 (b).

Land-use type	Area in 2003 (km²)	Area in 2011 (km²)	Growth (km²)	Growth rate
Construction land	2.60	3.70	1.10	42.41%
Cultivated land	83.67	81.74	-1.93	-2.30%
Forest land	3.12	3.17	0.04	1.36%
Grassland	14.03	14.76	0.73	5.18%
Road	0.89	0.86	-0.03	-3.45%
Unutilized land	0.72	1.07	0.34	47.41%
Urban land	79.29	99.95	20.66	26.05%
Water area	108.87	87.95	-20.92	-19.21%
Summary	293.20	293.20	—	—

Table 5. Statistical results of Zhan Jiang Port in 2003 and 2011

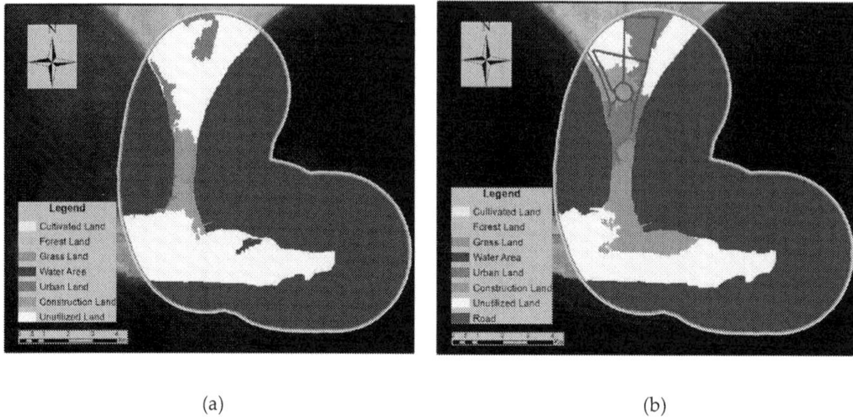

(a)   (b)

Figure 13. Landscape of Gwadar Port in 2003 (a) and 2011 (b).

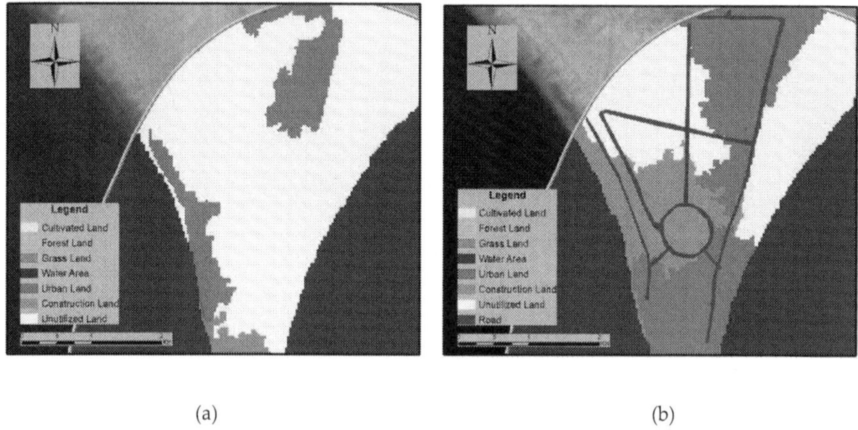

(a)   (b)

Figure 14. Landscape of partial enlargement of Gwadar Port in 2003 (a) and 2011 (b).

As shown in Fig. 13, the landscape of Gwadar Port changed significantly from 2003 to 2011. Construction land and urban land increased significantly by occupying the grass land and unutilized land. However, the most obvious change was the construction of the main road. In 2003, the main road of Gwadar Port was too small to be identified by the Landsat TM imagery, whereas, in 2011, it was clearly identifiable (see Fig. 14).

The landscape types of Gwadar Port were based primarily on unutilized land and water areas in both 2003 and 2011. From 2003 to 2011, construction land, urban land, and road increased and the unutilized land, grass land, and water area were decreased. From Table 5, we come to the conclusion that Gwadar Port was in the expansion stage, with a massive increase in construction land and urban land.

Landscape type	Area in 2003 (km^2)	Area in 2011 (km^2)	Growth (km^2)	Growth rate
Construction land	1.61	7.51	5.91	367.63%
Grassland	1.7	0.54	-1.16	-68.25%
Unutilized land	24.98	17.49	-7.49	-30.00%
Urban land	2.11	4.98	2.87	136.03%
Water area	63.61	63.49	-0.12	-0.19%
Road	0	1.08	1.08	—
Summary	94.01	94.01	—	—

Table 6. Statistical results of Gwadar Port in 2003 and 2011

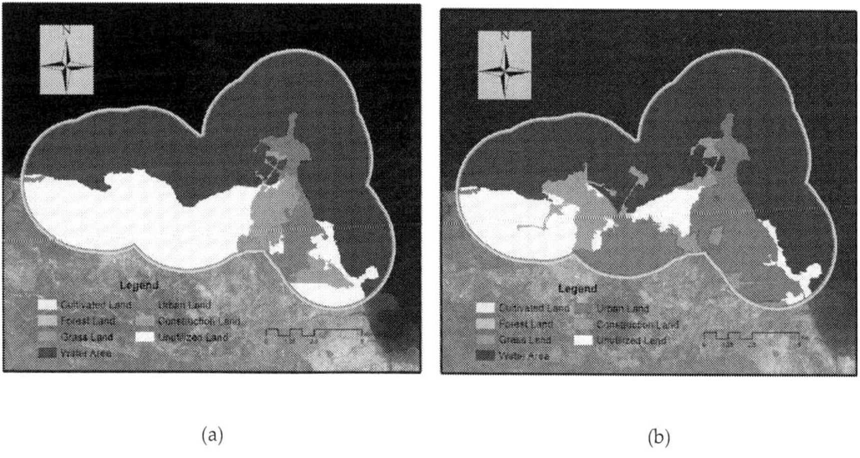

(a)   (b)

Figure 15. Landscape of Djibouti Port in 2000 (a) and 2011 (b).

As shown in Fig. 15, the landscape of Djibouti Port had a significant change from 2000 to 2011. Urban land, grass land, and construction land increased a lot, whereas unutilized land and water area were reduced. In the eastern part of Djibouti Port, the landscape-type changes were construction land to urban land, whereas, in the middle and western parts of Djibouti Port, the changes were primarily from unutilized land to urban land and construction land. The change pattern of landscape indicated that the expansion of Djibouti Port was from east to west. By 2011, the eastern part of Djibouti Port was developed, and the middle part was under construction, whereas the western part had not been developed yet.

Landscape type	Area in 2000 (km²)	Area in 2011 (km²)	Growth (km²)	Growth rate
Construction land	8.80	10.51	1.70	19.32%
Grass land	1.46	2.07	0.61	42.09%
Unutilized land	45.93	24.34	-21.60	-47.01%
Urban land	9.17	30.83	21.66	236.17%
Water area	85.08	82.70	-2.38	-2.79%
Summary	150.45	150.45	—	—

Table 7. Statistic results of Djibouti Port in 2000 and 2011

The unutilized land was the second largest area in 2000 in Djibouti Port, whereas urban land became the second in 2011, with an increase of 21.66 km². The great increase of urban land and great decrease of unutilized land indicated that Djibouti Port developed a lot from 2000 to 2011, and the increase of construction land showed that Djibouti Port was still expanding in 2011.

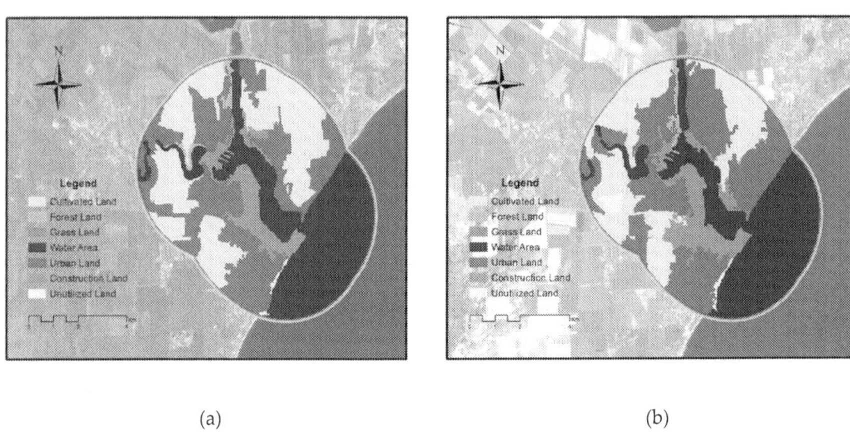

(a)   (b)

Figure 16. Landscape of Ilichevesk in 2003 (a) and 2011 (b).

From Fig. 16, we know that construction around the water area increased slightly, and urban land in the western part decreased. From Fig. 17, we can see that some cultivated land in this region changed to urban land, especially around the water area. All in all, the landscape structure remained stable from 2003 to 2011 in general, with urban expansion in northeast Ilichevesk.

The landscape structure of Ilichevsk Port was based primarily on water area, urban land, and cultivated land both in 2003 and 2011. From 2003 to 2011, urban land increased only 2.21 km².

Thus, the landscape structure of Ilichevsk Port remained stable from 2003 to 2011 in general, with the main trend of expansion in the northeast part.

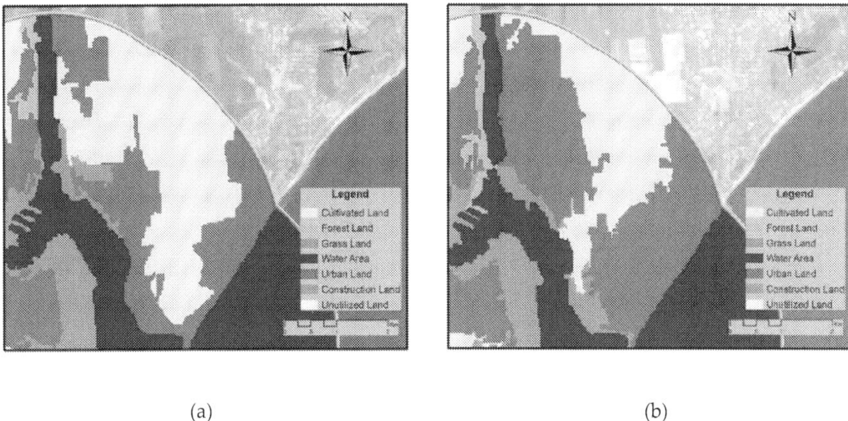

(a)   (b)

Figure 17. Landscape of northeast Ilichevesk in 2003 (a) and 2011 (b).

Landscape type	Area in 2003 (km²)	Area in 2011 (km²)	Growth (km²)	Growth rate
Construction land	5.63	6.96	1.33	23.66%
Cultivated land	19.86	16.28	-3.58	-18.02%
Forest land	0.68	0.64	-0.04	-6.35%
Unutilized land	0.20	0.72	0.52	260.37%
Urban land	20.56	22.77	2.21	10.76%
Water area	21.74	21.30	-0.44	-2.03%
Summary	68.66	68.66	—	—

Table 8. Statistic results of Ilichevsk Port in 2003 and 2011

4. Conclusions

In this chapter, we first discussed the automatic classification and change detection on a regional scale and the objective-based classification with high-resolution image data on a local scale for the landscape classification. We then retrieved the eco-environmental parameters based on remote sensing techniques, including solar radiation, ET, VIs, PAR, and NPP. On this basis, we introduced the mathematical models, including the AHP, the neural net, SVM, and ecosystem modeling, which can be combined with GIS for an integrated analysis of coastal environments. Finally, we used typical ports (Zhan Jiang, Gwadar, Djibouti, and Ilichevsk) as examples for landscape classification and analyzed the spatiotemporal variation from 2003 to 2011.

Zhan Jiang Port was at the stage of specialization. The landscape was relatively stable with a small trend of sea reclamation from 2003 to 2011. In the northern part of Zhan Jiang Port, the cultivated land, water area, and urban land were reduced significantly. The middle part shows a reduction of grassland and an increase in unutilized land. The southern part shows a reduction of construction land and a reduction of unutilized land. Gwadar Port was in the expansion stage, with a massive increase in construction land and urban land. Construction land and urban land in Gwadar Port increased significantly by occupying the grassland and unutilized land. However, the most obvious change was the construction of the main road. Djibouti Port developed considerably from 2000 to 2011, and the small increase in construction land in 2011 shows that Djibouti Port was still expanding. The urban land and construction land increased significantly, whereas unutilized land and water area were reduced. Ilichevsk Port remained generally stable from 2003 to 2011, with the slight trend of expansion in the northeast part.

The coastal zone represents a comparatively small but highly productive and extremely diverse system with a variety of ecosystems. Because of its characteristics of complex and diversity, the modern methods such as remote sensing and GIS become more and more important for the coastal environment. Remote sensing has the ability of wide and frequent information acquisition and GIS has the ability of spatial analysis. Therefore, they are very effective for monitoring the shoreline changes, understanding the physical processes in the coastal environment, planning and managing the coastal zone, and analyzing the interaction between coastal zone changes and the effects of human activities.

Author details

Dong Jiang*, Mengmeng Hao and Jingying Fu

*Address all correspondence to: jiangd@igsnrr.ac.cn

Institute of Geographical Sciences and Natural Resources Research, Chinese Academy of Sciences, Beijing, China

References

[1] Feng YQ, Niu J. Some problems about the coastal environment of china. Marine Geology Letters. 2004.

[2] Cracknell AP. Remote sensing techniques in estuaries and coastal zones an update. International Journal of Remote Sensing. 1999;20:485-96.

[3] Maselli F. Monitoring forest conditions in a protected Mediterranean coastal area by the analysis of multiyear NDVI data. Remote Sensing of Environment. 2004;89:423-33.

[4] Cracknell AP. Remote sensing techniques in estuaries and coastal zones - an update. International Journal of Remote Sensing. 1999;20:485-96.

[5] Chen J, Quan W, Cui T, Song Q. Estimation of total suspended matter concentration from MODIS data using a neural network model in the China eastern coastal zone. Estuarine Coastal & Shelf Science. 2015;155:104-13.

[6] Samee A, Yogesh A, Mohor B, Mugdha A, Inamdar AB. Monitoring and trend mapping of sea surface temperature (SST) from MODIS data: a case study of Mumbai coast. Environmental Monitoring & Assessment. 2015;187:1-13.

[7] Petropoulos GP, Kalivas DP, Griffiths HM, Dimou PP. Remote sensing and GIS analysis for mapping spatio-temporal changes of erosion and deposition of two Mediterranean river deltas: The case of the Axios and Aliakmonas rivers, Greece. International Journal of Applied Earth Observation & Geoinformation. 2015;35:217-28.

[8] Enaruvbe GO, Ige-Olumide O. Geospatial analysis of land-use change processes in a densely populated coastal city: the case of Port Harcourt, south-east Nigeria. Geocarto Int. 2015;30:441-56.

[9] Harvey ET, Kratzer S, Philipson P. Satellite-based water quality monitoring for im proved spatial and temporal retrieval of chlorophyll-a in coastal waters. Remote Sensing of Environment. 2015;158:417-30.

[10] Jusoff K. Malaysian mangrove forests and their significance to the coastal marine environment. Polish Journal of Environmental Studies. 2013;22:979-1005.

[11] Zhang H, Bo S. Analysis on the coastline change and erosion-accretion evolution of the Pearl River Estuary, China, based on remote-sensing images and nautical charts. Journal of Applied Remote Sensing. 2013;7:9558-.

[12] Nguyen DTH, Tripathi NK, Gallardo WG, Tipdecho T. Coastal and marine ecological changes and fish cage culture development in Phu Quoc, Vietnam (2001 to 2011). Geocarto International. 2013;29:486-506.

[13] Shaji J. Coastal sensitivity assessment for Thiruvananthapuram, west coast of India. Natural Hazards. 2014;73:1369-92.

[14] Drakopoulos P, Ghionis G, Lazogiannis K, Poulos S. Toward Precise Shoreline Detection and Extraction from Remotely Sensed Images with the Use of Wet and Dry Sand Spectral Signatures. Fresen Environ Bull. 2014;23:2809-13.

[15] GONG, Peng. Some essential questions in remote sensing science and technology. Journal of Remote Sensing (in Chinese) 2009:1-12.

[16] Wan H, Wang Q, Jiang D, Fu J, Yang Y, Liu X. Monitoring the Invasion of Spartina alterniflora Using Very High Resolution Unmanned Aerial Vehicle Imagery in Beihai, Guangxi (China). Scientific World Journal. 2014;2014:638296-.

[17] Chen J, Quan WT, Cui TW, Song QJ. Estimation of total suspended matter concentration from MODIS data using a neural network model in the China eastern coastal zone. Estuar Coast Shelf S. 2015;155:104-13.

[18] Filippi AM, Jensen JR. Fuzzy learning vector quantization for hyperspectral coastal vegetation classification. Remote Sensing of Environment. 2006;100:512-30.

[19] Zheng D, Xiang X, Tao J. Optimising the modelling of eutrophication for Bohai Bay based on the cellular automata-Support vector machine method. Journal of Hydroinformatics. 2014;16:1125-41.

[20] Chang NB, Xuan ZM, Yang YJ. Exploring spatiotemporal patterns of phosphorus concentrations in a coastal bay with MODIS images and machine learning models. Remote Sensing of Environment. 2013;134:100-10.

[21] Lu D, Weng Q. A survey of image classification methods and techniques for improving classification performance. International Journal of Remote Sensing. 2007;28:823-70.

[22] Jiang D, Huang Y, Zhuang D, Zhu Y, Xu X, Ren H. A Simple Semi-Automatic Approach for Land Cover Classification from Multispectral Remote Sensing Imagery. Plos One. 2012;7:e45889-e.

[23] Wang Y, Jiang D, Zhuang D, Huang Y, Wang W, Yu X. Effective key parameter determination for an automatic approach to land cover classification based on multispectral remote sensing imagery. Plos One. 2013;8:e75852-e.

[24] Baatz M, Schäpe A. Multiresolution segmentation: an optimization approach for high quality multi-scale image segmentation. Beiträge zum AGIT-Symposium2000. p. 12-23.

[25] Baatz M, Benz U, Dehghani S. eCognition User Guide. Definiens Imaging GmbH. 2000.

[26] Benz UC, Hofmann P, Willhauck G, Lingenfelder I, Heynen M. Multi-resolution, object-oriented fuzzy analysis of remote sensing data for GIS-ready information. Isprs Journal of Photogrammetry & Remote Sensing. 2004;58:239-58.

[27] Walker JS, Blaschke T. Object-based land-cover classification for the Phoenix metropolitan area: optimization vs. transportability. International Journal of Remote Sensing. 2008;29:2021-40.

[28] Ranzi R, Rosso R. Distributed estimation of incoming direct solar radiation over a drainage basin. Journal of Hydrology. 1995;166:461-78.

[29] Wloczyk C, Richter R. Estimation of incident solar radiation on the ground from multispectral satellite sensor imagery. International Journal of Remote Sensing. 2006;27:1253-9.

[30] Smith HJ. Of sunlight, water, tree. Science. 2005;310:19.

[31] Gentine P, Entekhabi D, Chehbouni A, Boulet G, Duchemin B. Analysis of evaporative fraction diurnal behavior. Agricultural & Forest Meteorology. 2007;143:13-29.

[32] Wilson KB, Baldocchi DD. Seasonal and interannual variability of energy fluxes over a broadleaved temperate deciduous forest in North America. Agricultural & Forest Meteorology. 2000;100:1-18.

[33] Bastiaanssen WGM, Molden DJ, Makin IW. Remote sensing for irrigated agriculture: examples from research and possible applications. Agricultural Water Management. 2000;46:137-55As the access to this document is restricted, you may want to look for a different version under "Related research" (further below) orfor a different version of it.

[34] Bannari A, Morin D, Bonn F, Huete AR. A review of vegetation indices. Remote Sensing Reviews. 1995;13:95-120.

[35] Haboudane D. Hyperspectral vegetation indices and novel algorithms for predicting green LAI of crop canopies: Modeling and validation in the context of precision agriculture. Remote Sensing of Environment. 2004;90:337-52.

[36] Nigam R, Bhattacharya BK, Gunjal KR, Padmanabhan N, Patel NK. Continental scale vegetation index from Indian geostationary satellite: algorithm definition and validation. Curr Sci India. 2011;100:1184-92.

[37] Govaerts YM, Verstraete MM, Pinty B, Gobron N. Designing optimal spectral indices: a feasibility and proof of concept study. Int J Remote Sens. 1999;20:1853-73.

[38] Rouse JW. Monitoring the vernal advancement and retrogradation (greenwave effect) of natural vegetation. Nasa. 1974.

[39] Kawashima S. Relation between vegetation, surface temperature, and surface composition in the tokyo region during winter. Remote Sensing of Environment. 1994;50:52-60.

[40] Jordan CF. Derivation of Leaf-Area Index from Quality of Light on Forest Floor. Ecology. 1969;50:663-&.

[41] Zhang XZ, Zhang YG, Zhoub YH. Measuring and modelling photosynthetically active radiation in Tibet Plateau during April-October. Agr Forest Meteorol. 2000;102:207-12.

[42] Alados I, Foyo-Moreno I, Alados-Arboledas L. Photosynthetically active radiation: measurements and modelling. Agricultural & Forest Meteorology. 1996;78:121-31.

[43] Hoyle CR, Myhre G, Isaksen ISA. Present-day contribution of anthropogenic emissions from China to the global burden and radiative forcing of aerosol and ozone. Tellus B. 2009;61:618-24.

[44] Wu C, Chen JM, Huang N. Predicting gross primary production from the enhanced vegetation index and photosynthetically active radiation: Evaluation and calibration. Remote Sensing of Environment. 2011;115:3424–35.

[45] Wang L, Wei G, Hu B, Zhu Z. Analysis of photosynthetically active radiation in Northwest China from observation and estimation. International Journal of Biometeorology. 2015;59:193-204.

[46] Liu JH, Wu HY, Xu XL, Wu HY. Photosynthetically active radiation interacts with UVR during photoinhibition and repair in the cyanobacterium Arthrospira platensis (Cyanophyceae).. Phycologia. 2014;53:508-12.

[47] Bai J. Photosynthetically active radiation loss in the atmosphere in North China. Atmospheric Pollution Research. 2013.

[48] Potter CS, Randerson JT, Field CB, Matson PA, Vitousek PM, Mooney HA, et al. Terrestrial Ecosystem Production - a Process Model-Based on Global Satellite and Surface Data. Global Biogeochem Cy. 1993;7:811-41.

[49] Prince SD, Haskett J, Steininger M, Strand H, Wright R. Net primary production of US Midwest croplands from agricultural harvest yield data. Ecol Appl. 2001;11:1194-205.

[50] Nemani R, White M, Thornton P, Nishida K, Reddy S, Jenkins J, et al. Recent trends in hydrologic balance have enhanced the terrestrial carbon sink in the United States. Geophys Res Lett. 2002;29.

[51] Crabtree R, Potter C, Mullen R, Sheldon J, Huang SL, Harmsen J, et al. A modeling and spatio-temporal analysis framework for monitoring environmental change using NPP as an ecosystem indicator. Remote Sensing of Environment. 2009;113:1486-96.

[52] Goetz SJ, Prince SD. Modelling terrestrial carbon exchange and storage: Evidence and implications of functional convergence in light-use efficiency. Adv Ecol Res. 1999;28:57-92.

[53] McCallum I, Wagner W, Schmullius C, Shvidenko A, Obersteiner M, Fritz S, et al. Satellite-based terrestrial production efficiency modeling. Carbon balance and management. 2009;4:8.

[54] Wang PJ, Xie DH, Zhou YY, Youhao E, Zhu QJ. Estimation of net primary productivity using a process-based model in Gansu Province, Northwest China. Environ Earth Sci. 2014;71:647-58.

[55] Bala G, Joshi J, Chaturvedi RK, Gangamani HV, Hashimoto H, Nemani R. Trends and Variability of AVHRR-Derived NPP in India. Remote Sens-Basel. 2013;5:810-29.

[56] Ruimy A, Dedieu G, Saugier B. TURC: A diagnostic model of continental gross primary productivity and net primary productivity. Global Biogeochem Cy. 1996;10:269-85.

[57] Prince SD, Goward SN. Global primary production: A remote sensing approach. J Biogeogr. 1995;22:815-35.

[58] Potter C, Klooster S, Myneni R, Genovese V, Tan PN, Kumar V. Continental-scale comparisons of terrestrial carbon sinks estimated from satellite data and ecosystem modeling 1982-1998. Global Planet Change. 2003;39:201-13.

[59] Cao MK, Prince SD, Small J, Goetz SJ. Remotely sensed interannual variations and trends in terestrial net primary productivity 1981-2000. Ecosystems. 2004;7:233-42.

[60] Nayak RK, Patel NR, Dadhwal VK. Estimation and analysis of terrestrial net primary productivity over India by remote-sensing-driven terrestrial biosphere model. Environ Monit Assess. 2010;170:195-213.

[61] Zhang YL, Qi W, Zhou CP, Ding MJ, Liu LS, Gao JG, et al. Spatial and temporal variability in the net primary production of alpine grassland on the Tibetan Plateau since 1982. J Geogr Sci. 2014;24:269-87.

[62] Field CB, Randerson JT, Malmstrom CM. Global Net Primary Production - Combining Ecology and Remote-Sensing. Remote Sensing of Environment. 1995;51:74-88.

[63] Lee C, Landgrebe DA. Decision boundary feature extraction for nonparametric classification. IEEE Transactions on Systems Man & Cybernetics. 1993;23:433-44.

[64] Fritzke B. Fritzke A growing neural gas network learns topologies. Advances in Neural Information Processing Systems. 1995;7:625--32.

[65] Filgueira R, Grant J, Bacher C, Carreau M. A physical–biogeochemical coupling scheme for modeling marine coastal ecosystems. Ecological Informatics. 2012;7:71–80.

[66] Filgueira R, Grant J, Stuart R, Brown MS. Ecosystem modelling for ecosystem-based management of bivalve aquaculture sites in data-poor environments. Aquacult Env Interac. 2013;4:117-33.

[67] Filgueira R, Grant J, Stuart R, Brown MS, Filgueira R, Stuart R, et al. Operational models for Ecosystem-Based Management of bivalve aquaculture sites in data-poor environments. Aquaculture Environment Interactions. 2013;4:117-33.

[68] Thomas Y, Mazurié J, Alunno-Bruscia M, Bacher C, Bouget JF, Gohin F, et al. Modelling spatio-temporal variability of Mytilus edulis (L.) growth by forcing a dynamic energy budget model with satellite-derived environmental data. Journal of Sea Research. 2011;66:308-17.

[69] Silva C, Ferreira JG, Bricker SB, Delvalls TA, Martín-Díaz ML, Yáñez E. Site selection for shellfish aquaculture by means of GIS and farm-scale models, with an emphasis on data-poor environments. Aquaculture. 2011;318:444–57.

[70] Ferreira JG, Hawkins AJS, Bricker SB. Management of productivity, environmental effects and profitability of shellfish aquaculture — the Farm Aquaculture Resource Management (FARM) model. Aquaculture. 2007;264:160-74.

[71] Collins MG, Steiner FR, Rushman MJ. Land-Use Suitability Analysis in the United States: Historical Development and Promising Technological Achievements. Environmental Management. 2001;28:611-21.

[72] Dai FC, Lee CF, Zhang XH. GIS-based geo-environmental evaluation for urban land-use planning: a case study. Engineering Geology. 2001;61:257–71.

[73] Liu Y, Lv X, Qin X, Guo H, Yu Y, Wang J, et al. An integrated GIS-based analysis system for land-use management of lake areas in urban fringe. Landscape & Urban Planning. 2007;82:233-46.

[74] Jiang D, Zhuang D, Xu X, Lei Y. Integrated Evaluation of Urban Development Suitability Based on Remote Sensing and GIS Techniques:A Case Study in Jingjinji Area, China. Sensors. 2008;8:5975-86.

Human History of Maritime Exploitation and Adaptation Process to Coastal and Marine Environmentsess to Coastal and Marine Environments

Human History of Maritime Exploitation and Adaptation Process to Coastal and Marine Environments – A View from the Case of Wallacea and the Pacific

Rintaro Ono

Additional information is available at the end of the chapter

http://dx.doi.org/10.5772/62013

Abstract

This chapter introduce the archaeological new findings and current outcomes for the past human marine exploitation and maritime or coastal adaptation particularly in the Wallacea region where I have studied for long time. One of the oldest and important data I discuss here is from Jerimalai Cave site from East Timor and Leang Sarru site from Talaud Islands. The finds from East Timor demonstrate the high level of maritime skills and technology possessed by the modern humans who colonized Wallacea. These skills would have made possible the occupation of the faunally depauperate islands of Wallacea and facilitated the early maritime colonization of Australia and Near Oceania. On the other hand, Leang Sarru site dated back to 35,000 years ago on Talaud Islands where located over 100 km away from neighbour islands. The site also produced large number of marine shells from the late Pleistocene via Last Glacial Maximum (LGM) to the early Holocene, then we can also discuss the past maritime exploitation and adaptation from the late Pleistocene to the early Holocene in the Talaud Islands, where located in Northern part of Wallacea. During the Holocene after 12,000 years ago, various capture technology invented, and great variety of fish and shellfish species had been exploited by modern human. Especially the fishing technology and Ocean navigation technology were developed after the Neolithic times in Wallacea to the Pacific. Most famous archaeological records related them are the Lapita migration and colonization to many islands in Melanesia to Western Polynesia where were mostly uninhabited islands before them. After the Lapita colonization, the Polynesians who are the descendant of Lapita people succeeded to colonize Hawaii, Easter Island, and New Zealand by the 12th to 13th centuries. The distance to these islands from their neighbour islands or continent is over 4000 km, hence the success of migration by the Polynesian clearly indicate their maritime adaption and navigation technology were highly developed. In fact, it is a dramatic event that modern human succeeded to migrate to all over the world except North and South Pole when the colonization to New Zealand was done by the Polynesians. It also shows that marine environment were our last target for migration and colonization in this world after the human birthed in and around inner forest environment over 600 million years ago. This chapter also discuss such developments of marine exploitation and maritime adap-

tation after the Holocene or Neolithic to modern times, then reviewing the human adaptation history to coastal and marine environments.

Keywords: Human history, maritime adaptation, fihing technology, Wallacea, Pacific

1. Introduction

Fish and shellfish are part of the major aquatic food resources and play significant role for human diet and tools for a long time, while its evolutional process of exploiting such aquatic species by hominins who originally birthed in forest and terrestrial environments is yet unclear. The archaeological traces of the first human aquatic exploitation now back to over 2 million years ago, while the active use of marine resources including shellfish, fish, and other marine animals seems much later and possibly started by *Homo sapiens* known as modern humans. This chapter firstly introduces a brief history of marine use and maritime adaptation by early humans, including *H. sapiens*, and then discusses further maritime adaptation to Wallacea and Oceania, where the largest archipelago modern humans have migrated and lived after around 50 ka years ago up to now.

The term "Wallacea" came from Alfred Russel Wallace, who voyaged through the Malay Archipelago to recognize the zoogeographic divide running between Lombok and Bali, Borneo/Kalimantan Island, and Sulawes [1], now known as Wallace Line. The major factor to make such zoogeographic divide is the existence of deep sea between Lombok and Bali and the whole Wallacea Archipelago; hence, this line is defined for Wallace two natural provinces that had evolved during a long history of separation and isolation. Now we know Wallace Line is exactly running the past border of Sunda subcontinent coastline, and many Eurasian or Asian animals, especially mammals, could not or were hard to migrate into far east beyond the line in the past.

In terms of the past climate and wind water masses around Wallacea and Oceania, we should remember that the history of human migration mainly has corresponded to the Pleistocene to Holocene times. The Pleistocene is now estimated to have started around 2.6 million years ago corresponding to the glacial and interglacial cycles up to now in the latest interglacial period known as the Holocene. During the longer Pleistocene, the Earth's temperature had been cycled and separated into two stages as (1) cooler and dry period (glacial period) and (2) warmer and humid period (interglacial period). The average time-span of each stage was around 100 ka years. During the glacial period, huge ice sheets expanded in high latitude area and the world average sea level decreased down to 150 to 140 m when 6 million km^3 water become ice sheet and more land area appeared. In the Southeast Asian region, Sunda subcontinent occurred as some major islands including Sumatra, Java, and Borneo (Kalimantan) were connected, while a huge Sahul Continent was formed by union of Australia and New Guinea in Oceania. Yet, the Wallacea region was never connected to these continents and formed as an archipelago even during the glacial period since their surrounding ocean is deeper than 150 m and hence become the water barrier to many animals for migration to Oceania.

However, Wallace recognized that it did not apply to the human occupants of the regions since he thought, "Man has means of transversing the sea which animals do not possess" [1:30]. Wallace inferred that maritime enterprise had allowed humans to travel between islands carrying with them their genes, language, cultural traits, domestic animals, and crops [2]. The current archaeological and anthropological evidences show that the earlier *Homo erectus* might not be able to go beyond the line. Only an exception is a human trace dated 84 ka in Flores Island, which is located in Wallacea. If this date is secure, the evidence shows that the later *H. erectus* might develop their maritime adaptation along the Sunda coast to cross the sea and migrate into islands in Wallacea.

However, all other archaeological traces of human maritime adaptation in Wallacea and Oceania are from the age of *Homo sapiens* and dated after around 50 ka years ago. One of the prominent evidences of our maritime adaption was the earlier modern human migration to Sahul Continent, since it was required to cross the sea over 80 km from Wallacea islands to Sahul even during the Late Pleistocene time with lower temperature and sea level. It also becomes the first migration to Oceania region by humans and long-distance ocean voyage in the world. Such archaeological evidences of modern human maritime adaptation far increased in both Wallacea and Oceania after the drastic sea rise and great increase of coastal distance during the Holocene with warmer temperature, which began around 12,000 years ago.

In Oceania, the islands in the South Pacific are used to be divided into (1) Near Oceania, which includes the past Sahul Continent and its neighboring islands between Bismarck Archipelago and Solomon Islands, and (2) Remote Oceania, which includes all other Melanesian Islands east beyond the Solomon Islands and whole Micronesia and Polynesian islands in the Pacific including Hawaii, New Zealand, and Easter Island (Rapa Nui). The highlight could be back to the first colonization by Austronesian-speaking people including Lapita people to the Remote Oceania Islands, which are located in remote and uninhabited islands before them around 3500 years ago. Some of their voyages for migration to new islands were required to cross the ocean over 800–2000 km; hence, they should have more sophisticated and high-level seafaring skills and maritime technology than their ancestors in the Late Pleistocene. In relation with the past climate and marine environment, the development of unstable sea conditions caused by ENSO from about 4–5000 years ago could be one of the factors for the development of more seaworthy craft and higher seafaring skills (e.g., [3]).

This chapter discusses such Holocene maritime adaptation and migration by modern humans based on our recent archaeological outcomes as well. With such discussions, we will see how the human maritime adaptation process has corresponded to the changes in the past marine environment including sea level, temperature, coastline, and marine resources.

1.1. Aquatic resource use by early hominins

Currently, the oldest trace of fish use by humans was found at the 1.95-million-year-old Oldowan site of FwJJ20 in Kenya (Fig. 1), and some early hominins possibly *Homo habiris* already exploited a varied diet including aquatic species such as freshwater fish singly occupied by catfish (*Clarias* sp.) family [4]. It is also reported that most of these catfish bones have cut marks and hence could be eaten by humans. However, most Oldowan-aged sites including FwJJ20 produced only freshwater fish and over 80% of them were catfish, which could be captured by hand [5].

Although it is yet unclear whether these aquatic species were captured and discarded by humans or not, some early Rift Valley sites in East Africa have also produced the bones of aquatic animals, including crocodiles, fish, frogs, and shellfish, dated around 2.5 to 1.7 million years ago [6, 7]. Similarly, most of these fish are occupied by catfish species. On the other hand, some evidences of aquatic use by *Homo erectus* dated between 1.1 and 0.4 million years ago were also discovered in Africa and Southeast Asia. At Olduvai in Africa, Leakey reported that the bones of catfish and hippos are ubiquitous in artifact-bearing sediments [8], which also produced remains of crocodiles, aquatic turtles, and freshwater shellfish [7].

The relatively unambiguous use of aquatic resources, mainly freshwater shellfish by *Homo erectus* in Southeast Asia, comes from the site of Kao Pah Nam, a limestone cave in northern Thailand occupied about 700 ka [9] and the site at Trinil in Java, Indonesia, occupied around 500 to 400 ka (e.g., [10]). Based on these evidences, it is now acceptable that capture and use of freshwater fish and shellfish species possibly begun by early hominins (cf. *H. habiris* and *H. erectus*) could be the initial stage of human aquatic exploitation, since our ancient ancestors were originally birthed in mountain-forest environments in Africa back to over 6 million years ago.

After 400 ka during the stage of archaic *Homo sapiens*, archaeological evidences for aquatic resource use increases [7]. For instance, the Lower Paleolithic site of Hoxne in England produced numerous specimens of freshwater fish and beaver, smaller numbers of otter and waterfowl along with Clactonian artifacts dated 350 to 300 ka, which mainly correspond to an interglacial period [11]. Several North African Middle Paleolithic or Aterian sites, including Haua Fteah in Libya (e.g., [12]), Mugharet el'Aliya in Morocco [13], and several sites in Morocco and Algeria (e.g., [14, 15]) produced shellfish remains as well. Also in southern Europe, Monte Circeo [16] and Grimaldi caves [17] in Italy, Ramandils in France [18], and Devil's Tower Rockshelter [19], Gorham's Cave (e.g., [20]), and Vanguard Cave [21] in Gibraltar produced shellfish along with Mousterian stone tools assemblage.

Another archaic *Homo sapiens* site at Terra Amata along the Mediterranean coast of France also produced mussels and other marine shells about 300 ka, although their context and quantity are poorly documented [22,23]. If their context could be confirmed as really dated back to 300 ka, the use of marine shellfish by humans may possibly be started during the archaic *H. sapiens* species, including Neanderthals who occupied along the coastal area in Europe to Middle East. However, there is little evidence for the marine fish capture and use by Neanderthals or other archaic *H. sapiens* in the regions (e.g., [18]).

1.2. Development of aquatic and marine resource use by early modern humans

During Middle Stone Age (MSA) after 160 ka, aquatic resources become regularly visible in the diets of some groups of hominins, including modern humans (*Homo sapiens*), in Africa (Fig. 1). One of the early and solid evidences for marine shellfish consumption so far was found in Pinnacle Point in the south coast of South Africa dated around 160 ka (e.g., [24,25]). Klaises

River Cave (e.g., [26]) and Blombos Cave (e.g., [27,28]) in the South African coastal caves produced over 18 spices of marine shellfish during the Middle Stone Age dated around 120 to 100 ka ([29]). Another early evidence for possible consumption of marine shells dated about 125 ka was also found near Abdur in Eritrea (Fig. 1) along the Red Sea coast ([30,31]). However, it is not clear whether these marine shells were consumed by modern humans or other hominids. It is also pointed that 10 km is the maximum transport distance for shellfish during the MSA stage (e.g. [32,33]).

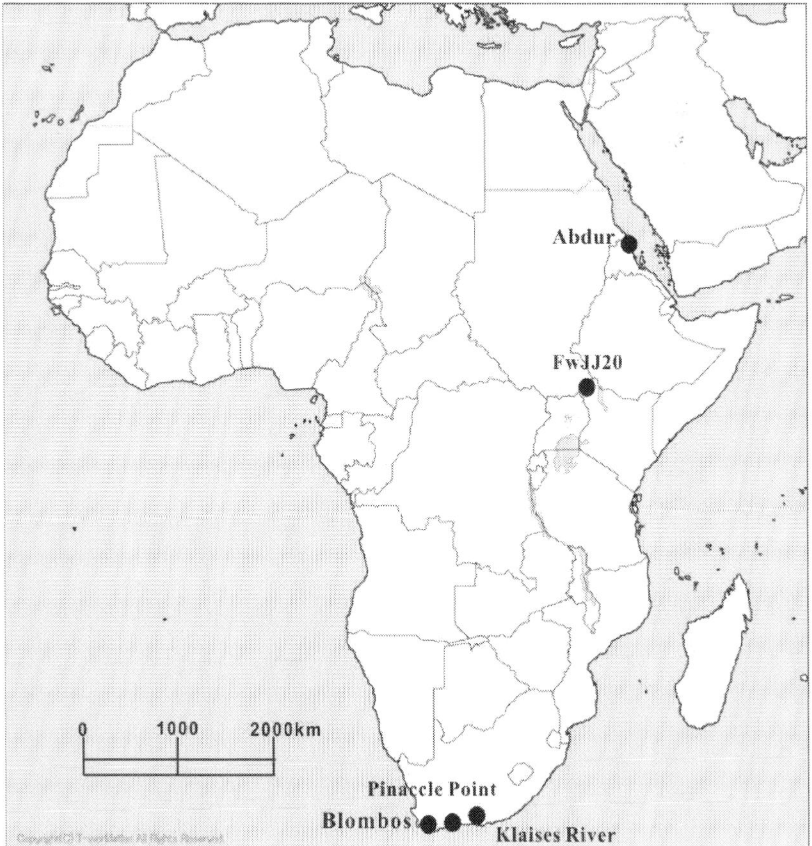

Figure 1. Locations of some important sites with traces of aquatic and maritime adaptation by early hominins and modern humans in Africa.

In terms of marine fish exploitation, one of the early evidences was also found at Blombos Cave. The site produced a number of marine-fish bones identified into 10 species of from MSA layers dated 100 to 70 ka years ago, yet inshore species were the majority [27]. Besides fish, the site also produced fur seals, penguins, and marine birds. A number of bone spears estimated as fishing and hunting tools were also found in MSA layers in Blombos Cave [28]. In north and east Africa, the major captured fish species were catfish taken in lake or river environments.

The capture and use of catfish seems like similar practice with the earlier use by early hominins, although the capture technique or skill could be much developed during MSA. For example, a number of sophisticated barbed bone spears with amount of catfish remains found from Katanda (Fig. 1) on the Semliki River in Zaire (e.g. [34,35,36]) dated back to 90 to 80 ka in east Africa (Fig. 1). Such evidences indicate that early modern humans (*Homo sapiens*) already invented spear technology during MSA to catch both freshwater and marine fish.

During the same ages, production and use of more sophisticated shell tools and ornaments seem to have started. Blombos produced perforated *Nassarius* shell beads dated back to 75 ka [37], while Skhul in Israel and Oued Djebbana in Algeria also produced similar *Nassarius* beads back to 100–135 ka [38]. Since the Middle East including Israel to Europe was dominated by Neanderthals during this stage, these beads could be also made and used by them. Another evidence for use of shell tools by Neanderthals was also found at the Grotta Moscerini Cave in Italy dated 60–80 ka [16].

Recently, however, another new evidence of possible shellfish use for tool or ornament and engraving by *Homo erectus* at Trinil in Indonesia dated 540 to 450 ka was discovered [39]. According to their report, the well-preserved Pseudodon shells with a small hole (5–10 mm in diameter) around the anterior adductor muscle scar inside the shell (n=73) could be the ornament produced by *H. erectus*. Among them, five shells were dated by luminescence dating and back to around 500 ka. The finding of possible engraving by *H. erectus* may be a more surprising one, since the earliest previously known undisputable engravings are at least 300,000 years younger and by modern humans (cf. some engraved ochre pieces dated 70 ka at Blombos). If such evidence will be added, the possible use of shell tool and invention of engraving art could be traced back to later *H. erectus* level in near future. In any case, we shall consider that the use of shellfish for various tools could be another new tradition and technology occasionally practiced both by modern humans and Neanderthals rather than *H. erectus* so far.

In Late Stone Age (LSA) after 50 ka, further development of fish capture methods occurred during migration process by modern humans into Eurasia and via the Wallacea Archipelago to the Pacific. For example, the oldest trace of fast-swimming fish including tunas was found at Jerimalai Cave site on East Timor in Wallacea dated back to 42 ka. The site also produced shell fish-hooks after 23 ka [40]. Early fish-hooks appeared in Europe by Late to Final Palaeolithic around 20 ka [41] and East Asia to North Pacific after 10 to 8 ka. Beads or other ornaments made from marine shells or artistic depictions of aquatic animals were also produced from a number of coastal Upper Paleolithic sites in Europe and southwest Asia (e.g., [7, 18, 42, 43]).

All these evidence show that our adaptation to coastal and marine environments had been dramatically developed particularly after modern humans birthed in Africa and migrated into Eurasian continent and insular Pacific region including the Wallacea Archipelago (eastern Indonesia and East Timor regions). Based on such understanding, the next section introduces the archaeological new findings and current outcomes for the past human marine exploitation and maritime or coastal adaptation, particularly in the Wallacea Archipelago, where the author has conducted archaeological fieldworks for over 10 years. One of the oldest and important data discussed here is from Jerimalai Cave site on East Timor and Leang Sarru Rockshelter site on Talaud Islands. The author also briefly introduces other important sites in and around Wallacea and the Pacific for comparative purpose to see the evidences of human use of marine resources and maritime adaptation in this region.

2. The late pleistocene marine exploitation and adaptation in Wallacea

2.1. Human migration and dispersal into Wallacea and the Pacific

A few million years later since humans birthed in African forest or subforest environments, a few groups of our ancestors succeeded out of the African continent and migrated into Eurasian continent or Old World during the stage of early hominins. One of the earliest hominins outside of Africa were recently found at Dmanisi Cave in the Republic of Georgia. The Dmanisi specimens dated 1.75–1.8 million years ago. Although they were reported to show clear affinities to African *Homo ergaster* rather than to more typical Asian *Homo erectus* or to any European hominid (e.g., [44, 45]), it is dominantly considered that Dmanisi men are the early type of *H. erectus* that possibly evolved from *H. ergaster*.

Other early hominin remains discovered outside of Africa are all identified as *Homo erectus*. The oldest aged *H. erectus* now is Sangiran 2 discovered in Java, Indonesia, and possibly dated around 1.66 to 1.49 million years ago (e.g., [46, 47]). Since the Java and Southeast Asian region is basically a tropic zone with rich floral and faunal resources for human food consumption, and Java (and also Sumatra and Borneo/Kalimantan Island) has been connected to the Asian mainland to form Sunda continent, except during the interglacial period like now in Holocene with much warmer temperature and higher sea level, early *H. erectus* might target this region soon after they left Africa for some reasons.

Following the current major hypothesis, *Homo erectus* possibly birthed around 2.4 million years ago and succeeded out of Africa around 2 million years ago into Eurasian continent, and they seemed to prefer the tropic or subtropic area in early stage. Their first migration into Java could be done during the glacial period when Java was connected to the Asian continent (Malay Peninsula), Sumatra, and Borneo/Kalimantan Island, which formed the Sunda subcontinent. They could disperse into Java by walking and without the need to cross ocean. In fact, all the old *H. erectus* bones and stone tools possibly made by them are only found in Java or in the islands that belonged to the last Sunda subcontinent but not in further eastern islands in Indonesia that are exactly located in the Wallacea Archipelago.

The earliest possible human trace in Wallacea so far is from Flores Island dated 840 ka. The evidence are some stone tools possibly produced by *Homo erectus*. Although the bones themselves are not discovered yet, if it could be the earliest human trace in Flores Island, it is the earliest evidence of sea crossing by humans in *H. erectus* level, since Flores Island is never connected to the Sunda subcontinent or the neighboring islands. There have been at least 19 km distance of sea gap between Bali and Lombok even during glacial period, and they needed to cross this sea gap as well as other much shorter gaps to reach Flores Island. Therefore, the earlier human trace in Flores Island tentatively indicates the possible development of maritime adaptation by *H. erectus* or early hominins.

In terms of the Flores Island case, however, we should not forget the recent discovery of another new human species, *Homo floresiensis,* whose partial skeleton including a cranium, mandible, and several lower limb elements named as LB1 were excavated at Liang Bua in 2003 (e.g., [48, 49]). Although the direct dating of the bones only dated around 17 ka and surprisingly younger, the morphological character and size of cranial and other bones indicate that *H. floresiensis* could be one of the early hominin species and differs from the *Homo erectus* group (e.g., [50, 51]). Later excavations at Liang Bua added more bones possibly belonging to LB1 and also other individuals (LB2-6), while all of these bones could be identified as *H. floresiensis* (e.g., [52]). Among these bones, a few bones belong to LB 2 and 3 dated 74 ka, while this date is also much younger than other early hominin bones in the world.

In any case, if they are another early hominin species that differ from *Homo erectus*, there is another possibility that the Flores Island stone tools dated 840 ka can be also produced by *Homo floresiensis*, since they are only early hominins found in Flores Island so far, even their dates are very young. If so, this newly found human species might also develop maritime adaptation in the past to cross sea gaps and reach Flores Island. Although the truth is yet unclear, the Flores Island case tentatively shows that human maritime adaptation and migration could be developed in Wallacea since the early hominin age as far older than the age of modern humans (*Homo sapiens*). However, the possible evidences of human maritime adaptation and migration in early hominin level are yet limited and all the other evidences belong to the age of modern humans dated after around 50 ka in Wallacea.

The birth of modern humans in Africa is now estimated around 200 ka based on DNA model, past-environment changing model, and several archaeological traces in Africa (e.g., [30,42,53, 54,55]. As discussed in above, Middle Stone Age (MSA) after 160 ka may correspond to the age of modern humans, and their use of aquatic and marine resources became regularly visible in Africa, particularly in the coast of South Africa represented by some sites including Pinnacle Point, Klaises River Cave, and Blombos Cave.

After some extent of such maritime and aquatic adaptation, single or a few groups of modern humans succeeded to migrate into Eurasia possibly by the route of over the mouth of the Red Sea and subsequently dispersed into Arabia and Southern Asia. Possible dates for out of Africa by modern humans are now estimated around 100 to 60 ka (e.g., [30,56,57,58, 59,60,61], and one of the oldest evidences for modern human migration out of Africa is from Australia dated back to 60 to 40 ka (e.g., [62,63]) but most potentially around 50 to 45 ka (e.g., [7,64,65]. Since Australia is located in Oceania further east from Eurasia, the existence of modern humans there clearly indicates that *Homo sapiens* should have migrated to Arabia, South Asia, and Southeast Asia, including the Wallacea Archipelago, before 50 to 45 ka.

It is also worth to mention that this earlier modern human migration to Australia, which was connected to New Guinea and formed the larger Sahul Continent, is so far the oldest evidence of human sea crossing over 80 km, since there had been over 80–100 km sea gap between Sahul or Australia and its neighboring islands in Wallacea including Timor or northern and southern Maluku Islands even during the Late Pleistocene at the last glacial period with much lower sea level and temperature (e.g., [65, 66,67,68]).

2.2. Major late pleistocene sites in Wallacea and the evidences

While the colonization of Sahul including Australia and New Guinea represents the earliest evidence of intentional and relatively long-distance seafaring in the world, there is relatively little known about the antiquity and nature of seafaring and coastal occupation in Wallacea, the nursery grounds for these early seafarers. However, recent excavations at Late Pleistocene sites in Wallacea provide evidence of early aquatic culture and marine exploitation. For instance, excavations in Maluku Islands, Aru Islands, and East Timor have uncovered evidence of modern human colonization dated 40 to 30 ka.

The major Late Pleistocene to Early Holocene sites in the Wallacea Archipelago (Fig. 2) are Jerimalai Cave, Lene Hara Cave, and Matja Kuru 2 Cave in East Timor (e.g. [40, 64,69, 70,71]), Golo Cave in northern Maluku Islands (e.g. [72-75]), Leang Sarru Rockshelter in Talaud Islands (e.g. [65,76,77,78]), Leang Burung 2 in southern Sulawesi (e.g. [79,80]), and Leang Lembudu Cave in Aru Islands (e.g., [71, 81]).

2.2.1. Evidence from Jerimalai cave site

Among them, the Jerimalai Cave site is currently the oldest prehistoric site by modern humans dated back to 42 to 38 ka in Wallacea. It is located at the eastern end of East Timor (see Fig. 2), where Pleistocene-raised coralline terraces run parallel to the present coastline with many caves and shelters. The site also plays significant roles as producing the oldest evidence of fast-swimming pelagic fish exploitation back to 42 to 38 ka and oldest shell fish-hooks possibly back to 23 to 16 ka [40]. The excavation was conducted by Sue O'Connor and her team from the Australian National University in 2005 and the author analyzed these excavated fish bones during 2009 and 2010.

Although Jerimalai Cave was excavated by only two 1×1 m test pits (squares A and B) in 2005, the site produced a rich assemblage of cultural material, including well-preserved faunal remains including marine fish and shellfish, marine turtles, murid rodents, bats, birds, and various terrestrial reptiles, as well as stone artifacts (n=9752), bone points, shell fish-hooks, and shell beads dated to the Terminal Pleistocene. The excavation encountered only three cultural layers in Jerimalai Cave [64].

The radiocarbon determinations on marine shell indicate that the lowest Layer 3 accumulated during the Late Pleistocene between 42 and 38 ka and the lower part of Layer 2 accumulated during the terminal stage of the LGM around 17 to 16 ka. Layer 1 with heaviest volume of sediment formed during the Early Holocene, around 10,000 to 5000 years ago. The site appears to have little evidence of occupation between ~38 and 24 ka. This may be due to sea-level retreat

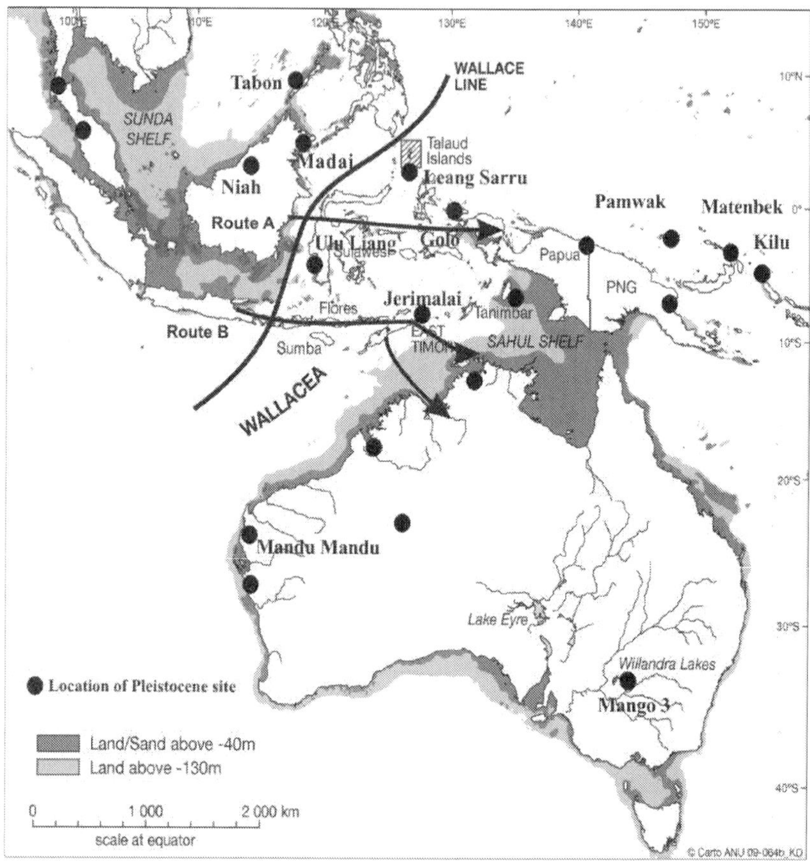

Figure 2. Map showing Sunda Shelf, Wallacea, and Sahul Shelf and major Pleistocene sites, with associated chronometric ages (after [65]).

rendering the shelter farther from the coast during this time; however, research at other nearby coastal sites such as Lene Hara Cave suggests that it more likely reflects limited sampling due to the small size of the excavation [40].

Among the faunal remains, the number and weight of fish bones far exceeds those of other fauna, averaging 56% by weight of the total vertebrate remains from the upper to lower layers. Especially, relatively larger bones from pelagic fish including tunas (Scombridae) and trevallies (Carangidae) compose almost 50% of the total fish assemblage in the earliest occupation levels. A total of 38,687 fish bones from square B and 23 taxa were identified. The total MNI (minimum number of individual) is 796 (= at least 796 fish were captured and discarded at the site) and the total NISP (number of identified specimen) is 2822 (= 2822 bones were identified into family or species level). Twenty-three taxa were identified, including 22 families and 1 species as *Monotaxis granoculis* (see [40:1119]).

The MNI of Scombrids (mainly identified as yellow-fin tuna and skipjack tuna) is greatest in most of the spits down to the base of the site. Scaridae (parrotfish), Carangidae (trevallies), Balistidae (triggerfish), and Serranidae (groupers) fish families follow in MNI and NISP. These species make up about 12% to 15% of the total MNI. Other major fish families identified are Lethrinids (emperors), Lutjanids (snappers), Acanthurids (unicornfish), Labrids (wrasses), and Tetradontids (puffers). For Elasmobranchi (rays and sharks), both are recognized, and some shark vertebra were identified as Carcharhinidae (see [40: 1118-1119]).

Among the three cultural layers in Jerimalai Cave, at least 15 fish taxa were exploited from the lowest layer (Layer 3), which may correspond to the earliest period of occupation, dated 42 to 38 ka, yet the quantity of fish bones from this layer is still limited. However, the tuna bones from this layer is so far the oldest evidence of human capture and use of tunas or fast-swimming offshore marine fish species in the world and tentatively indicates that modern humans migrated into Wallacea might have developed their maritime adaptation to have the high level of fish capture skills including hook and line fishing. Many of the current ethno-archaeological or fishery data in Wallacea to the Pacific islands (e.g., [65,82-91]) show the major fishing method to capture fast-swimming fish such as tunas are (1) fishing with hook and line or (2) trolling with lure (traditionally pearl shell lures have been used in the Pacific).

Since lure fishing could be developed in much later times possibly during the Neolithic times as discussed in below, it is highly plausible that hook and line fishing could be practiced to capture tunas at Jerimalai Cave. Yet, it should be noted that this evidence does not clearly show the modern humans at Jerimalai Cave practiced pelagic fishing in offshore sea zone with the use of a larger and faster boat or canoe as discussed by Anderson [92]. In fact, some tuna species including small-sized yellow-fin and skipjack tunas can also be captured near-shore or inside bay sometimes. However, wherever they caught tunas, they still might need to use hook and line for capturing tunas, as they are fast-swimming fish and hard to capture by other methods such as spearing, poisoning, trapping, and netting (see [93]). In recent or modern times, the large size of the net has been also used for capturing schools of tunas by large-scale motorized fishing vessel, although such modernized netting is hardly accepted in the prehistoric or even ethnographic fishing in Wallacea and the Pacific.

Therefore, the existence of tuna bones from Layer 3 in Jerimalai Cave shows the human use of fish hooks and line to capture marine fish dated back to around 40 ka, and the site also did produce some shell fish-hooks, although the oldest one from Layer 2 dated between ~23 and 16 ka, while later one in square A dated ~11 ka [40]. All are *Trochus* shell single-piece baited hooks and do not seem suitable for pelagic fishing. However, it is possible that other types of hooks were also developed at the same time. Bone points made on the spines of large fish first appear at Jerimalai Cave in Layer 2. Their function is uncertain, but they clearly represent a component of a composite tool, such as fine barbs for fish spears.

In terms of temporal change of marine fish exploitation in Jerimalai Cave, the prominent feature during this earliest occupation phase is the intensive focus on pelagic fish species, particularly Scombrids (Fig. 3). The ratio of pelagic species (49%) and inshore species (51%) based on MNI is almost equal, and sharks and rays were also exploited in Layer 3. In Layer 2, the ratio of pelagic species in the total MNI still remained high (46%), but the number of Carangids (trevallies) dramatically increased, whereas the exploitation of Scombrids (tunas) remained similar. The exploitation of inshore fish species also shows the drastic changes, as parrotfish were still intensively exploited, but the number of groupers and triggerfish also increased in Layer 2, which possibly corresponded to LGM. In Layer 1 during the mid-Holocene time, the most prominent change is the dramatic increase (66–76%) in the total MNI and lower ratio of pelagic fish species (34–24%). Pelagic species were still targeted, but inshore fish dominate the assemblage. The increase of inshore species may reflect the stabilization of sea level, warmer temperatures, and the development of coral reef environment along the coast (see [40:1119]).

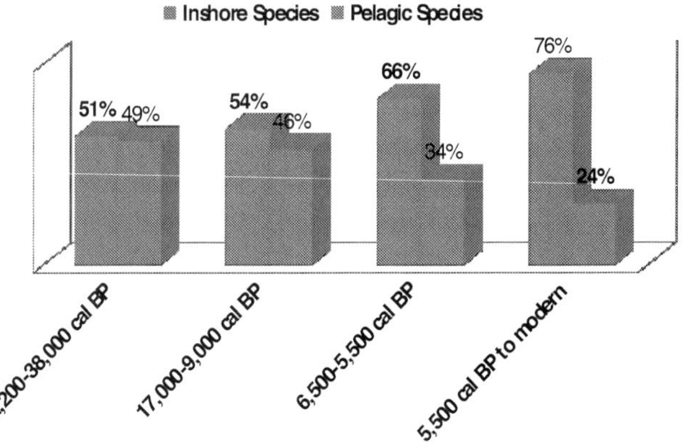

(Original map from Ono et al. 2010; Fig. 1)

Figure 3. Changing ratios of pelagic species vs inshore species in Jerimalai Cave (source: O'Connor et al. 2011; Fig. 2, which originally made by Ono).

The finds from East Timor demonstrate the high level of maritime skills and technology possessed by the modern humans who colonized Wallacea. These skills would have made possible the occupation of the faunally depauperate islands of Wallacea and facilitated the early maritime colonization of Australia and Near Oceania. Here, the author also introduces other archaeological evidences from Leang Sarru Rockshelter site, where the author re-excavated in 2005 and dated back to 35 ka in next.

2.2.2. Evidence from Leang Sarru Rockshelter site

Leang Sarru Rockshelter is a small limestone rockshelter located about 400 m from the eastern coast of Salibabu Island in the Talaud Islands, which is located over 100 km away from neighbor islands (see Fig. 1). It is situated in an uplifted coral limestone block about 15 m above sea level in the middle of a clove plantation. The shelter faces northeast and is about 5×3 m in area. It has a curving ceiling about 2.5 m high at the drip line [65: 320].

The site was previously excavated by Tanudirjo [76, 77] in 1995, and he dug two 1 m² pits in 10 cm spits to a depth of about 80–90 cm below ground surface. This excavation revealed four sedimentary layers and unearthed thousands of chert artifacts and shell remains but no animal or fish bones. This excavation was relatively small, about 2 m² (1.8 m³), and the front area of the shelter was not examined. For this reason and to collect further archaeological data, Ono and Balai Arkeologi Manado (Institute for Archaeological Research in Manado) re-excavated the site in 2004 [65, 78, 94]. This excavation uncovered a further area of about 6 m² (3.6 m³) and comprised areas both inside and outside the shelter.

The excavation encountered three cultural layers (corresponding to Tanudirjo's Layers 1–3) before reaching the hard calcareous deposit (corresponding to Tanudirjo's Layer 4), and we inferred that the hard deposit shelved downward toward the back wall of the shelter where Tanudirjo had excavated. The radiocarbon determinations on marine shell indicate that Layer 3 (and Layer 4 in Tanudirjo's excavation) accumulated during the Late Pleistocene possibly between 35 and 32 ka, and the lower part of Layer 2 accumulated during the final stage of the LGM around 21 to 18 ka. The upper part of Layer 2 and possibly the lower part of Layer 1 formed during the Early Holocene around 10,000 to 8000 years ago [65: 322-323].

There were no conventional ¹⁴C or AMS determinations for the periods between 27–21 ka and 17 ka to 10,000 years ago; hence, it is possible that the shelter may not have been inhabited during these periods, other than perhaps very occasionally with little or no cultural discard. All the evidence possibly show the shelter had been occupied during at least three different periods in the Late Pleistocene and Early Holocene, an interpretation consistent with the tentative conclusion reached by Tanudirjo[77].

In total, 9465 stone artifacts, including flake tools, flakes, cores, chips, chunks, and hammer-stones, together with 3371 NISP (number of identified specimen) of marine shell, land snail, crustacean, and sea urchin, and 580 earthenware sherds (only from the upper layer and top soil) were excavated. This range of cultural materials was generally similar to those from the 1995 excavation by Tanudirjo, with no fish or animal bones. The lack of animal bones possibly indicates that edible animals were scarce in the Talaud Islands. In fact, the Talaud Islands in modern times have no land mammals other than about 14 species of bat, 5 species of rat, 4 species of flying fox (*Pteropus* spp.), and 2 species of cuscus (*Ailurops ursinus* and *Strigocuscus celebensis*). Some introduced animals include chicken, dog, cattle (*Bos javanicus*), and pig (*Sus celebensis*) (see [95]). There is no archaeological evidence for the existence of large or mid-sized animals either [65: 323].

Shell remains are another major material from Leang Sarru Rockshelter. A total NISP of 3281 marine shell and land snail (26 kg) recovered from the excavation were sorted into 53 taxa: 23 of these taxa were identified to species level and the remainder to genus and family levels. One species each of crustacean (Brachyura) and sea urchin (*Heterocentrotus mammillatus*) were also identified as marine resources. With only minor differences, the shell density demon-strates a consistent pattern. When the 4135 in NISP and 15.2 kg/1.8 m³ of shell remains from the previous excavation (identified as 40 taxa; see Tanudirjo 2001) are added, a grand total of 7416 (NISP) shell and land snail remains and 42.1 kg/5.4 m³ was retrieved from Leang Sarru Rockshelter [65: 325-326].

The analysis of the shell remains confirmed that *Turbo* spp. (e.g., *Turbo marmoratus* and *Turbo setosus*), *Nerita* spp. (e.g., *Nerita balteata* and *Nerita undata*), and *Trochus* spp. (e.g., *Trochus maculatus* and *Trochus niloticus*) were the predominant faunal species at Leang Sarru Rockshelter. Among them, *Turbo* spp. and *Nerita* spp. were the most abundant in number, whereas, in terms of size and actual meat value, *Turbo* spp. and *Trochus* spp., which were much larger than *Nerita* spp., were more important in terms of food and protein sources [65: 327]. These shell families are also the major exploited shells in other Late Pleistocene to Holocene sites in Wallacea and Near Oceanic islands (e.g., [64,96]).

In terms of temporal change of excavated shell and land snail, 33 species of them, with one species each of crustacean and sea urchin, were excavated from Layer 3 dated between 35 and 32 ka. The intertidal to subtidal rocky shore species such as Neritidae (*Nerita balteata*), Patellidae, Muricidae, Haliotidae, and Chitonidae were dominant, while *Turbo* spp., a subtidal species, were also exploited. The number of land snails such as Ellobiidae (*Pythia pantherina*) was also predominantly exploited during this period. However, the total amount of marine shell are yet small in number (NISP=859) during this early period.

In Layer 2B, dated between 21 and 17 ka, which corresponds to LGM, the total number and variety of species dramatically increased (NISP=1456; 42 species), as did crustacean and sea urchin (NISP=56 compared with NISP=12 in the earlier period). For instance, the major marine shell species of Neritidae, Turbinidae (Fig. 4A), and Trochidae (Fig. 4B) greatly increased in number (Fig. 4) and species variety, while the number of intertidal shell species such as *Haliotis varia* and land mollusc such as *Pythia* spp. slightly decreased. For *Nerita* spp., the number of *Nerita balteata*, *Nerita undata*, and *Nerita albicilla* dramatically increased, and for *Turbo* spp., the large to mid-sized species such as *Turbo setosus* and *Turbo marmoratus* also increased.

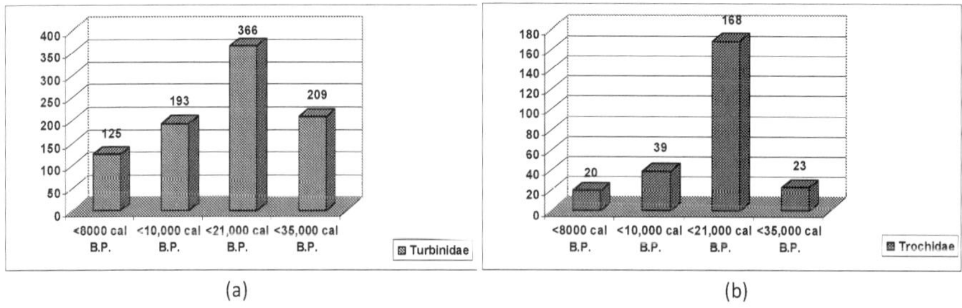

(a)   (b)

Figure 4. Number (MNI) of Turbinidae (a) and Trochidae (b) shell families in each layer/age at Leang Sarru Rockshelter.

For other species, shell belonging to Trochidae (particularly *Trochus maculatus*) and Fasciolariidae also dramatically increased. The increase in the total amount of shell remains in Layer 2B indicates more active exploitation of shell resources during the LGM. The exact reasons for such an increase is not completely clear, although one of the crucial factors may have been a decline in the available food resources in the vicinity of the rockshelter during the LGM; alternatively, the site might have been more intensively used during this period.

However, the total amount of shell remains dramatically decreased in Layer 2A (NISP=516) and Layer 1 (NISP=450), which dated around 8000–10,000 years ago. In upper layers, however, some subtidal and coral rubble-dwelling species belong to Fasciolariidae (*Latirus nagasakiensis*) and Tridacnidae (*Tridacna maxima* and *Tridacna crocea*) increased. The increase of such coral rubble-dwelling species possibly indicates warming of sea and air temperatures and renewed growth of coastal corals after the Holocene. In contrast, the major subtidal species such as *Turbo* spp. and *Trochus* spp. as well as major intertidal species such as *Nerita* spp., *Cellana* spp., and *Chiton* dramatically decreased during this Early Holocene stage, although *Turbo* spp. were still dominant even in this stage. Regarding land snails, Ellobidae (*Pythia* spp.) and other snails (Zonitidae and Helicarionidae) increased, particularly in Layer 1, although it is unclear whether these were collected by humans or naturally accumulated in the sediments [65: 327-330].

2.3. Late pleistocene maritime exploitation and adaptation in Wallacea and adjacent regions

Previously excavated Pleistocene sites (Fig. 2) at the present coast of the former Sunda Shelf region reveal very little or no evidence for human use of marine resources during the Late Pleistocene, as all of these sites were located a considerable distance from the coast at that time. For instance, Lang Rogrien Rockshelter, which is currently located in the western coast of southern Thailand but was about 75–140 km from the coast during the Late Pleistocene (e.g., [97]), produced no marine faunal remains other than a single fish bone identified as a probable marine species [98].

Similarly, there are very little or no marine shell and fish remains in the cave sites of Niah (e.g., [99, 100]), Leang Burung 2 [80], and Tabon in the west coast of Palawan Island [101] (Fox 1970) or from the open-air site of Tingkayu in northeastern part of Borneo Island [102], all of which were a distance of 30–400 km from the coast during the Late Pleistocene; in general, these sites do have remains of freshwater mollusc and fish along with a variety of terrestrial animals (e.g., [101,103,104]).

Such archaeological results demonstrate the difficulty of investigating Pleistocene marine exploitation and adaption in the Sunda Shelf region where former coastal sites are now submerged on the sea bed. On the other hand, island coasts in the Wallacea region were far less affected by the rapid rise in sea level since the Late Pleistocene, because they shelve steeply into deep water. For this reason, coastal sites in Wallacea provide good potential for identifying human exploitation of marine resources during the Late Pleistocene. For instance, Golo Cave, located 60 m inland from the northwestern coast of Gebe Island in Maluku Islands, produced shell tools and a variety (47 species) of marine shell that mainly inhabit rocky subtidal zones [75:703-704].

Similarly, in East Timor, Lene Hara Cave and Matja Kuru 2 Cave, located less than 10 km from the coast during the LGM, produced a larger amount of marine shell [70.72], while Jerimalai

Cave, located only a few kilometers from the shore, produced a number of *Nerita* spp., *Strombus* spp., *Trochus* spp., *Turbo* sp., and *Chiton* sp. as well as a large amount of fish remains including tuna and a variety of inshore fishes [64: 530].

In the Pacific region, next to Wallacea, Matenkupkum and Buang Merabak on New Ireland in the Bismarck Archipelago also produced amounts of marine shell, mainly rocky platform species such as *Strombus luhuanus, Trochus niloticus, Lambis lambis, Turbo* spp., *Nerita* spp., and *Chiton* dated around 40 to 32 ka (e.g. [105-108]). Wickler [109] noted a similar pattern at Kilu Cave on Buka Island, in which *Chiton* spp. comprised a relatively common component of the shell assemblages until the Early Holocene.

Although evidence for intensive marine exploitation in Australia during the Late Pleistocene is lacking, several sites in Western Australia have produced limited amounts of marine shell from strata dated between about 36 and 20 ka (e.g., [110-113]). For instance, at Mandu Mandu Creek Rockshelter (Fig. 2), located only about 4–5 km from the coast just prior to the LGM, a low-density midden deposit includes the remains of shell, crab, fish, and terrestrial fauna [110]. These sites could be interpreted as evidence for limited Pleistocene use of marine resources, but sea-level and shoreline reconstructions show a strong correlation between the presence and density of marine resources and the variable distance of each site from the sea [7:316].

These results tentatively indicate that coastal people in the Wallacea and Sahul regions (and probably in the Sunda Shelf region as well) actively exploited marine resources, especially shell, during the Late Pleistocene. The most important shell species exploited at these sites during the Late Pleistocene were *Turbo* spp., *Nerita* spp., *Trochus* spp., *Strombus* spp., and *Chiton*. In terms of actual meat value, *Turbo* spp. might have been the most important followed by *Trochus* spp. and *Strombus* spp. After 30 ka, the total volume and number of shell from Matja Kuru 2 Cave in East Timor dramatically decreased [72]. In contrast, the amount and variety of marine shell resources at Leang Sarru Rockshelter dramatically increased, especially during the LGM. These differences between the two sites may possibly relate to their distances from the coast.

The distance to the coast from each site was greater during the LGM. Even so, Leang Sarru Rockshelter was located within 2.5 km of the nearest coast [76:264], while Matja Kuru 2 Cave was over 10 km distant. Similarly, a heavy reliance on marine shell and other marine resources continued at Jerimalai Cave, which was located within 5 km of the coast during the Late Pleistocene [40:530]. The dramatic decrease in number and volume of marine shell in Matja Kuru 2 Cave during the LGM suggests the possible movement of the inhabitants to other locations closer to the coast or to other areas of the island. The absent of any human trace at most of LGM times during 24 to 17 ka in Jerimalai Cave also indicates the possible movement of inhabitants to other locations, but truth is yet unclear now with the very limited number of archaeological sites we have investigated.

In Sundaland, Niah and Tabon caves (Fig. 1) had been continually used, although their distances from the coast increased during the LGM, and very little or no marine shells were brought to these sites. Similarly, all other inland sites, such as Hagop Bilo Cave (17 to 12 ka) in northern Borneo and Ulu Leang Cave (Fig. 1) in southern Sulawesi, produce very little or no marine shell but instead freshwater and land mollusc species. Overall, it is plausible that the access to marine resources depended on distance from the coast (see also [7]).

The fact that marine shell remains are abundant at Leang Sarru Rockshelter indicates that extensive marine exploitation was conducted by the people who lived close to the coast even during the LGM. Furthermore, the dramatic increase in volume and variety of shell species in Leang Sarru Rockshelter during the LGM strongly indicates that there was heavy reliance on marine and coastal resources particularly during cold conditions. It is possible that maritime and coastal adaptation in remote islands such as Talaud during the LGM was a consequence of very limited terrestrial resources.

For capture and use of marine fish, both pelagic and inshore species in the Late Pleistocene Wallacea, we only have a solid data from Jerimalai Cave so far. In the Pacific, a more convincing case for systematic fishing comes from Kilu Cave (Fig. 1) in the northern Solomon Islands dated 32 ka. Although Kilu Cave contains a larger quantity of fish bone, including some pelagic species, overall fishing at Kilu Cave seems to have been inshore fish species.

As similar to other early modern human sites in Africa, shellfish could be more important marine resources for human food consumption in Wallacea and the Pacific, and such possibility may be the main factor of very limited evidence of fish capture during the Late Pleistocene compared with the evidence of shellfish use. However, the Jerimalai Cave and Kilu Cave cases show that the modern humans migrated into the archipelagos also exploited fish and developed their fishing skills and techniques to capture both inshore and pelagic species far before the LGM, and the importance of marine resources as food and tool source should become much important after the Holocene with a dramatic rise of sea level and the coastal area and line far expanded as we see in next section.

Lastly, in terms of seafaring or voyaging skills as another aspect of maritime adaptation, the current archaeological evidences in Australia, New Guinea and Wallacea show that modern humans had ability to cross ocean over 80 km distance by 50 to 45 ka. Identifying evidence of early seafaring tradition can be difficult, yet evidence for the settlement of offshore inlands not connected to adjacent mainland tentatively indicates the use of watercraft by humans during the Pleistocene [114].

Although no evidence for reconstructing the early seafaring technology including their vessels is found yet, it is assumed that its progenitors arrived by raft possibly made of bamboo (e.g., [3, 115]). The trial by Thorne and Raymond [115] to build a 15×2 m bamboo raft with a 2 m² square sail of matting on a 2 m short mast revealed that such raft could travel 4–5 knots in a light breeze. It is reasonable to assume that bamboo rafts or rafts of other materials were the means of the earliest ocean passages [3, 66].

In any case, the fact is that modern humans could cross ocean to migrate into new lands or island with rather new environment and terrestrial resources from their home lands, including Africa. The exact departing location(s) where the earlier group(s) of modern humans tried to cross sea for migrating to Sahul is not sure, although it should be somewhere in the Wallacea Archipelago. In hypothetically, the two major migration routes from Sunda subcontinent via Wallacea to Sahul Continent are estimated (e.g., [81, 116]).

The first major route (Route A, see Fig. 2) is from Borneo/Kalimantan Island through Sulawesi, Sula and then into north or south Maluku Islands (e.g., Halmahera, Buru, and Seram) and into the Bird's Head in New Guinea or directly to Sahul Shelf where the present Aru Islands lie. The second one (Route B, see Fig. 2) is along the Lesser Sundas to Timor then via Maluku Islands to Aru Islands or directly onto the Sahul Shelf near the Kimberley region of Western Australia. When based on the current evidences as the oldest modern human sites located in East Timor, the direct migration from Timor is the most potential and supported route so far (e.g., [64,68,81]). However, the current archaeological evidences we have are yet very limited since many islands or area are not archaeologically investigated in Wallacea especially along the first route, and we need to increase more data to reconstruct the initial and past human migration route(s) in the near future.

After the initial migration(s) to Sahul (both Australia and New Guinea), the seafaring skills could be further developed during 35 to 20 ka in Wallacea and the Pacific. As already discussed, the excavations at Leang Sarru Rockshelter have shown that humans colonized the Talaud Islands by sea crossing of over 100 km at least by around 35 ka in Wallacea. Although the exact route of this migration to the islands is not known, it appears likely that many islands in Wallacea were already colonized by modern humans at the time the relatively remote islands of the Talaud group were settled.

Correspondingly, voyages to the Bismarck Archipelago and Solomon Islands from Sahul occurred after 40 to 30 ka (e.g., [108, 117,119]). While the initial settlement of New Guinea, New Britain, and New Ireland required voyages of up to 100 km, colonization of Buka Island in the Solomon Islands at least 28 ka required a minimum sea voyage of 140 km and possibly 175 km [67:20]. Furthermore, the archaeological evidences on Kilu Cave on Buka Island show the use and possible production of some plant food, including *Colocasia* taro and *Alocasia* taro. The residues of these plant species were discovered on the earliest stone tools from the site [120].

During the LGM, Pamwak site on Manus Island (Fig. 2) was newly settled around 21 ka. It should require at least 230 km voyaging to reach Manus and Admiralty islands from the north coast of New Guinea or the northwestern tip of New Ireland [120]. So the people's seafaring skills might be much developed by LGM around 20 ka. The Manus Island case is the earliest example of literally sailing into the unknown and beyond the range of one-way intervisibility [96]. Migration or movement to Manus Island required an uninterrupted voyage of 200–230 km, 60–90 km of which would have been completely out of sight of land [67:21].

Furthermore, the archaeological evidences at Matenbek and Buang Merabak on the New Island (Fig. 2) also show the intentional transfer of obsidian as tool material and cuscus as food source around 20 to 16 ka. Obsidian is mainly from the New Britain source of Mopir or Talasea, c. 55 km from Mopir [122]. Two quite different source distributions might imply different linkage between New Ireland sites and New Britain sources, with implications for canoe travel [96:154]. The cuscus is a kind of small- or mid-sized marsupial originally not habited in the Bismarcks; hence, they could be introduced by humans into these islands possibly from the main island of New Guinea as a half-dozen domesticated and nondomesticated animals. The cuscus might have been taken across the 30 km strait separating New Britain and New Island, while the direct distance between the obsidian sources and both sites could be around 300–350 km. It is yet unknown whether the transfer of obsidian was done directly by crossing over 300 km sea or by hopping the coastal of New Britain to New Ireland coast. In the latter case, the maximum distance to cross sea could be about 30 km same as the estimated cuscus transfer.

It has often been argued that initial colonization of the many islands of Wallacea and its adjacent regions must have been facilitated by a maritime adaptation and that coastal lowland regions would therefore have been the logical focus of early settlement (e.g. [123, 124]. All of these evidences clearly show that the modern human seafaring skill did develop during the Late Pleistocene via LGM to Early Holocene in Wallacea and the Pacific.

However, as with the case of Leang Sarru Rockshelter, many of these early sites or remote islands were not continually used or habited during the Late Pleistocene, and it remains uncertain whether the early phase colonization of Talaud Islands was successful or not. The ^{14}C dates indicate that the initial occupation of the site was rather short, lasting only about 3000 years, from 35 to 32 ka. It is not known why people stopped using Leang Sarru Rockshelter after 32 ka. Did they move to other locations or islands to be close to better resources or did they just die out? Certainly, terrestrial resources were very limited in the Talaud Islands, as there were no land mammals, except for a few species of bat and rat, possibly with cuscus and flying fox (e.g., [95]). This paucity of terrestrial resources must be one of the major factors for the short occupation of Leang Sarru Rockshelter in the Late Pleistocene.

The Leang Sarru Rockshelter case tentatively shows that modern humans in Wallacea might not have had the strategies and skills prior to the advent of agriculture to sustain continual colonization of remote islands such as Talaud, which had limited terrestrial resources during the Late Pleistocene and Holocene. The difficulty of sustaining hunter-gathering subsistence in a remote island with limited natural resources has been discussed in relation to islands of the coast of southern Australia, such as Kangaroo Island (e.g., [125]) and Hunter Island (e.g., [126]), and Ryukyu (Okinawa) Islands (e.g. [127]) in south Japan.

On the other hand, in the Pacific, as with the case of Matenbek and Buang Merabak on New Island, people seemed to transfer animal resources and other materials from outside, possibly for supply of the limited resources in such remote islands especially during the LGM. However, the Matenkupkum site had been only occupied between 21 and around 14 ka, while Matenbek located 70 m south in the same cliff line had been occupied only between 20 and 18 ka and then abandoned. The site seems not re-occupied until 8000 years ago. Also, Buang Merabak on New Island and Kilu Cave on Buka Island were abandoned around 20 ka and re-occupied around Early Holocene. Therefore, the sites in Bismarcks and Solomon Islands were also not occupied continually the same as the sites in Wallacia.

The excavations of Golo Cave on Gebe Island (Fig. 2) in Maluku Islands (e.g. [74,75] and Jerimalai Cave (e.g. [40,64]) in East Timor tentatively show that these sites also had not been used continually during the Late Pleistocene to Holocene. Golo Cave, for instance, was only occupied intermittently during the Late Pleistocene around 32 to 28 ka, and 21 to 19 ka, then later during the Holocene around 12 to 10 ka and possibly most recently during 7000 to 3000 years ago [74,75].

Similarly, Jerimalai Cave was occupied intermittently, during the Late Pleistocene from about 40 to 38 ka and again around 14 ka and during the Holocene around 6000 to 4000 years ago (see [64:528-529]). As with these sites, particularly Golo Cave in Maluku Islands, the second occupation phase of Leang Sarru Rockshelter occurred during 21 to 18 ka corresponding to the LGM. Although it is unknown whether people voyaged from other islands in the Talaud group or from adjacent regions such as the Sangihe Islands or Mindanao, it is clear that marine exploitation was practiced in the Talaud Islands during the LGM.

3. Holocene maritime exploitation and adaptation in Wallacea

3.1. The early to Mid-Holocene maritime exploitation and adaptation in Wallacea

After the LGM, the climate rapidly warmed and the sea level rose progressively until about 6500 years ago. The number of archaeological sites in Wallacea and adjacent regions postdating the Early Holocene dramatically increased. In East Timor, human habitation of Matja Kuru 1 Cave (e.g., [72]) and Uai Bobo 2 Cave [69] began at around 13 ka and there is evidence of increased human presence and activity from 8000 to 5000 years ago. During this latter period, human habitation of Bui Ceri Uato Cave in East Timor also commenced [69].

Matja Kuru 1 Cave and Bui Ceri Uato Cave, which are close to the coast, produced a variety of marine shellfish, while Uai Bobo 2 Cave, which is located in the inland hill country, has no marine shell. In northern Borneo, Hagop Bilo Cave was abandoned about 10 ka, while human activity at Madai caves is evident from 11 ka until 7000 years ago [102]. Madai caves produced a variety of fresh and brackish water shellfish but no marine shellfish. The distance of these caves from the coast during the Early Holocene was over 15 km, which, as for Uai Bobo 2 Cave in East Timor, would explain why marine shellfish are absent. Other sites such as Niah Cave, Lang Rogrien, and Leang Burung 2 were continually inhabited during the Early Holocene but no marine exploitation occurred there either.

On the other hand, total number and volumes of marine fish and shellfish remains dramatically increased at Jerimalai Cave during the Early to Middle Holocene dated around 6500 to 5500 years ago. For marine fish, numbers and percentage of inshore coral-dwelling fish species including Scarids (parrotfish), Balistids (triggerfish), and Serranids (grouper) increased. The heavy reliance on marine resources, especially shell, was continually evident in Lene Hara Cave [70] and Bui Ceri Uato Cave [69] in East Timor.

Similarly, some marine shellfish species increased in Leang Sarru Rockshelter during the Early Holocene. This is especially the case for Tridacnidae and Fasciolariidae, which are subtidal or coral rubble-dwelling species. As noted above, the increase of coral rubble-dwelling species such as Tridacnidae possibly relates to the rise in sea temperature and the growth of coral reefs. Similarly, at Golo Cave in Maluku Islands, the Holocene sediment prior to 3000 years ago contains a high density of shells [75:704]. In Buang Merabak on New Ireland, subtidal species such as *Trochus* spp. and *Limpet* suddenly increased after 10 ka, while the coral rubble-dwelling species, such as *Cypraea* spp. and *Strombus* spp., increased around 5000 years ago [106,128]. All this evidence indicates that the possible change in target species corresponds to the development of Early Holocene coastal environments, although shellfish continued to be one of the major marine resources exploited by humans in Talaud Islands as well as in other island coasts in Wallacea and its adjacent regions.

In another aspect of maritime adaptation process during this age, the evidences both in the Wallacea Archipelago and the Pacific islands show a new tradition of shell use for tool and ornament. In Wallacea, Jerimalai Cave and Lene Hara Cave on East Timor produced some *Trochus* single-piece fish-hooks around 10 ka, especially the one from Lene Hara Cave [129], which is almost complete with a rather larger size possibly for capturing larger-sized fish. Although Jerimalai Cave has a possibly much older *Trochus* fish-hook as discussed above, much clear evidences are all after 11 to 10 ka during the Terminal Pleistocene to Early Holocene. Jerimalai Cave also produced a number of shell beads and ornaments during the Middle Holocene age around 6500 to 5500 years ago [40,64].

In northern Maluku Islands, Golo Cave produced a number of *Tridacna* shell adzes after 10 ka and *Cassis* shell adzes later around 6000 years ago [73,74], while the site also produced possible *Turbo* shell scrapers from the Late Pleistocene periods back to 35 ka [75]. The *Tridacna* and *Cassis* shell adzes from Golo Cave are so far one of the oldest shell adzes in the world. Since early modern humans in Africa and Arabia already produced and used marine shell ornaments and tools dated 130 to 70 ka (or possibly even *Homo erectus* after 500 to 400 ka), it is not so surprising that the modern humans migrated into Wallacea also used such shell tools from the initial stage during the Late Pleistocene. More visibility of shell use for tools and ornaments after the Holocene in the region might indicate that such use was more actively practiced during the Holocene.

In the Pacific, Pamwak site on Manus Island, stone and *Tridacna* shell edge-ground tools started to occur after 10 ka, while single edge-ground axe fragment, shark teeth, and cut *Trochus* shell occurred in Balof, another site at Manus Island from 10 ka as well. Matenbek on New Island produced a decorate shell bead, while Kilu Cave on Buka Island produced drilled shark teeth [96]. All of these archaeological evidences in Wallacea and the Pacific clearly show the human maritime adaptation and more elaborate techniques both for fishing and tool-making had been developed by this time.

Furthermore, some evidences of possible use of food plant resources in the Pacific may indicate that some kind of agricultural practices had been started during the Terminal Pleistocene to Early Holocene in the region. For instance, the analysis of plant residues on stone tools and shell scrapers from Balof on Manus Island indicates that these tools were used for processing starchy plants including yam (*Dioscorea* sp.), which are regarded as introduced plants from New Guinea after 14 ka [130]. For other evidences, *Canarium* nutshells, which are also estimated as introduced plants by humans from Wallacea or New Guinea, occur in Pamwak site before 11 ka and on Kilu Cave before 10 ka [131]. The fruit of such introduced trees has also been archaeologically presented on the north coast of New Guinea, where wild forms were endemic [132].

The possible development of agriculture or subsistent activities and technologies can be also supported by another evidence that human use of caves as habitation or camping site decrease after 6–5000 years ago. Both in Wallacea and the Pacific, most of the sites discussed here were abandoned by 5000 years and seemed to be seldom used as habitation sites, except for some case of use for secondly burial practices mainly after Metal Age in Wallacea or for temporal occupation in much later times (cf. use by Japanese soldiers during WWII). Possible development process of agricultural and subsistent technologies could correspond to the development process of people's maritime adaptation and seafaring skill during the ages, and then much-skilled colonizers known as Austronesian language group with their highly maritime and agricultural skills appeared in later times after around 3500 years ago in both Wallacea and the Pacific.

3.2. The Late Holocene marine exploitation and adaptation in Wallacea and Oceania

During the Late Holocene after 5000 years ago, various capture technology invented, and great variety of fish and shellfish species had been exploited by modern humans. Especially, the fishing and seafaring technology were further developed after the Neolithic times in Wallacea and the Pacific after the possible new migration by the Austronesian-speaking people who were originally birthed along the coast of southern China to Taiwan region before 5000 years ago and then dispersed into Southeast Asian islands including the Wallacea Archipelago and to the Pacific islands dated around 3500 to 2000 years ago (e.g., [73]). Since they had knowledge and skill of pottery production, animal husbandry of pig, dog, and chicken, as well as agriculture of some food plants, the early age of them is archaeologically recognized as Neolithic times.

In Wallacea and adjacent islands in Southeast Asia (Fig. 2), several major sites dated to this age are known as Leang Two Manae site on the Talaud Islands (e.g., [73,76]), Uattamdi site on Kayoa, northern Maluku Islands (e.g., [74, 133]), Bukit Tengkorak site on the eastern coast of Borneo/Kalimantan Island (e.g., [134,135,136,137.138]), Karumpang site on Sulawesi Island (e.g., [73]), and Duyong Cave on Palawan Island (e.g., [101]).

Among them, only two sites, including Uattamdi on Kayoa (e.g. [74,133,139,140]) and Bukit Tengkorak on Borneo/Kalimantan Island produced a number of marine fish and shellfish remains (e.g. [136-138]) and then have traces of modern human maritime exploitation and adaptation in the Neolithic times. Both sites had been mainly used during 3500 to 1800 years ago, although Bukit Tengkorak site was used more heavily during the Neolithic times dated 3400 to 2800 years ago [136]), while Uattamdi site was used as a habitation or camping site during Neolithic dated 3500 to 2800 years ago and as secondly burial site during Early Metal Age dated 2200 to 1800 years ago (e.g. [74,133,139,140]).

The Bukit Tengkorak site was first excavated by the Sabah Museum under the direction of Peter Bellwood in 1987 [134,141], and then by a joint team from the University of Science Malaysia and the Sabah Museum in 1994 and 1995 [135] and in 2001 [136,137,138,142]. The total area excavated during these investigations by 2001 was 10.5 m^2, and each excavation unearthed a number of fish remains. From these excavations, the author earlier analyzed fish remains excavated during 1994–1995 (from the southern part of the site) and in 2001 (from the northern

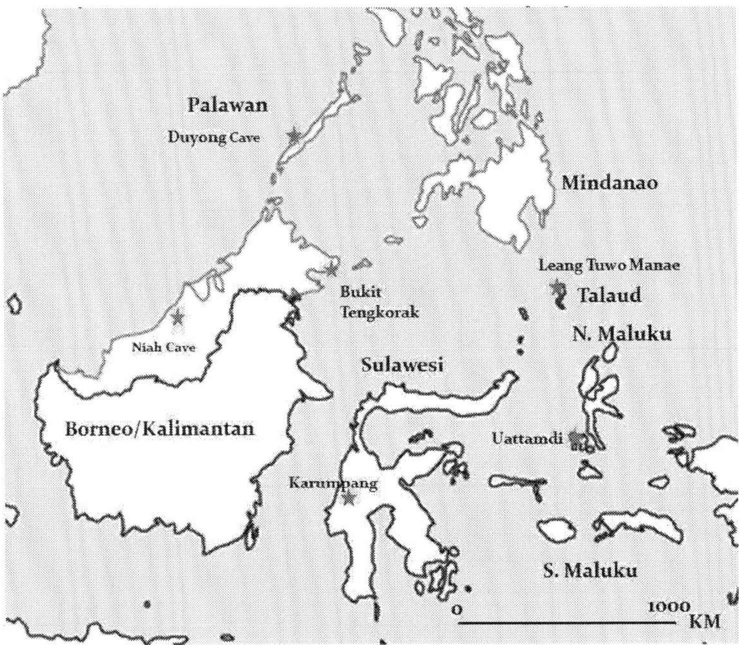

Figure 5. Major Neolithic sites in and around the Wallacea Archipelago.

part of the site), identifying a total of 28 fish taxa to family level (NISP=4132, MNI=1976) from 3 m^2 selected units [138,142].

Over 99% of the identified bones were inshore or coastal coral fish species such as groupers (Serranids), emperors (Lethrinids), parrotfish (Scarids), wrasses (Labrids), snappers (Lutjanids), and triggerfish (Balistids). The number of pelagic fish was very limited: only two bones (vertebrae) were identified as tuna (Scombrids) and one bone as barracuda (Sphyraenids). Although no materials related to fishing (cf. net sinker, hook, and lure) were recovered, the variety of excavated fish species and the author's intensive ethnoarchaeological study on recent Sama maritime people fishing around the site indicate that the Bukit Tengkorak people practiced netting, hook and line fishing, spearing, and possibly trapping and poisoning but possibly not trolling and offshore fishing [138]. Since the small peninsula, where the site is located, is surrounded by a large coral reef over a few kilometers, the heavy reliance on inshore fishing to capture coral-dwelling inshore fish species could be their strategic decision rather than lack of skills of offshore fishing.

On the other hand, the Uattamdi site is a complex of limestone rockshelters on a much smaller island but again fringing coral reef developed on the western coast where the site is located. The site was also excavated by Peter Bellwood and his team during the 1990s to find a number of potsherds, stone tools, shell tools or ornaments, glass beads, shellfish, and a few of fish, animal, and human bones. Human bones and glass beads were only excavated from upper layers, which belong to Early Metal Age burial. Styles and types of potsherds also show the clear difference between upper Metal-aged layers and lower Neolithic layers [74,140].

However, the details of excavated shellfish and fish remains as well as shell tools are not fully reported and remain unclear. Ono and the National Archaeological Research Centre of Indonesia conducted new excavations in two rockshelters at Uattamdi in 2014 [133]. One of them is same larger rockshelter where Bellwood excavated. Our excavations also unearthed a number of potsherds, stone and shell tools, glass beads, marine fish, shellfish, animal bones, and human remains in both shelters at Uattamdi. Our excavations confirm over 150 species of shellfish mainly marine species, over 10 families of marine fish including tunas, groupers, emperors, parrotfish, wrasses, snappers, triggerfish, trevallies, unicorn fish, porcupine fish (Diodontidae), moray eels (Muraenidae), Haemurids, and sharks (mainly Carcharhinidae) as well as sea-turtle bones. A shark tooth from the Neolithic layers had a drilled hole and could be used as tool or ornament.

Although the total number and volume of fish remains is not so large like Bukit Tengkorak site, most of tunas mainly occupied by skipjack (*Katsuwonus* sp.) were only from Neolithic layers, and it seemed both offshore and inshore fishing were engaged in Uattamdi site during the Neolithic times. Also, we excavated over 20 specimens of possible shell fish-hook blanks both made from some bivalve species and gastropod species, mainly *Nerita* sp. Since they are all blanks and not complete hooks, it is yet unclear whether they are really hooks or not, although a number of tuna and grouper bones from the site tentatively show the Uattamdi people did hook and line fishing at least.

The location of Bukit Tengkorak and Uattamdi also indicate that these Neolithic people, possibly newly migrated into the region, developed maritime skills and knowledge and strategically selected such coastal environment with large coral reefs as one of their initial settlements. Such migration strategies and preference of coastal settlement can be seen much clearly in the case of Lapita sites in the Pacific.

Lapita people are usually identified as the early Austronesian-speaking people originally from Asian region with Neolithic material culture, including pottery making and using tradition and more systematic agriculture and animal husbandry practices, and migrated into the Pacific islands after around 3300 years ago. They migrated into many remote islands in Melanesia to Western Polynesia, which had been mostly uninhabited before them. Therefore, it is believed that the Lapita people are the first modern humans who succeeded to reach and migrate to these islands in Remote Oceania including Polynesia; hence, they are also estimated as the direct ancestors of Polynesians who migrated to most of all the islands in Polynesia including Hawaii, Rapa Nui (Easter Island), and New Zealand in later times.

However, the Lapita culture seemed not singly from the Asian region as well. One of the most specific materials that belong to Lapita is their pottery assemblage, which includes diagnostic motif of human face and geometric patterns made by dentate-stamping technique on its pottery surface and some of them are impressed by white-lime material. The technique of making pottery, dentate stamping, and lime impressing could have originated from Island

Southeast Asia or Taiwan to southern China since much older potteries made by such technique were found in these regions. However, when reducing pottery to a single element, there are 30 Lapita elements, of which 17 (57%) have been recovered from pre-Lapita Melanesian sites, 21 (70%) from Neolithic Southeast Asian sites, and only 8 (27%) from pre-Neolithic Southeast Asian sites [96,143].

Even for the pottery itself, the diagnostic motif of human face and mix of various patterns in one single pottery is a very specific character of Lapita pot and never seen in other regions including Wallacea yet. With such evidences, it is also believed that the Lapita is basically of Southeast Asian Austronesian origin, but with added elements innovated in pre-Lapita times in Melanesian Islands. If so, Lapita culture is more like a complex synergy involving both Austronesian immigrants and the Melanesian descendants of the Pleistocene colonizers of New Guinea and the Bismarcks (e.g., [96,144,145]). It also means that the Lapita people could be human group(s) mixed with both Asian originated newcomers and indigenous Melanesian islanders, yet the truth is unclear and we need more archaeological, anthropological, and genetic studies in the near future.

In terms of maritime exploitation and adaptation of the Lapita people, they seemed to have more developed maritime skills both for fishing and seafaring. Firstly, for fishing, most of the Lapita sites produced more inshore fish remains like in the Bukit Tengkorak site, including parrotfish, wrasses, snappers, triggerfish, groupers, emperors, unicorn fish, and porcupine fish. Although the number and volume are smaller, they also produced pelagic fish remains including tunas and barracuda (e.g., [83,136,137,145-151].

Furthermore, several Lapita sites produced large to small-sized shell fish-hooks and *Trochus* lure shanks (e.g., [145,147,151,152]). Such lure shanks clearly show the possible employment of trolling offshore to catch skipjack tunas or other fast-swimming fish species, and the Lapita *Trochus* lure shanks are possibly the oldest evidence for such type of hook so far. Various sizes and styles of shell fish-hooks are another evidence of active line fishing. Most of these hooks are made from *Trochus* sp. or *Turbo* sp. shell, and the former could have been originally from the *Trochus* fish-hook tradition back to the Late Pleistocene to Early Holocene in Wallacea to the Pacific regions.

On the other hand, the absence of shell trolling lures and variety of fish hooks in Island Southeast Asia Neolithic sites, including Bukit Tengkorak and Uattamdi, implies a discontinuity between this region and the Pacific in the aspect of fishing technology. It is risky to make any conclusive statement at this stage, however, given the small number of investigated sites and paucity of data. Nonetheless, the present information could suggest that offshore fishing techniques with trolling lures might be an innovation of Lapita fishers in Melanesia where more islands are surrounded by deep sea.

Secondly, for seafaring skills, we do not have any archaeological evidences of vessels used by Lapita people, although it is estimated that their colonization could have been carried by outrigger canoes using a primitive Oceanic spritsail [3:38]. Outrigger canoes are major canoe types in almost the whole area where the Austronesian language dispersed from Islands Southeast Asia to the Pacific and also in Madagascar in Africa, where the later migrations by the Austronesian-speaking people were done possibly from Indonesian islands somewhere after 2000 years ago. The only exception is Taiwan, which is considered as the potential original place of the Austronesian language group, and no outrigger canoes have been used in the island. It is yet unknown where the original outrigger canoes were birthed, but it is sure that they are the major type of canoe in the Austronesian-speaking area.

Another fact that the past Lapita colonization of the islands in Remote Oceania shows is that the capable distance of sea crossing by the Lapita people should be much longer than the modern human voyage during the Late Pleistocene to Middle Holocene. The Lapita successful migration beyond the sea gap in the Solomon Islands or between Near Oceania and Remote Oceania and then into many remote islands including New Caledonia, Vanuatu, Fiji, Tonga, Samoa, and so on sometimes need over 800 km navigation. Furthermore, the corresponding Austronesian migration into Mariana Islands in Micronesia dated about 3500 years ago possibly from the islands of the Philippines required over 2000 km navigation (e.g., [153,154]). Thus, the facts that these remote islands were colonized by the Austronesian including Lapita people clearly indicate that their seafaring technology and skills were developed around the Neolithic times.

After the Lapita colonization to Samoa, a western border of Polynesia dated around 2800 years ago, their migration movement seemed to stop in this volcanic island, and they did not try to migrate further east into a whole Polynesia. However, voyaging clearly continued between the known islands after the Lapita boundary was established and later reached several new islands to the east including Niue and perhaps Pukapuka by about 2000 years ago [3]. Similarly, after East Polynesia was reached, the Polynesians who are the descendants of the Lapita people also succeeded to colonize Hawaii, Easter Island, and New Zealand by the 12th to 13th centuries. The distance to these islands from their neighbor islands or continent is over 4000 km; hence, the success of migration by the Polynesians clearly indicate that their maritime adaption and navigation technology were highly developed.

In fact, it is a dramatic event that modern humans did succeed to migrate to all over the world, except for the North Pole and South Pole, when the colonization to New Zealand was done by the Polynesians. Such evidence also shows that marine environment could be our last target for migration and colonization in this world after the human birthed in and around inner forest environment over 6 million years ago. In this regard, the human history of maritime exploitation and adaptation process to coastal and marine environments in the Wallacea Archipelago and the Pacific islands should be important to think about the past relationships between coastal and marine environment and humans and thus is worth to be continually investigated by future studies as well.

4. Conclusion

This chapter overviews over 2 million years of aquatic and maritime adaptation history by human. It seems now that the early aquatic adaptation and fish use had been started in lake and river environments in Africa by early Hominins, while the maritime adaptation and marine use had been started much later by *Homo sapiens*. The progress of our maritime adaptation had been far exceeded when *H. sapiens* reached various marine and coastal environments after out of Africa, particularly in Wallacea and Oceania islands dated around 50 ka years. Currently, the oldest traces of long-distance seafaring, possible pelagic fishing or line fishing, and use of shell fish-hooks by modern humans were all discovered in Wallacea and Near Oceania regions back to the Late to Terminal Pleistocene.

Such facts strongly indicate that their archipelagic environment with many small and remote islands in Wallacea and Near Oceania could be one of the major backgrounds for such high maritime adaptation of the people in this maritime world. Another possible factor is the effect of temperature and sea-level change in the past. In fact, many of the Late Pleistocene to Early Holocene sites in Wallacea and Near Oceania show the drastic change in the variety and volume of fish and shellfish for exploitation; usually, the high peak of shellfish use was possibly practiced during the LGM period with the coldest and driest times in the Late Pleistocene.

After the Holocene, modern human maritime adaptation had been accelerated as corresponding to warmer temperature, rapid rise of sea level, and expansion of coastal area including the possible development of coral reefs after around 6000 years ago. In Wallacea and Near Oceania, Austronesian-speaking people who were originally from Taiwan or Southern China having developed farming fishing and animal husbandry skills and knowledge newly migrated after 4000 years ago. Among these groups, Lapita people migrated to the remote islands in Near Oceania at first and then rapidly and firstly colonized into many Melanesian Islands in Remote Oceania east to Samoa and Tonga, while other groups migrated to Mariana or Palau Islands in Micronesia. Such new migration to these remote islands in Remote Oceania were required over 2000 to 500 km nonstop sea crossing; hence, these facts clearly show that modern human seafaring skills had been further developed. All of these cases discussed and introduced in this chapter tentatively indicate that the human maritime adaptation process has corresponded to rather long-term temporal changes in the past marine environment including sea level, temperature, coastline, and marine resources.

Author details

Rintaro Ono*

Address all correspondence to: onorintaro@gmail.com

School of Marine Science and Technology, Tokai University, Japan

References

[1] Wallace, A. 1869 *The Malay Archipelago: The Land of the Orang-utan and the Bird of Paradise. A Narrative of Travel with Studies of Man and Nature.* 2 vols. London: Macmillan and Co. Reprinted version 1986: Oxford: Oxford University Press.

[2] O'Connor, S., and P. Veth 2000a East of Wallace's Line: An introduction. In S. O'Connor and P. Veth (eds.), *East of Wallace's Line: Studies of Past and Present Maritime Culture of the Indo-Pacific Region. Modern Quaternary Research in Southeast Asia* 16:1–11.

[3] Anderson, A. 2000 Slow boats from China: Issues in the prehistory of Indo-Pacific seafaring. In S. O'Connor and P. Veth (eds.), *East of Wallace's Line: Studies of Past and Present Maritime Culture of the Indo-Pacific Region. Modern Quaternary Research in Southeast Asia* 16:13–50.

[4] Braun, D., et al. 2010 Early hominin diet included diverse terrestrial and aquatic animals 1.95 Maagoin East Turkana, Kenya. *Proc. Natl. Acad. Sci. U. S. A.* 107:10002–10007

[5] Stewart, K. 1994 Early hominid utilization of fish resources and implications for seasonality and behaviour. *J. Hum. Evol.* 27:229–245.

[6] Leakey, M. D. 1971 *Olduvai Gorge, Vol. 3: Excavations in Beds I and II, 1960–1963.* Cambridge: Cambridge University Press.

[7] Erlandson, J. 2001 The archaeology of aquatic adaptations: Paradigms for a new millennium. *J. Archaeol. Res.* 9(4):287–350.

[8] Leakey, M. D. 1994 *Olduvai Gorge, Vol. 5: Excavations in Beds III, IV and the Masek Beds, 1968–1971.* Cambridge: Cambridge University Press.

[9] Pope, G. 1989 Bamboo and human evolution. *Nat. Hist.* (October):49–57.

[10] Choi, K. and D. Driwantoro 2007 Shell tool use by early members of *Homo erectus* in Sangiran, central Java, Indonesia: Cut mark evidence. *J. Archaeol. Sci.* 34:48–58.

[11] Singer, R., B. Gladfelter, and J. Wymer 1993 *The Lower Paleolithic Site at Hoxne, England.* Chicago: University of Chicago Press.

[12] Klein, R. G., and K. Scott 1986 Re-analysis of faunal assemblages from the Haua Fteah and other Late Quaternary sites in Cyrenaican Libya. *J. Archaeol. Sci.* 13:515–542.

[13] Howe, B. 1967 *The Palaeolithic of Tangier, Morocco: Excavations at Cape Ashakar, 1939–1947, Bulletin 22.* Cambridge: American School of Prehistoric Research.

[14] Roubet, F. 1969 Le niveau Aterian dans la stratigraphie cotiere a l'ouest d'Alger. *Paleoecol. Afr.* 4:124–128.

[15] Roche, J., and Texier, J.-P. 1976 Decouverte de restes humains dans un niveau aterien superieur de la grotte des Contrebandiers, a Temara (Maroc). *C. R. S. Acad. Sci.* 282:45–47.

[16] Stiner, M. 1994 *Honor Among Thieves: A Zooarchaeological Study of Neandertal Ecology.* Princeton, NJ: Princeton University Press.

[17] Stiner, M. 1999 Paleolithic mollusc exploitation at Riparo Mochi (Balzi Rossi, Italy): Food and ornaments from the Aurignacian through Epigravettian. *Antiquity* 73:735–754.

[18] Cleyet-Merle, J., and Madelaine, S. 1995 Inland evidence of human sea coast exploitation in Palaeolithic France. In A. Fischer (ed.), *Man and Sea in the Mesolithic*, Oxbow Monographs 53, Oxford, pp. 303–308.

[19] Garrod, D. A. E., L. H. D. Buxton, G. E. Smith, and D. M. A. Bate 1928 Excavation of a Mousterian rock-shelter at Devil's Tower, Gibraltar. J. *R. Anthropol. Inst. Great Britain Ireland* 58:33–113.

[20] Waechter, J. D. 1964 The excavation of Gorham's Cave, Gibraltar, 1951–54. *Bull. Inst. Archaeol.* 4:189–221.

[21] Barton, R. N. E., Currant, A. P., Fernandez-Jalvo, Y., Finlayson, J. C., Goldberg, P., MacPhail, R., Pettitt, P. B., and Stringer, C. B. 1999 Gibraltar Neanderthals and results of recent excavations in Gorham's, Vanguard and Ibex caves. *Antiquity* 73:13–23.

[22] de Lumley, H. 1969 A Paleolithic camp at Nice. *Sci. Am.* 220:42–50.

[23] Villa, P. 1983 *Terra Amata and the Middle Pleistocene Archaeological Record of Southern France*. Publications in Anthropology 13. Berkeley: University of California.

[24] Marean, C. W. 2010. Pinnacle Point Cave 13B (Western Cape Province, South Africa) in context: The Cape floral kingdom, shellfish, and modern human origins. *J. Hum. Evol.* 59:425–443.

[25] Marean, C. W., M. Bar-Matthews, J. Bernatchez, E. Fisher, P. Goldberg, A. I. R. Herries, Z. Jacobs, A. Jerardino, P. Karkanas, T. Minichillo, P. J. Nilssen, E. Thompson, I. Watts, and H. M. Williams 2007. Early human use of marine resources and pigment in South Africa during the Middle Pleistocene. *Nature* 449:905–908.

[26] Thackeray, J. F. 1988 Molluscan fauna from Klasies River, South Africa. *S. Afr. Archaeol. Bull.* 43:27–32.

[27] Henshilwood, C., J. Sealy, R. Yates, K. Cruz-Uribe, P. Goldberg, F. Grine, R. Klein, C. Poggenpoel, K. van Niekerk, and I. Watts 2001a Blombos Cave, southern Cape, South Africa: Preliminary report on the 1992–1999 excavations of the Middle Stone Age levels. *J. Archaeol. Sci.* 28:421–448.

[28] Henshilwood, C., F. d'Errico, C. Marean, R. Milo, and R. Yates 2001b An early bone tool industry from the Middle Stone Age, Blombos Cave, South Africa: Implications for the origins of modern human behaviour, symbolism and language. *J. Hum. Evol.* 41:631–678.

[29] Geeske, H. J. L., K. L. van Niekerk, G. L. Dusseldorp, and J. F. Thackeray 2012 Middle Stone Age shellfish exploitation: Potential indications for mass collecting and re-

source intensification at Blombos Cave and Klasies River, South Africa. *Quat. Int.* 270:80–94.

[30] Stringer, C. B. 2000 Coasting out of Africa. *Nature* 405:24–27.

[31] Walter, R. C., R. T. Buffler, J. H. Bruggemann, M. M. M. Guillaume, S. M. Berhe, B. Negasi, Y. Libsekal, H. Cheng, R. L. Edwards, R. von Cosel, D. Neraudeau, and M. Gagnon 2000 Early human occupation of the Red Sea coast of Eritrea during the last interglacial. *Nature* 405:65–69.

[32] Avery, G., D. Halkett, J. Orton, T. Steele, M. Tusenius, and R. Klein 2008 The Ysterfontein 1 Middle Stone Age rock shelter and the evolution of coastal foraging. *S. Afr. Archaeol. Soc. Goodwin Ser.* 10:66–89.

[33] Dusseldorp, G. L., and G. H. J. Langejans 2013 Carry that weight: Coastal foraging and transport of marine resources during the South African Middle Stone Age. *S. Afr. Humanities* 25:105–135.

[34] Brooks, A., D. Helgren, J. Cramer, A. Franklin, W. Hornyak, J. Keating, R. Klein, W. Rink, H. Schwarcz, J. Smith, K. Stewart, N. Todd, J. Verniers, and J. Yellen 1995 Dating and context of three Middle Stone Age sites with bone points in the Upper Semliki Valley, Zaire. *Science* 268:548–552.

[35] Yellen, J. E., A. S. Brooks, E. Cornelissen, M. J. Mehiman, and K. Stewart 1995 A Middle Stone Age worked bone industry from Katanda, Upper Semliki Valley, Zaire. *Science* 268:553–556.

[36] Yellen, J. E. 1998 Barbed bone points: Tradition and continuity in Saharan and sub-Saharan Africa. *Afr. Archaeol. Rev.* 15:173–198.

[37] Henshilwood, C., F. d'Errico, M. Vanhaeren, K. van Niekerk, and Z. Jacobs 2004 Middle Stone Age shell beads from South Africa. *Science* 304:404.

[38] Vanhaeren, M., F. d'Errico, C. Stringer, S. L. James, J. A. Todd, and H. K. Mienis 2007 Middle paleolithic shell beads in Israel and Algeria. *Science* 312:1785–1788.

[39] Joorden, J., et al. 2014 *Homo erectus* at Trinil on Java used shells for tool production and engraving. *Nature* 2014. DOI:10.1038/nature13962.

[40] O'Connor, S., R. Ono, and C. Clarkson 2011 Pelagic fishing at 42,000 years before the present and the maritime skills of modern humans. *Science* 334:1117–1121.

[41] Bernhard, G., J. Beran, S. Hanik, and R. S. Sommer 2013 A Palaeolithic fishhook made of ivory and the earliest fishhook tradition in Europe. *J. Archaeol. Sci.* 40:2458–2463.

[42] White, T. D., et al. 2003 Pleistocene *Homo sapiens* from Middle Awash, Ethiopia. *Nature* 423:742–747.

[43] Clottes, J., and J. Courtin 1996 *The Cave Beneath the Sea: Paleolithic Images at Cosquer.* New York: H. N. Abrams.

[44] Gabunia, L., et al. 2000. Earliest Pleistocene hominid cranial remains from Dmanisi Republic of Georgia: Taxonomy, geological setting, and age. *Science* 288:1019–1025.

[45] Velua, A., et al. 2002 A new skull of early *Homo* from Dmanisi, Georgia. *Science* 297:85–89.

[46] Huffman, O. F., et al. 2006. Relocation of the 1936 Mojokerto skull discovery site near Perning, East Java. *J. Hum. Evol.* 50:431–451.

[47] Dennell, R. 2009 *The Palaeolithic Settlement of Asia*. London: Cambridge Univ. Press.

[48] Brown, P., et al. 2004 A new small-bodied hominin from Late Pleistocene of Flores, Indonesia. *Nature* 431:1055–1061.

[49] Morwood, M. J., R. J. Soejono, R. G. Roberts, T. Sutikna, C. S. M. Turney, K. E. Westaway, W. J. Rink, J.-x. Zhao, G. D. van den Bergh, R. Awe Due, D. R. Hobbs, M. W. Moore, M. I. Bird, and L. K. Fifield 2004 Archaeology and age of a new hominin species from Flores in eastern Indonesia. *Nature* 431:1087–1091.

[50] Tocheri, M. W., et al. 2007 The primitive wist of *H. floresiensis* and its implications for hominin evolution. *Science* 317:1743–1745.

[51] Jungers, W. L., W. E. Harcourt-Smith, R. E. Wunderlich, M. W. Tocheri, S. G. Larson, T. Sutikna, R. A. Due, and M. J. Morwood 2009 The foot of *Homo floresiensis*. *Nature* 459(7243):81-4.

[52] Larson, S., et al. 2008 Descriptions of upper limb skeleton of *Homo floresiensis*. *J. Hum. Evol.* DOI: 10.1016/ji jhevol.2008.06.007.

[53] McDougall, I., et al. 2005 Stratigraphic placement and age of modern humans from Kibish, Ethiopia. *Nature* 433:733–736.

[54] Liu, H., F. Prugnolle, A. Manica, and F. Balloux 2006 A geographically explicit genetic model of worldwide human-settlement history. *Am. J. Hum. Genet.* 79(2):230–237.

[55] *Ghirotto, S., L. Penso-Dolfin, and G. Barbujani 2011* Genomic evidence for an African expansion of anatomically modern humans by a Southern route. *Hum. Biol. 83(4): 477–489.*

[56] Lahr, M. and R. Foley 1994 Multiple dispersals and modern human origins. *Evol. Anthropol.* 3:48–60.

[57] Lahr, M. 1996 *The Evolution of Modern Human Diversity: A Study of Cranial Variation*. Cambridge: Cambridge University Press.

[58] Oppenheimer, S. J. 2003 *Out of Eden: The Peopling of the World*. London: Constable.

[59] Oppenheimer, S. J. 2008 The great arc of dispersal of modern humans: Africa to Australia. *Quat. Int.* 2008. DOI 10.1016/j.quaint.2008.05.015.

[60] Foster, A. 2004 Ice Ages and the mitochondrial DNA chronology of human dispersals: A review. *Philos. Trans. R. Soc. Lond. B* 359:243–254.

[61] Foster, A. and Matsumura, S. 2005 Did early humans go north or south? *Science* 308:965–966.

[62] Mulvaney, J., and J. Kamminga 1999 *Prehistory of Australia*. Washington, DC: Smithsonian Institution Press.

[63] O'Connor, S., and J. Chappell 2003 Colonisation and coastal subsistence in Australia and Papua New Guinea: Different timing, different modes. In *Pacific Archaeology: Assessments and Prospects. Proceedings of the International Conference for the 50th Anniversary of the First Lapita Excavation (July 1952), Koné-Nouméa 2002: 15–32, ed. C. Sand. Les Cahiers de l'Archéologie en Nouvelle Calédonie 11*. Nouméa: Département d'Archéologie, Service des Musées et du Patrimoine.

[64] O'Connor, S 2007 New evidence from East Timor contributes to our understanding of earliest modern human colonization east of the Sunda Shelf. *Antiquity* 81:523–535.

[65] Ono, R., S. Soegondho, and M. Yoneda 2010 Changing marine exploitation during Late Pleistocene in northern Wallacea: Shellfish remains from Leang Sarru Rockshelter in Talaud Islands. *Asian Perspect.* 48(2):318–341.

[66] Clark, J. 1991 Early settlement of the Indo-Pacific. J. *Anthropol. Archaeol.* 10:27–53.

[67] Irwin, G. 1992 *The Prehistoric Exploration and Colonisation of the Pacific*. Cambridge: Cambridge University Press.

[68] Chappell, J. 2000 Pleistocene seedbeds of western Pacific maritime cultures and the importance of chronology. In S. O'Connor and P. Veth (eds.), *East of Wallace's Line: Studies of Past and Present Maritime Culture of the Indo-Pacific Region. Modern Quaternary Research in Southeast Asia* 16:77–98.

[69] Glover, I. 1986 *Archaeology in Eastern Timor, 1966–67*. Terra Australis 11, Department of Prehistory, Research School of Pacific Studies. Canberra: The Australian National University.

[70] O'Connor, S., M. Spriggs, and P. Veth 2002 Excavation at Lene Hara Cave establishes occupation in East Timor at least 30,000–35,000 years ago. *Antiquity* 76:45–50.

[71] O'Connor, S., M. Spriggs, and P. Veth, eds. 2005 *The Archaeology of the Aru Islands, Eastern Indonesia*. Terra Australis 23. Canberra: Pandanus Books.

[72] Veth, P, M. Spriggs, and S. O'Connor 2005 The continuity of cave use in the tropics: Examples from East Timor and the Aru Islands, Maluku. *Asian Perspect.* 44(1):180–192.

[73] Bellwood, P. 1997 *Prehistory of the Indo-Malaysian Archipelago* (revised edition). Honolulu: University of Hawaii Press.

[74] Bellwood, P., G. Nitihaminoto, G. Irwin, A. Waluyo, and D. Tanudirjo 1998 35,000 years of prehistory in the northern Moluccas. In J.-C. Galipaud and I. Lilley (eds.) *Bird's Head Approaches*, pp. 233–275. Modern Quaternary Research in Southeast Asia 15. Rotterdam: Balkema.

[75] Szabó, K., A. Brumm, and P. Bellwood 2007 Shell artefact production at 32,000–28,000 B.P. in Island Southeast Asia: Thinking across media? *Curr. Anthropol.* 48(5): 701–723.

[76] Tanudirjo, D. 2001 *Islands in Between: Prehistory of the Northeastern Indonesian Archipelago*. Ph.D. dissertation. Canberra: The Australian National University.

[77] Tanudirjo, D. 2005 Long-continuous or short-occasional occupation?: The human use of Leang Sarru Rockshelter in the Talaud Islands, north eastern Indonesia. *Bull. Indo-Pac Prehistory Assoc* 25:15–19.

[78] Ono, R., N. Nakajima, H. Nishizawa, S. Oda, and S. Soegondho 2015 Maritime adaptation and development of lithic technology in Talaud Islands during the Late Pleistocene to the Early Holocene. In Y. Kaifu, M. Izuho, T. Goebel, H. Sato, and A. Ono (eds.), *Emergence and Diversity of Modern Human Behavior in Paleolithic Asia*. Texas A&M University Press, pp. 201–213.

[79] Glover, I. 1976 Ulu Leang Cave, Maros: A preliminary sequence of post-Pleistocene cultural development in south Sulawesi. *Archipel* 11:113–154.

[80] Glover, I. 1981 Leang Burung 2: An upper Paleolithic rock shelter in south Sulawesi, Indonesia. *Mod. Quat. Res. Southeast Asia* 6:1–38.

[81] O'Connor, S. and P. Veth 2000b The world's first mariners: Savannah dwellers in an island continent. In S. O'Connor and P. Veth (eds.), *East of Wallace's Line: Studies of Past and Present Maritime Culture of the Indo-Pacific Region. Modern Quaternary Research in Southeast Asia* 16:99–137.

[82] Kirch, P. 1976 Ethno-archaeological investigations in Futuna and Uvea (Western Polynesia): A preliminary report. *J. Polynes. Soc.* 85:27–69.

[83] Kirch, P. and T. Dye 1979 Ethno-archaeology and the development of Polynesian fishing strategies. *J. Polynes. Soc.* 88:53–76.

[84] Dye, T. 1983 Fish and fishing on Niuatoputapu, Tonga. *Oceania* 53:242–271.

[85] Masse, B. 1986 A millennium of fishing in the Palau Islands, Micronesia. In A. Anderson (ed.), *Traditional Fishing in the Pacific: Ethnographic and Archaeological Papers from the 15th Pacific Science Congress. Pacific Anthropological Records 37*. Honolulu: Bernice P. Bishop Museum, pp. 85–117.

[86] Masse, B. 1990 *The Archaeology and Ecology of Fishing in the Belau Islands, Micronesia*. Ph.D. Thesis, Southern Illinois University.

[87] Chikamori, M. 1988 *Ethno-Archaeology in Coral Reefs.* Tokyo: Yuzankaku Press (in Japanese).

[88] Goto, A. 1990 Prehistoric Hawaiian fishing lore: An integrated approach. *Man Culture Oceania* 6:1–34.

[89] Goto, A. 1996 Lagoon life among the Langalanga, Malaita Island, Solomon Islands. In T. Akimichi (ed.), *Coastal Foragers in Transition.* Osaka: Senri Ethnological Studies 42, pp. 11–54.

[90] Rolett, B. 1989 *Hanamiai: Changing Subsistence and Ecology in the Prehistory of Tahuata, Marquesas Islands, French Polynesia.* Unpublished Ph.D. thesis, Stanford University.

[91] Ono, R, and D. Addison 2009 Ethno-ecology and Tokelauan fishing lore from Atafu atoll, Tokelau. *SPC Traditional Marine Resource Management Knowledge Information Bulletin.* 26: 3-22.

[92] Anderson, A. 2013 Inshore or offshore? Boating and fishing in the Pleistocene. *Antiquity* 87:879–885.

[93] O'Connor, and R. Ono 2013 The case for complex fishing technologies: a response to Anderson. *Antiquity* 87: 886-890.

[94] Ono, R., and S. Soegondho 2004 A short report for the re-excavation at Leang Sarru site, Talaud Islands. *Jejak-Jejak Arkeologi* 4:37–50.

[95] Riley, J. 2002 Mammals on the Sangihe and Talaud Islands, Indonesia, and the impact of hunting and habitat loss. *Oryx* 36(3):288–296.

[96] Allen, J. 2000 From beach to beach: The development of maritime economies in prehistoric Melanesia. In S. O'Connor and P. Veth (eds.), *East of Wallace's Line: Studies of Past and Present Maritime Culture of the Indo-Pacific Region. Modern Quaternary Research in Southeast Asia* 16:139–176.

[97] Anderson, D. 1997 Cave archaeology in Southeast Asia. Geoarchaeology 12:607–638.

[98] Muder, K, and D. Anderson 2007 New evidence for Southeast Asian Pleistocene foraging economies: Faunal remains from the early levels of Lang Rongrien rockshelter, Krabi, Thailand. Asian Perspectives 46(2):298–334.

[99] Zuraina, M. 1982 The west mouth, Niah: In the prehistory of Southeast Asia. *Sarawak Museum Journal* 31. Kuchin: Sarawak Museum.

[100] Barker, G. 2005 The archaeology of foraging and farming at Niah Cave, Sarawak. Asian Perspectives 44:90–106.

[101] Fox, R. 1970 *The Tabon Caves: Archaeological Explorations and Excavations on Palawan Island, Philippines. Monograph of the National Museum 1.* Manila: The Philippines National Museum.

[102] Bellwood, P. 1988 Archaeological research in southern Sabah. Sabah Museum Monograph 2, Kota Kinabaru: Sabah Museum.

[103] Medway, L. 1976 The Niah excavations: An assessment of the impact of early man on mammals in Borneo. *Asian Perspectives* 21:51–69.

[104] Cranbrook, E. 2000 Northern Borneo environments of the past 40,000 years: Archaezoological evidences. *Sarawak Museum Journal* 55:61–109.

[105] Gosden, C and N. Robertson 1991 Models for Matenkupkum: interpreting a late Pleistocene site from southern New Ireland, Papua New Guinea. In J. Allen and C. Gosden (eds), *Report of the Lapita Homeland Project. Occasional Papers in Prehistory 20.* Canberra: The Australian National University, pp. 20-45.

[106] Swadling, P. 1994 Changing shell resources and their exploitation for food and artifact production in Papua New Guinea. *Man Culture Oceania*10:127–150.

[107] Leavesley, M, and J. Allen 1998 Dates, disturbance and artifact distributions: Another analysis of Buang Merabak, a Pleistocene site on New Ireland, Papua New Guinea. *Archaeol. Oceania* 33(2):63–82.

[108] Specht, J. 2005 Revisiting the Bismarcks: Some alternative views. In A. Pawley, R. Attenborough, J. Golson, and R. Hide (eds.). *Papuan Past: Cultural, Linguistic and Biological Histories of Papuan-Speaking Peoples.* Pacific Linguistics 572, Research School of Pacific and Asian Studies. Canberra: The Australian National University, pp. 235–288.

[109] Wickler, S. 1995 *Twenty-nine Thousand Years on Buka: Long Term Cultural Change in the Northern Solomon Islands.* Unpublished Ph.D. thesis. University of Hawai'i, Honolulu.

[110] Morse, K. 1993 *West Side Story: towards a prehistory of the Cape Range Peninsula.* Western Australia. Unpublished PhD thesis, University of Western Australia.

[111] O'Connor, S. 1999 30,000 years in the Kimberley: results of excavation of three rock-shelters in the coastal west Kimberley, W.A. In P. Veth and P. Hiscock (eds), The Archaeology of Northern Australia. Tempus 4. St. Lucia, Queensland: Anthropology Museum, University of Queensland, pp. 26-29.

[112] Bowdler, S. 1990 The Silver Dollar site, Shark Bay: An interim report. *Australian Aboriginal Studies* 2:60– 63.

[113] Veth,P. 1993 The Aboriginal occupation of the Montebello Islands, northwest Australia. *Australian Aboriginal Studies* 2: 39-47.

[114] Erlandson, J., and S. Fitzpatrick 2006 Oceans, islands, and coasts: Current perspectives on the role of the sea in human prehistory. *J. Island Coast. Archaeol.* 1:5–32.

[115] Thorne, A., and R. Raymond 1989 *Man on the Rim: The Peopling of the Pacific.* North Ryde: Angus and Robertson.

[116] Birdsell, J. 1977 The recalibration of a paradigm for the first peopling of Greater Australia. In J. Allen. J. Golson, and R. Jones (eds.), *Sunda and Sahul: Prehistoric Studies in Southeast Asia, Melanesia and Australia.* London: Academic Press, pp. 113–167.

[117] Allen, J., C. Gosden, and J. P. White 1989 Pleistocene New Island. *Antiquity* 63:548–561.

[118] Wickler, S., and M. Spriggs 1988 Pleistocene human occupation of the Solomon Islands, Melanesia. *Antiquity* 62:703–707.

[119] Leavesley, M, and J. Chappell 2004 Buang Merabak: Additional early radiocarbon evidence of the colonisation of the Bismarck Archipelago, Papua New Guinea. *Antiquity Project Gallery* 78(301). http://antiquity.ac.uk/ProjGall/leavesley/index.html#leavesley

[120] Loy, T., M. Spriggs, and S. Wickler 1998 Dates, disturbance and aretefact distributions: Another analysis of Buanbg Merabak, a Pleistocene site on New Ireland, Papua New Guinea. *Archaeol. Oceania* 33(2):63–82.

[121] Frederichsen, C., M. Spriggs, and W. Ambrose 1993 Pamwak Rockshelter: A Pleistocene site on Manus Island. Papua New Guinea. In M. A. Smith, M. Spriggs, and B. Fankhauser (eds.), *Sahul in Review. Pleistocene Archaeology in Australia, New Guinea and Island Melanesia.* Canberra: The Australian National University, pp. 144–152.

[122] Summerhayes, G. and J. Allen 1993 The transport of Mopir obsidian to Pleistocene New Ireland. *Archaeol. Oceania* 29(3):144–148.

[123] Bulbeck, D. 2003 Hunter-gatherer occupation of the Malay Peninsula from the Ice Age to the Iron Age. In J. Mercader (ed.), *Under the Canopy: The Archaeology of Tropical Rain Forests.* New Brunswick, NJ: Rutgers University Press, pp. 119–160.

[124] Bulbeck, D., I. Sumantri, and P. Hiscock 2004 Leang Sakapao 1: A second dated Pleistocene site from south Sulawesi, Indonesia, in quaternary research in Indonesia. *Mod. Quat. Res. Southeast Asia* 18:111– 129.

[125] Draper, N. 1988 Stone tools and cultural landscapes: Investigating the archaeology of Kangaroo Island. *S. Aust. Geo. J.* 88:15–36.

[126] Bowdler, S. 1984 *Hunter Hill, Hunter Island. Terra Australis 8.* Canberra: Department of Prehistory, Research School of Pacific Studies, The Australian National University.

[127] Takamiya, H. 2006 An unusual case? Hunter-gatherer adaptations to an island environment: A case study from Okinawa, Japan. *J. Island Coastal Archaeol.* 1:49–66.

[128] Rosenfeld, A. 1997 Excavation of Buang Merabak, Central New Ireland. *Bull. Indo-Pac. Prehistory Assoc.* 16:213–224.

[129] O'Connor, S. and P. Veth 2005 Early Holocene shell fish hooks from Lene Hara Cave, East Timor, establish complex fishing technology was in use in Island Southeast Asia five thousand years before Austronesian settlement. *Antiquity* 79:1–8.

[130] Barton, H. and J. P. White 1993Use of stone and shell artefacts at Balof 2, New Ireland, Papua New Guinea. *Asian Perspectives* 32 (2): 169-181.

[131] Spriggs, M. 1997 *The Island Melanesians*. Oxford: Blackwell Publishers.

[132] Yen, D. E. 1995 The development of Sahul agriculture with Australia as bystander. In J. Allen and J.F. O'Connell (eds.), Transitions: Pleistocene to Holocene in Australia and New Guinea. *Antiquity* 69(Special Number 265):831–847.

[133] Ono, R., F. Aziz, I. B. Zesse, M. Yoneda, and K. Tanaka in preparation *Neolithic and Early Metal Aged Culture in Northern Maluku: Preliminary Report of Archaeological Excavations of Uattamdi Rockshelter on Kayoa Island, Indonesia.*

[134] Bellwood, P. 1989 Archaeological investigations at Bukit Tengkorak and Segarong, southeastern Sabah. *Bull. Indo-Pacific Prehistory Assoc.* 9:122–162.

[135] Chia, S. 2003 *The Prehistory of Bukit Tengkorak as a Major Pottery Making Site in Southeast Asia. Sabah Museum Monograph No. 8*. Kota Kinabalu: Sabah Museum.

[136] Ono, R. 2003 Prehistoric Austronesian fishing strategies: A comparison between Island Southeast Asia and the Lapita Cultural Complex. In C. Sand (ed.), *Pacific Archaeology: Assessments and Prospects*. Noumea: Service des Musées et du Patrimoine du Nouvélle Calédonie, Département Archéologie, pp. 191–201.

[137] Ono, R. 2004 Prehistoric fishing at Bukit Tengkorak, east coast of Borneo Island. *N. Z. J. Archaeol.* 24:77–106.

[138] Ono, R. 2010 Ethno-Archaeology and the early Austronesian fishing strategies in near-shore environments. *J. Polynes. Soc.* 119(3):269–314.

[139] Bellwood, P., A. Waluyo, G. Nitihaminoto, and G. Irwin 1993 Archaeological research in the northern Moluccas; interim results, 1991 field season. *Bull. Indo-Pac. Prehistory Assoc.* 13:20–33.

[140] Bellwood, P. in preparation *The Spice Islands in Prehistory: Archaeology in the Northern Moluccas, Indonesia.*

[141] Bellwood, P and P. Koon 1989 Lapita colonists leave boats unburned, *Antiquity* 63: 613-622.

[142] Ono, R. 2011 *Maritime Exploitation and Fishing Strategies in the Celebes Sea: An Ethno-Archaeological Study of Maritime People*. Kyoto: The Kyoto University Press (in Japanese).

[143] Spriggs, M. 1996 What is Southeast Asian about Lapita? In T. Akazawa and E. J. E. Szathmary (eds.), *Prehistoric Mongoloid Dispersals*. Oxford: Oxford University Press, pp. 324–348.

[144] Green, R. 1991 The Lapita cultural complex: Current evidence and proposed models. In P. Bellwood (ed.), Indo-Pacific Prehistory 1990. *Bull. Indo-Pac. Prehistory Assoc.* 11:295–305.

[145] Kirch, P. 1997 *The Lapita Peoples: Ancestors of the Oceanic World.* London: Blackwell.

[146] Kirch, P. 1988 *Niuatoputapu: The Prehistory of a Polynesian Chiefdom.* Seattle: Burke Museum.

[147] Kirch, P. 2000 *On the Road of the Winds: An Archaeological History of the Pacific Islands Before European Contacts.* Los Angeles: University of California Press.

[148] Green, R. 1986 Lapita fishing: The evidence of site SE-RF-2 from the main Reef Islands, Santa Cruz Group, Solomons. In A. Anderson (ed.), *Traditional Fishing in the Pacific: Ethnographic and Archaeological Papers from the 15th Pacific Science Congress. Pacific Anthropological Records 37.* Honolulu: Bernice P. Bishop Museum, pp. 19–35.

[149] Butler, V. 1988 Lapita fishing strategies: The faunal evidence. In P. V. Kirch and T. L. Hunt (eds.), *Archaeology of the Lapita Cultural Complex: A Critical Review. Research Report 5.* Seattle: Thomas Burke Memorial Washington State Museum, pp. 99–115.

[150] Butler, V. 1994 Fish feeding behaviour and fish capture: The case for variation in Lapita fishing strategies. *Archaeol. Oceania* 29:81–90.

[151] Ono, R., S. Hawkins, and S. Bedford 2015 *Lapita fish use and development of fishing technology: A view from northern Vanuatu.* A paper presented at the 8th International Lapita Conference, Port Vila, Vanuatu (7th July).

[152] Szabó, K., 2010 Shell Artefacts and Shell-Working within the Lapita Cultural Complex. *Journal of Pacific Archaeology* 1(2): 115-127.

[153] Bellwood, P. 2005 *First Farmers: The Origins of Agricultural Societies.* Blackwell Publishing Ltd.

[154] Bellwood, P. 2013 *First Migrants: Ancient Migration in Global Perspective.* John Wiley and Sons.